CONTROL SYSTEMS ENGINEERING

CONTROL SYSTEMS ENGINEERING

NORMAN S. NISE

California State Polytechnic University, Pomona

THE BENJAMIN/CUMMINGS PUBLISHING COMPANY, INC.

Redwood City, California • Menlo Park, California • Reading, Massachusetts
New York • Don Mills, Ontario • Wokingham, U.K. • Amsterdam
Bonn • Sydney • Singapore • Tokyo • Madrid • San Juan

Sponsoring Editor: Alan Apt

Associate Editor: Mary Ann Telatnik

Production Supervisor: Laura Kenney

Freelance Production Management: Kirsten Stigberg,
    Publication Services, Inc.

Copyeditor: Nicholas Murray

Cover Design: Rudy Zehntner

Cover Photo: Uniphoto, Inc.

Technical Art: Publication Services, Inc.

Composition: Publication Services, Inc.

Library of Congress Cataloging-in-Publication Data

Nise, Norman S.
    Control systems engineering / Norman S. Nise
        p.   cm.
    Includes index.
    ISBN 0-8053-5420-4
        1. Automatic control.   2. Systems engineering.   I. Title.
    TJ213.N497   1991   91-22066
    629.8—dc20   CIP

ISBN 0-8053-5420-4
1 2 3 4 5 6 7 8 9 10—DO—95 94 93 92 91
The Benjamin/Cummings Publishing Company, Inc.
390 Bridge Parkway
Redwood City, California 94065

To my wife, Ellen; sons, Benjamin and Alan; and daughter, Sharon.

PREFACE

The control systems course that the student is about to begin is the basis of many scientific and engineering design solutions. If the student is an electrical, mechanical, aerospace, or chemical engineer, this course is important because its content plays a major role in the design of robots, spacecraft, aircraft, and process control systems. This book is dedicated to control systems engineering concepts, theory, and practice with an emphasis on the practical application of control systems engineering to the design and analysis of feedback systems.

Typically, control systems is taught in universities and colleges in the junior or senior year of an engineering curriculum. This textbook is suitable for use in either of these applications. The student using this text should have typical lower-division courses in physics and mathematics through differential equations. Other required background material, including Laplace transforms and linear algebra, is covered in the chapters or appendixes as review topics. This background material may be skipped without loss of continuity if the student's preparation so warrants it.

Key Features

This textbook is written for the student! It is written for both the student under the tutelage of a professor at a university and the student who wishes to learn the subject through self-study. With these objectives in mind, the following are the key features of this textbook:

- Qualitative and quantitative explanations
- Numerous examples
- Numerous illustrations
- Numerous homework problems
- Standard chapter organization
- Emphasis on design
- Flexible coverage
- Emphasis on computational assistance

Qualitative and Quantitative Explanations

Explanations are readable and complete. Where appropriate, they include review material as well as a discussion of where we were and where we are going. The text consists of topics that build upon and support one another in a logical way. Topics are introduced qualitatively and intuitively; quantitative analysis and design follow, supported by appropriate mathematics. The writing style is user-friendly. Groundwork for new concepts and terminology is carefully and logically laid out so as not to overwhelm the student. During classroom testing, students indicated that they used the book successfully for self-study when they missed class.

Numerous Examples

Explanations are supported by examples, which are focused. They tightly and completely fit the explanation that they are demonstrating, thus forming a clear picture of the material presented. Broader examples that tie together most of the developed concepts and show practical applications are at the ends of the chapters. An antenna azimuth control system is used at the end of many chapters to demonstrate the objectives of the entire chapter. Thus, the student can see how the chapter's material can be applied to the same system that they see over and over again.

Numerous Illustrations

Any explanation that requires an illustration has one. The student is not left to create an illustration in order to visualize a concept. For example, after a design has been completed, illustrations or tables are used to compare the performance of the system before and after the design.

Numerous Homework Problems

Every chapter ends with homework problems that allow the student to practice the ideas covered. Some are simple and demonstrate a single concept, and others are more challenging and require the interweaving of several ideas. Many problems are expressed in terms of practical systems so the student can experience real applications.

Standard Chapter Organization

The introduction to each chapter, except the first, contains a statement of the objective in terms of a problem that describes what the student is expected to be able to do after studying the chapter. The concluding section of the chapter illustrates the objective with a real-life problem, which, for many chapters, includes the antenna azimuth control system.

Each chapter has homework problems as well as review questions requiring short answers. Some homework problems are clearly stated as chapter-objective problems.

Emphasis on Design

In design problems, a desired response is specified and the student must then evaluate gain or other system parameters or specify a system configuration along with parameter values. Design is emphasized in Chapters 8, 9, 11, 12, and 13. Even in earlier chapters, such as Chapters 4 through 7, where the emphasis is on analysis, some simple design examples are included.

This textbook emphasizes design using the following format:

- Design objectives stated
- Methodology explained
- Simplifying assumptions explained
- Examples given that demonstrate each design methodology
- Improvement in performance shown with tables or plots; assumptions checked

Design objectives focus on *stability, steady-state error,* and *transient response* as specified by *percent overshoot, settling time, peak time,* and *rise time.* In the chapters using frequency response design methods, frequency response characteristics such as phase margin that do not yield a vivid picture of the design objectives for first-time students are carefully and deliberately tied back to more vivid design objectives as itemized above.

For each type of design problem, methodology is presented—in many cases in the form of a step-by-step procedure. Example problems serve to demonstrate the method by following the procedure, making necessary assumptions, and presenting the results of the design through tables or plots that compare the performance of the original system to the improved system. This comparison also affords a check on the simplifying assumptions. For an illustration of the approach see Chapters 9 and 11 and their examples.

Transient response design topics are covered comprehensively. They include:

- Design via gain adjustment using the root locus
- Design of compensation and controllers via the root locus
- Design via gain adjustment using sinusoidal frequency response methods
- Design of compensation via sinusoidal frequency response methods
- Design of controllers in state space using pole-placement techniques
- Design of observers in state space using pole-placement techniques
- Design of digital control systems via gain adjustment on the root locus

Steady-state error design is similarly covered comprehensively in this textbook and includes:

- Gain adjustment
- Design of compensation and controllers via the root locus
- Design of compensation via sinusoidal frequency response methods
- Design of integral control in state space

Finally, the design of gain to yield stability is covered from the following points of view:

- Routh-Hurwitz criterion
- Root locus
- Nyquist criterion
- Bode plots

Although quantitative solutions are extremely important, a qualitative "feel" for the problem and the methods of solution yields the insight required for design. At every opportunity where qualitative insight can be demonstrated, the material has been included. For example, in Chapter 8, qualitative information about the transient response that can be obtained from the root locus is related to a previous chapter on second-order system responses. The student is then able to look at a root locus and describe, qualitatively, the changes in the transient response that occur as a system parameter is varied. This insight occurs without any calculations other than those required to rapidly sketch the root locus.

Flexible Coverage

The amount of material that can be covered in a one-quarter course and a one-semester course is different. The book lends itself well to the selection of material to fit the allotted time. State-space methods are covered along with the classical approach where the state-space material is functionally the same. For example, the chapter on stability discusses the subject from both the classical and the state-space perspectives. Throughout the book, chapters and sections (as well as examples, questions, and problems) that cover state space are marked with an asterisk (*) and can be omitted without loss of continuity. Thus, those wishing to study just classical methods can skip all chapters and sections marked with an asterisk (*). Those wishing to include just a basic introduction to state-space modeling can include Chapter 3. Semester-system schools can include state-space analysis, in Chapters 4, 5, and 6, as well as state-space design, covered in Chapter 12. If preferred, state space can also be taught separately by gathering those chapters and sections marked with an asterisk (*) into a single unit that follows the classical approach. In a one-quarter course, Chapter 13, Digital Control Systems, could also be eliminated.

Emphasis on Computational Assistance

Control systems problems can be tedious, particularly analysis and design problems using the root locus, since trial and error must be used. To solve the problems, the students should be provided with access to programs found on mainframe or personal computers. The author's experience in teaching control systems has shown that a small, hand-held programmable calculator will adequately perform the following control systems computations:

- Conversions between the various descriptions of second-order systems. For example:

  Pole location in polar coordinates
  Pole location in Cartesian coordinates
  Characteristic polynomial
  Natural frequency and damping ratio
  Settling time and percent overshoot
  Peak time and percent overshoot
  Settling time and peak time

- Determining whether a point on the *s*-plane is on the root locus
- Finding the magnitude and phase frequency response data for Nyquist and Bode plots

The student can program the second-order system conversions on a hand-held calculator using the relationships derived in Chapter 4. A program written in Microsoft® QuickBASIC in Appendix C and referred to at appropriate places in the text satisfies the requirements of the last two items in the above list. Some modification of the program may be required depending upon the form of BASIC and the machine used. The problems of students' logging on to mainframes and waiting for access are eliminated for these simple applications. Furthermore, the hand-held machine can be used at home as well as in the classroom for examinations.

For more computational-intensive applications, such as finding the time response, plotting root loci, finding state-transition matrices, etc., mainframes or personal computers can be used. Students can write their own programs or use other available programs such as MATLAB™, MacLocus, or Program CC. One such program, written in Microsoft® QuickBASIC to plot the time response for a system represented in state space, is provided in Appendix C. Again, some modification of the program may be required depending upon the form of BASIC and the machine used. Without the computational assistance to perform actual calculations, the students would be hard pressed to obtain meaningful analysis and design results and the experience gained could be limited.

Chapter Organization

Many times it is helpful to understand an author's reasoning behind the organization of the course material. The following paragraphs may shed light on this topic.

Chapter 1 begins with motivational material. The student becomes aware of not only the applications of control systems but also the advantages of study and a career choice in this field. Control systems engineering design objectives such as transient response, steady-state error, and stability are described as well as the path to obtain these objectives. Many students have trouble with the first step in the design and analysis sequence—transforming a physical system into a schematic. Many simplifying assumptions must be made based upon experience that the student does not at present possess. Identifying some of these assumptions in Chapter 1 helps break this difficult barrier.

Chapters 2, 3, and 5 cover the representation of physical systems. Chapters 2 and 3 cover open-loop systems using frequency response techniques and state-space techniques, respectively. Chapter 5 continues with the discussion of the representation of closed-loop and other systems that consist of the interconnection of open-loop subsystems. Only a representative sample of physical systems can be covered in this type of book. Electrical, mechanical (both translational and rotational), and electromechanical systems will serve as samples of the physical systems that are modeled, analyzed, and designed. Linearization of a nonlinear system, one technique used by the engineer to simplify a system in order to represent it mathematically, is also introduced.

Chapter 4 provides the student with the first introduction to system analysis—finding and describing the output response of a system. It may appear more logical to put this material after Chapter 5 along with other chapters covering analysis; however, many years of teaching control systems has shown that the quicker the student can see an application for the study of system representation, the greater the chance for continued motivation.

Chapters 6, 7, and 8 return to control systems analysis and design with the study of stability, steady-state errors, and transient response of higher-order systems using root locus techniques, respectively. Chapter 9 covers design of compensators and controllers using the root locus.

Chapters 10 and 11 are to sinusoidal frequency analysis and design what Chapters 8 and 9 are to root locus analysis and design. Chapter 10, like Chapter 8, covers basic concepts for stability, transient, and steady-state error analysis. However, Nyquist and Bode methods are used rather than root locus. Chapter 11, like Chapter 9, covers the design of compensators, but from the point of view of sinusoidal frequency techniques rather than root locus.

State-space and digital control systems analysis and design concepts complete the text in Chapters 12 and 13, respectively. Although this material can be used as an introduction for students continuing their studies in control systems engineering, these topics are useful by themselves and supplement the analysis and design covered in the previous chapters. Although all aspects of these topics cannot be covered in just two chapters, the emphasis is clearly stated, scoped, and logically linked to the rest of the book.

Acknowledgments

The author would like to acknowledge the contributions of many people to this book. I want to thank my colleagues at California State Polytechnic University, Pomona, for their help and encouragement: Mohamed Rafiquzzaman, for his inspiration to begin the project in the first place; Samy M. El-Sawah, who error-checked the manuscript; as well as Elhami T. Ibrahim, Robert G. Irvine, Herbert A. Johnson, R. Frank Smith, and Bob Kennerknecht for their suggestions and contributions. I would especially like to thank my students who used the manuscript over the last three years. Their suggestions were invaluable and their eagerness to find and report errors was inspiring.

The staff at Benjamin/Cummings Publishing Company must receive kudos for their untiring attention to the project through all phases. Their knowledge and expertise were always available to me, freely and timely. In particular, I want to thank the editors who worked with me at various times on the project: Alan R. Apt, Craig S. Bartholomew, Mary Ann Telatnik, and Jake Warde.

The several drafts of the manuscript were scrutinized by numerous reviewers whose mastery of the subject and laudable attention to details yielded many suggestions and improvements. In particular, I want to acknowledge Jeffrey E. Froyd, Rose-Hulman Institute of Technology, whose input through all the drafts helped shaped the pedagogy of the book; Francois E. Cellier, University of Arizona, whose early suggestions made major contributions to the topic selection; as well as Gordon Lee, North Carolina State University; Hun H. Sun, Drexel University; Carlos Tavora, Gonzaga University; Glenn R. Widmann, Colorado State University; Jessy W. Grizzle, University of Michigan; David Talwar, University of Virginia; Robert O. Barr, Jr., Michigan State University; Charles P. Neuman, Carnegie-Mellon University; and Wiley Thompson, New Mexico State University. Other contributors include Menachem Rafaelof, who provided software that was invaluable to the project, and Ben Nise for his suggestions about technical content and writing style.

Norman S. Nise

CONTENTS

*3 SYSTEM REPRESENTATION IN THE TIME DOMAIN 111

4 TIME RESPONSE FOR THE SYSTEM MODEL 149

5 REPRESENTATION AND REDUCTION OF MULTIPLE SUBSYSTEMS 211

6 STABILITY 277

13 DIGITAL CONTROL SYSTEMS 661

APPENDIX A LIST OF SYMBOLS 707

APPENDIX B MATRICES, DETERMINANTS, AND SYSTEMS OF EQUATIONS 710

APPENDIX C COMPUTER PROGRAMS 719

1

INTRODUCTION

1.1 Control Systems Applications

Control systems are an integral part of modern society. We can see numerous applications all around us. The rockets fire, and the space shuttle lifts off towards earth orbit (Figure 1.1); amid splashing cooling water, a metallic part is automatically machined; a vehicle delivering material to various work stations in an aerospace assembly plant, automatically and unattended, glides along the floor seeking its destination. These are just a few examples of the automatically controlled, physical systems that we can create.

Figure 1.1 Space Shuttle *Discovery* Launch, April 24, 1990 (Courtesy of NASA)

We are not the only creators of automatically controlled systems, however; these systems also exist in nature. Within our own bodies are numerous control systems such as the pancreas, which regulates our blood sugar. In time of flight or fright, our adrenalin automatically increases along with our heart rate, causing more oxygen to be delivered to our cells. Our eyes follow a moving object to keep it in view; our hands grasp the object and place it precisely at a predetermined location.

Even the nonphysical world appears to be automatically regulated. Models have been suggested showing automatic control of student performance. The input to the model is the student's available study time, and the output is the grade. The model can be used to predict the time required for the grade to rise if a sudden increase in the study time is available. Using this model, you can determine if increased study is worth the effort during the last week of the semester.

With control systems we can move large equipment with precision that would otherwise be impossible. We can point large antennas toward the farthest reaches of the universe to pick up faint radio signals. Moving these antennas by hand would be impossible. Elevators automatically stop at the right floor, taking us there in comfort (Figure 1.2). We alone could not provide the power required for

Figure 1.2 (*a*) Rope is Cut to Demonstrate Safety in an Early Elevator; (*b*) Modern Elevators that Use Control Systems to Regulate Position and Velocity (Courtesy of Otis Elevator)

(*a*)   (*b*)

the load and the speed; motors provide the power, and control systems regulate the position and speed.

Control systems find widespread application in the steering of missiles, planes, spacecraft, and ships at sea. For example, modern ships use a combination of electrical, mechanical, and hydraulic components to develop rudder commands in response to desired heading commands (Figure 1.3). The rudder commands, in turn, produce a rudder angle, which steers the ship. An example of the application of control systems to the process-control industry is a thickness-control system for a steel plate finishing mill (Figure 1.4). Steel enters the finishing mill and passes through rollers. In the finishing mill, X-rays can be used to measure the actual thickness of the steel plate, which is compared to the desired thickness. Any difference is adjusted by a screw-down position control that changes the roll gap at the rollers through which the steel passes. This change in roll gap regulates the thickness.

The home is not without its own control systems. In a video disc or compact disc machine, microscopic pits representing the information are cut into the disc by a laser during the recording process (Figure 1.5). During playback, a reflected laser beam focused on the pits changes intensity. The changes in light intensity are then converted to an electrical signal and processed as sound or pic-

Figure 1.3 CG 47 Class Cruiser Uses Control Systems for Steering
(Provided by the Engineering Department, Ingalls Shipbuilding Division
of Litton, Pascagoula, Mississippi)

Figure 1.4 Steel Plate Rolls through Finishing Mill. Screw-Down Motors
on Top Are Part of a Control System that Regulates Thickness (Courtesy of
Bethlehem Steel)

Figure 1.5 (*a*) Video Laser Disc Player (Courtesy of
Pioneer Video)

(*a*)

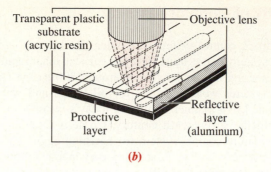

(b)

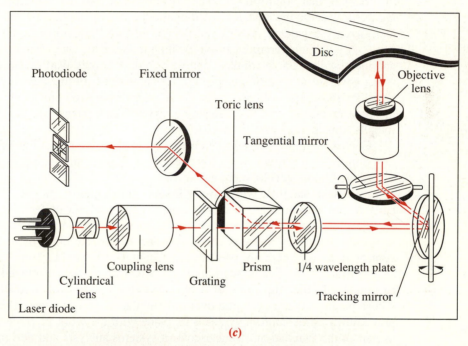

(c)

Figure 1.5 (continued) (*b*) Objective Lens Reading Pits on a Laser Disc; (*c*) Optical Path for Playback Showing Tracking Mirror that Is Rotated by a Control System to Keep the Laser Beam Positioned on the Pits (Courtesy of Pioneer Video)

ture. A control system keeps the laser beam positioned on the pits, which are cut as concentric circles. In a home heating system is a simple control system consisting of a bimetallic material that expands or contracts with changing temperature. This expansion or contraction moves a vial of mercury that acts as a switch, turning the heater on or off. The amount of expansion or contraction required to move the mercury switch is determined by our setting of the temperature.

Now that we have seen some applications of control systems, let us trace some of their history and see how you will benefit from the study of control systems.

1.2 Your Study of Control Systems

Feedback control systems are older than human beings. Numerous biological control systems were built into the earliest inhabitants of our planet. People began engineering feedback systems around 300 B.C. in Greece, where liquid-level control systems were designed for oil lamps and water clocks. These mechanisms operated in a manner similar to today's water-level control in our flush toilets.

In the seventeenth century, Cornelis Drebbel invented a purely mechanical temperature control system for hatching eggs. The device used a vial of alcohol and mercury with a floater inserted in the mixture. The floater was connected to a damper that controlled a flame. A portion of the vial was inserted into the incubator to sense the heat generated by the fire. As the heat increased, the alcohol and mercury expanded, raising the floater, closing the damper, and reducing the flame. A lower temperature caused the float to descend, which opened the damper and increased the flame. In the eighteenth century, James Watt invented the flyball speed governor to control the speed of steam engines. In this device two spinning flyballs rise as rotational speed increases. A steam valve connected to the flyball mechanism closes with the ascending flyballs and opens with the descending flyballs, thus regulating the speed.

Control systems theory as we know it today began to crystallize in the last half of the nineteenth century. J. C. Maxwell, E. J. Routh, and A. M. Lyapunov each contributed to the development and formulation of today's theories and practice in control system stability, which we will study in Chapter 6. In the late 1920s and early 1930s at Bell Telephone Laboratories, the analysis of feedback amplifiers was developed by H. W. Bode and H. Nyquist. These contributions evolved into sinusoidal frequency analysis and design techniques currently used for feedback control systems, which are presented in Chapters 10 and 11.

In 1948, W. R. Evans, working in the aircraft industry, developed a technique to graphically plot the roots of a characteristic equation of a feedback system whose parameters are changing over a particular range of values. This technique, now known as the *root locus,* takes its place with the work of Bode and Nyquist as part of the foundation of linear control systems analysis and design theory. We will study root locus in Chapters 8, 9, and 13.

Control systems engineering is an exciting field in which to apply your engineering talents, because it cuts across numerous disciplines and numerous functions within those disciplines. The control engineer can be found at the top level of large projects, engaged at the conceptual phase in determining or implementing overall system requirements. These requirements include total system performance specifications, subsystem functions, and the interconnection of these functions, including interface requirements, hardware and software design, and test plans and procedures.

Many engineers find themselves engaged in only one of the above areas, such as circuit design or software development. However, as a control systems engineer, you may find yourself working in a broad and diversified arena. You can interact with many people from numerous branches of engineering and the sciences. For example, if you are working on a biological system, you will need to interact with colleagues in the biological sciences, mechanical engineering, and electrical engineering, not to mention perhaps mathematics and physics. You will be working with these engineers at all levels of project development from concept through design and, finally, testing. At the design level, the control systems engineer can be performing hardware selection, design, and interface, including total subsystem design to meet specified requirements. The control engineer can be working with sensors and motors, as well as electronic, pneumatic, and hydraulic circuits. Let us look in more detail at an example of a control systems application and see the diversity required of the control engineer.

Space exploration would be impossible without control systems. America's space shuttle contains numerous control systems operated by an on-board computer on a time-shared basis. Without control systems, it would be impossible to guide the shuttle to and from earth orbit or to adjust the orbit itself and support the life on board. Navigation functions programmed into the shuttle's computers use data from the shuttle's hardware to estimate vehicle position and velocity. This information is fed to the guidance equations that calculate commands for the shuttle's flight control systems, which steer the spacecraft. In space, the flight control system gimbals (rotates) the orbital maneuvering system (OMS) engines into a position that provides thrust in the commanded direction to steer the spacecraft. Within the earth's atmosphere, the shuttle is steered by commands sent from the flight control system to the aerosurfaces, such as the elevons.

Within this large control system represented by navigation, guidance, and control, there are numerous subsystems relegated to the control of the vehicle's functions. For example, the elevons themselves require a control system to ensure that their position is indeed that which was commanded, since disturbances such as wind could certainly rotate the elevons away from the commanded position. Similarly, in space, the gimbaling of the orbital maneuvering engines requires a similar control system to ensure that the rotating engine can accomplish its function with speed and accuracy. Control systems are also used to control and stabilize the vehicle during its descent from orbit. Numerous small jets that compose the reaction control system (RCS) are used initially in the exoatmosphere, where the aerosurfaces are ineffective. Later, control is passed to the aerosurfaces as the orbiter descends into the atmosphere.

Inside the shuttle, numerous control systems are required for power and life support. For example, the orbiter has three fuel-cell powerplants that convert hydrogen and oxygen (reactants) into electricity and water for use by the crew. The fuel cells involve the use of control systems to regulate temperature and pressure. The reactant tanks are kept at constant pressure as the quantity of reactant diminishes. Sensors in the tanks send signals to the control systems to turn heaters on or off to keep the tank pressure constant (Rockwell International, 1984).

You can see that the shuttle's control systems cut across many branches of science: orbital mechanics and propulsion, aerodynamics, electrical engineering, and mechanical engineering, including hydraulics, temperature, and pressure. As a control engineer, you apply a broad-based knowledge to the solution of engineering control problems. You have the opportunity to expand your engineering horizons beyond your university curriculum.

You are now aware of the future. But for now, what advantages does this course offer to a student of control systems (other than the fact that you need this course to graduate)? Engineering curricula tend to emphasize *bottom-up* design. That is, you start from the components, develop circuits, and then assemble a product. In *top-down* design, a high-level picture of the requirements is first formulated; then the functions and hardware required to implement the system are determined.

A major reason for not teaching top-down design is the high level of mathematics initially required for the systems approach. For example, control systems theory, which requires differential equations, could not be taught as a lower-division course. However, while progressing through bottom-up design, it is difficult to see how the courses you are taking fit logically into the large picture of the product development cycle. After completing the control systems course, you will be able to stand back and see how your previous studies fit into the large picture. Your amplifier course or your vibrations course will take on new meaning as you begin to see the role they play as part of product development. For example, as engineers, we want to describe the physical world mathematically so that we can create systems that will benefit humanity. You will find that you have indeed acquired, through your previous courses, the ability to model physical systems mathematically, although at the time you may not have understood where in the product development cycle the modeling fits. This course will clarify the analysis and design procedure and show you how the knowledge you acquired fits into the total picture of system design. Understanding control systems enables students from all branches of engineering to speak a common language and develop an appreciation and working knowledge of the other branches. You will find that there really isn't much difference among the branches of engineering as far as the goals and applications are concerned. As you study control systems, you will see this commonality.

We now will describe what a control system is, its characteristics, and its advantages. We will also discuss two major classifications for control systems.

1.3 Describing Control Systems

The Control System

A control system consists of *subsystems* and *processes* (or *plants*) assembled for the purpose of controlling the output of the processes. For example, a furnace is a process that produces heat as a result of the flow of fuel. This process, assembled with subsystems called *fuel valves* and *fuel-valve actuators,* regulates the temperature of a room by controlling the heat output from the furnace. Other subsystems such as thermostats, which act as sensors, measure the room temperature. In its simplest form, a control system provides an output or response for a given input or stimulus, as shown in Figure 1.6.

Figure 1.6 Simplified Description of a Control System

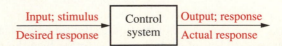

Description of the Input and Output

The input represents a desired response; the output is the actual response. For example, when the fourth-floor button of an elevator is pushed on the ground floor, the elevator rises to the fourth floor with a speed and floor-leveling accuracy designed for passenger comfort. Figure 1.7 shows the input and output for the elevator system. The push of the fourth-floor button is the input and is represented by a step command. Note that in the interest of passenger comfort, not to mention the limited power available, we would not want the elevator to mimic the suddenness of the input. The input represents what we would like the output to be after the elevator has stopped; the elevator itself follows the displacement described by the curve marked *elevator response*.

Two factors make the output different from the input. First, compare the instantaneous change of the input against the gradual change of the output in Figure 1.7. Physical entities cannot change their state (e.g., position or velocity) instantaneously. The state changes through a path that is related to the physical device and the way it acquires or dissipates energy. Thus, the elevator undergoes a gradual change as it rises from the first floor to the fourth floor. We call this part of the response the *transient response*.

After the transient response, a physical system approaches its *steady-state* response, which is its approximation to the commanded or desired response. For the elevator example, this response occurs when the elevator reaches the fourth floor. The accuracy of the elevator's final leveling with the floor is a second factor that could make the output different from the input. We call this difference, shown in Figure 1.7, *steady-state error*. Steady-state error need not exist only in defective control systems. Many times, steady-state errors are inherent in the designed system, and the control systems engineer determines whether or not that error leads to significant degradation of system functions. For example, if a system is tracking a satellite, some steady-state error can be tolerated as long as it is small enough to keep the satellite close to the center of the tracking radar beam. However, if a robot is inserting a memory chip onto a board, the steady-state error must be zero. Why do we use control systems that yield these differences between input and output?

Figure 1.7 Elevator Input and Output

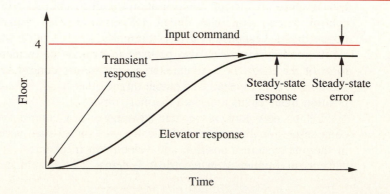

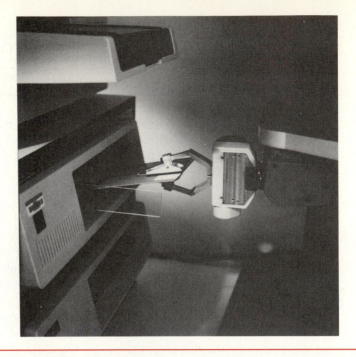

Figure 1.8 Responding to Voice Commands, a Robot Arm
Changes Floppy Diskettes at a Work Station for Quadriplegic
Programmers (Courtesy of Boeing Computer Services)

Advantages of Control Systems

We build systems for specific reasons and advantages—because, for instance, they can yield a degree of power amplification. For example, a radar antenna, positioned by the low-power rotation of a knob at the input, requires a large amount of power for its output rotation. Power amplification, or power *gain,* is one reason to build a control system. A system can also be used for control in a remote or dangerous location. The remote control of a robotic arm used for picking up dangerous material in a radioactive environment is one example. Handicapped people can compensate for their disabilities with robots designed by control systems principles. Figure 1.8 shows a robot arm changing a computer floppy disk in response to a voice command.

Control systems can also be used to provide convenience by changing the form of the input. For example, in a temperature control system, the input is a *position* on a thermostat. The output on the other hand, is *heat.* Thus a convenient position input yields a desired thermal output.

Let us now look at another advantage of a control system, the ability to compensate for disturbances. Typically, we control such variables as temperature in thermal systems, position and velocity in mechanical systems, and voltage, current, or frequency in electrical systems. The system must be able to yield

Figure 1.9 Goldstone 210-ft. Antenna Used for Deep Space
Communications (Courtesy of Jet Propulsion Laboratory)

the correct output even with a disturbance. For example, in Figure 1.9, consider
an antenna system that points in a commanded direction. If a wind forces the
antenna from its commanded position, or if noise enters internally, the system must
be able to detect this disturbance and correct the antenna's position. Obviously,
the system's input will not change to make the correction. Consequently, the
system itself must (1) measure the amount that the disturbance has repositioned
the antenna and (2) return the antenna to the position commanded by the input.
What then is the basic architecture of a control system?

Although Figure 1.6 is a total view of a control system from its input to its
output, we previously mentioned that it consists of a process and subsystems in
some configuration. We now discuss two specific internal architectures.

Open-Loop Systems

A generic *open-loop* system is shown in Figure 1.10(*a*). It consists of a subsystem called a *controller,* which drives a process or plant. The input is sometimes
called the *reference,* while the output can be called the *controlled variable.* Other
signals, such as *disturbances,* are shown added to the controller and process outputs via *summing junctions* that yield the algebraic sum of their input signals
using associated signs. For example, the plant can be a furnace or air condition-

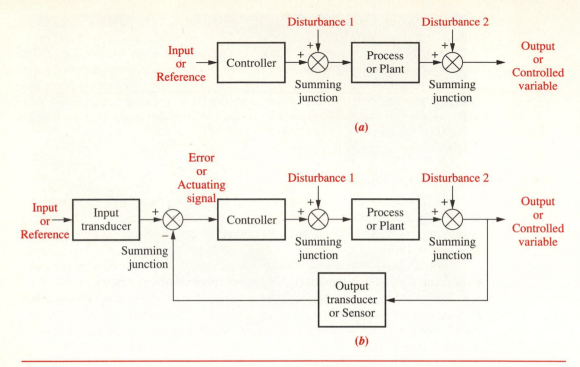

Figure 1.10 Block Diagrams of Control Systems: (*a*) Open-Loop; (*b*) Closed-Loop

ing system where the output variable is temperature. The controller in a heating system consists of fuel valves and the electrical system that operates the valves.

The distinguishing characteristic of an open-loop system is that it cannot compensate for any disturbances that add to the controller's driving signal (Disturbance 1 in Figure 1.10(*a*)). For example, if the controller is an electronic amplifier and Disturbance 1 is noise, then any additive amplifier noise at the first summing junction will also drive the process, so that the output will be corrupted by the effect of the noise. The output of an open-loop system is corrupted not only by signals that add to the controller's commands, but also by disturbances at the output (Disturbance 2 in Figure 1.10(*a*)). The system cannot correct for these disturbances either.

Open-loop systems, then, do not correct for disturbances and are simply commanded by the input. For example, toasters are open-loop systems, as anyone with burnt toast can attest. The controlled variable (output) of a toaster is the color of the toast. The device is designed with the assumption that the toast will be darker the longer it is subjected to heat. The toaster does not measure the color of the toast; it does not correct for the fact that the toast is rye, white, or sourdough, nor does it correct for the fact that toast comes in different thicknesses.

Other examples of open-loop systems are mechanical systems consisting of a mass, spring, and damper with a constant force positioning the mass. The greater the force, the greater the displacement. Again, the system position will change with a disturbance such as an additional force, and the system will not detect or correct for the disturbance. Or suppose that you calculate the amount of time required to study for an examination that covers three chapters in order to get an A. If the professor adds a fourth chapter, the disturbance, you are an open-loop system if you do not detect the disturbance, the increased material, and add study time to that previously calculated. The result of this oversight will be a grade lower than the one you expected.

How can we improve the open-loop system?

Closed-Loop Systems

The disadvantages of open-loop systems, namely, sensitivity to disturbances and the system's inability to correct for these disturbances, may be overcome in closed-loop systems. The generic architecture of a closed-loop system is shown in Figure 1.10(*b*).

Here, an input transducer, which was included in the controller of the open-loop system, is shown explicitly. The input transducer converts the form of the input to the form used by the controller. An output transducer, or sensor, measures the output response and also converts this response into the form used by the controller. For example, if the controller uses electrical signals to operate the valves of a temperature control system, the input position and the output temperature are converted to electrical signals. The input position can be converted to a voltage by a *potentiometer,* a variable resistor, and the output temperature can be converted to a voltage by a *thermistor,* a device that changes its electrical resistance with temperature.

The first summing junction algebraically adds the signal from the input to the signal from the output, which arrives via the *feedback path,* the return path from the output to the summing junction. In Figure 1.10(*b*), the output signal is subtracted from the input signal. The result is generally called the *actuating signal*. However, in systems where both the input and output transducers have *unity gain* (i.e., the transducer amplifies its input by 1), the actuating signal's value is equal to the actual difference between the input and the output. Under this condition, the actuating signal is called the *error*.

The closed-loop system compensates for disturbances by measuring the output response, feeding that measurement back through a feedback path, and comparing that response to the input at the summing junction. If there is any difference between the two responses, the system drives the plant, via the actuating signal, to make a correction. If there is no difference, the system does not drive the plant, since the plant's response is already the desired response.

Closed-loop systems, then, have the obvious advantage of greater accuracy than open-loop systems. They are less sensitive to noise, disturbances, and changes in the environment. Transient response and steady-state error can be controlled more conveniently and with greater flexibility in closed-loop systems, often by a simple adjustment of gains in the loop, and sometimes by redesigning the controller. We refer to the redesign as *compensating* the system and to the resulting hardware as a *compensator*. On the other hand, closed-loop systems are more complex and expensive than open-loop systems. A standard, open-loop toaster serves as an example: it is simple and inexpensive. A closed-loop toaster oven is more complex and more expensive since it has to measure both color (through light reflectivity) and the humidity inside the toaster oven. Thus, the control systems engineer must consider the trade-off between the simplicity and low cost of an open-loop system and the accuracy and higher cost of a closed-loop system.

In summary, systems that perform the previously described measurement and correction are called *closed-loop,* or *feedback control, systems*. Systems that do not have this property of measurement and correction are called *open-loop systems*.

Computer-Controlled Systems

In many modern systems, the controller (or *compensator*) is a digital computer. The advantage of using a computer in the loop is that many loops can be controlled or compensated by the same computer through time sharing. Further, any adjustments of the compensator parameters required to yield a desired response can be made by changes in software rather than hardware. The computer can also perform supervisory functions, such as the scheduling of many required applications. For example, the space shuttle main engine (SSME) controller (Fig-

Figure 1.11 Space Shuttle Main Engine (SSME) Controller
(Courtesy of Honeywell, Inc. Space Systems Group)

ure 1.11), which contains two digital computers, alone controls numerous engine functions. For example, it monitors engine sensors that provide pressures, temperatures, flow rates, turbopump speed, valve positions, and engine servovalve actuator positions. The controller further provides closed-loop control of thrust and propellant mixture ratio, sensor excitation, valve actuators, spark igniters, as well as control of other functions (Rockwell International, 1984).

Now that we have described control systems, let us define our analysis and design objectives.

1.4 Control Systems Analysis and Design Objectives

Control systems are *dynamic* systems: they respond to an input by undergoing a transient response prior to reaching a steady-state response that generally resembles the input. We have already identified these two responses and cited a position control system (an elevator) as an example. We now discuss the importance of transient and steady-state response and then establish our objectives. We will also introduce a new concept called *stability*.

Transient Response

Transient response is important. In the case of an elevator, a slow transient response will make passengers impatient, whereas an excessively rapid response will make them uncomfortable. If the elevator oscillates about the floor for more than a second, a disconcerting feeling can result. Transient response is also important for structural reasons: too fast a transient response could cause permanent physical damage. In a computer, transient response contributes to the time required to read from or write to a computer's disk storage (see Figure 1.12). Since reading and writing cannot take place until the head stops, the speed of the read/write head's movement from one track on the disk to another will influence the overall speed of the computer.

In this book we establish quantitative definitions for transient response. We then analyze the system for its *existing* transient response. Finally, we will adjust parameters or design components to yield a *desired* transient response.

We have discussed how a system arrives at its destination; let us now look at how close it gets to the desired location.

Steady-State Response

Another analysis and design goal focuses on the steady-state response. As we have seen, this response resembles the input and is usually what remains after the transients have decayed to zero. For example, this response may be an elevator stopped near the fourth floor, or the head of the disk drive finally stopped at the correct track. We are concerned about the accuracy of the steady-state solution. An elevator must be level enough with the floor for the passengers to exit, and a read/write head yields errors if its head is not positioned over the commanded track. An antenna tracking a satellite must keep the satellite well

Figure 1.12 A Computer Hard Disk Drive (Courtesy of
Rodime Systems)

within its beamwidth in order not to lose track. In this text, we define steady-state errors quantitatively, analyze a system's steady-state error, and then design corrective action to reduce the steady-state error.

Discussion of transient response and steady-state error is a moot point if the system does not have the next characteristic.

Stability

Another control system specification that we have not yet described is *stability*. In order to explain stability, we start from the fact that the total response of a system is the sum of the natural response and the forced response. When you studied linear differential equations, you probably referred to these responses as the *homogeneous* and the *particular* solutions, respectively. *Natural response* describes the way the system dissipates or acquires energy. The form or nature of this response is dependent only upon the system, not the input. On the other hand, the form or nature of the *forced response* is dependent upon the input. Thus, for a linear system, we can write,

$$\text{Total response} = \text{Natural response} + \text{Forced response} \qquad (1.1)$$

In order for a control system to be useful, the natural response must (1) eventually approach zero, thus leaving only the forced response, or (2) oscillate.

In some systems, however, the natural response grows without bound rather than diminishing to zero or oscillating. Eventually, the natural response is so much greater than the forced response that the system is no longer controlled. This condition, called *instability*, could lead to self-destruction of the physical device if limit stops are not provided as part of the design. For example, the elevator would crash through the floor or exit through the ceiling. A time plot of an unstable system would show a transient response that grows without bound and without any evidence of a steady-state response.

Control systems must be designed to be stable. That is, their natural response must decay to zero as time approaches infinity or oscillate. In many systems, the transient response you see on a time response plot can be directly related to the natural response. Thus, if the natural response decays to zero as time approaches infinity, the transient response will also die out, leaving only the forced response. If the system is stable, then the proper transient response and steady-state error characteristics can be designed.

In this book we will define, as well as analyze and design, systems for three specifications: (1) transient response, (2) steady-state errors, and (3) stability. Now that our objectives are defined, how do we meet them? We first look at an example of a feedback control system to become familiar with the problem and our objectives. Next, we will look at the design and analysis sequence that will help us arrive at our goals.

1.5 Example of a Closed-Loop System: The Position Control

In this section we look at a physical example of a feedback control system. We will see how it works and how we can effect changes in its performance. The discussion will be on a qualitative level with the objective of giving us an intuitive feel for the systems that we will be dealing with throughout the book. The example will also illustrate the principles discussed in many of the ensuing chapters.

Position control systems find widespread applications. A position control system converts a position input command to a position output response. Conceptually, such a system is incorporated into the caulking robot shown in Figure 1.13.

A Detailed Example: Antenna Azimuth Position Control

Another example of a position control system is the antenna azimuth position control system shown in Figure 1.14. The azimuth angle output of the antenna, $\theta_o(t)$, follows the input angle of the potentiometer, $\theta_i(t)$. This system will serve as a model for our discussion throughout the book. Figure 1.15 shows a *functional block diagram* of the system. The functions are shown above the blocks, and the required hardware is indicated inside the blocks. Consider the input command to be an angular displacement. The potentiometer, converts the angular displace-

Figure 1.13 A Robot Caulking an Aircraft Window
Structure (Courtesy of Boeing Computer Services)

Figure 1.14 An Antenna Azimuth Position Control System

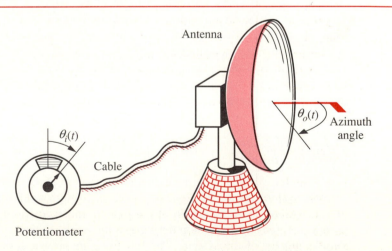

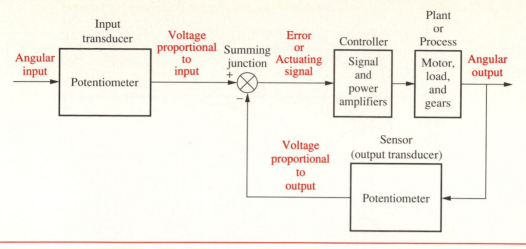

Figure 1.15 Functional Block Diagram of a Position Control System

ment into a voltage. Similarly, the output angular displacement is converted to a voltage by the potentiometer in the feedback path. The signal and power amplifiers boost the difference between the input and output voltages. This amplified actuating signal drives the plant.

The system normally operates to drive the error to zero. When the input and output match, the error will be zero, and the motor will not turn. Thus, the motor is only driven when the output and the input do not match. The greater the difference between the input and the output, the larger the motor input voltage, and the faster the motor will turn.

If we increase the gain of the signal amplifier, will there be an increase in the steady-state value of the output? If the gain is increased, then for a given actuating signal, the motor will be driven harder, but the motor will still stop when the actuating signal reaches zero, that is, when the output matches the input. The difference in the response, however, will be in the transients. Since the motor is driven harder, it turns faster toward its final position. Also, because of the increased speed, increased momentum could cause the motor to overshoot the final value and be forced by the system to return to the commanded position. Thus, the possibility exists for a transient response that consists of *damped oscillations* (i.e., a sinusoidal response whose amplitude diminishes with time) about the steady-state value if the gain is high. The responses for low gain and high gain are shown in Figure 1.16.

We have discussed the transient response of the position control system. Let us now direct our attention to the steady-state position to see how closely the output matches the input after the transients disappear. Figure 1.16 shows zero error in the steady-state response; that is, after the transients have disappeared, the output position equals the commanded input position. In some systems, the steady-state error will not be zero; for these systems, a simple gain adjustment to

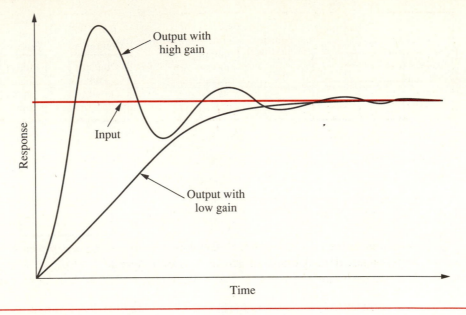

Figure 1.16 Response of a Position Control Showing Effect of High and Low Controller Gain on the Output Response

regulate the transient response is either not effective or leads to a trade-off between the desired transient response and the desired steady-state accuracy.

To solve this problem, a controller with a dynamic response, such as an electrical filter, is used along with an amplifier. With this type of controller, it is possible to design both the required transient response and the required steady-state accuracy without the trade-off required by a simple setting of gain. However, the controller is now more complex. The filter in this case is called a *compensator*. Many systems also use dynamic elements in the feedback path along with the output transducer to improve system performance.

In summary, then, our design objectives and the system's performance revolve around the transient response, the steady-state error, and stability. Gain adjustments can affect performance and sometimes lead to trade-offs among the performance criteria. Compensators can often be designed to achieve performance specifications without the need for trade-offs. Now that we have stated our objectives and some of the methods available to meet those objectives, we describe the orderly progression that leads us to the final system design.

1.6 The Design and Analysis Sequence

In this section, we establish an orderly sequence for the analysis and design of feedback control systems that will be followed as we progress through the rest of the book. Figure 1.17 shows the described analysis and design sequence as

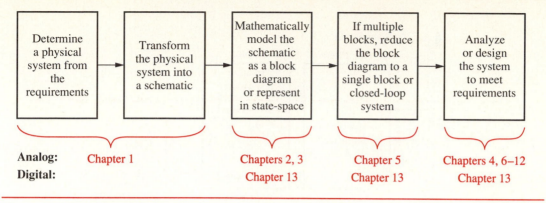

Figure 1.17 The Control System Design and Analysis Sequence

well as the chapters in which the steps are discussed. The position control system discussed in the last section is representative of control systems that must be analyzed and designed. Let us now elaborate on each block of Figure 1.17.

Determine a Physical System from the Requirements

We begin by transforming the requirements into a physical system. For example, in the antenna azimuth control system, the requirements would state the desire to position the antenna from a remote location and describe such features as weight, physical dimensions, desired transient response, and steady-state accuracy, to mention a few. First, a functional description of the subsystems may be developed, followed by a description of the required hardware. Figure 1.15 is an example of a block diagram that defines functions and also suggests possible hardware to perform the functions. For example, the function of the input transducer is implemented with a potentiometer in the figure.

Transform the Physical System into a Schematic

As we have seen, position control systems consist of electrical, mechanical, and electromechanical components. After producing the description of a physical system, the control systems engineer next transforms the physical system into a schematic diagram. The control system designer can begin with the physical description contained in the block diagram of Figure 1.15 and derive a schematic. The engineer must make approximations about the system and neglect certain phenomena or else the schematic will be unwieldy, making it difficult to extract a useful mathematical model during the next phase of the design and analysis sequence. The designer starts with a simple schematic representation and, at subsequent phases of the analysis and design sequence, checks the assumptions made about the physical system through analysis and computer simulation. If the schematic is too simple and does not adequately account for observed behavior, the control systems engineer adds additional phenomena to the schematic that were previously assumed negligible. A schematic diagram for the antenna azimuth position control system is shown in Figure 1.18.

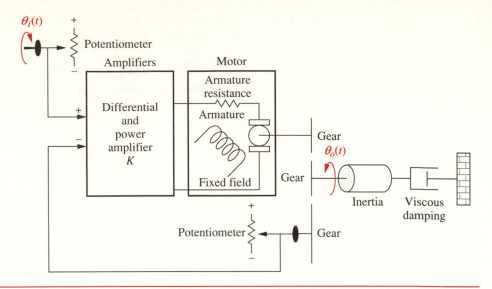

Figure 1.18 Schematic Diagram of a Position Control
System

When we draw the potentiometers, we make our first simplifying assumption by neglecting the friction or inertia of the potentiometers. These mechanical characteristics yield a dynamic rather than an instantaneous response in the output voltage. We assume that these mechanical effects are negligible and that the voltage across a potentiometer changes instantaneously as the potentiometer shaft turns.

A differential amplifier and a power amplifier are used as the controller to yield gain and power amplification, respectively, to drive the motor. Again, we assume that the dynamics of the amplifiers are rapid compared to the response time of the motor; thus we model them as a pure gain, K.

A motor and equivalent load produce the output angular displacement. An *armature-controlled* dc motor produces torque by setting up two opposing magnetic fields. One field is constant and is called the *fixed field;* the other field, which can be varied, is produced by a current flowing through a rotating member called an *armature*. The armature rotates within the magnetic field set up by the fixed field. If we change the armature voltage, the speed or torque of the motor can be controlled. Both inductance and resistance are part of the armature circuit. By showing just the armature resistance in Figure 1.18, we are assuming that the effect of the armature inductance is negligible for a dc motor.

The designer makes further assumptions about the load. The load consists of a rotating mass and bearing friction. Thus, the model consists of *inertia* and *viscous damping* whose resistive torque increases with speed, as in an automobile's shock absorber or screen-door damper.

The decisions that we made to develop the schematic stem from a knowledge of the physical system, the physical laws governing the system's behavior, and *practical experience*. These decisions are not easy; however, as you acquire more design experience, you will gain the insight required for this difficult task. This simplified schematic is representative of a vast array of position control systems. Antennas, robot arms, and computer disk drives are just a few applications that can be simplified to the position control system of Figure 1.18. We will use this position control system throughout this book as an example to show the application of each control system concept.

Mathematical Models for the Schematic

Once the schematic is drawn, the designer uses physical laws, such as Kirchhoff's laws for electrical networks and Newton's law for mechanical systems, along with simplifying assumptions, to model the system mathematically. These laws are:

Kirchhoff's voltage law: The sum of voltages around a closed path equals zero.

Kirchhoff's current law: The sum of electric currents flowing from a node equals zero.

Newton's laws: The sum of forces on a body equals zero;[1] the sum of torques on a body equals zero.

Kirchhoff's and Newton's laws lead to a differential equation that describes the relationship between the input and output of a dynamic system.

Simplifying assumptions are made in the process of obtaining a mathematical model. Without these assumptions, the model becomes complicated and clouds the answers the designer is looking for. These assumptions not only yield adequate models for the system, but can also yield insight to the analysis and design problem. We examine some of these simplifying assumptions in Chapter 2.

The designer uses the physical system, its laws, and approximations to create a mathematical model. One such model is the *linear, time-invariant differential equation,* Equation 1.2:

$$\frac{d^n c(t)}{dt^n} + a_{n-1}\frac{d^{n-1}c(t)}{dt^{n-1}} + \cdots + a_0 c(t) = b_m \frac{d^m r(t)}{dt^m} + b_{m-1}\frac{d^{m-1}r(t)}{dt^{m-1}}$$

$$+ \cdots + b_0 r(t) \qquad (1.2)$$

Many systems can be approximately described by this equation, which relates the output, $c(t)$, to the input, $r(t)$, by way of the system parameters, a_i and b_j. The reader is assumed to be familiar with differential equations. Problems and a bibliography are provided at the end of the chapter for you to review your knowledge of differential equations.

---

[1]Alternately, $\sum$ forces $= Ma$. In this text, the force, Ma, will be brought to the left-hand side of the equation to yield $\sum$ forces $= 0$ (D'Alembert's principle). We can then have a consistent analogy between force and voltage, and Kirchhoff's and Newton's laws; i.e., $\sum$ forces $= 0$; $\sum$ voltages $= 0$.

In addition to the differential equation, the *transfer function* is another way of mathematically modeling a system. The model is derived from the linear time-invariant differential equation using what we call the *Laplace transform*. Although the transfer function can be used only for linear systems, it yields more intuitive information than the differential equation. We will be able to change system parameters and rapidly sense the effect of these changes on the system response. The transfer function also is useful in modeling the interconnection of subsystems by forming a block diagram similar to Figure 1.15, but with a mathematical function inside each block.

Still another model is the *state-space* representation. One advantage of state-space methods is that they can also be used for systems that cannot be described by linear differential equations. Further, state-space methods are used to model systems for simulation on the digital computer. Basically, this representation turns an *n*th-order differential equation into *n* simultaneous first-order differential equations. Let this description suffice for now; we describe this approach in more detail in Chapter 3.

We now discuss the fourth block of the analysis and design sequence of Figure 1.17, block diagram reduction.

Block Diagram Reduction

Subsystem models are interconnected to form block diagrams of larger systems. In order to evaluate system response, you will need to reduce a large system's block diagram to a single block that represents the system from its input to its output. Hence, the next step in the design and analysis sequence is to reduce the block diagram to a single block, so that the response can be analyzed. Once the block diagram is reduced, we are ready to analyze and design the system.

Analysis and Design

If you are interested only in the performance of an individual subsystem, you can skip the block diagram reduction and move immediately into the next phase, analysis and design. We have already discussed some of the characteristics of control systems that are of interest to the designer. Once the system is represented by a single block, characteristics such as transient response, steady-state error, and stability can be analyzed and designed by the control systems engineer. It is not necessarily practical or illuminating to choose complicated input signals to analyze a system's performance. Thus, the engineer usually selects standard test inputs. These inputs are impulses, sinusoids, steps, ramps, and parabolas, as shown in Table 1.1.

An *impulse* is infinite over time from $0-$ to $0+$ and zero elsewhere. The area under the unit impulse is 1. An approximation of this type of waveform is used to place initial energy into a system so that the resulting response, due to that initial energy, is only the transient response of a system. From this response, the designer can derive a mathematical model of the system. *Sinusoidal* inputs can also be used to test a physical system to arrive at a mathematical model. A *step* input represents a constant commanded position input and typically would be used

Table 1.1 Test Waveforms Used in Control Systems

Input	Function	Description	Sketch	Use
Impulse	$\delta(t)$	$\delta(t) = \infty$ for $0- < t < 0+$ $= 0$ elsewhere $\int_{0-}^{0+} \delta(t)\,dt = 1$		Transient response Modeling
Sinusoid	$\sin \omega t$			Transient response Modeling Steady-state error
Step	$u(t)$	$u(t) = 1$ for $t \geq 0$ $= 0$ elsewhere		Transient response Steady-state error
Ramp	$tu(t)$	$tu(t) = t$ for $t \geq 0$ $= 0$ elsewhere		Steady-state error
Parabola	$\frac{1}{2}t^2 u(t)$	$\frac{1}{2}t^2 u(t) = \frac{1}{2}t^2$ for $t \geq 0$ $= 0$ elsewhere		Steady-state error

as the input to the position control system that we have previously discussed. The designer can clearly see both the transient response and the steady-state response with a step input. The *ramp* input represents an input of constant velocity, such as that found when tracking a vehicle moving at constant velocity. The response from this test signal yields information about the steady-state error. Similarly, for systems where the target is accelerating, the *parabolic* input represents constant acceleration and also yields information about the steady-state error.

Choosing control system components, which are assembled into a viable system, is based upon requirements, such as speed and power. After choosing the system configuration, the engineer must analyze the system to see if the response requirements can be met. If these requirements cannot be met, the designer then designs additional hardware or makes adjustments to the system's parameters in order to achieve the desired performance.

The control systems engineer must take into consideration other characteristics about feedback control systems that we have not yet mentioned. For example, control system behavior changes with changes in component values or system parameters. These changes can be caused by temperature, pressure, or other environmental changes. Systems must be built so that expected changes do not degrade performance beyond specified bounds. A *sensitivity* analysis can yield the percent of change in a specification as a function of a change in a system parameter. One of the designer's goals, then, is to build a system with minimum sensitivity over an expected range of environmental changes. Finally, one of the basic analysis and design requirements is to evaluate the time response of a system for a given input. Throughout the book you will learn numerous methods for accomplishing this goal.

In this section, we looked at some control systems design and analysis considerations. We showed that the designer is concerned about transient response, steady-state error, stability, and sensitivity. We pointed out that although the basis of evaluating system performance is the differential equation, other methods such as transfer functions and state-space, which have distinct advantages over differential equations, will be used. The advantages of these new techniques will become apparent as we discuss them in later chapters. Now that we have discussed the analysis and design sequence, we are ready to study control systems.

1.7 Summary

Control systems contribute to every aspect of modern society. You can find these systems, for example, in automobiles, airplanes, spacecraft, and home appliances. Control systems also exist naturally; our bodies contain numerous control systems. Even economic and psychological system representations have been proposed based upon control system theory. Control systems are used where power gain, remote control, or conversion of the form of the input are required.

A control system has an input, a process, and an output. Open-loop systems do not monitor or correct the output for disturbances; however, they are simpler and less expensive than closed-loop systems. Closed-loop systems, on the other hand, monitor the output and compare it to the input. If an error is detected, the closed-loop system corrects the output; thus, they can correct the effects of disturbances.

Control systems design and analysis focuses on three criteria: (1) transient response, (2) steady-state errors, and (3) stability. A system must first be stable in order to produce the proper transient and steady-state response. Transient response is important because it affects the speed of a system and influences human patience and comfort, not to mention mechanical stress. Steady-state response deals with the accuracy of a control system; it determines how closely the output matches the desired response.

The design of a control system follows these steps: (1) determine a physical system from requirements, (2) represent the physical system as a schematic, (3) mathematically model the schematic, (4) if the system consists of multiple blocks, reduce the mathematical model to a form that can be analyzed, and (5) analyze and design the system to meet specified requirements. During the last step, the engineer performs analysis and design to meet stability, transient, and steady-state requirements.

In the next chapter, we continue through the design and analysis sequence and see how to model the schematic mathematically.

REVIEW QUESTIONS

1. Name three applications for feedback control systems.
2. Name three reasons for using feedback control systems and at least one reason for not using them.
3. Give three examples of open-loop systems.
4. Functionally, how do closed-loop systems differ from open-loop systems?
5. Physically, what happens to a system that is unstable?
6. State one condition under which the error signal of a feedback control system would not be the difference between the input and the output.
7. If the error signal is not the difference between input and output, by what general name can we describe the error signal?
8. Adjustments of the forward path gain can cause changes in the transient response. True or false?
9. Adjustments in the forward path gain never affect the steady-state value. True or false?
10. Name two advantages of having a computer in the loop.
11. Name the three major design criteria for control systems.
12. Name three approaches to the mathematical modeling of control systems.
13. Briefly describe each of your answers to Question 12.

PROBLEMS

1. A variable resistor, called a *potentiometer,* is shown in Figure P1.1. The resistance is varied by moving a wiper arm along a fixed resistance. The resistance from A to C is fixed, but the resistance from B to C varies with the position of the wiper arm. If it takes 10 turns to move the wiper arm from A to C, draw a block diagram of the potentiometer showing the input variable, the output variable, and (inside the block) the gain, which is a constant and is the amount by which the input is multiplied to obtain the output.

Figure P1.1 Potentiometer

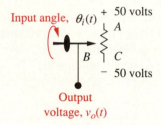

2. A temperature control system operates by sensing the difference between the thermostat setting and the actual temperature and then opening a fuel valve an amount proportional to this difference. Draw a functional closed-loop block diagram similar to Figure 1.15 identifying the input and output transducers, the controller, and the plant. Further, identify the input and output signals of all subsystems previously described.

3. An aircraft's attitude varies in roll, pitch, and yaw as defined in Figure P1.2. Draw a functional block diagram for a closed-loop system that stabilizes the roll as follows: the system measures the actual roll angle with a gyro and compares the actual roll angle with the desired roll angle. The ailerons respond to the roll-angle error by undergoing an angular deflection. The aircraft responds to this angular deflection, producing a roll angle rate. Identify the input and output transducers, the controller, and the plant. Further, identify the nature of each signal.

Figure P1.2 Aircraft Attitude Defined

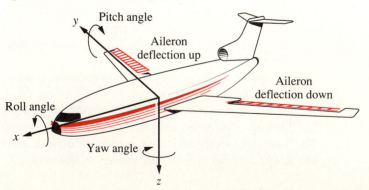

Figure P1.3 Winder (Adapted from Ayers, J. Taking the Mystery Out of Winder Controls. *Power Transmission Design,* April 1988. Penton Publishing Inc., Cleveland, Ohio)

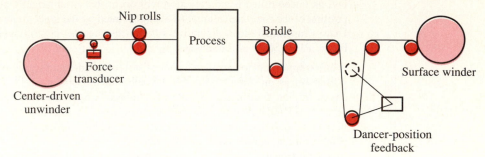

4. Many processes operate on rolled material that moves from a supply reel to a takeup reel. Typically, these systems, called *winders,* control the material so that it travels at a constant velocity. Beside velocity, complex winders also control tension, compensate for roll inertia while accelerating or decelerating, and regulate acceleration due to sudden changes. A winder is shown in Figure P1.3. The force transducer measures tension; the winder pulls against the nip rolls, which provide an opposing force; and the bridle provides slip. In order to compensate for changes in speed, the material is looped around a *dancer.* The loop prevents rapid changes from yielding excessive slack or damaging the material. If the dancer position is sensed by a potentiometer or other device, speed variations due to build-up on the take-up reel or other causes can be controlled by comparing the potentiometer voltage to the commanded speed. The system then corrects the speed and resets the dancer to the desired position (Ayers, 1988). Draw a functional block diagram for the speed control system showing each component and signal.

5. In a nuclear power generating plant, heat from a reactor is used to generate steam for the turbines. The rate of the fission reaction determines the amount of heat generated, and this rate is controlled by rods inserted into the radioactive core.

Figure P1.4 Control of a Nuclear Reactor

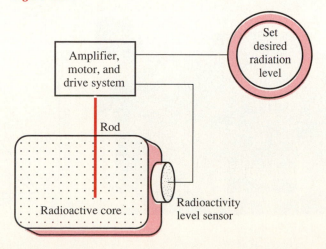

The rods regulate the flow of neutrons. If the rods are inserted completely into the core, the fission process will cease altogether; if the rods are pulled completely out, an uncontrolled fission reaction will occur. By automatically controlling the position of the rods, the amount of heat generated by the reactor can be regulated. Draw a functional block diagram for the nuclear reactor control system shown in Figure P1.4. Show all blocks and signals.

6. A university wants to establish a control systems model that represents the student population as an output with the desired student population as an input. The administration determines the rate of admissions by comparing the current and desired student population. The admissions office then uses the rate established by the administration to admit students. Draw a functional block diagram showing the administration and the admissions office as blocks of the system. Also show the following signals: the desired student population, the actual student population, the desired student rate as determined by the administration, the actual student rate as generated by the admissions office, the drop-out rate, and the net rate of influx.

7. A control system that will automatically adjust a motorcycle's radio volume as the noise generated by the motorcycle changes is part of the radio shown in Figure P1.5. The noise generated by the motorcycle increases with speed. As the noise increases, the system increases the volume of the radio. Assume that the amount of noise can be represented by a voltage generated by the speedometer cable and the volume of the radio is controlled by a dc voltage (Hogan, 1988). If the dc voltage represents the desired volume disturbed by the motorcycle noise, draw

Figure P1.5 (*a*) Voyager XII Radio and Cassette Player with Self-Adjusting Volume Mounted on a Motorcycle; (*b*) Variation of Speaker Volume with Vehicle Speed (Courtesy of Kawasaki Motors Corp.)

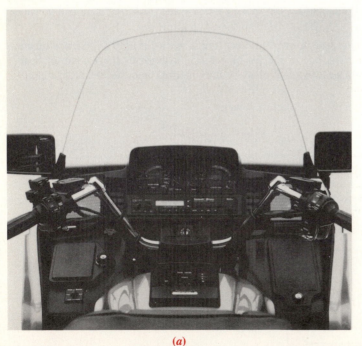

(*a*)

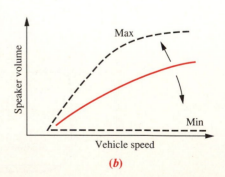

(*b*)

the functional block diagram of the automatic volume control system, showing the input transducer, the volume control circuit, and the speed transducer as blocks. Also show the following signals: the desired volume as an input, the actual volume as an output, and voltages representing speed, desired volume, and actual volume.

8. Given the electric network shown in Figure P1.6,
   a. Write the differential equation for the network if $v(t) = u(t)$, a unit step.
   b. Solve the differential equation for the current, $i(t)$, if all initial conditions are zero.
   c. Make a plot of your solution if $R/L = 1$.

Figure P1.6 Electric Network

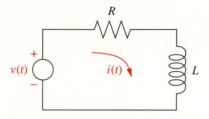

9. Solve the following differential equations using classical methods. Assume zero initial conditions.

   a. $\dfrac{dx}{dt} + 7x = 5\cos 2t$

   b. $\dfrac{d^2x}{dt^2} + 8\dfrac{dx}{dt} + 15x = 7u(t)$

   c. $\dfrac{d^2x}{dt^2} + 2\dfrac{dx}{dt} + 5x = 3\sin 5t$

10. Solve the following differential equations using classical methods and the given initial conditions.

   a. $\dfrac{d^2x}{dt^2} + 2\dfrac{dx}{dt} + 2x = \sin 2t$

   $x(0) = 1; \quad \dfrac{dx}{dt}(0) = -1$

   b. $\dfrac{d^2x}{dt^2} + 2\dfrac{dx}{dt} + x = 5e^{-2t} + t$

   $x(0) = 2; \quad \dfrac{dx}{dt}(0) = 1$

   c. $\dfrac{d^2x}{dt^2} + 4x = t^2$

   $x(0) = 1; \dfrac{dx}{dt}(0) = 2$

BIBLIOGRAPHY

Ayers, J. Taking the Mystery Out of Winder Controls. *Power Transmission Design,* April 1988, pp. 27–34.

Bahill, A. T. *Bioengineering: Biomedical, Medical, and Clinical Engineering.* Prentice-Hall, Englewood Cliffs, N.J., 1981.

Bode, H. W. *Network Analysis and Feedback Amplifier Design.* Van Nostrand, Princeton, N.J., 1945.

Cannon, R. H., Jr. *Dynamics of Physical Systems.* McGraw-Hill, New York, 1967.

D'Azzo, J. J., and Houpis, C. H. *Feedback Control System Analysis and Synthesis.* 2d ed. McGraw-Hill, New York, 1966.

Dorf, R. C. *Modern Control Systems.* 5th ed. Addison-Wesley, Reading, Mass., 1989.

Franklin, G. F.; Powell, J. D.; and Emami-Naeini, A. *Feedback Control of Dynamic Systems.* Addison-Wesley, Reading, Mass., 1986.

Heller, H. C.; Crawshaw, L. I.; and Hammel, H. T. The Thermostat of Vertebrate Animals. *Scientific American,* August 1978, pp. 102–113.

Hogan, B. J. As Motorcycle's Speed Changes, Circuit Adjusts Radio's Volume. *Design News,* 18 August 1988, pp.118–119.

Hostetter, G. H.; Savant, C. J., Jr.; and Stefani, R. T. *Design of Feedback Control Systems.* 2d ed. Saunders College Publishing, New York, 1989.

Klapper, J., and Frankle, J. T. *Phase-Locked and Frequency-Feedback Systems.* Academic Press, New York, 1972.

Martin, R. H., Jr. *Elementary Differential Equations with Boundary Value Problems.* McGraw-Hill, New York, 1984.

Mayr, O. The Origins of Feedback Control. *Scientific American,* October 1970, pp. 110–118.

Novosad, J. P. *Systems, Modeling, and Decision Making.* Kendall/Hunt, Dubuque, Ia. 1982.

Nyquist, H. Regeneration Theory. *Bell System Technical Journal,* January 1932.

Rockwell International. *Space Shuttle Transportation System.* 1984 (press information).

Shaw, D. A., and Turnbull, G. A. Modern Thickness Control for a Generation III Hot Strip Mill. *The International Steel Rolling Conference—The Science & Technology of Flat Rolling Vol. 1.* Association Technique de la Siderurgie Francaise, Deauville, France, 1–3 June 1987.

2

SYSTEM MODELING IN THE FREQUENCY DOMAIN

2.1 Introduction

In Chapter 1 we discussed the design and analysis sequence that included obtaining the system's schematic and demonstrated this step for a position control system. To obtain a schematic, the control systems engineer must often make many simplifying assumptions in order to keep the ensuing model manageable and still approximate physical reality.

The next step is to develop mathematical models from schematics of physical systems. We will discuss two methods: (1) transfer functions in the frequency domain and (2) state equations in the time domain. These topics are covered in this chapter and in Chapter 3, respectively. As we proceed, we will notice that in every case the first step in developing a mathematical model is to apply the appropriate fundamental principles of science and engineering. For example, when we model electrical networks, Ohm's law and Kirchhoff's laws, which are basic principles of electric networks, will be applied initially. We will sum voltages in a loop or sum currents at a node. When we study mechanical networks, we will use Newton's laws as the fundamental guiding principles. Here, we will sum forces or torques. From these equations we will obtain the relationship between the system's output and input.

In Chapter 1 we saw that a differential equation can describe the relationship between the input and output of a system. The form of the differential equation as well as its coefficients are a formulation or description of the system. Although the differential equation relates the system to its input and output, it is not a satisfying representation from a system perspective. Looking at Equation 1.2, a general, nth-order, linear, time-invariant differential equation, we see that the system parameters, which are the coefficients, as well as the output, $c(t)$, and the input, $r(t)$, appear throughout the equation.

We would prefer a mathematical representation such as that shown in Figure 2.1(a), where the input, output, and system are distinct and separate parts. Also, we would like to conveniently represent the interconnection of several subsystems. For example, we would like to represent *cascaded* interconnections, as shown in Figure 2.1(b), where a mathematical function is inside each block and

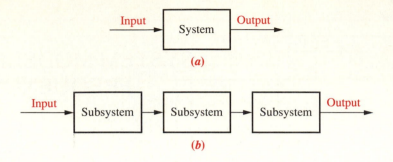

Figure 2.1 (**a**) Representation of a System; (**b**) Representation of an Interconnection of Subsystems

block functions can be easily combined to yield Figure 2.1(*a*) for ease of analysis and design. This convenience cannot be obtained with the differential equation. After stating the chapter objective, we will develop the representation shown in Figure 2.1.

Chapter Objective

Given the antenna azimuth position control system of Figure 2.2(*a*) and the subsequently derived schematic of Figure 2.2(*b*), the student will be able to identify each subsystem and derive for each a mathematical model, similar to Figure 2.2(*c*), that relates a mathematical representation of the subsystem's input, $r(t)$, to a mathematical representation of the subsystem's output, $c(t)$. Before we can meet our objective, a review of the Laplace transform is necessary.

Figure 2.2 Chapter Objective—Antenna Azimuth Position Control System: (**a**) Physical System

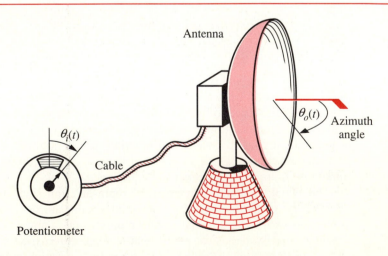

(*a*)

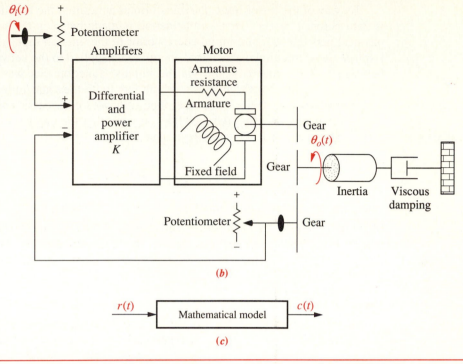

Figure 2.2 (continued) Chapter Objective—Antenna Azimuth Position Control System: (**b**) Schematic; (**c**) Mathematical Representation of Each Subsystem

2.2 Laplace Transform Review

Since a system represented by a differential equation is difficult to visualize, we now lay the groundwork for the Laplace transform. With this transform, not only can we represent the input, output, and system as separate entities, but their interrelationship will be simply algebraic. Let us first define the Laplace transform and then show how it simplifies the representation of physical systems (Nilsson, 1986).

The Laplace transform is defined as

$$\mathscr{L}[f(t)] = F(s) = \int_{0-}^{\infty} f(t)e^{-st}\, dt \tag{2.1}$$

where $s = \sigma + j\omega$, a complex variable. Thus, knowing $f(t)$ and that the integral in Equation 2.1 exists, we can find a function, $F(s)$, that is called the *Laplace transform* of $f(t)$.[1]

---

[1]The Laplace transform exists if the integral of Equation 2.1 converges. The integral will converge if $\int_{0-}^{\infty} |f(t)|e^{-\sigma_1 t}\, dt < \infty$. If $|f(t)| < Me^{\sigma_2 t}$, $0 < t < \infty$, the integral will converge if $\infty > \sigma_1 > \sigma_2$. We call σ_2 the *abscissa of convergence*.

By way of the lower limit, the transform assumes that the time function, $f(t)$, is zero before $t = 0-$. Here, a distinction is made between $0-$ and $0+$. Many physical networks acquire initial energy prior to the time an input is applied. The Laplace transform allows this initial energy to be applied to the network all at once between $t = 0-$ and $t = 0+$ via the impulse function that was introduced in Chapter 1. Recall that the impulse, which has finite area and infinite amplitude, is nonzero only for $0- < t < 0+$. For example, consider a capacitor that is charged between $t = 0-$ and $t = 0+$. Assume that the capacitor has a current, $i(t) = 5\delta(t)$, applied. The voltage is given by

$$v_C(t) = \frac{1}{C} \int_{0-}^{0+} 5\delta(t)\,dt = \frac{5}{C} \tag{2.2}$$

since the area under a unit impulse is 1. This is the voltage after the application of the impulse, so we may call it $v(0+)$. Our conclusion is that the impulse of current applied an initial voltage of $5/C$ to the capacitor between $t = 0-$ and $t = 0+$.

The inverse Laplace transform, which allows us to find $f(t)$ given $F(s)$, is

$$\mathcal{L}^{-1}[F(s)] = \frac{1}{2\pi j} \int_{\sigma - j\infty}^{\sigma + j\infty} F(s)e^{st}\,ds$$

$$= f(t)u(t) \tag{2.3}$$

where

$$u(t) = 1 \qquad t > 0$$
$$= 0 \qquad t < 0$$

is the unit step function. Multiplication of $f(t)$ by $u(t)$ yields a time function that is zero for $t < 0$.

Using Equation 2.1, it is possible to derive a table relating $f(t)$ to $F(s)$ for specific cases. Table 2.1 shows the results for a representative sample of functions. If we use the tables, we do not have to use Equation 2.3, which requires complex integration, to find $f(t)$ given $F(s)$. In the following example, we demonstrate the use of Equation 2.1 to find the Laplace transform of a time function.

Example 2.1   Find the Laplace transform of $f(t)$.

Problem   Find the Laplace transform of $f(t) = Ae^{-at}u(t)$.

Solution   Since the time function does not contain an impulse function, we can replace the lower limit of Equation 2.1 with $0+$ or simply 0. Hence,

$$F(s) = \int_0^\infty f(t)e^{-st}\,dt = \int_0^\infty Ae^{-at}e^{-st}\,dt = A\int_0^\infty e^{-(s+a)t}\,dt$$

$$= -\frac{A}{s+a}e^{-at}\bigg|_{t=0}^\infty = \frac{A}{s+a} \tag{2.4}$$

Table 2.1 Laplace Transform Table

Laplace transform pairs		
Item no.	$f(t)$	$F(s)$
1.	$\delta(t)$	1
2.	$u(t)$	$\dfrac{1}{s}$
3.	$tu(t)$	$\dfrac{1}{s^2}$
4.	$t^n u(t)$	$\dfrac{n!}{s^{n+1}}$
5.	$e^{-at}u(t)$	$\dfrac{1}{s+a}$
6.	$\sin \omega t\, u(t)$	$\dfrac{\omega}{s^2 + \omega^2}$
7.	$\cos \omega t\, u(t)$	$\dfrac{s}{s^2 + \omega^2}$

In the next example, we demonstrate the use of the Laplace transform theorems shown in Table 2.2 to find $f(t)$ given $F(s)$.

Example 2.2   Find the inverse Laplace transform of $F(s)$.

Problem   Find the inverse Laplace transform of $F_1(s) = \dfrac{1}{(s+3)^2}$.

Solution   For this example, we make use of the frequency shift theorem, Item 4 in Table 2.2, and the Laplace transform of $f(t) = tu(t)$, Item 3 of Table 2.1. If the inverse transform of $F(s) = \dfrac{1}{s^2}$ is $tu(t)$, the inverse transform of $F(s+a) = \dfrac{1}{(s+a)^2}$ is $e^{-at}t\,u(t)$. Hence, $f_1(t) = e^{-3t}t\,u(t)$.

Partial Fraction Expansion

To find the inverse Laplace transform of more complicated functions, we must be able to convert these functions to a sum of simpler terms for which we know the Laplace transform. The result is called a *partial fraction expansion*. If $F_1(s) = N(s)/D(s)$, where the order of $N(s)$ is less than the order of $D(s)$, then a partial fraction expansion can be made. If the order of $N(s)$ is greater than or

Table 2.2 Laplace Transform Theorems

Laplace transform theorems

Item no.	Theorem		Name
1.	$\mathcal{L}[f(t)] = F(s) =$	$\displaystyle\int_{0-}^{\infty} f(t)e^{-st}\,dt$	Definition
2.	$\mathcal{L}[kf(t)]$	$= kF(s)$	Linearity theorem
3.	$\mathcal{L}[f_1(t) + f_2(t)]$	$= F_1(s) + F_2(s)$	Linearity theorem
4.	$\mathcal{L}[e^{-at}f(t)]$	$= F(s + a)$	Frequency shift theorem
5.	$\mathcal{L}[f(t - T)]$	$= e^{-sT}F(s)$	Time shift theorem
6.	$\mathcal{L}[f(at)]$	$= \dfrac{1}{a}F\left(\dfrac{s}{a}\right)$	Scaling theorem
7.	$\mathcal{L}\left[\dfrac{df}{dt}\right]$	$= sF(s) - f(0-)$	Differentiation theorem
8.	$\mathcal{L}\left[\dfrac{d^2f}{dt^2}\right]$	$= s^2F(s) - sf(0-) - \dot{f}(0-)$	Differentiation theorem
9.	$\mathcal{L}\left[\dfrac{d^nf}{dt^n}\right]$	$= s^nF(s) - \displaystyle\sum_{k=1}^{n} s^{n-k}f^{k-1}(0-)$	Differentiation theorem
10.	$\mathcal{L}\left[\displaystyle\int_{0-}^{t} f(\tau)\,d\tau\right]$	$= \dfrac{F(s)}{s}$	Integration theorem
11.	$f(\infty)$	$= \displaystyle\lim_{s \to 0} sF(s)$	Final value theorem[1]
12.	$f(0+)$	$= \displaystyle\lim_{s \to \infty} sF(s)$	Initial value theorem[2]

[1] For this theorem to be valid, all roots of the denominator of $F(s)$ must have negative real parts or be at the origin.
[2] For this theorem to be valid, $f(t)$ must be continuous or have a step discontinuity at $t = 0$ (i.e., no impulses or their derivatives at $t = 0$).

equal to the order of $D(s)$, then $N(s)$ must be divided by $D(s)$ successively until the result has a remainder whose numerator is of order less than its denominator. For example, if

$$F_1(s) = \frac{s^3 + 2s^2 + 6s + 7}{s^2 + s + 5} \tag{2.5}$$

we must perform the indicated division until we obtain a remainder whose numerator is of order less than its denominator. Hence,

$$F_1(s) = s + 1 + \frac{2}{s^2 + s + 5} \tag{2.6}$$

Taking the inverse Laplace transform, using Item 1 of Table 2.1, along with the differentiation theorem, Item 7 and the linearity theorem, Item 3 of Table 2.2, we obtain

$$f_1(t) = \frac{d\delta(t)}{dt} + \delta(t) + \mathcal{L}^{-1}\left[\frac{2}{s^2 + s + 5}\right] \tag{2.7}$$

Using partial fraction expansion, we will be able to expand functions like $F(s) = \frac{2}{s^2 + s + 5}$ into a sum of terms and then find the inverse Laplace transform for each term. We will now consider three cases and show for each case how an $F(s)$ can be expanded into partial fractions.

Case 1: Roots of the Denominator of $F(s)$ Are Real and Distinct

An example of an $F(s)$ with real and distinct roots in the denominator is

$$F(s) = \frac{2}{(s + 1)(s + 2)} \tag{2.8}$$

The roots of the denominator are distinct, since each factor is raised only to unity power. We can write the partial fraction expansion as a sum of terms where each factor of the original denominator forms the denominator of each term, and constants, called *residues,* form the numerators. Hence,

$$F(s) = \frac{2}{(s + 1)(s + 2)} = \frac{K_1}{(s + 1)} + \frac{K_2}{(s + 2)} \tag{2.9}$$

To find K_1, we first multiply Equation 2.9 by $(s + 1)$, which isolates K_1. Thus,

$$\frac{2}{(s + 2)} = K_1 + \frac{(s + 1)K_2}{(s + 2)} \tag{2.10}$$

Letting s approach -1 eliminates the last term and yields $K_1 = 2$. Similarly, K_2 can be found by multiplying Equation 2.9 by $(s + 2)$ and then letting s approach -2; hence, $K_2 = -2$.

Each component part of Equation 2.9 is an $F(s)$ in Table 2.1. Hence, $f(t)$ is the sum of the inverse Laplace transform of each term, or

$$f(t) = (2e^{-t} - 2e^{-2t})u(t) \tag{2.11}$$

In general, then, given an $F(s)$ whose denominator has real and distinct roots, a partial fraction expansion,

$$F(s) = \frac{N(s)}{D(s)} = \frac{N(s)}{(s + p_1)(s + p_2)\cdots(s + p_m)\cdots(s + p_n)}$$

$$= \frac{K_1}{(s + p_1)} + \frac{K_2}{(s + p_2)} + \cdots + \frac{K_m}{(s + p_m)} + \cdots + \frac{K_n}{(s + p_n)}$$

$$\tag{2.12}$$

can be made if the order of $N(s)$ is less than the order of $D(s)$. To evaluate each residue, K_i, we multiply Equation 2.12 by the denominator of the corresponding partial fraction. Thus, if we want to find K_m, we multiply Equation 2.12 by $(s + p_m)$ and get

$$(s + p_m)F(s) = \frac{(s + p_m)N(s)}{(s + p_1)(s + p_2)\cdots(s + p_m)\cdots(s + p_n)}$$

$$= (s + p_m)\frac{K_1}{(s + p_1)} + (s + p_m)\frac{K_2}{(s + p_2)} + \cdots + K_m \quad (2.13)$$

If we let s approach $-p_m$, all terms on the right-hand side of Equation 2.13 go to zero except the term K_m, leaving

$$\left. \frac{(s + p_m)N(s)}{(s + p_1)(s + p_2)\cdots(s + p_m)\cdots(s + p_n)} \right|_{s \to -p_m} = K_m \quad (2.14)$$

The following example demonstrates the use of the partial fraction expansion to solve a differential equation. We will see that the Laplace transform reduces the task of finding the solution to simple algebra.

Example 2.3   Solve a second-order, linear, time-invariant differential equation using the Laplace transform and the partial fraction expansion.

Problem   Given the following differential equation, solve for $y(t)$ if all initial conditions are zero. Use the Laplace transform.

$$\frac{d^2y}{dt^2} + 12\frac{dy}{dt} + 32y = 32u(t) \quad (2.15)$$

Solution   Substitute the corresponding $F(s)$ for each term in Equation 2.15, using Item 2 in Table 2.1, Items 7 and 8 in Table 2.2, and the initial conditions of $y(t)$ and $dy(t)/dt$, given by $y(0-) = 0$ and $\dot{y}(0-) = 0$, respectively. Hence, the Laplace transform of Equation 2.15 is

$$S^2Y(s) + 12sY(s) + 32Y(s) = \frac{32}{s} \quad (2.16)$$

Solving for the response, $Y(s)$, yields

$$Y(s) = \frac{32}{s(s^2 + 12s + 32)} = \frac{32}{s(s + 4)(s + 8)} \quad (2.17)$$

To solve for $y(t)$, we notice that Equation 2.17 does not match any of the terms in Table 2.1. Thus, we form the partial fraction expansion of the right-hand term and match each of the resulting terms with $F(s)$ in Table 2.1. Therefore,

$$Y(s) = \frac{32}{s(s + 4)(s + 8)} = \frac{K_1}{s} + \frac{K_2}{(s + 4)} + \frac{K_3}{(s + 8)} \quad (2.18)$$

where, from Equation 2.14,

$$K_1 = \left.\frac{32}{(s+4)(s+8)}\right|_{s\to0} = 1 \tag{2.19a}$$

$$K_2 = \left.\frac{32}{s(s+8)}\right|_{s\to-4} = -2 \tag{2.19b}$$

$$K_3 = \left.\frac{32}{s(s+4)}\right|_{s\to-8} = 1 \tag{2.19c}$$

Since each of the three component parts of Equation 2.18 is represented as an $F(s)$ in Table 2.1, $y(t)$ is the sum of the inverse Laplace transforms of each term. Hence,

$$y(t) = (1 - 2e^{-4t} + e^{-8t})u(t) \tag{2.20}$$

The $u(t)$ in Equation 2.20 shows that the response is zero until $t = 0+$. Unless otherwise specified, all inputs to systems in the text will not start until $t = 0+$.[2] Thus, output responses will also be zero until $t = 0+$. For the sake of convenience, we will leave off the $u(t)$ notation from now on. Accordingly, we write the output response as

$$y(t) = 1 - 2e^{-4t} + e^{-8t} \tag{2.21}$$

Case 2: Roots of the Denominator of F(s) Are Real and Repeated

An example of an $F(s)$ with real and repeated roots in the denominator is

$$F(s) = \frac{2}{(s+1)(s+2)^2} \tag{2.22}$$

The roots of $(s+2)^2$ in the denominator are repeated, since the factor is raised to an integer power higher than 1. In this case, the denominator root at -2 is a *multiple root* of *multiplicity* 2.

We can write the partial fraction expansion as a sum of terms, where each factor of the denominator forms the denominator of each term. In addition, each multiple root generates additional terms consisting of denominator factors of reduced multiplicity. For example, if

$$F(s) = \frac{2}{(s+1)(s+2)^2} = \frac{K_1}{(s+1)} + \frac{K_2}{(s+2)^2} + \frac{K_3}{(s+2)} \tag{2.23}$$

then $K_1 = 2$, which can be found as previously described. K_2 can be isolated by multiplying Equation 2.23 by $(s+2)^2$, yielding

$$\frac{2}{s+1} = (s+2)^2\frac{K_1}{(s+1)} + K_2 + (s+2)K_3 \tag{2.24}$$

---

[2] With the exception of the delta function and its derivatives, which exist before $t = 0+$.

Letting s approach -2, $K_2 = -2$. To find K_3, we see that if we differentiate Equation 2.24 with respect to s,

$$\frac{-2}{(s+1)^2} = \frac{(s+2)s}{(s+1)^2}K_1 + K_3 \tag{2.25}$$

K_3 is isolated and can be found if we let s approach -2. Hence, $K_3 = -2$.

Each component part of Equation 2.23 is an $F(s)$ in Table 2.1; hence, $f(t)$ is the sum of the inverse Laplace transform of each term, or

$$f(t) = 2e^{-t} - 2te^{-2t} - 2e^{-2t} \tag{2.26}$$

If the denominator root is of higher multiplicity than 2, successive differentiation would isolate each residue in the expansion of the multiple root.

In general, then, given an $F(s)$ whose denominator has real and repeated roots, a partial fraction expansion,

$$F(s) = \frac{N(s)}{D(s)}$$

$$= \frac{N(s)}{(s+p_1)^r(s+p_2)\cdots(s+p_n)}$$

$$= \frac{K_1}{(s+p_1)^r} + \frac{K_2}{(s+p_1)^{r-1}} + \cdots + \frac{K_r}{(s+p_1)}$$

$$+ \frac{K_{r+1}}{(s+p_2)} + \cdots + \frac{K_n}{(s+p_n)} \tag{2.27}$$

can be made if the order of $N(s)$ is less than the order of $D(s)$ and the repeated roots are of multiplicity r at $-p_1$. To find K_1 through K_r for the roots of multiplicity greater than unity, first multiply Equation 2.27 by $(s+p_1)^r$ getting $F_1(s)$, which is

$$F_1(s) = (s+p_1)^r F(s)$$

$$= \frac{(s+p_1)^r N(s)}{(s+p_1)^r(s+p_2)\cdots(s+p_n)}$$

$$= K_1 + (s+p_1)K_2 + (s+p_1)^2 K_3 + \cdots + (s+p_1)^{r-1}K_r$$

$$+ \frac{K_{r+1}(s+p_1)^r}{(s+p_2)} + \cdots + \frac{K_n(s+p_1)^r}{(s+p_n)} \tag{2.28}$$

Immediately, we can solve for K_1 if we let s approach $-p_1$. We can solve for K_2 if we differentiate Equation 2.28 with respect to s first and then let s approach $-p_1$. Subsequent differentiation will allow us to find K_3 through K_r. The general expression for K_1 through K_r for the multiple roots is

$$K_i = \frac{1}{(i-1)!}\frac{d^{i-1}F_1(s)}{ds^{i-1}}\bigg|_{s\to-p_1} \qquad i = 1, 2, \ldots, r; \quad 0! = 1 \tag{2.29}$$

Case 3: Roots of the Denominator of F(s) Are Complex or Purely Imaginary

An example of $F(s)$ with complex roots in the denominator is

$$F(s) = \frac{3}{s(s^2 + 2s + 5)} \tag{2.30}$$

This function can be expanded in the following form:

$$\frac{3}{s(s^2 + 2s + 5)} = \frac{K_1}{s} + \frac{K_2 s + K_3}{s^2 + 2s + 5} \tag{2.31}$$

K_1 is found in the usual way to be $3/5$. K_2 and K_3 can be found by first multiplying Equation 2.31 by the lowest common denominator, $s(s^2 + 2s + 5)$, and clearing the fractions. After simplification, with $K_1 = 3/5$, we obtain

$$3 = \left(K_2 + \frac{3}{5}\right)s^2 + \left(K_3 + \frac{6}{5}\right)s + 3 \tag{2.32}$$

Balancing coefficients, $\left(K_2 + \frac{3}{5}\right) = 0$, and $\left(K_3 + \frac{6}{5}\right) = 0$. Hence $K_2 = -\frac{3}{5}$ and $K_3 = -\frac{6}{5}$. Thus,

$$F(s) = \frac{3}{s(s^2 + 2s + 5)} = \frac{3/5}{s} - \frac{3}{5}\frac{s + 2}{s^2 + 2s + 5} \tag{2.33}$$

The last term can be shown to be the sum of the Laplace transforms of an exponentially damped sine and cosine. Using Item 7 in Table 2.1 and Items 2 and 4 in Table 2.2, we get

$$\mathscr{L}[Ae^{-at}\cos\omega t] = \frac{A(s + a)}{(s + a)^2 + \omega^2} \tag{2.34}$$

Similarly,

$$\mathscr{L}[Be^{-at}\sin\omega t] = \frac{B\omega}{(s + a)^2 + \omega^2} \tag{2.35}$$

Adding Equations 2.34 and 2.35, we get

$$\mathscr{L}[Ae^{-at}\cos\omega t + Be^{-at}\sin\omega t] = \frac{A(s + a) + B\omega}{(s + a)^2 + \omega^2} \tag{2.36}$$

We now convert the last term of Equation 2.33 to the form suggested by Equation 2.36 by completing the squares in the denominator and adjusting terms in the numerator without changing its value. Hence,

$$F(s) = \frac{3/5}{s} - \frac{3}{5}\frac{(s + 1) + (1/2)(2)}{(s + 1)^2 + 2^2} \tag{2.37}$$

Comparing Equation 2.37 to Table 2.1 and Equation 2.36, we find

$$f(t) = \frac{3}{5} - \frac{3}{5}e^{-t}\left(\cos 2t + \frac{1}{2}\sin 2t\right) \qquad (2.38)$$

In order to visualize the solution, an alternate form of $c(t)$, obtained by trigonometric identities, is preferable. Using the amplitudes of the cos and sin terms, we factor out $\sqrt{1^2 + (1/2)^2}$ from the term in parentheses and obtain

$$c(t) = \frac{3}{5} - \frac{3}{5}\sqrt{1^2 + (1/2)^2}e^{-t}\left(\frac{1}{\sqrt{1^2 + (1/2)^2}}\cos 2t + \frac{1/2}{\sqrt{1^2 + (1/2)^2}}\sin 2t\right)$$

$$(2.39)$$

Letting $\dfrac{1}{\sqrt{1^2 + (1/2)^2}} = \cos\phi$ and $\dfrac{1/2}{\sqrt{1^2 + (1/2)^2}} = \sin\phi$,

$$c(t) = \frac{3}{5} - \frac{3}{5}\sqrt{1^2 + (1/2)^2}e^{-t}(\cos\phi\cos 2t + \sin\phi\sin 2t) \qquad (2.40)$$

or

$$c(t) = 0.6 - 0.671e^{-t}\cos(2t - \phi) \qquad (2.41)$$

where $\phi = \arctan 0.5 = 26.57°$. Thus, $c(t)$ is a constant plus an exponentially damped sinusoid.

In general, then, given an $F(s)$ whose denominator has complex or purely imaginary roots, a partial fraction expansion,

$$F(s) = \frac{N(s)}{D(s)} = \frac{N(s)}{(s + p_1)(s^2 + as + b)\cdots}$$

$$= \frac{K_1}{(s + p_1)} + \frac{(K_2 s + K_3)}{(s^2 + as + b)} + \cdots \qquad (2.42)$$

can be made if the order of $N(s)$ is less than the order of $D(s)$, p_1 is real, and $(s2 + as + b)$ has complex or purely imaginary roots. The complex or imaginary roots are expanded with $(K_2 s + K_3)$ terms in the numerator rather than just simply K_i, as in the case of real roots. The K_i's in Equation 2.42 are found through balancing the coefficients of the equation after clearing fractions. After completing the squares on $(s^2 + as + b)$ and adjusting the numerator, $\dfrac{K_2 s + K_3}{(s^2 + as + b)}$ can be put into the form shown on the right-hand side of Equation 2.36.

Finally, the case of purely imaginary roots arises if $a = 0$ in Equation 2.42. The calculations are the same.

Another method that follows the technique used for the partial fraction expansion of $F(s)$ with real roots in the denominator can be used for complex and imaginary roots. However, the residues of the complex and imaginary roots are themselves complex conjugates. Then, after taking the inverse Laplace transform, the resulting terms can be identified as

$$\frac{e^{j\theta} + e^{-j\theta}}{2} = \cos\theta \tag{2.43}$$

and

$$\frac{e^{j\theta} - e^{-j\theta}}{2j} = \sin\theta \tag{2.44}$$

For example, the previous $F(s)$ can also be expanded in partial fractions as

$$F(s) = \frac{3}{s(s^2 + 2s + 5)} = \frac{3}{s(s + 1 + j2)(s + 1 - j2)}$$

$$= \frac{K_1}{s} + \frac{K_2}{s + 1 + j2} + \frac{K_3}{s + 1 - j2} \tag{2.45}$$

Finding K_2,

$$K_2 = \left.\frac{3}{s(s + 1 - j2)}\right|_{s \to -1 - j2} = -\frac{3}{20}(2 + j1) \tag{2.46}$$

Similarly, K_3 is found to be the complex conjugate of K_2, and K_1 is found as previously described. Hence,

$$F(s) = \frac{3/5}{s} - \frac{3}{20}\left(\frac{2 + j1}{s + 1 + j2} + \frac{2 - j1}{s + 1 - j2}\right) \tag{2.47}$$

from which

$$f(t) = \frac{3}{5} - \frac{3}{20}\left[(2 + j1)e^{-(1+j2)t} + (2 - j1)e^{-(1-j2)t}\right]$$

$$= \frac{3}{5} - \frac{3}{20}e^{-t}\left[4\left(\frac{e^{j2t} + e^{-j2t}}{2}\right) + 2\left(\frac{e^{j2t} - e^{-j2t}}{2j}\right)\right] \tag{2.48}$$

Using Equations 2.43 and 2.44, we get

$$f(t) = \frac{3}{5} - \frac{3}{5}e^{-t}\left(\cos 2t + \frac{1}{2}\sin 2t\right) = 0.6 - 0.671e^{-t}\cos(2t - \phi) \tag{2.49}$$

where $\phi = \arctan 0.5 = 26.57°$.

In this section, we defined the Laplace transform and its inverse. We presented the idea of the partial fraction expansion and applied the concepts to the solution of differential equations. We are now ready to formulate the system representation shown in Figure 2.1.

2.3 The Transfer Function

We now establish a viable definition for a function that algebraically relates a system's output to its input. This function will allow separation of the input, system, and output into three separate and distinct parts, unlike the differential equation. The function will also allow us to *algebraically* combine mathematical representations of subsystems to yield a total system representation.

Let us begin by first writing a general nth-order, linear, time-invariant differential equation,

$$a_n \frac{d^n c(t)}{dt^n} + a_{n-1} \frac{d^{n-1} c(t)}{dt^{n-1}} + \cdots + a_0 c(t)$$

$$= b_m \frac{d^m r(t)}{dt^m} + b_{m-1} \frac{d^{m-1} r(t)}{dt^{m-1}} + \cdots + b_0 r(t) \qquad (2.50)$$

where $c(t)$ is the output, $r(t)$ is the input, and the a_1's and the form of the differential equation represent the system. Taking the Laplace transform of both sides,

$$a_n s^n C(s) + a_{n-1} s^{n-1} C(s) + \cdots + a_0 C(s) + \text{Initial condition}$$
$$\text{terms involving } c(t)$$

$$= b_m s^m R(s) + b_{m-1} s^{m-1} R(s) + \cdots + b_0 R(s) + \text{Initial condition}$$
$$\text{terms involving } r(t)$$
$$(2.51)$$

Equation 2.51 is a purely algebraic expression. If we assume that *all initial conditions are zero,* Equation 2.51 reduces to

$$(a_n s^n + a_{n-1} s^{n-1} + \cdots + a_0) C(s)$$
$$= (b_m s^m + b_{m-1} s^{m-1} + \cdots + b_0) R(s) \qquad (2.52)$$

Now form the ratio of the output transform, $C(s)$, divided by the input transform, $R(s)$:

$$\frac{C(s)}{R(s)} = G(s) = \frac{(b_m s^m + b_{m-1} s^{m-1} + \cdots + b_0)}{(a_n s^n + a_{n-1} s^{n-1} + \cdots + a_0)} \qquad (2.53)$$

Notice that Equation 2.53 separates the output, $C(s)$, the input, $R(s)$, and the system, the ratio on the right. We call this ratio, $G(s)$, the *transfer function,* and it is evaluated with *zero initial conditions*.

The transfer function can be represented as a block diagram, as shown in

Figure 2.3 Block Diagram of a Transfer Function

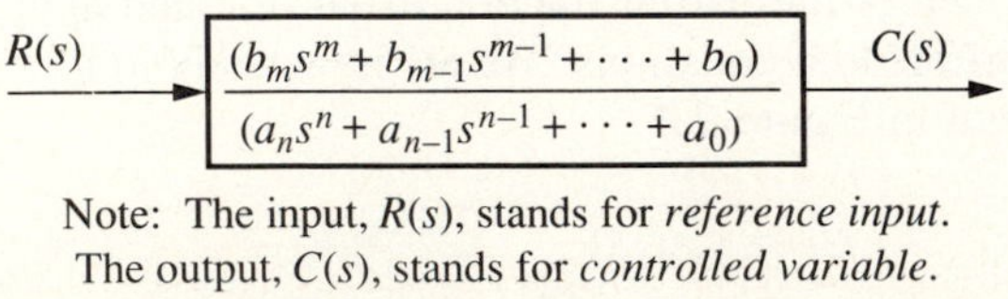

Note: The input, $R(s)$, stands for *reference input.*
The output, $C(s)$, stands for *controlled variable.*

Figure 2.3, with the input on the left, the output on the right, and the system transfer function inside the block. Notice that the denominator of the transfer function is identical to the characteristic polynomial of the differential equation. Also, we can find the output, $C(s)$, by using

$$C(s) = R(s)G(s) \tag{2.54}$$

Let us apply the concept to an example.

Example 2.4   Find the transfer function represented by a differential equation.

Problem   Find the transfer function represented by Equation 2.55.

$$\frac{dx}{dt} + 2x = r(t) \tag{2.55}$$

Solution   Taking the Laplace transform of both sides, assuming zero initial conditions, we have

$$sX(s) + 2X(s) = R(s) \tag{2.56}$$

The transfer function, $G(s)$, is

$$G(s) = \frac{X(s)}{R(s)} = \frac{1}{s + 2} \tag{2.57}$$

The transfer function can be used to find the response of a system, using Equation 2.54. The following example demonstrates the method.

Example 2.5   Find the response of a system from the transfer function.

Problem   Use the result of Example 2.4 to find the response, $x(t)$, to an input, $r(t) = u(t)$, a unit step, assuming zero initial conditions.

Solution   To solve the problem, we use Equation 2.54, where $G(s) = \dfrac{1}{s + 2}$ as found in Example 2.4. Since $r(t) = u(t)$, $R(s) = 1/s$, from Table 2.1. Since the initial conditions are zero,

$$C(s) = R(s)G(s) = \frac{1}{s(s + 2)} \tag{2.58}$$

Expanding by partial fractions, we get

$$C(s) = \frac{1/2}{s} - \frac{1/2}{s + 2} \tag{2.59}$$

Finally, taking the inverse Laplace transform of each term yields

$$c(t) = \frac{1}{2} - \frac{1}{2}e^{-2t} \tag{2.60}$$

In general, physical systems that can be represented by a linear, time-invariant differential equation can be modeled as a transfer function. The rest of this chapter will be devoted to the task of modeling individual subsystems. We will learn how to represent electrical networks, translational mechanical systems, rotational mechanical systems, and electromechanical systems as transfer functions. As the need arises, the reader can consult the Bibliography at the end of the chapter for discussions of other types of systems, such as pneumatic, hydraulic, and heat-transfer systems (Cannon, 1967).

2.4 Transfer Functions for Electrical Networks

In this section we formally apply the transfer function to the mathematical modeling of electrical networks. Subsequent sections cover mechanical and electromechanical systems.

Equivalent circuits for electrical networks that we will work with will consist of three passive, linear components: resistors, capacitors, and inductors.[3] Table 2.3 summarizes the components and the relationships between voltage and current, and voltage and charge under zero initial conditions.

We form a transfer function for each of the electrical components by first taking the Laplace transform of the equations in the second column. For the capacitor,

$$V(s) = \frac{1}{Cs}I(s) \qquad (2.61)$$

For the resistor,

$$V(s) = RI(s) \qquad (2.62)$$

For the inductor,

$$V(s) = LsI(s) \qquad (2.63)$$

Let us now define a particular kind of transfer function:

$$\frac{V(s)}{I(s)} = Z(s) \qquad (2.64)$$

Notice that this function is similar to the definition of resistance, that is, the ratio of voltage to current. But, unlike resistance, this function is applicable to capacitors and inductors and carries with it information on the dynamic behavior of the component, since it represents an equivalent differential equation. We will call this particular transfer function *impedance*. The impedance for each of the electrical elements is shown in the last column of Table 2.3.

We now combine electrical components into circuits, decide upon the input and output, and find the transfer function. Our guiding principles are Kirchhoff's laws. We will sum voltages around loops, or sum currents at nodes, depending

---

[3] *Passive* means that there is no internal source of energy.

Table 2.3 Voltage-Current, Voltage-Charge, and Impedance Relationships for a Capacitor, a Resistor, and an Inductor

Component	Voltage-current	Voltage-charge	Impedance $Z(s) = V(s)/I(s)$
⊣⊢ Capacitor	$v(t) = \dfrac{1}{C}\displaystyle\int_0^t i(\tau)\,d\tau$	$v(t) = \dfrac{1}{C}q(t)$	$\dfrac{1}{Cs}$
⟋⋁⋁⋁⟍ Resistor	$v(t) = Ri(t)$	$v(t) = R\dfrac{dq(t)}{dt}$	R
⟋0000⟍ Inductor	$v(t) = L\dfrac{di(t)}{dt}$	$v(t) = L\dfrac{d^2q(t)}{dt^2}$	Ls

Note: The following set of symbols and units are used throughout this book: $v(t)$ = V (volts), $i(t)$ = A (amperes), $q(t)$ = Q (coulombs), C = F (farads), $R = \Omega$ (ohms), L = H (henries).

upon which technique yields the least effort in algebraic manipulation, and then equate the result to zero. From these relationships, we can write differential equations for the circuit. Then we can take the Laplace transform of the differential equations and finally solve for the transfer function.

Another technique is to replace the circuit elements with their impedance and write Kirchhoff's law directly, bypassing the need to write differential equations first. The following examples demonstrate the use of the various techniques. We solve the problem in three different ways, so that you can gain insight as to the easiest method.

Example 2.6

Find the transfer function of a single-loop electrical network.

Problem   Find the transfer function relating the capacitor voltage, $V_C(s)$, to the input voltage, $V(s)$, in Figure 2.4.

Solution   In any problem, the designer must first decide what the input and output should be. In this network several variables could be chosen to be the output, for example, the inductor voltage, the capacitor voltage, the resistor voltage, or the cur-

Figure 2.4 *RLC* Network

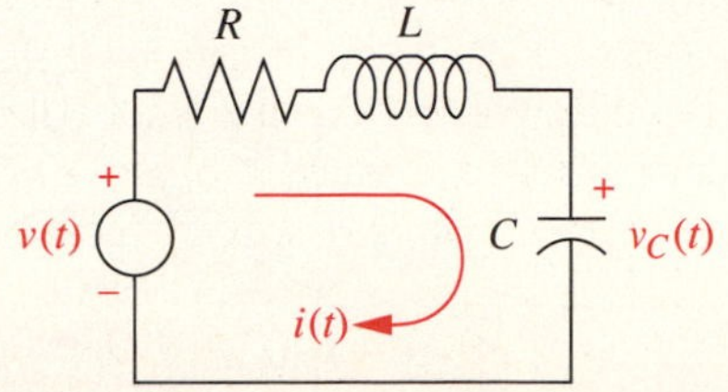

rent. The problem statement, however, is quite clear in this case: we are to treat the capacitor voltage as the output and the applied voltage as the input.

Method 1: Loop Equations Using Kirchhoff's Voltage Law Assuming zero initial conditions, summing the voltages around the loop yields the following integro-differential equation for this network:

$$Ri(t) + L\frac{di}{dt} + \frac{1}{C}\int_0^t i(\tau)\,d\tau = v(t) \tag{2.65}$$

Taking the Laplace transform, again assuming zero initial conditions, yields

$$RI(s) + LsI(s) + \frac{1}{Cs}I(s) = V(s) \tag{2.66}$$

Rewriting Equation 2.66, we get

$$V(s) = \left(R + Ls + \frac{1}{Cs}\right)I(s) \tag{2.67}$$

Notice, from Equation 2.67, that the applied voltage equals the sum of impedances around the loop times the current in the loop.

Forming the transfer function, $I(s)/V(s)$, from Equation 2.67 yields

$$\frac{I(s)}{V(s)} = \frac{1}{\left(R + Ls + \dfrac{1}{Cs}\right)} = \frac{\dfrac{1}{L}s}{s^2 + \dfrac{R}{L}s + \dfrac{1}{LC}} \tag{2.68}$$

From Equation 2.61, the voltage across the capacitor, $V_C(s)$, is $\dfrac{1}{Cs}I(s)$. Thus, multiplying the transfer function of Equation 2.68 by $\dfrac{1}{Cs}$ yields the final result:

$$\frac{V_C(s)}{V(s)} = \frac{\dfrac{1}{LC}}{s^2 + \dfrac{R}{L}s + \dfrac{1}{LC}} \tag{2.69}$$

To simplify the solution for future problems, we notice that Equation 2.67 suggests the transformed circuit shown in Figure 2.5. This circuit can also be obtained by replacing the component values by their impedances. Thus, rather than writing the

Figure 2.5 Laplace Transformed Network

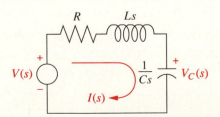

differential equation first and then taking the Laplace transform, we can immediately draw a transformed circuit as follows:

1. Redraw the original circuit showing all time variables, such as $v(t)$, $i(t)$, and $v_C(t)$, as Laplace transforms $V(s)$, $I(s)$, and $V_C(s)$, respectively.
2. Replace the component values with their impedance values. This replacement is similar to the case of dc circuits, where we represent resistors with their resistance values.

Method 2: Node Equations Using Kirchhoff's Current Law   The transfer function can also be obtained by summing currents flowing out of the node whose voltage is $V_C(s)$ in Figure 2.5. We assume that currents leaving the node are positive and currents entering the node are negative. The currents consist of the current through the capacitor and the current flowing through the series resistor and inductor. From Equation 2.64, each $I(s) = V(s)/Z(s)$. Hence,

$$\frac{V_C(s)}{1/Cs} + \frac{V_C(s) - V(s)}{R + Ls} = 0 \qquad (2.70)$$

where $\dfrac{V_C(s)}{1/Cs}$ is the current flowing out of the node through the capacitor, and $\dfrac{V_C(s) - V(s)}{R + Ls}$ is the current flowing out of the node through the series resistor and inductor. Solving Equation 2.70 for the transfer function, we arrive at the same result as Equation 2.69.

Method 3: Voltage Division   The problem also can be solved directly by using voltage division on the transformed network. That is, the voltage across the capacitor is some proportion of the input voltage, namely, the impedance of the capacitor divided by the sum of the impedances. Thus,

$$V_C(s) = \frac{\dfrac{1}{Cs}}{R + Ls + \dfrac{1}{Cs}} \qquad (2.71)$$

which, after simplification, yields the same result as Equation 2.69.

Which method is the easiest for this circuit?

The preceding example involves a simple, single-loop electrical network. Many electrical networks consist of multiple loops and nodes; this requires that we write and solve simultaneous differential equations in order to find the transfer function or solve for the output.

In order use loop equations to handle multiple loops and nodes in electrical networks, we can perform the following steps:

1. Replace passive element values with their impedance.
2. Assume a current direction in each *mesh*.[4]
3. Write Kirchhoff's voltage law around each mesh.

---

[4]A particular loop that resembles the spaces in a screen or fence is called a *mesh*.

4. Solve the simultaneous equations for the output.

5. Form the transfer function.

Let us look at an example.

Example 2.7   Find the transfer function of an electrical network, using mesh equations.

Problem   Given the network of Figure 2.6(*a*), find the transfer function $I_2(s)/V(s)$.

Solution   The first step in the problem solution is to convert the network into Laplace transforms for impedances and circuit variables, assuming zero initial conditions. The result is shown in Figure 2.6(*b*). The circuit with which we are dealing requires two simultaneous equations to solve for the transfer function. These equations can be found by summing voltages around each mesh through which the assumed currents, $I_1(s)$ and $I_2(s)$, flow. Around Mesh 1, where $I_1(s)$ flows,

$$R_1I_1(s) + LsI_1(s) - LsI_2(s) = V(s) \tag{2.72}$$

Around Mesh 2, where $I_2(s)$ flows,

$$LsI_2(s) + R_2I_2(s) + \frac{1}{Cs}I_2(s) - LsI_1(s) = 0 \tag{2.73}$$

Figure 2.6 (*a*) Two-Loop Electrical Network; (*b*) Transformed Two-Loop Electrical Network; (*c*) Transfer Function Representation

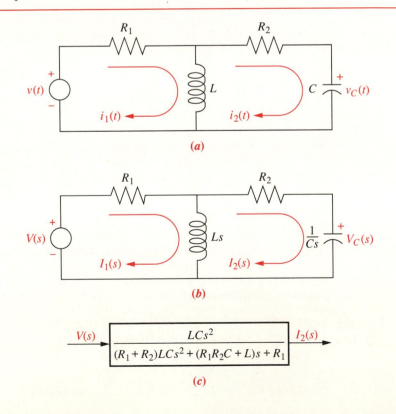

(*a*)

(*b*)

(*c*)

Combining terms, Equations 2.72 and 2.73 become simultaneous equations in $I_1(s)$ and $I_2(s)$:

$$(R_1 + Ls)I_1(s) \qquad\qquad - LsI_2(s) = V(s) \qquad (2.74a)$$

$$-LsI_1(s) + \left(Ls + R_2 + \frac{1}{Cs}\right)I_2(s) = 0 \qquad (2.74b)$$

We can use Cramer's rule (or any other method for solving simultaneous equations) to solve Equations 2.74 for $I_2(s)$. [5] Hence,

$$I_2(s) = \frac{\begin{vmatrix} (R_1 + Ls) & V(s) \\ -Ls & 0 \end{vmatrix}}{\Delta} = \frac{LsV(s)}{\Delta} \qquad (2.75)$$

where

$$\Delta = \begin{vmatrix} (R_1 + Ls) & -Ls \\ -Ls & \left(Ls + R_2 + \dfrac{1}{Cs}\right) \end{vmatrix}$$

Forming the transfer function, $G(s)$, yields

$$G(s) = \frac{I_2(s)}{V(s)} = \frac{Ls}{\Delta} = \frac{LCs^2}{(R_1 + R_2)LCs^2 + (R_1R_2C + L)s + R_1} \qquad (2.76)$$

We have succeeded in modeling a physical network as a transfer function: the network of Figure 2.6(a) is now modeled as the transfer function of Figure 2.6(c). Before leaving the example, we notice a pattern first illustrated in Example 2.6. The form that Equations 2.74 take is

$$\begin{bmatrix} \text{Sum of} \\ \text{impedances} \\ \text{around Mesh 1} \end{bmatrix} I_1 - \begin{bmatrix} \text{Sum of} \\ \text{impedances} \\ \text{common to the} \\ \text{two meshes} \end{bmatrix} I_2 = \begin{bmatrix} \text{Sum of applied} \\ \text{voltages around} \\ \text{Mesh 1} \end{bmatrix} \qquad (2.77a)$$

$$-\begin{bmatrix} \text{Sum of} \\ \text{impedances} \\ \text{common to the} \\ \text{two meshes} \end{bmatrix} I_1 + \begin{bmatrix} \text{Sum of} \\ \text{impedances} \\ \text{around Mesh 2} \end{bmatrix} I_2 = \begin{bmatrix} \text{Sum of applied} \\ \text{voltages around} \\ \text{Mesh 2} \end{bmatrix} \qquad (2.77b)$$

Recognizing the form will help us write such equations rapidly; for example, mechanical equations of motion (covered in Section 2.5) have the same form.

Many times the easiest way to find the transfer function is to use node equations rather than loop equations. The number of simultaneous differential equations that must be written is equal to the number of nodes whose voltage is unknown. In the previous example, we wrote simultaneous mesh equations using Kirchhoff's

---

[5] See Appendix B for Cramer's rule (section B.4).

voltage law. For multiple nodes, we use Kirchhoff's current law and sum currents flowing from each node. Again, as a convention, currents flowing from the node are assumed to be positive, and currents flowing into the node are assumed to be negative.

Before working an example, let us first define the term *admittance, $Y(s)$,* as the reciprocal of impedance,

$$Y(s) = \frac{1}{Z(s)} = \frac{I(s)}{V(s)} \tag{2.78}$$

When writing nodal equations, it is more convenient to represent circuit elements by their admittance.

Another simplifying step in the solution of node equations is to replace voltage sources by current sources. Since a voltage source presents a constant voltage to any load, conversely, a current source delivers a constant current to any load. Practically, a current source can be constructed from a voltage source by placing a large resistance in series with the voltage source. Thus, any variations in the load do not appreciably change the current, since the current is determined approximately by the large series resistor and the voltage source. Theoretically, we rely on Norton's theorem which states that a voltage source, $V(s)$, in series with an impedance, $Z_s(s)$, can be replaced by a current source, $I(s) = V(s)/Z_s(s)$ in parallel with $Z_s(s)$.

In order to handle multiple-node electrical networks, we perform the following steps:

1. Replace passive element values with their admittances.
2. Replace voltage sources with current sources (optional).
3. Write Kirchhoff's current law at each node.
4. Solve the simultaneous equations for the output.
5. Form the transfer function.

Let us look at an example.

Example 2.8   Find the transfer function of an electrical network using node equations.

Problem   For the network of Figure 2.6, find the transfer function, $G(s) = V_C(s)/V(s)$, using nodal analysis.

Solution   For this problem, there will be less algebraic manipulation if we sum currents at the nodes rather than sum voltages around the meshes. Further, the equations follow a pattern like that of the mesh equations if we use admittances and current sources rather than impedances and voltage sources. Therefore, we will convert all impedances to admittances and all voltage sources in series with an impedance to current sources in parallel with an admittance, using Norton's theorem.

Redrawing Figure 2.6(*b*) to reflect the above changes, we obtain Figure 2.7, where $G_1 = 1/R_1$, and the node voltages—the voltages across the inductor and the

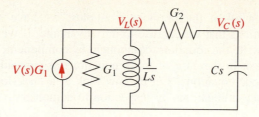

Figure 2.7 A Transformed Network Ready for Nodal Analysis

capacitor—have been identified as $V_L(s)$ and $V_C(s)$, respectively.[6] Using the general relationship $I(s) = Y(s)V(s)$ and summing currents at the node $V_L(s)$, we get

$$G_1 V_L(s) + \frac{1}{Ls}V_L(s) + G_2[V_L(s) - V_C(s)] = V(s)G_1 \qquad (2.79)$$

Summing the currents at the node $V_C(s)$ yields

$$C s V_C(s) + G_2[V_C(s) - V_L(s)] = 0 \qquad (2.80)$$

Combining terms, Equations 2.79 and 2.80 become simultaneous equations in $V_C(s)$ and $V_L(s)$:

$$\left(G_1 + G_2 + \frac{1}{Ls}\right)V_L(s) \qquad\qquad - G_2 V_C(s) = V(s)G_1 \qquad (2.81a)$$

$$-G_2 V_L(s) + (G_2 + C s)V_C(s) = 0 \qquad (2.81b)$$

Solving for the transfer function, $V_C(s)/V(s)$, we get

$$\frac{V_C(s)}{V(s)} = \frac{\dfrac{G_1 G_2}{C}s}{(G_1 + G_2)s^2 + \dfrac{G_1 G_2 L + C}{LC}s + \dfrac{G_2}{LC}} \qquad (2.82)$$

Notice again the format of Equations 2.81:

$$\begin{bmatrix} \text{Sum of} \\ \text{admittances} \\ \text{connected to} \\ \text{Node 1} \end{bmatrix}V_L - \begin{bmatrix} \text{Sum of} \\ \text{admittances} \\ \text{common to} \\ \text{the two nodes} \end{bmatrix}V_C = \begin{bmatrix} \text{Sum of} \\ \text{applied} \\ \text{currents at} \\ \text{Node 1} \end{bmatrix} \qquad (2.83a)$$

$$-\begin{bmatrix} \text{Sum of} \\ \text{admittances} \\ \text{common to} \\ \text{the two nodes} \end{bmatrix}V_L + \begin{bmatrix} \text{Sum of} \\ \text{admittances} \\ \text{connected to} \\ \text{Node 2} \end{bmatrix}V_C = \begin{bmatrix} \text{Sum of} \\ \text{applied} \\ \text{currents} \\ \text{at Node 2} \end{bmatrix} \qquad (2.83b)$$

[6]In general, admittance is complex. The real part is called *conductance,* and the imaginary part is called *susceptance*. But when we take the reciprocal of resistance to obtain admittance, a purely real quantity results. The reciprocal of resistance is called *conductance*.

In all of the previous examples, we have seen a repeating pattern in the equations that we can use to our advantage. If we recognize this pattern, we need not write the equations component by component; we can sum impedances around a mesh in the case of mesh equations or sum admittances at a node in the case of node equations. Let us now look at a three-loop electrical network and write the mesh equations by inspection to demonstrate the process.

Example 2.9   Write the mesh equations by inspection.

Problem   Write, but do not solve, the mesh equations for the network shown in Figure 2.8.

Solution   We have pointed out in each of the previous problems that the mesh equations and nodal equations have a predictable form. We use that knowledge to solve this three-loop problem. The equation for Mesh 1 will have the following form:

$$
\begin{bmatrix} \text{Sum of} \\ \text{impedances} \\ \text{around Mesh 1} \end{bmatrix} I_1(s) - \begin{bmatrix} \text{Sum of} \\ \text{impedances} \\ \text{common to} \\ \text{Mesh 1 and} \\ \text{Mesh 2} \end{bmatrix} I_2(s)
$$

$$
- \begin{bmatrix} \text{Sum of} \\ \text{impedances} \\ \text{common to} \\ \text{Mesh 1 and} \\ \text{Mesh 3} \end{bmatrix} I_3(s) = \begin{bmatrix} \text{Sum of applied} \\ \text{voltages around} \\ \text{Mesh 1} \end{bmatrix} \qquad (2.84)
$$

Similarly, Meshes 2 and 3, respectively, are

$$
- \begin{bmatrix} \text{Sum of} \\ \text{impedances} \\ \text{common to} \\ \text{Mesh 1 and} \\ \text{Mesh 2} \end{bmatrix} I_1(s) + \begin{bmatrix} \text{Sum of} \\ \text{impedances} \\ \text{around Mesh 2} \end{bmatrix} I_2(s)
$$

$$
- \begin{bmatrix} \text{Sum of} \\ \text{impedances} \\ \text{common to} \\ \text{Mesh 2 and} \\ \text{Mesh 3} \end{bmatrix} I_3(s) = \begin{bmatrix} \text{Sum of applied} \\ \text{voltages around} \\ \text{Mesh 2} \end{bmatrix} \qquad (2.85)
$$

and

$$
- \begin{bmatrix} \text{Sum of} \\ \text{impedances} \\ \text{common to} \\ \text{Mesh 1 and} \\ \text{Mesh 3} \end{bmatrix} I_1(s) - \begin{bmatrix} \text{Sum of} \\ \text{impedances} \\ \text{common to} \\ \text{Mesh 2 and} \\ \text{Mesh 3} \end{bmatrix} I_2(s)
$$

$$
+ \begin{bmatrix} \text{Sum of} \\ \text{impedances} \\ \text{around Mesh 3} \end{bmatrix} I_3(s) = \begin{bmatrix} \text{Sum of applied} \\ \text{voltages around} \\ \text{Mesh 3} \end{bmatrix} \qquad (2.86)
$$

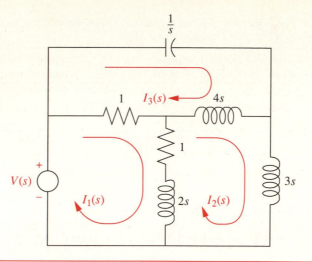

Figure 2.8 A Three-Loop Electrical Network

Substituting the values from Figure 2.8 into Equations 2.84 through 2.86 yields

$$+(2s + 2)I_1(s) - (2s + 1)I_2(s) \qquad\qquad - I_3(s) = V(s) \qquad (2.87a)$$

$$-(2s + 1)I_1(s) + (9s + 1)I_2(s) \qquad\qquad - 4sI_3(s) = 0 \qquad (2.87b)$$

$$- I_1(s) \qquad - 4sI_2(s) + \left(4s + 1 + \frac{1}{s}\right)I_3(s) = 0 \qquad (2.87c)$$

which can be solved simultaneously for any desired transfer function, for example, $I_3(s)/V(s)$.

In this section we found transfer functions for multiple-loop and multiple-node electrical networks. Repeating patterns in the simultaneous equations were used to our advantage to simplify writing them. In the next section, we begin our work with mechanical systems. We will see that many of the concepts applied to electrical networks also apply to mechanical systems. This revelation will give you the confidence to move beyond this textbook to study systems not covered here, such as hydraulic or pneumatic systems.

2.5 Transfer Functions for Translational Mechanical Systems

We have shown that electrical networks can be modeled by a transfer function, $G(s)$, that algebraically relates the Laplace transform of the output to the Laplace transform of the input. Now we will do the same for mechanical systems. In this

Table 2.4 Force-Velocity, Force-Displacement, and Impedance Translational Relationships for a Spring, a Viscous Damper, and a Mass

Component	Force-Velocity	Force-Displacement	Impedance $Z_M(s) = F(s)/X(s)$
Spring K	$f(t) = K \int_0^t v(\tau)\, d\tau$	$f(t) = K x(t)$	K
Viscous damper f_v	$f(t) = f_v v(t)$	$f(t) = f_v \dfrac{dx(t)}{dt}$	$f_v s$
Mass M	$f(t) = M \dfrac{dv(t)}{dt}$	$f(t) = M \dfrac{d^2 x(t)}{dt^2}$	$M s^2$

Note: The following set of symbols and units are used throughout this book: $f(t)$ = N (newtons), $x(t)$ = m (meters), $v(t)$ = m/s (meters/second), K = N/m (newtons/meter), f_v = N-s/m (newton-seconds/meter), M = kg (kilograms = newton-seconds^2/meter).

section we concentrate on translational mechanical systems. In the next section we extend the concepts to rotational mechanical systems. Notice that the end product, shown in Figure 2.3, will be mathematically indistinguishable from an electrical network. Hence, an electrical network can be interfaced to a mechanical system by cascading their transfer functions, provided that one system is not loaded by the other.[7]

Mechanical systems parallel electrical networks to such an extent that there are analogies between electrical and mechanical components and variables. Mechanical systems, like electrical networks, have three passive, linear components. Two of them, the spring and the mass, are energy-storage elements; one of them, the viscous damper, dissipates energy. The two energy-storage elements are analogous to the two electrical energy-storage elements, the inductor and capacitor. The energy dissipator is analogous to electrical resistance. Let us take a look at these mechanical elements, which are shown in Table 2.4. In the table, K, f_v, and M are called *spring constant*, *coefficient of viscous friction*, and *mass*, respectively.

---

[7]The concept of loading is explained further in Chapter 5.

We can see immediately the analogy between mechanical and electrical components by comparing Table 2.4 for mechanical components to Table 2.3 for electrical components. The applied mechanical force is analogous to the applied electrical voltage; the mechanical response of velocity is analogous to the electrical response of current; and the mechanical response of displacement is analogous to the electrical response of charge. Another analogy can be drawn if a different form of the electrical relationships is used. For example, if the electrical equations in Table 2.3 are written with current as a function of voltage (i.e., assuming that current is applied and voltage is the response), what analogy would you draw? This second analogy, which draws a parallel between variables that go *through* the component (force and current) and variables that appear *across* the component (velocity and voltage) is particularly useful when analyzing a mechanical system by first converting it into an electrical network, solving the electrical network, and then relating the results back to the mechanical system. However, in this book we will work directly with the mechanical system; therefore, we need not use any analogy except our original one, where voltage and force, respectively, are typically the applied variables.

We have previously defined impedance as the ratio of the Laplace transform of voltage to the Laplace transform of current—the ratio of the applied variable to a response variable. Let us now define impedance for mechanical components by selecting force as the applied variable and displacement as the response variable. Thus, mechanical impedance is defined as

$$Z_M(s) = \frac{F(s)}{X(s)} \qquad (2.88)$$

In order to evaluate the impedance for each mechanical component, we must first look at the force-displacement column in Table 2.4 and take the Laplace transform. For the spring,

$$F(s) = KX(s) \qquad (2.89)$$

For the viscous damper,

$$F(s) = f_v sX(s) \qquad (2.90)$$

For the mass,

$$F(s) = Ms^2X(s) \qquad (2.91)$$

Table 2.4 summarizes the impedances as calculated from Equation 2.88, using Equations 2.89 through 2.91.

We are now ready to find transfer functions for translational mechanical systems. Our first example, shown in Figure 2.9(a), is similar to the simple *RLC* network of Example 2.6 (see Figure 2.4). The mechanical system requires just one differential equation, called the *equation of motion,* to describe it. We will begin by assuming a positive direction of motion, for example, to the right. This assumed positive direction of motion is similar to assuming a current direction in an electrical loop. Using our assumed direction of positive motion, we first

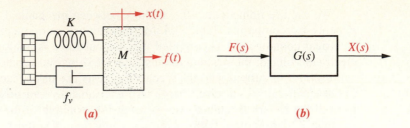

Figure 2.9 (*a*) A Mass, Spring, and Damper System;
(*b*) A Representation

draw a free-body diagram, placing on the body all forces that act on the body either in the direction of motion or opposite to it. The forces can be represented either in the time domain or by their Laplace transforms, assuming zero initial conditions. Next, we use Newton's law by summing the forces and setting the sum equal to zero. Using the Laplace transformed differential equation with zero initial conditions, we finally separate variables and arrive at the transfer function. An example follows.

Example 2.10   Find the transfer function of a translational mechanical system described by one equation of motion.

Problem   Find the transfer function, $X(s)/F(s)$, for the system of Figure 2.9.

Solution   Begin the solution by drawing the free-body diagram shown in Figure 2.10(*a*). Place on the mass all forces felt by the mass. We assume the mass is traveling toward the right. Thus, only the applied force points to the right; all other forces impede the motion and act to oppose it. Hence, the spring, viscous damper, and the force due to acceleration point to the left.

Figure 2.10 (*a*) Free-Body Diagram of Mass, Spring, and Damper System; (*b*) Transformed Free-Body Diagram

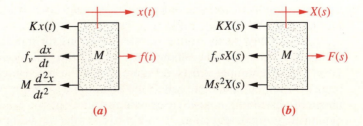

At this point, the forces in Figure 2.10(a) could be summed and equated to zero. We would then take the Laplace transform of that differential equation, assuming zero initial conditions, and find the required transfer function. Let us instead take a lead from our work with electrical networks and immediately show the forces on the free-body diagram as Laplace transforms with zero initial conditions, creating the transformed free-body diagram of Figure 2.10(b). Now the forces can be summed directly in the s-domain, saving us the step of taking the Laplace transform of the differential equation. Hence,

$$M s^2 X(s) + f_v s X(s) + K X(s) = F(s) \qquad (2.92)$$

or

$$(M s^2 + f_v s + K) X(s) = F(s) \qquad (2.93)$$

Notice that the left-hand side of Equation 2.93 contains a term that is the sum of the mechanical impedances associated with the motion.

Solving for the transfer function yields

$$G(s) = \frac{X(s)}{F(s)} = \frac{1}{M s^2 + f_v s + K} \qquad (2.94)$$

Many mechanical systems are similar to multiple-loop and multiple-node electrical networks, where more than one simultaneous differential equation is required to describe the system. In mechanical systems, the number of equations of motion required is equal to the number of *linearly independent* motions. Linear independence implies that a point of motion in a system can still move if all other points of motion are held still. Another name for the number of linearly independent motions is *degrees of freedom*. This discussion is not meant to imply that these motions are not coupled to one another; in general, they are. For example, in a two-loop electrical network, each loop current depends upon the other loop current, but if we open-circuit just one of the loops, the other current can still exist if there is a voltage source in that loop. Similarly, in a mechanical system with two degrees of freedom, one point of motion can be held still while the other point of motion moves under the influence of an applied force.

In order to work such a problem, we draw the free-body diagram for each point of motion and then we use superposition. For each free-body diagram, we begin by holding all other points of motion still and finding the forces acting on the body due only to its own motion. Then we hold the body still and activate the other points of motion one at a time, placing on the original body the forces created by the adjacent motion.

Using Newton's law, we sum the forces on each body and set the sum to zero. The result is a system of simultaneous equations of motion. As Laplace transforms, these equations are then solved for the output variable of interest in terms of the input variable from which the transfer function is evaluated. We demonstrate with an example.

Example 2.11

Find the transfer function of a two-degrees-of-freedom translational mechanical system.

Problem Find the transfer function, $X_2(s)/F(s)$, for the system of Figure 2.11.

Solution The system has two degrees of freedom, since each mass can be moved in the horizontal direction while the other is held still. Thus, two simultaneous equations of motion will be required to describe the system. The two equations come from free-body diagrams of each mass. Superposition is used to draw the free-body diagrams. For example, the forces on M_1 are due to (1) its own motion and (2) the motion of M_2 transmitted to M_1 through the system. We will consider these two sources separately.

If we hold M_2 still and move M_1 to the right, we see the forces shown in Figure 2.12(a). If we hold M_1 still and move M_2 to the right, we see the forces shown in Figure 2.12(b). The total force on M_1 is the superposition, or sum, of the forces just discussed. This result is shown in Figure 2.12(c). For M_2, we proceed in a similar fashion: first we move M_2 to the right while holding M_1 still; then we move M_1 to the right and hold M_2 still. For each case, we evaluate the forces on M_2. The results appear in Figure 2.13.

The Laplace transform of the equations of motion can now be written from Figures 2.12(c) and 2.13(c) as

$$[(M_1s^2 + (f_{v_1} + f_{v_3})s + (K_1 + K_2)]X_1(s) \quad - (f_{v_3}s + K_2)X_2(s) = F(s) \quad (2.95a)$$

$$-(f_{v_3}s + K_2)X_1(s)$$
$$+ [M_2s^2 + (f_{v_2} + f_{v_3})s + (K_2 + K_3)]X_2(s) = 0 \qquad (2.95b)$$

From this, the transfer function $X_2(s)/F(s)$ is

$$\frac{X_2(s)}{F(s)} = G(s) = \frac{(f_{v_3}s + K_2)}{\Delta} \qquad (2.96)$$

where

$$\Delta = \begin{vmatrix} [M_1s^2 + (f_{v_1} + f_{v_3})s + (K_1 + K_2)] & -(f_{v_3}s + K_2) \\ -(f_{v_3}s + K_2) & [M_2s^2 + (f_{v_2} + f_{v_3})s + (K_2 + K_3)] \end{vmatrix}$$

Figure 2.11 Two-Degrees-of-Freedom Translational Mechanical System

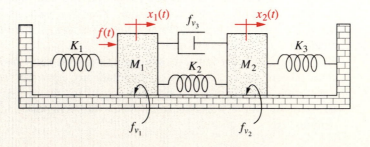

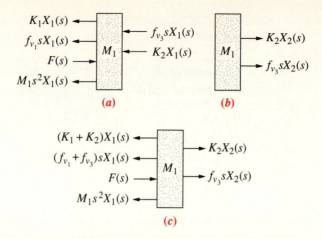

Figure 2.12 (*a*) Forces on M_1 Due Only to Motion of M_1;
(*b*) Forces on M_1 Due Only to Motion of M_2; (*c*) All Forces
on M_1

Figure 2.13 (*a*) Forces on M_2 Due Only to Motion of M_2;
(*b*) Forces on M_2 Due Only to Motion of M_1; (*c*) All Forces
on M_2

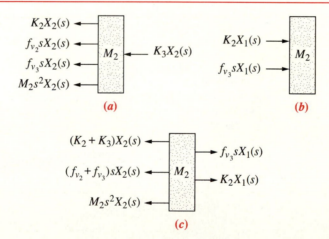

Notice again, in Equations 2.95, that the form of the equations is similar to that found in electrical networks:

$$
\begin{bmatrix} \text{Sum of} \\ \text{impedances} \\ \text{connected} \\ \text{to the motion} \\ \text{at } x_1 \end{bmatrix} X_1(s) - \begin{bmatrix} \text{Sum of} \\ \text{impedances} \\ \text{between} \\ x_1 \text{ and } x_2 \end{bmatrix} X_2(s) = \begin{bmatrix} \text{Sum of} \\ \text{applied forces} \\ \text{at } x_1 \end{bmatrix} \quad (2.97a)
$$

$$
- \begin{bmatrix} \text{Sum of} \\ \text{impedances} \\ \text{between} \\ x_1 \text{ and } x_2 \end{bmatrix} X_1(s) + \begin{bmatrix} \text{Sum of} \\ \text{impedances} \\ \text{connected} \\ \text{to the motion} \\ \text{at } x_2 \end{bmatrix} X_2(s) = \begin{bmatrix} \text{Sum of} \\ \text{applied forces} \\ \text{at } x_2 \end{bmatrix} \quad (2.97b)
$$

The pattern shown in Equation 2.97 should now be familiar to us. Let us use the concept to write the equations of motion of a three-degrees-of-freedom mechanical network by inspection, without drawing the free-body diagram.

Example 2.12 Write the equations of motion of a three-degrees-of-freedom translational mechanical system.

Problem Write, but do not solve, the equations of motion for the mechanical network of Figure 2.14.

Solution The system has three degrees of freedom, since each of the three masses can be moved independently while the others are held still. The form of the equations will be, for M_1,

$$
\begin{bmatrix} \text{Sum of} \\ \text{impedances} \\ \text{connected} \\ \text{to the motion} \\ \text{at } x_1 \end{bmatrix} X_1(s) - \begin{bmatrix} \text{Sum of} \\ \text{impedances} \\ \text{between} \\ x_1 \text{ and } x_2 \end{bmatrix} X_2(s)
$$

$$
- \begin{bmatrix} \text{Sum of} \\ \text{impedances} \\ \text{between} \\ x_1 \text{ and } x_3 \end{bmatrix} X_3(s) = \begin{bmatrix} \text{Sum of} \\ \text{applied forces} \\ \text{at } x_1 \end{bmatrix} \quad (2.98)
$$

Similarly, for M_2 and M_3, respectively,

$$
- \begin{bmatrix} \text{Sum of} \\ \text{impedances} \\ \text{between} \\ x_1 \text{ and } x_2 \end{bmatrix} X_1(s) + \begin{bmatrix} \text{Sum of} \\ \text{impedances} \\ \text{connected} \\ \text{to the motion} \\ \text{at } x_2 \end{bmatrix} X_2(s)
$$

$$
- \begin{bmatrix} \text{Sum of} \\ \text{impedances} \\ \text{between} \\ x_2 \text{ and } x_3 \end{bmatrix} X_3(s) = \begin{bmatrix} \text{Sum of} \\ \text{applied forces} \\ \text{at } x_2 \end{bmatrix} \quad (2.99)
$$

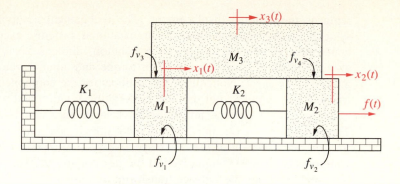

Figure 2.14 A Three-Degrees-of-Freedom Translational Mechanical System

$$-\begin{bmatrix} \text{Sum of} \\ \text{impedances} \\ \text{between} \\ x_1 \text{ and } x_3 \end{bmatrix} X_1(s) - \begin{bmatrix} \text{Sum of} \\ \text{impedances} \\ \text{between} \\ x_2 \text{ and } x_3 \end{bmatrix} X_2(s)$$

$$+ \begin{bmatrix} \text{Sum of} \\ \text{impedances} \\ \text{connected} \\ \text{to the motion} \\ \text{at } x_3 \end{bmatrix} X_3(s) = \begin{bmatrix} \text{Sum of} \\ \text{applied forces} \\ \text{at } x_3 \end{bmatrix} \qquad (2.100)$$

M_1 has two springs, two viscous dampers, and mass associated with its motion. There is one spring between M_1 and M_2 and one viscous damper between M_1 and M_3. Thus, using Equation 2.98,

$$\left[M_1 s^2 + (f_{v_1} + f_{v_3})s + (K_1 + K_2)\right] X_1(s) - K_2 X_2(s) - f_{v_3} s X_3(s) = 0 \qquad (2.101)$$

Similarly, for M_2,

$$-K_2 X_1(s) + \left[M_2 s^2 + (f_{v_2} + f_{v_4})s + K_2\right] X_2(s) - f_{v_4} s X_3(s) = F(s) \qquad (2.102)$$

and for M_3,

$$-f_{v_3} s X_1(s) - f_{v_4} s X_2(s) + \left[M_3 s^2 + (f_{v_3} + f_{v_4})s\right] X_3(s) = 0 \qquad (2.103)$$

Equations 2.101 through 2.103 are the equations of motion. We can solve them for any displacement, X_1, X_2, or X_3, or transfer function.

Having covered electrical and translational mechanical systems, we now move on to consider rotational mechanical systems.

2.6 Transfer Functions for Rotational Mechanical Systems

Rotational mechanical systems are handled the same way as translational mechanical systems, except that torque replaces force, and angular displacement replaces translational displacement. The mechanical components for rotational systems are the same as those for translational systems, except that the components undergo rotation instead of translation. Table 2.5 shows the components along with the relationships between torque and angular velocity, as well as angular displacement. Notice that the symbols for the components look the same as translational symbols, but they are undergoing rotation and not translation.

Also notice that the term associated with the mass is replaced by inertia. The values of K, D, and J are called *spring constant, coefficient of viscous friction,* and *moment of inertia,* respectively. The impedances of the mechanical components are also summarized in the last column of Table 2.5. The values can be found by taking the Laplace transform, assuming zero initial conditions, of the torque-angular displacement column of Table 2.5.

The concept of degrees of freedom carries over to rotational systems, except that we test a point of motion by *rotating* it while holding all other points of motion still. The number of points of motion that can be rotated while all others are held still equals the number of equations of motion required to describe the system.

Table 2.5 Torque-Angular Velocity, Torque-Angular Displacement, and Impedance Rotational Relationships for a Spring, a Viscous Damper, and an Inertia

Component	Torque-angular velocity	Torque-angular displacement	Impedance $Z_M(s) = T(s)/\theta(s)$
Spring $T(t)$ $\theta(t)$ K	$T(t) = K \int_0^t \omega(\tau)\, d\tau$	$T(t) = K\theta(t)$	K
Viscous damper $T(t)$ $\theta(t)$ D	$T(t) = D\omega(t)$	$T(t) = D\dfrac{d\theta(t)}{dt}$	Ds
Inertia $T(t)$ $\theta(t)$ J	$T(t) = J\dfrac{d\omega(t)}{dt}$	$T(t) = J\dfrac{d^2\theta(t)}{dt^2}$	Js^2

Note: The following set of symbols and units are used throughout this book: $T(t)$ = N-m (newton-meters), $\theta(t)$ = rad (radians), $\omega(t)$ = rad/s (radians/second), K = N-m/rad (newton-meters/radian), D = N-m-s/rad (newton-meters-seconds/radian), J = kg-m^2 (kilogram-meters^2 = newton-meters-seconds^2/radian).

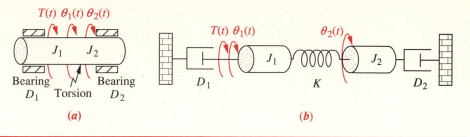

Figure 2.15 (*a*) Physical System; (*b*) Schematic

Writing the equations of motion for rotational systems is similar to writing them for translational systems; the only difference is that the free-body diagram consists of torques rather than forces. We obtain these torques using superposition. First, we rotate a body while holding all other points still and place on its free-body diagram all torques due to the body's own motion. Then, holding the body still, we rotate adjacent points of motion one at a time and add the torques due to the adjacent motion to the free-body diagram. The process is repeated for each point of motion. For each free-body diagram, these torques are summed and set equal to zero to form the equations of motion.

Two examples will demonstrate the solution of rotational systems. The first one uses free-body diagrams; the second uses the concept of impedances to write the equations of motion by inspection.

Example 2.13

Find the transfer function of a two-degrees-of-freedom rotational system.

Problem Find the transfer function, $\theta_2(s)/T(s)$, for the rotational system shown in Figure 2.15(*a*). The rod is supported by bearings at either end and is undergoing torsion. A torque is applied at the left, and the displacement is measured at the right.

Solution First, obtain the schematic from the physical system. Even though torsion occurs throughout the rod in Figure 2.15(*a*),[8] we approximate the system by assuming that the torsion acts like a spring concentrated at one particular point in the rod, with an inertia, J_1, to the left, and an inertia, J_2 to the right.[9] We also assume that the damping inside the flexible shaft is negligible. The schematic is shown in Figure 2.15(*b*). There are two degrees of freedom, since each inertia can be rotated while the other is held still. Hence it will take two simultaneous equations to solve the system.

Next, draw a free-body diagram of J_1, using superposition. Figure 2.16(*a*) shows the torques on J_1 if J_2 is held still and J_1 rotated. Figure 2.16(*b*) shows the torques

---

[8]In this case the parameter is referred to as a *distributed* parameter.
[9]The parameter is now referred to as a *lumped* parameter.

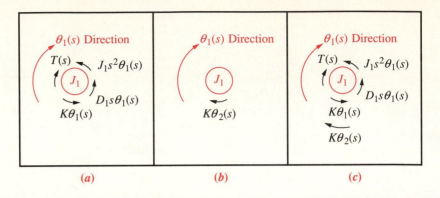

(a)   (b)   (c)

Figure 2.16 (**a**) Torques on J_1 Due Only to the Motion of J_1; (**b**) Torques on J_1 Due Only to the Motion of J_2; (**c**) Final Free-Body Diagram for J_1

on J_1 if J_1 is held still and J_2 rotated. Finally, the sum of Figures 2.16(*a*) and 2.16(*b*) is shown in Figure 2.16(*c*), the final free-body diagram for J_1. The same process is repeated in Figure 2.17 for J_2.

Summing torques respectively from Figures 2.16(*c*) and 2.17(*c*) we obtain the equations of motion,

$$(J_1 s^2 + D_1 s + K)\theta_1(s) \qquad\qquad - K\theta_2(s) = T(s) \qquad (2.104a)$$

$$-K\theta_1(s) + (J_2 s^2 + D_2 s + K)\theta_2(s) = 0 \qquad (2.104b)$$

from which the required transfer function is found to be

$$\frac{\theta_2(s)}{T(s)} = \frac{K}{\Delta} \qquad (2.105)$$

Figure 2.17 (**a**) Torques on J_2 Due Only to the Motion of J_2; (**b**) Torques on J_2 Due Only to the Motion of J_1; (**c**) Final Free-Body Diagram for J_2

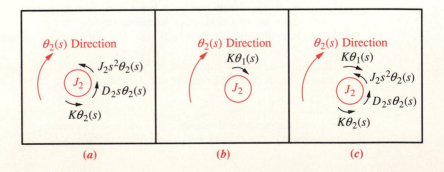

(a)   (b)   (c)

where

$$\Delta = \begin{vmatrix} (J_1 s^2 + D_1 s + K) & -K \\ -K & (J_2 s^2 + D_2 s + K) \end{vmatrix}$$

Notice, again, that the Equations 2.104 have that now well-known form,

$$\begin{bmatrix} \text{Sum of} \\ \text{impedances} \\ \text{connected} \\ \text{to the motion} \\ \text{at } \theta_1 \end{bmatrix} \theta_1(s) - \begin{bmatrix} \text{Sum of} \\ \text{impedances} \\ \text{between} \\ \theta_1 \text{ and } \theta_2 \end{bmatrix} \theta_2(s) = \begin{bmatrix} \text{Sum of} \\ \text{applied torques} \\ \text{at } \theta_1 \end{bmatrix} \quad (2.106a)$$

$$- \begin{bmatrix} \text{Sum of} \\ \text{impedances} \\ \text{between} \\ \theta_1 \text{ and } \theta_2 \end{bmatrix} \theta_1(s) + \begin{bmatrix} \text{Sum of} \\ \text{impedances} \\ \text{connected} \\ \text{to the motion} \\ \text{at } \theta_2 \end{bmatrix} \theta_2(s) = \begin{bmatrix} \text{Sum of} \\ \text{applied torques} \\ \text{at } \theta_2 \end{bmatrix} \quad (2.106b)$$

Let us now write the equations of motion for a three-degrees-of-freedom rotational problem. Although we will bypass the free-body diagrams, you should also be able to do the problem using free-body diagrams.

Example 2.14

Write the equations of motion for a three-degrees-of-freedom rotational system.

Problem Write, but do not solve, the Laplace transform of the equations of motion for the system shown in Figure 2.18.

Solution The equations will take on the following form:

$$\begin{bmatrix} \text{Sum of} \\ \text{impedances} \\ \text{connected} \\ \text{to the motion} \\ \text{at } \theta_1 \end{bmatrix} \theta_1(s) - \begin{bmatrix} \text{Sum of} \\ \text{impedances} \\ \text{between} \\ \theta_1 \text{ and } \theta_2 \end{bmatrix} \theta_2(s)$$

$$- \begin{bmatrix} \text{Sum of} \\ \text{impedances} \\ \text{between} \\ \theta_1 \text{ and } \theta_3 \end{bmatrix} \theta_3(s) = \begin{bmatrix} \text{Sum of} \\ \text{applied torques} \\ \text{at } \theta_1 \end{bmatrix} \quad (2.107)$$

Figure 2.18 A Three-Degrees-of-Freedom Rotational System

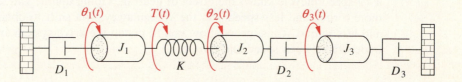

$$-\begin{bmatrix} \text{Sum of} \\ \text{impedances} \\ \text{between} \\ \theta_1 \text{ and } \theta_2 \end{bmatrix}\theta_1(s) + \begin{bmatrix} \text{Sum of} \\ \text{impedances} \\ \text{connected} \\ \text{to the motion} \\ \text{at } \theta_2 \end{bmatrix}\theta_2(s)$$

$$-\begin{bmatrix} \text{Sum of} \\ \text{impedances} \\ \text{between} \\ \theta_2 \text{ and } \theta_3 \end{bmatrix}\theta_3(s) = \begin{bmatrix} \text{Sum of} \\ \text{applied torques} \\ \text{at } \theta_2 \end{bmatrix} \quad (2.108)$$

$$-\begin{bmatrix} \text{Sum of} \\ \text{impedances} \\ \text{between} \\ \theta_1 \text{ and } \theta_3 \end{bmatrix}\theta_1(s) - \begin{bmatrix} \text{Sum of} \\ \text{impedances} \\ \text{between} \\ \theta_2 \text{ and } \theta_3 \end{bmatrix}\theta_2(s)$$

$$+\begin{bmatrix} \text{Sum of} \\ \text{impedances} \\ \text{connected} \\ \text{to the motion} \\ \text{at } \theta_3 \end{bmatrix}\theta_3(s) = \begin{bmatrix} \text{Sum of} \\ \text{applied torques} \\ \text{at } \theta_3 \end{bmatrix} \quad (2.109)$$

Hence,

$$(J_1 s^2 + D_1 s + K)\theta_1(s) \qquad - K\,\theta_2(s) \qquad - 0\theta_3(s) = T(s)$$

$$-K\theta_1(s) + (J_2 s^2 + D_2 s + K)\theta_2(s) \qquad - D_2 s\theta_3(s) = 0$$

$$-0\theta_1(s) \qquad - D_2 s\theta_2(s) + (J_3 s^2 + D_3 s + D_2 s)\theta_3(s) = 0$$

$$(2.110a, b, c)$$

Now that we are able to find the transfer function for rotational systems, we realize that these systems, especially those driven by motors, are rarely seen without associated gear trains driving the load. The next section covers this important topic.

2.7 Transfer Functions for Systems with Gears

Gears provide mechanical advantage to rotational systems. Anyone who has ridden a 10-speed bicycle knows the effect of gearing. Going uphill, you shift to provide more torque and less speed. On the straightaway, you shift to obtain more speed and less torque. Thus, gears allow you to match the drive system and the load—a trade-off between speed and torque.

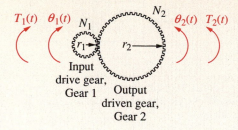

Figure 2.19 A Gear System

For many applications, gears exhibit *backlash,* which occurs because of the loose fit between two meshed gears. The drive gear rotates through a small angle before making contact with the meshed gear. The result is that the angular rotation of the output gear does not occur until a small angular rotation of the input gear has occurred. In this section, we idealize the behavior of gears and assume that there is no backlash.

The linearized interaction between two gears is depicted in Figure 2.19. An input gear with radius r_1 and N_1 teeth is rotated through angle $\theta_1(t)$ due to a torque, $T_1(t)$. An output gear with radius r_2 and N_2 teeth responds by rotating through angle $\theta_2(t)$ and delivering a torque, $T_2(t)$. Let us now find the relationship that exists between the rotation of Gear 1, θ_1, and Gear 2, θ_2.

From Figure 2.19, as the gears turn, the distance traveled along each gear's circumference is the same. Thus,

$$r_1\theta_1 = r_2\theta_2 \tag{2.111}$$

or

$$\frac{\theta_2}{\theta_1} = \frac{r_1}{r_2} = \frac{N_1}{N_2} \tag{2.112}$$

since the ratio of the number of teeth along the circumference is in the same proportion as the ratio of the radii. We conclude that the ratio of the angular displacement of the gears is inversely proportional to the ratio of the number of teeth.

What is the relationship between the input torque, T_1, and the delivered torque, T_2? If we assume the gears do not absorb or store energy, the energy into Gear 1 equals the energy out of Gear 2.[10] Since the translational energy of force times displacement becomes the rotational energy of torque times angular displacement,

$$T_1\theta_1 = T_2\theta_2 \tag{2.113}$$

---

[10]This is equivalent to saying that the gears have negligible inertia and damping.

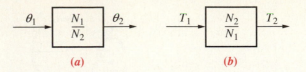

(a)   (b)

Figure 2.20 (*a*) Transfer Function for Angular Displacement in Lossless Gears; (*b*) Transfer Function for Torque in Lossless Gears

Solving Equation 2.113 for the ratio of the torques, and using Equation 2.112, we get

$$\frac{T_2}{T_1} = \frac{\theta_1}{\theta_2} = \frac{N_2}{N_1} \tag{2.114}$$

Thus, the torques are directly proportional to the ratio of the number of teeth. All results are summarized in Figure 2.20.

Let us see what happens to mechanical impedances that are driven by gears. Figure 2.21(*a*) shows gears driving a rotational inertia, spring, and viscous damper. For clarity, the gears are shown by an end-on view. We want to represent Fig-

Figure 2.21 (*a*) A Rotational System Driven by Gears; (*b*) Equivalent System at the Output After Reflection of Input Torque; (*c*) Equivalent System at the Input After Reflection of Impedances

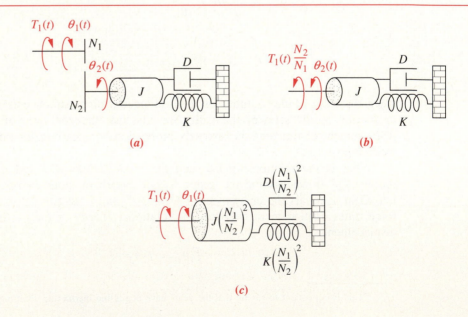

(a)   (b)

(c)

ure 2.21(a) as an equivalent system at θ_1 without the gears. In other words, can the mechanical impedances be reflected from the output to the input, thereby eliminating the gears?

From Figure 2.20(b), T_1 can be reflected to the output by multiplying by N_2/N_1. The result is shown in Figure 2.21(b), from which we write the equation of motion as

$$(Js^2 + Ds + K)\theta_2(s) = T_1(s)\frac{N_2}{N_1} \tag{2.115}$$

Now convert $\theta_2(s)$ into an equivalent $\theta_1(s)$, so that Equation 2.115 will look as if it were written at the input. Using Figure 2.20(a) to obtain $\theta_2(s)$ in terms of $\theta_1(s)$, we get

$$(Js^2 + Ds + K)\frac{N_1}{N_2}\theta_1(s) = T_1(s)\frac{N_2}{N_1} \tag{2.116}$$

After simplification,

$$\left[J\left(\frac{N_1}{N_2}\right)^2 s^2 + D\left(\frac{N_1}{N_2}\right)^2 s + K\left(\frac{N_1}{N_2}\right)^2 \right]\theta_1(s) = T_1(s) \tag{2.117}$$

which suggests the equivalent system at the input and without gears shown in Figure 2.21(c). Thus, the load can be thought of as having been reflected from the output to the input.

Generalizing the results, we can make the following statement: **Rotational mechanical impedances can be reflected through gear trains by multiplying the mechanical impedance by the ratio**

$$\left(\frac{\textbf{Number of teeth of } \textit{destination} \textbf{ gear}}{\textbf{Number of teeth of } \textit{source} \textbf{ gear}} \right)^{2}$$

where the impedance to be reflected is attached to the source gear and is being reflected to the destination gear. The next example demonstrates the application of the concept of reflected impedances as we find the transfer function of a rotational mechanical system with gears.

Example 2.15

Find the transfer function of a mechanical rotational system with gears.

Problem Find the transfer function, $\theta_2(s)/T_1(s)$, for the system of Figure 2.22(a).

Solution It may be tempting at this point to search for two simultaneous equations corresponding to each inertia. The inertias, however, do not undergo linearly inde-

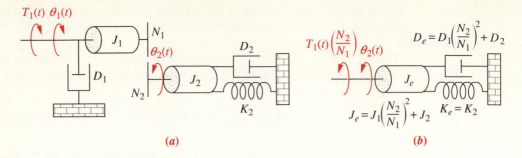

Figure 2.22 (*a*) Rotational Mechanical System with Gears; (*b*) System After Reflection of Torques and Impedance to the Output Shaft

pendent motion, since they are tied together by the gears. Thus, there is only one degree of freedom and hence one equation of motion.

Let us first reflect the impedances (J_1 and D_1) and torque (T_1) on the input shaft to the output as shown in Figure 2.22(*b*), where the impedances are reflected by $(N_2/N_1)^2$ and the torque is reflected by (N_2/N_1). The equation of motion can now be written as

$$(J_e s^2 + D_e s + K_e)\theta_2(s) = T_1(s)\frac{N_2}{N_1} \tag{2.118}$$

where

$$J_e = J_1\left(\frac{N_2}{N_1}\right)^2 + J_2; \qquad D_e = D_1\left(\frac{N_2}{N_1}\right)^2 + D_2; \qquad K_e = K_2$$

Solving for $\theta_2(s)/T_1(s)$, the transfer function is found to be

$$G(s) = \frac{\theta_2(s)}{T_1(s)} = \frac{N_2/N_1}{J_e s^2 + D_e s + K_e} \tag{2.119}$$

In order to eliminate gears with large radii, a *gear train* is used to implement large gear ratios by cascading smaller gear ratios. A schematic diagram of a gear train is shown in Figure 2.23. Next to each rotation, the angular displacement relative to θ_1 has been calculated. From Figure 2.23,

$$\theta_4 = \frac{N_1 N_3 N_5}{N_2 N_4 N_6}\theta_1 \tag{2.120}$$

For gear trains, we conclude that the equivalent gear ratio is the product of the individual gear ratios. We now apply this result to solve for the transfer function of a system that does not have lossless gears.

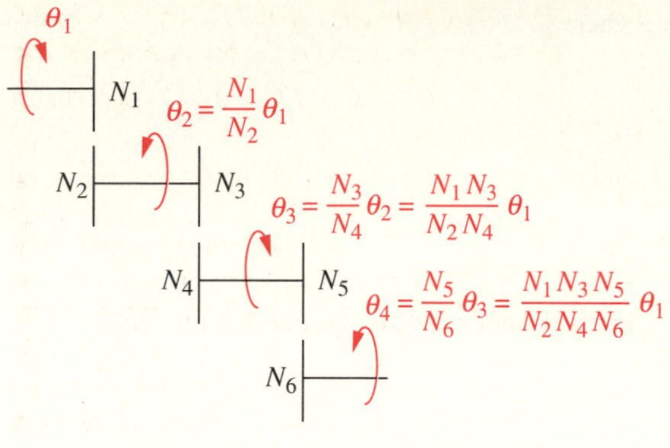

Figure 2.23 A Gear Train

Example 2.16

Find the transfer function of a system with gear trains that have inertia and damping.

Problem   Find the transfer function, $\theta_1(s)/T_1(s)$, for the system of Figure 2.24(a).

Solution   This system, which uses a gear train, does not have lossless gears. All of the gears have inertia and for some shafts there is viscous friction. To solve the problem, we want to reflect all of the impedances to the input shaft, θ_1. The gear ratio is not the same for all impedances. For example, D_2 is reflected only through one gear ratio as $D_2(N_1/N_2)^2$ whereas J_4 plus J_5 is reflected through two gear ratios as $(J_4 + J_5)\,[(N_3/N_4)(N_1/N_2)]^2$. The result of reflecting all impedances to θ_1 is shown in Figure 2.24(b), from which the equations of motion are

$$(J_e s^2 + D_e s)\theta_1(s) = T_1(s) \tag{2.121}$$

Figure 2.24 (**a**) A System Using a Gear Train; (**b**) Equivalent System at the Input

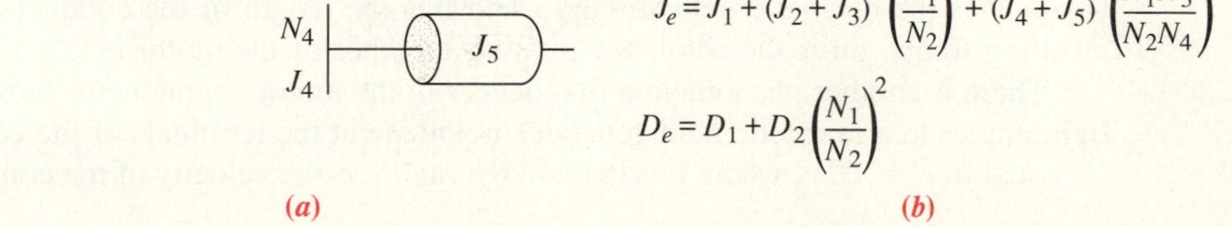

$$J_e = J_1 + (J_2 + J_3)\left(\frac{N_1}{N_2}\right)^2 + (J_4 + J_5)\left(\frac{N_1 N_3}{N_2 N_4}\right)^2$$

$$D_e = D_1 + D_2\left(\frac{N_1}{N_2}\right)^2$$

(*a*)   (*b*)

where

$$J_e = J_1 + (J_2 + J_3)\left(\frac{N_1}{N_2}\right)^2 + (J_4 + J_5)\left(\frac{N_1 N_3}{N_2 N_4}\right)^2$$

and

$$D_e = D_1 + D_2\left(\frac{N_1}{N_2}\right)^2$$

From Equation 2.121, the transfer function is

$$G(s) = \frac{\theta_1(s)}{T_1(s)} = \frac{1}{J_e s^2 + D_e s} \tag{2.122}$$

In this section we talked about rotational systems with gears, and our discussion of purely mechanical systems is now complete. Next, we move to systems that are hybrids of electrical and mechanical, the electromechanical systems.

2.8 Transfer Functions for Electromechanical Systems

Electrical and mechanical networks are combined to form what are called *electromechanical systems*. We have seen one application of an electromechanical system in Chapter 1, the antenna azimuth control system. Other applications for systems that use electromechanical components are robot controls, sun and star trackers, and computer tape and disk-drive position controls. An example of a control system that uses electromechanical components is shown in Figure 2.25.

A motor is an electromechanical component that yields a displacement output for a voltage input, that is, a mechanical output generated by an electrical input. We will derive the transfer function for one particular kind of electromechanical system, the armature-controlled dc servomotor (Mablekos, 1980). The motor's schematic is shown in Figure 2.26(a), and the transfer function we will derive appears in Figure 2.26(b).

In Figure 2.26(a), a magnetic field is developed by stationary permanent magnets or a stationary electromagnet called the *fixed field*. A rotating circuit called the *armature,* through which current $i_a(t)$ flows, passes through this magnetic field at right angles and feels a force that follows the motor law, $F = Bli_a(t)$, where B is the magnetic field strength, and l is the length of the conductor. The resulting torque turns the *rotor,* the rotating member of the motor.

There is another phenomenon that occurs in the motor: a conductor moving at right angles to a magnetic field generates a voltage at the terminals of the conductor equal to $e = Blv$, where e is the voltage, and v is the velocity of the conductor

Figure 2.25 A Piper Seneca III Flight Simulator Uses Electromechanical Control System Components (Courtesy of Frasca International)

Figure 2.26 (**a**) A DC Motor; (**b**) Transfer Function Representation

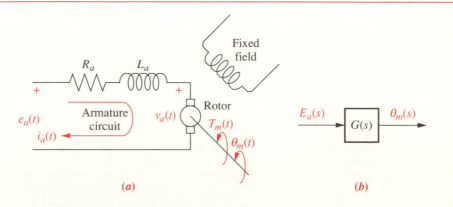

normal to the magnetic field. Since the current-carrying armature is rotating in a magnetic field, its voltage is proportional to speed. Thus,

$$v_a(t) = K_b \frac{d\theta_m(t)}{dt} \tag{2.123}$$

We call $v_a(t)$ the *back electromotive force (back emf)*; K_b is a constant of proportionality called the back emf constant; and $d\theta_m(t)/dt = \omega_m(t)$ is the angular velocity of the motor. Taking the Laplace transform, we get

$$V_a(s) = K_b s \theta_m(s) \tag{2.124}$$

The relationship between the armature current, $i_a(t)$, the applied armature voltage, $e_a(t)$, and the back emf, $v_a(t)$, is found by writing a loop equation around the Laplace transformed armature circuit (see Figure 2.26(a)):

$$R_a I_a(s) + L_a s I_a(s) + V_a(s) = E_a(s) \tag{2.125}$$

The torque developed by the motor is proportional to the armature current; thus, we can write

$$T_m(s) = K_t I_a(s) \tag{2.126}$$

where T_m is the torque developed by the motor, and K_t is a constant of proportionality, called the motor torque constant, which depends upon the motor and magnetic field characteristics. In a consistent set of units, the value of K_t is equal to the value of K_b. Rearranging Equation 2.126 yields

$$I_a(s) = \frac{1}{K_t} T_m(s) \tag{2.127}$$

To find the transfer function of the motor, we first substitute Equations 2.124 and 2.127 into 2.125, yielding

$$\frac{(R_a + L_a s) T_m(s)}{K_t} + K_b s \theta_m(s) = E_a(s) \tag{2.128}$$

Now we must find $T_m(s)$ in terms of $\theta_m(s)$ if we are to separate the input and output variables and obtain the transfer function, $\theta_m(s)/E_a(s)$.

Figure 2.27 shows a typical equivalent mechanical loading on a motor. J_m is the equivalent inertia at the armature and includes both the armature inertia and,

Figure 2.27 Typical Equivalent Mechanical Loading on a Motor

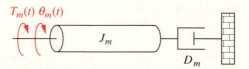

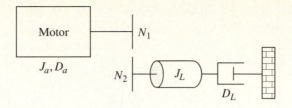

Figure 2.28 DC Motor Driving a Rotational Mechanical Load

as we will see later, the load inertia reflected to the armature. D_m is the equivalent viscous damping at the armature and includes both the armature viscous damping and, as we will see later, the load viscous damping reflected to the armature. From Figure 2.27,

$$T_m(s) = (J_m s^2 + D_m s)\theta_m(s) \tag{2.129}$$

Substituting Equation 2.129 into Equation 2.128 yields

$$\frac{(R_a + L_a s)(J_m s^2 + D_m s)\theta_m(s)}{K_t} + K_b s\theta_m(s) = E_a(s) \tag{2.130}$$

If we assume that the armature inductance, L_a, is small compared to the armature resistance, R_a, which is usual for a dc motor, Equation 2.130 becomes

$$\left[\frac{R_a}{K_t}(J_m s + D_m) + K_b\right]s\theta_m(s) = E_a(s) \tag{2.131}$$

After simplification, the desired transfer function, $\theta_m(s)/E_a(s)$, is found to be,[11]

$$\frac{\theta_m(s)}{E_a(s)} = \frac{K_t/R_a J_m}{s\left[s + \frac{1}{J_m}\left(D_m + \frac{K_t K_b}{R_a}\right)\right]} \tag{2.132}$$

Even though the form of Equation 2.132 is relatively simple, namely,

$$\frac{\theta_m(s)}{E_a(s)} = \frac{K}{s(s + \alpha)} \tag{2.133}$$

the reader may be concerned about how to evaluate the constants.

Let us first discuss the mechanical constants, J_m and D_m. Consider Figure 2.28, which shows a motor with inertia J_a and damping D_a at the armature driving a load consisting of inertia J_L and damping D_L. Assuming that all inertia

---

[11]The units for the electrical constants are K_t = N-m/A (newton-meters/ampere), and K_b = V-s/rad (volt-seconds/radian).

and damping values shown are known, J_L and D_L can be reflected back to the armature as some equivalent inertia and damping to be added to J_a and D_a, respectively. Thus, the equivalent inertia, J_m, and equivalent damping, D_m, at the armature are

$$J_m = J_a + J_L \left(\frac{N_1}{N_2}\right)^2 ; \qquad D_m = D_a + D_L \left(\frac{N_1}{N_2}\right)^2 \qquad (2.134)$$

Now that we have evaluated the mechanical constants, J_m and D_m, what about the electrical constants in the transfer function of Equation 2.132? We will show that these constants can be obtained through a *dynamometer* test of the motor, where a dynamometer measures the torque and speed of a motor under the condition of a constant applied voltage. Let us first develop the relationships that dictate the use of a dynamometer.

Substituting Equations 2.124 and 2.127 into Equation 2.125, with $L_a = 0$, yields

$$\frac{R_a}{K_t}T_m(s) + K_b s\theta_m(s) = E_a(s) \qquad (2.135)$$

Taking the inverse Laplace transform, we get

$$\frac{R_a}{K_t}T_m(t) + K_b\omega_m(t) = e_a(t) \qquad (2.136)$$

where the inverse Laplace transform of $s\theta_m(s)$ is $d\theta_m(t)/dt$ or, alternately, $\omega_m(t)$.

If a dc voltage, e_a, is applied, the motor will turn at a constant angular velocity, ω_m, with a constant torque, T_m. Hence, dropping the functional relationship based on time from Equation 2.136, the following relationship exists when the motor is operating at steady-state with a dc voltage input:

$$\frac{R_a}{K_t}T_m + K_b\omega_m = e_a \qquad (2.137)$$

Solving for T_m yields

$$T_m = -\frac{K_b K_t}{R_a}\omega_m + \frac{K_t}{R_a}e_a \qquad (2.138)$$

Equation 2.138 is a straight line, T_m vs ω_m, and is shown in Figure 2.29. This plot is called the *torque-speed curve*. The torque axis intercept occurs when the angular velocity reaches zero. That value of torque is called the *stall torque*, T_{stall}. Thus,

$$T_{\text{stall}} = \frac{K_t}{R_a}e_a \qquad (2.139)$$

The angular velocity occurring when the torque is zero is called the *no-load speed*,

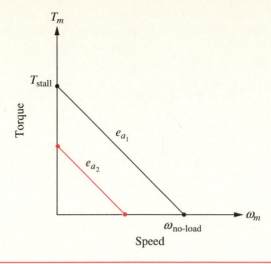

Figure 2.29 Torque-Speed Curves with an Armature Voltage, e_a, as a Parameter

$\omega_{\text{no-load}}$. Thus,

$$\omega_{\text{no-load}} = \frac{e_a}{K_b} \tag{2.140}$$

The electrical constants of the motor's transfer function can now be found from Equations 2.139 and 2.140 as

$$\frac{K_t}{R_a} = \frac{T_{\text{stall}}}{e_a} \tag{2.141}$$

and

$$K_b = \frac{e_a}{\omega_{\text{no-load}}} \tag{2.142}$$

The electrical constants K_t/R_a and K_b, can be found from a dynamometer test of the motor, which would yield T_{stall} and $\omega_{\text{no-load}}$ for a given e_a.

Example 2.17   Find the transfer function of an armature-controlled dc motor and load.

Problem   Given the system and torque-speed curve of Figure 2.30, find the transfer function, $\theta_L(s)/E_a(s)$.

Solution   Begin by finding the mechanical constants, J_m and D_m, in Equation 2.132. From Equation 2.134, the total inertia at the armature of the motor is

$$J_m = J_a + J_L\left(\frac{N_1}{N_2}\right)^2 = 5 + 700\left(\frac{1}{10}\right)^2 = 12 \tag{2.143}$$

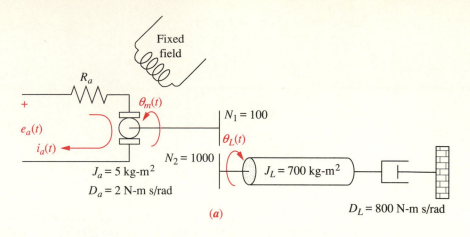

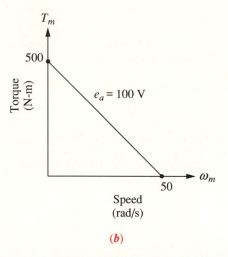

Figure 2.30 (**a**) DC Motor and Load; (**b**) Torque-Speed Curve

and the total damping at the armature of the motor is

$$D_m = D_a + D_L \left(\frac{N_1}{N_2}\right)^2 = 2 + 800 \left(\frac{1}{10}\right)^2 = 10 \tag{2.144}$$

Now we will find the electrical constants, K_t/R_a and K_b. From the torque-speed curve of Figure 2.30(**b**),

$$T_{\text{stall}} = 500 \tag{2.145}$$

$$\omega_{\text{no-load}} = 50 \tag{2.146}$$

$$e_a = 100 \tag{2.147}$$

Hence the electrical constants are

$$\frac{K_t}{R_a} = \frac{T_{stall}}{e_a} = \frac{500}{100} = 5 \tag{2.148}$$

and

$$K_b = \frac{e_a}{\omega_{no\text{-}load}} = \frac{100}{50} = 2 \tag{2.149}$$

Substituting Equations 2.143, 2.144, 2.148, and 2.149 into Equation 2.132 yields

$$\frac{\theta_m(s)}{E_a(s)} = \frac{5/12}{s\left\{s + \dfrac{1}{12}\big[10 + (5)(2)\big]\right\}} = \frac{0.417}{s(s + 1.667)} \tag{2.150}$$

In order to find $\theta_L(s)/E_a(s)$, we use the gear ratio, $N_1/N_2 = 1/10$, and find

$$\frac{\theta_L(s)}{E_a(s)} = \frac{0.0417}{s(s + 1.667)} \tag{2.151}$$

The models thus far are developed from systems that can be described approximately by linear, time-invariant differential equations. An assumption of *linearity* was implicit in the development of these models. In the next section, we formally define the terms *linear* and *nonlinear* and show how to distinguish between the two. Then we show how to approximate a nonlinear system as a linear system, so that we can use the modeling techniques previously covered in this chapter.

2.9 Nonlinearities

In this section we examine the meanings of *linear* and *nonlinear,* since these properties of a system affect our ability to use the modeling techniques covered so far or to extract useful information from more complex models (Hsu, 1968).

A linear system possesses two properties: (1) superposition and (2) homogeneity. The property of *superposition* means that the output response of a system to the sum of inputs is the sum of the responses to the individual inputs. Thus, if an input of $r_1(t)$ yields an output of $c_1(t)$, and an input of $r_2(t)$ yields an output of $c_2(t)$, then an input of $r_1(t) + r_2(t)$ yields an output of $c_1(t) + c_2(t)$. The property of *homogeneity* describes the response of the system to a multiplication of the input by a scalar. Specifically, in a linear system, the property of homogeneity is demonstrated if for an input of $r_1(t)$ that yields an output of $c_1(t)$, an input of $Ar_1(t)$ yields an output of $Ac_1(t)$; that is, multiplication of an input by a scalar yields a response that is multiplied by the same scalar.

We can visualize linearity as shown in Figure 2.31. Figure 2.31(a) is a linear system where the output is always 1/2 the input, or $f(x) = 0.5x$, regardless of the value of x. Thus each of the two properties of linear systems applies. For

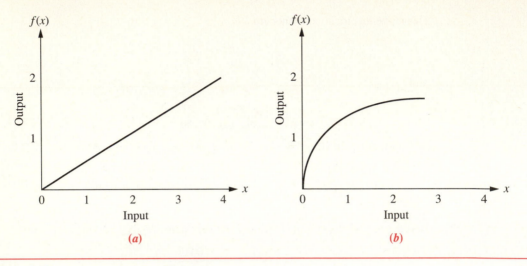

Figure 2.31 (*a*) Linear System; (*b*) Nonlinear System

example, an input of 1 yields an output of $1/2$ and an input of 2 yields an output of 1. Using superposition, an input that is the sum of the original inputs, or 3, should yield an output that is the sum of the individual outputs, or 1.5. From Figure 2.31(*a*), an input of 3 does indeed yield an output of 1.5.

To test the property of homogeneity, assume an input of 2, which yields an output of 1. Multiplying this input by 2 should yield an output of twice as much, or 2. From Figure 2.31(*a*), an input of 4 does indeed yield an output of 2. The reader can verify that the above properties certainly do not apply to the relationship shown in Figure 2.31(*b*).

Figure 2.32 shows some examples of physical nonlinearities. An electronic amplifier is linear over a specific range but exhibits the nonlinearity called *saturation* at high input voltages. A motor that does not respond at very low input voltages due to frictional forces exhibits a nonlinearity called *dead zone*. Gears that do not fit tightly exhibit a nonlinearity called *backlash:* the input moves over a small range without the output responding. The reader should verify that the curves shown in Figure 2.32 do not fit the definitions of linearity over their entire range. Another example of a nonlinear subsystem is a phase detector, used in a phase-locked loop in an FM radio receiver, whose output response is the sine of the input.

A designer can often make a linear approximation to a nonlinear system. Linear approximations simplify the analysis and design of a system and are used as long as the results yield a good approximation to reality. For example, a linear relationship can be established at a point on the nonlinear curve if the range of input values about that point is small and the origin is translated to that point. Electronic amplifiers are an example of physical devices that perform linear amplification with small excursions about a point.

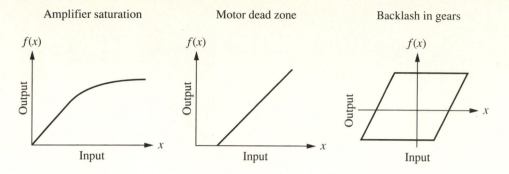

Figure 2.32 Some Physical Nonlinearities

2.10 Linearization

The electrical and mechanical systems covered thus far were assumed to be linear. However, if any nonlinear components are present, we must linearize the system before we can find the transfer function. In the last section we defined and discussed nonlinearities; in this section we show how to obtain linear approximations to nonlinear systems in order to obtain transfer functions.

The first step is to recognize the nonlinear component and write the nonlinear differential equation. When we linearize a nonlinear differential equation, we linearize it for small-signal inputs about the steady-state solution when the small-signal input is equal to zero. This steady-state solution is called *equilibrium* and is selected as the second step in the linearization process. For example, when a pendulum is at rest, it is at equilibrium. The angular displacement is described by a nonlinear differential equation, but it can be expressed with a linear differential equation for small excursions about this equilibrium point.

Next, we linearize the nonlinear differential equation, and then we take the Laplace transform of the linearized differential equation, assuming zero initial conditions. Finally, we separate input and output variables and form the transfer function. Let us first see how to linearize a function; later, we will apply the method to the linearization of a differential equation.

If we assume a nonlinear system operating at point A, $[x_0, f(x_0)]$ in Figure 2.33, small changes in the input can be related to changes in the output about the point by way of the slope of the curve at the point A. Thus, if the slope of the curve at point A is m_a, then small excursions of the input about point A, δx, yield small changes in the output, $\delta f(x)$, related by the slope at point A. Thus,

$$[f(x) - f(x_0)] \approx m_a(x - x_0) \qquad (2.152)$$

from which

$$\delta f(x) \approx m_a \delta x \qquad (2.153)$$

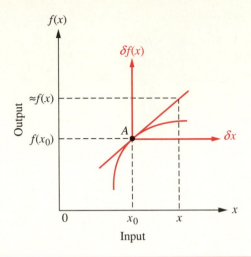

Figure 2.33 Linearization about a Point, A

and

$$f(x) \approx f(x_0) + m_a(x - x_0) \approx f(x_0) + m_a \delta x \qquad (2.154)$$

This relationship is shown graphically in Figure 2.33, where a new set of axes, δx and $\delta f(x)$, is created at the point A, and $f(x)$ is approximately equal to $f(x_0)$, the ordinate of the new origin, plus small excursions, $m_a \delta x$, away from point A. Let us look at an example.

Example 2.18

Linearize a function about a point.

Problem Linearize $f(x) = 5 \cos x$ about $x = \pi/2$

Solution We first find that the derivative of $f(x)$ is $df/dx = (-5 \sin x)$. At $x = \pi/2$, the derivative is -5. Also $f(x_0) = f(\pi/2) = 5 \cos(\pi/2) = 0$. Thus, from Equation 2.154, the system can be represented as $f(x) = -5 \delta x$ for small excursions of x about $\pi/2$. The process is shown graphically in Figure 2.34, where the cosine curve does indeed look like a straight line of slope -5 near $\pi/2$.

The previous discussion can be formalized using the Taylor series expansion, which expresses the value of a function in terms of the value of that function at a particular point, the excursion away from that point, and derivatives evaluated at that point. The Taylor series is shown in Equation 2.155.

$$f(x) = f(x_0) + \left.\frac{df}{dx}\right|_{x=x_0} \frac{(x - x_0)}{1!} + \left.\frac{d^2 f}{dx^2}\right|_{x=x_0} \frac{(x - x_0)^2}{2!} + \cdots \qquad (2.155)$$

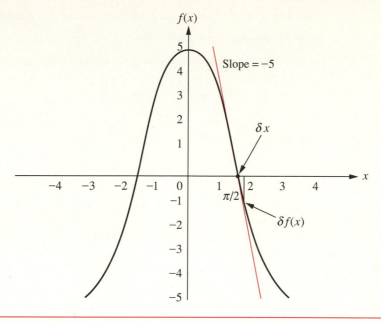

Figure 2.34 Linearization of $5 \cos x$ about $x = \pi/2$

For small excursions of x from x_0, we can neglect higher-order terms. The resulting approximation yields a straight-line relationship between the change in $f(x)$ and the excursions away from x_0. Neglecting the higher-order terms in Equation 2.155, we get

$$f(x) - f(x_0) \approx \left. \frac{df}{dx} \right|_{x = x_0} (x - x_0) \qquad (2.156)$$

or

$$\delta f(x) \approx m|_{x = x_0}\, \delta x \qquad (2.157)$$

which is a linear relationship between $\delta f(x)$ and δx for small excursions away from x_0. It is interesting to note that Equations 2.156 and 2.157 are identical to Equations 2.152 and 2.153, which we derived intuitively. The following example illustrates the application of linearization to differential equations.

Example 2.19   Linearize a differential equation.

   Problem   Linearize Equation 2.158 for small excursions about $x = \pi/4$.

$$\frac{d^2x}{dt^2} + 2\frac{dx}{dt} + \cos x = 0 \qquad (2.158)$$

Solution   The presence of the term $\cos x$ makes this equation nonlinear. Since we want to linearize the equation about $x = \dfrac{\pi}{4}$, we let $x = \delta x + \dfrac{\pi}{4}$, where δx is the small excursion about $\dfrac{\pi}{4}$, and substitute x into Equation 2.158:

$$\frac{d^2\left(\delta x + \dfrac{\pi}{4}\right)}{dt^2} + 2\frac{d\left(\delta x + \dfrac{\pi}{4}\right)}{dt} + \cos\left(\delta x + \frac{\pi}{4}\right) = 0 \tag{2.159}$$

But

$$\frac{d^2\left(\delta x + \dfrac{\pi}{4}\right)}{dt^2} = \frac{d^2\delta x}{dt^2} \tag{2.160}$$

and

$$\frac{d\left(\delta x + \dfrac{\pi}{4}\right)}{dt} = \frac{d\delta x}{dt} \tag{2.161}$$

Finally, the term $\cos\left(\delta x + \dfrac{\pi}{4}\right)$ can be linearized with the truncated Taylor series. Substituting $f(x) = \cos\left(\delta x + \dfrac{\pi}{4}\right)$, $f(x_0) = f\left(\dfrac{\pi}{4}\right) = \cos\left(\dfrac{\pi}{4}\right)$, and $(x - x_0) = \delta x$ into Equation 2.156 yields

$$\cos\left(\delta x + \frac{\pi}{4}\right) - \cos\left(\frac{\pi}{4}\right) = \left.\frac{d\cos x}{dx}\right|_{x=\frac{\pi}{4}} \delta x = -\sin\left(\frac{\pi}{4}\right)\delta x \tag{2.162}$$

Solving Equation 2.162 for $\cos\left(\delta x + \dfrac{\pi}{4}\right)$, we get

$$\cos\left(\delta x + \frac{\pi}{4}\right) = \cos\left(\frac{\pi}{4}\right) - \sin\left(\frac{\pi}{4}\right)\delta x = \frac{\sqrt{2}}{2} - \frac{\sqrt{2}}{2}\delta x \tag{2.163}$$

Substituting Equations 2.160, 2.161, and 2.163 into Equation 2.159 yields the following linearized differential equation:

$$\frac{d^2\delta x}{dt^2} + 2\frac{d\delta x}{dt} - \frac{\sqrt{2}}{2}\delta x = -\frac{\sqrt{2}}{2} \tag{2.164}$$

This equation can now be solved for δx, from which we can obtain $x = \delta x + \dfrac{\pi}{4}$.

Even though the nonlinear equation 2.158 is homogeneous, the linearized Equation 2.164 is not homogeneous. Equation 2.164 has a forcing function on its right-hand side. This additional term can be thought of as an input to a system represented by Equation 2.158.

Another observation about Equation 2.164 is the negative sign on the left-hand side. The study of differential equations tells us that since the roots of the characteristic equation are positive, the homogeneous solution grows without bound instead of diminishing to zero. Thus, this system linearized around $x = \dfrac{\pi}{4}$ is not stable.

As a final example, let us apply linearizing techniques to find the transfer function of a nonlinear system.

Example 2.20   Find the transfer function of a nonlinear electrical network.

Problem   Find the transfer function, $V_L(s)/V(s)$, for the electrical network shown in Figure 2.35, which contains a nonlinear resistor whose voltage-current relationship is defined by $i_r = 2e^{0.1v_r}$, where i_r and v_r are the resistor current and voltage, respectively. Also, $v(t)$ in Figure 2.35 is a small signal source.

Solution   We will use Kirchhoff's voltage law to sum the voltages in the loop to obtain the nonlinear differential equation, but first, we must solve for the voltage across the nonlinear resistor. Taking the natural log of the resistor's current-voltage relationship, we get $v_r = 10 \ln \frac{1}{2} i_r$. Applying Kirchhoff's voltage law around the loop, where $i_r = i$, yields

$$\frac{di}{dt} + 10 \ln \frac{1}{2} i - 20 = v(t) \tag{2.165}$$

Next, let us evaluate the equilibrium solution. First, set the small signal source, $v(t)$, equal to zero. Now evaluate the steady-state current. With $v(t) = 0$, the circuit consists of a 20 V battery in series with the inductor and nonlinear resistor. In the steady state, the voltage across the inductor will be zero, since $v_L(t) = L\frac{di}{dt}$ and $\frac{di}{dt}$ is zero in the steady state, given a constant battery source. Hence, the resistor voltage, v_r, is 20 V. Using the characteristics of the resistor, $i_r = 2e^{0.1v_r}$, we find that $i_r = i = 14.78$ amps. This current, i_0, is the equilibrium value of the network current. Hence $i = i_0 + \delta i$. Substituting this current into Equation 2.165 yields

$$L\frac{d(i_0 + \delta i)}{dt} + 10 \ln \frac{1}{2}(i_0 + \delta i) - 20 = v(t) \tag{2.166}$$

Figure 2.35 Nonlinear Electrical Network

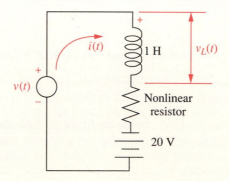

Using Equation 2.156 to linearize $\ln \frac{1}{2}(i_0 + \delta i)$, we get

$$\ln \frac{1}{2}(i_0 + \delta i) - \ln \frac{1}{2}i_0 = \left. \frac{d\left(\ln \frac{1}{2}i\right)}{di}\right|_{i=i_0} \delta i = \left.\frac{1}{i}\right|_{i=i_0} \delta i = \frac{1}{i_0}\delta i \qquad (2.167)$$

or

$$\ln \frac{1}{2}(i_0 + \delta i) = \ln \frac{i_0}{2} + \frac{1}{i_0}\delta i \qquad (2.168)$$

Substituting into Equation 2.166, the linearized equation becomes

$$L\frac{d\delta i}{dt} + 10\left(\ln \frac{i_0}{2} + \frac{1}{i_0}\delta i\right) - 20 = v(t) \qquad (2.169)$$

Letting $L = 1$, and $i_0 = 14.78$, the final, linearized differential equation is

$$\frac{d\delta i}{dt} + 0.677\delta i = v(t) \qquad (2.170)$$

Taking the Laplace transform with zero initial conditions and solving for $\delta i(s)$, we get

$$\delta i(s) = \frac{V(s)}{s + 0.667} \qquad (2.171)$$

But the voltage across the inductor about the equilibrium point is $v_L(t) = L\frac{d}{dt}(i_0 + \delta i) = L\frac{d\delta i}{dt}$. Taking the Laplace transform,

$$V_L(s) = Ls\delta i(s) = s\delta i(s) \qquad (2.172)$$

Substituting Equation 2.171 into Equation 2.172 yields

$$V_L(s) = s\frac{V(s)}{s + 0.667} \qquad (2.173)$$

from which the final transfer function is

$$\frac{V_L(s)}{V(s)} = \frac{s}{s + 0.667} \qquad (2.174)$$

for small excursions about $i = 14.78$ or, equivalently, about $v(t) = 0$.

In this section, we learned how to find transfer functions for nonlinear systems. Let us now summarize the essential concepts emphasized in this chapter; we present some demonstration problems, followed by a written summary.

2.11 Chapter-Objective Demonstration Problems

This chapter showed us that physical systems can be modeled mathematically with transfer functions. Typically, systems are composed of subsystems of different types, such as electrical, mechanical, and electromechanical.

The first demonstration problem relates to the chapter objective and uses our ongoing example of the antenna azimuth control system to show how to mathematically represent each of the azimuth control's subsystems as a transfer function.

Example 2.21 Find the transfer function of each subsystem composing an antenna azimuth control system.

Problem Find the transfer function of each subsystem for the antenna azimuth control system shown in Figure 2.36.

Solution First, we identify the individual subsystems for which we must find transfer functions; they are summarized in Table 2.6. We proceed to find the transfer function for each subsystem.

Input Potentiometer; Output Potentiometer Since both input and output potentiometers are configured the same, their transfer functions will be the same. We *neglect* the dynamics for the potentiometers and simply find the relationship between the output voltage and the input angular displacement. In the center position, the output voltage is zero. Five turns either toward the positive 10 volts or the negative 10 volts yields a voltage change of 10 volts. Thus, the transfer function, $V_i(s)/\theta_i(s)$, for the potentiometers is found by dividing the voltage change by the angular displacement. Hence,

$$\frac{V_i(s)}{\theta_i(s)} = \frac{10}{10\pi} = \frac{1}{\pi} \qquad (2.175)$$

Preamplifier; Power Amplifier The transfer functions of the amplifiers are given in the problem statement. Two phenomena are *neglected*. First, we *assume* that saturation is never reached. Second, the dynamics of the preamplifier are *neglected,* since its speed of response is typically much greater than that of the power amplifier. The transfer functions of both amplifiers are given in the problem statement and are the ratio of the Laplace transforms of the output voltage divided by the input voltage. Hence, for the preamplifier,

$$\frac{V_p(s)}{V_i(s)} = K \qquad (2.176)$$

and for the power amplifier,

$$\frac{E_a(s)}{V_p(s)} = \frac{100}{s + 100} \qquad (2.177)$$

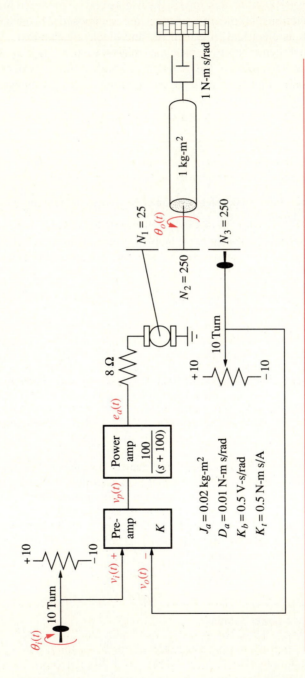

Figure 2.36 Antenna Azimuth Control System for the Chapter-Objective
Demonstration Problem

Table 2.6 Subsystems of the Antenna Azimuth Control System

Subsystem	Input	Output
Input potentiometer	Angular rotation from user $\theta_i(t)$	Voltage to preamp $v_i(t)$
Preamp	Voltage from potentiometers $v_i(t)$, $v_o(t)$	Voltage to power amp $v_p(t)$
Power amp	Voltage from preamp $v_p(t)$	Voltage to motor $e_a(t)$
Motor	Voltage from power amp $e_a(t)$	Angular rotation to load $\theta_0(t)$
Output potentiometer	Angular rotation from load $\theta_0(t)$	Voltage to preamp $v_0(t)$

Motor and Load   The motor and its load are next. The transfer function relating the armature displacement to the armature voltage is given in Equation 2.132. The equivalent inertia, J_m, is,

$$J_m = J_a + J_L \left(\frac{25}{250}\right)^2 = 0.02 + 1\frac{1}{100} = 0.03 \qquad (2.178)$$

where $J_L = 1$ is the load inertia at θ_0. The equivalent viscous damping, D_m, at the armature is

$$D_m = D_a + D_L \left(\frac{25}{250}\right)^2 = 0.01 + 1\frac{1}{100} = 0.02 \qquad (2.179)$$

where D_L is the load viscous damping at θ_0. From the problem statement, $K_t = 0.5$ N-m/A, $K_b = 0.5$ V-s/rad, and the armature resistance $R_a = 8$ ohms. These quantities along with J_m and D_m are substituted into Equation 2.132, yielding the transfer function of the motor from the armature voltage to the armature displacement, or

$$\frac{\theta_m(s)}{E_a(s)} = \frac{K_t/(R_a J_m)}{s\left[s + \frac{1}{J_m}\left(D_m + \frac{K_t K_b}{R_a}\right)\right]} = \frac{2.083}{s(s + 1.71)} \qquad (2.180)$$

To complete the transfer function of the motor, we multiply by the gear ratio to arrive at the transfer function relating load displacement to armature voltage:

$$\frac{\theta_0(s)}{E_a(s)} = 0.1\frac{\theta_m(s)}{E_a(s)} = \frac{0.2083}{s(s + 1.71)} \qquad (2.181)$$

In the next example, we demonstrate the first two steps in the design and analysis sequence by finding a transfer function of a biological system. The system is that of a human leg, which pivots from the hip joint. In this problem, the component of weight is nonlinear, and thus the system requires linearization prior to the evaluation of the transfer function.

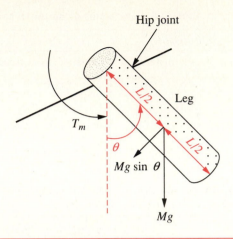

Figure 2.37 Cylinder Model of a Human Leg (Adapted from Milsum, J. H. *Biological Control Systems Analysis*. McGraw-Hill, New York, 1966, p. 182)

Example 2.22   Find the transfer function of a human leg.

Problem   The transfer function of a human leg relates the output angular rotation about the hip joint to the input torque supplied by the leg muscle. A simplified model for the leg is shown in Figure 2.37. The model *assumes* an applied muscular torque, $T_m(t)$, viscous damping, D, at the hip joint, and inertia, J, around the hip joint.[12] Also, a component of the weight of the leg, Mg, where M is the mass of the leg and g is the acceleration due to gravity, creates a nonlinear torque. If we *assume* that the leg is of uniform density, then the weight can be applied at $L/2$, where L is the length of the leg (Milsum, 1966). Do the following:

a. Evaluate the nonlinear torque.
b. Find the transfer function, $\theta(s)/T_m(s)$, for small angles of rotation, where $\theta(s)$ is the angular rotation of the leg about the hip joint.

Solution   First, calculate the torque due to the weight. The total weight of the leg is Mg acting vertically. The component of the weight in the direction of rotation is $Mg \sin \theta$. This force is applied at a distance $L/2$ from the hip joint. Hence the torque in the direction of rotation, $T_W(t)$, is $Mg\dfrac{L}{2}\sin \theta$. Next, draw a free-body diagram of the leg, showing the applied torque, $T_m(t)$, the torque due to the weight, $T_W(t)$, and the opposing torques due to inertia and viscous damping (see Figure 2.38).
Summing torques, we get

$$J \frac{d^2\theta}{dt^2} + D \frac{d\theta}{dt} + Mg \frac{L}{2} \sin \theta = T_m(t) \qquad (2.182)$$

---

[12]For emphasis, J is not around the center of mass, as we previously assumed for inertia in mechanical rotation.

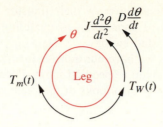

Figure 2.38 Free-Body Diagram of Leg Model

We linearize the system about the equilibrium point, $\theta = 0$, the vertical position of the leg. Using Equation 2.156, we get

$$\sin\theta - \sin 0 = (\cos 0)\,\delta\theta \tag{2.183}$$

From which, $\sin\theta = \delta\theta$. Also, $J\dfrac{d^2\theta}{dt^2} = J\dfrac{d^2\delta\theta}{dt^2}$ and $D\dfrac{d\theta}{dt} = D\dfrac{d\delta\theta}{dt}$. Hence Equation 2.182 becomes

$$J\frac{d^2\delta\theta}{dt^2} + D\frac{d\delta\theta}{dt} + Mg\frac{L}{2}\delta\theta = T_m(t) \tag{2.184}$$

Notice that the torque due to the weight approximates a spring torque on the leg. Taking the Laplace transform with zero initial conditions yields

$$\left(Js^2 + Ds + Mg\frac{L}{2}\right)\delta\theta(s) = T_m(s) \tag{2.185}$$

from which the transfer function is

$$\frac{\delta\theta(s)}{T_m(s)} = \frac{1/J}{s^2 + \dfrac{D}{J}s + \dfrac{MgL}{2J}} \tag{2.186}$$

for small excursions about the equilibrium point, $\theta = 0$.

This section showed two examples of how to arrive at the transfer function of a subsystem from the physical system by first making simplifying assumptions to arrive at a schematic representation of the system and then calculating the transfer function.

2.12 Summary

Having completed our discussion of electrical, mechanical, and electromechanical systems, we are prepared to find transfer functions for these physical systems. We realize that the physical world consists of more than the systems we chose to illustrate. There are also thermal, pneumatic, and hydraulic systems, to mention a few. The evaluation of their transfer functions is handled in a similar way. Using inductive approaches to modeling, we can obtain transfer functions even for economic and biological systems.

Now that we have our transfer function, we move on in Chapter 4 to evaluate its response given a specified input. For those pursuing the state-space approach, the next chapter also looks at the mathematical modeling of systems in the time domain.

REVIEW QUESTIONS

1. Name three performance criteria important to the design of control systems.
2. What mathematical model permits the easy interconnection of physical systems?
3. To what classification of systems can the transfer function be best applied?
4. What are the physical implications of an unstable system?
5. Name the two parts of a system's response.
6. To what part of the total response is instability attributable?
7. What transformation turns the solution of differential equations into algebraic manipulations?
8. Define the transfer function.
9. What assumption is made concerning initial conditions when dealing with transfer functions?
10. What do we call the mechanical equations written in order to evaluate the transfer function?
11. If we understand the form that the mechanical equations take, what step do we avoid in the evaluation of the transfer function?
12. Why do transfer functions for mechanical networks look identical to transfer functions for electrical networks?
13. What function do gears perform?
14. What are the component parts of the mechanical constants of a motor's transfer function?
15. Since the motor's transfer function relates armature displacement to armature voltage, how can the transfer function that relates load displacement and armature voltage be determined?
16. Summarize the steps taken to linearize a nonlinear system.

PROBLEMS

1. Derive the Laplace transform for the following time functions:
   a. $u(t)$
   b. $tu(t)$
   c. $\sin \omega t\, u(t)$
   d. $\cos \omega t\, u(t)$

2. Using the Laplace transform pairs of Table 2.1 and the Laplace transform theorems of Table 2.2, derive the Laplace transforms for the following time functions:

a. $e^{-at} \sin \omega t \, u(t)$

b. $e^{-at} \cos \omega t \, u(t)$

c. $t^3 u(t)$

3. Repeat Problem 9 in Chapter 1, using Laplace transforms. Assume that the forcing functions are zero prior to $t = 0-$.

4. Repeat Problem 10 in Chapter 1, using Laplace transforms. Assume that the forcing functions are zero prior to $t = 0-$.

5. A system is described by the following differential equation:

$$\frac{d^3y}{dt^3} + 5\frac{d^2y}{dt^2} + 7\frac{dy}{dt} + y = \frac{d^3x}{dt^3} + 2\frac{d^2x}{dt^2} + 3\frac{dx}{dt} + 7x$$

Find the expression for the transfer function of the system, $Y(s)/X(s)$.

6. For each transfer function below, write the corresponding differential equation.

a. $\dfrac{X(s)}{F(s)} = \dfrac{1}{s^2 + 2s + 7}$

b. $\dfrac{X(s)}{F(s)} = \dfrac{s + 2}{s^3 + 8s^2 + 9s + 15}$

7. Write the differential equation for the system shown in Figure P2.1.

Figure P2.1

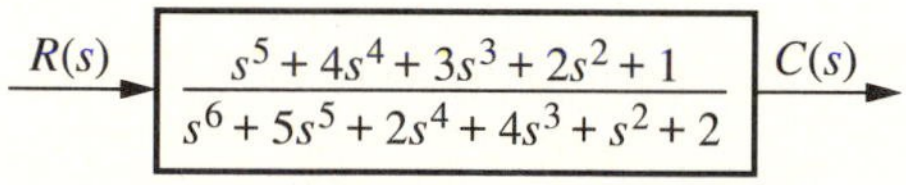

$$R(s) \longrightarrow \boxed{\frac{s^5 + 4s^4 + 3s^3 + 2s^2 + 1}{s^6 + 5s^5 + 2s^4 + 4s^3 + s^2 + 2}} \longrightarrow C(s)$$

8. Write the differential equation that is mathematically equivalent to the block diagram shown in Figure P2.2. Assume that $r(t) = t^3$.

Figure P2.2

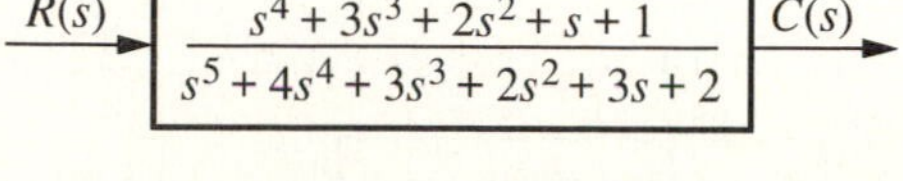

$$R(s) \longrightarrow \boxed{\frac{s^4 + 3s^3 + 2s^2 + s + 1}{s^5 + 4s^4 + 3s^3 + 2s^2 + 3s + 2}} \longrightarrow C(s)$$

9. A system is described by the following differential equation:

$$\frac{dx^2}{dt^2} + 2\frac{dx}{dt} + 3x = 1$$

with the following initial conditions: $x(0) = 1$; $\dot{x}(0) = -1$. Show a block diagram of the system, giving its transfer function and all pertinent inputs and outputs. (Hint: the initial conditions will show up as added inputs to an effective system with zero initial conditions.)

10. Find the transfer function, $G(s) = V_o(s)/V_i(s)$, for the network shown in Figure P2.3.

Figure P2.3

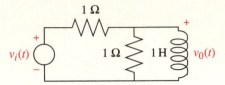

11. Find the transfer function, $G(s) = V_o(s)/V_i(s)$, for the network shown in Figure P2.4.

Figure P2.4

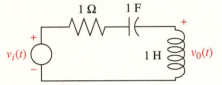

12. Find the transfer function, $G(s) = V_o(s)/V_i(s)$, for the network shown in Figure P2.5. Solve the problem using mesh analysis.

Figure P2.5

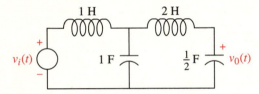

13. Repeat Problem 12 using nodal equations.

14. Find the transfer function, $G(s) = V_o(s)/V_i(s)$, for the network shown in Figure P2.6. Solve the problem using mesh analysis.

Figure P2.6

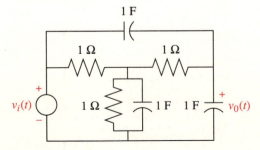

15. Repeat Problem 14 using nodal equations.

16. Find the transfer function, $G(s) = X_1(s)/F(s)$, for the translational mechanical system shown in Figure P2.7.

Figure P2.7

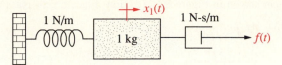

17. Find the transfer function, $G(s) = X_2(s)/F(s)$, for the translational mechanical network shown in Figure P2.8.

Figure P2.8

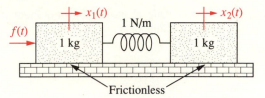

18. For the translational mechanical system shown in Figure P2.9, find the transfer function, $G(s) = X_1(s)/F(s)$.

Figure P2.9

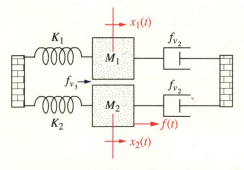

19. For the translational mechanical system shown in Figure P2.10, find the transfer function, $G(s) = X_1(s)/F(s)$.

Figure P2.10

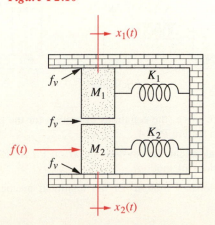

20. Find the transfer function, $G(s) = X_2(s)/F(s)$, for the translational mechanical system shown in Figure P2.11. (Hint: Place a zero mass at x_2.)

Figure P2.11

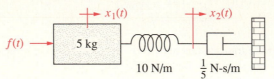

21. For the translational mechanical system shown in Figure P2.12, write, but do not solve, the system's equations of motion.

Figure P2.12

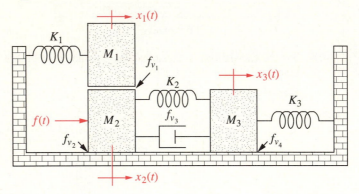

22. Find the transfer function, $G(s) = X_3(s)/F(s)$, for the translational mechanical system shown in Figure P2.13.

Figure P2.13

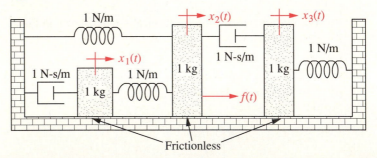

23. Write, but do not solve, the equations of motion for the translational mechanical system shown in Figure P2.14.

24. For the rotational mechanical system shown in Figure P2.15, write, but do not solve, the equations of motion.

25. Write the equations of motion for the rotational mechanical system shown in Figure P2.16.

Figure P2.14

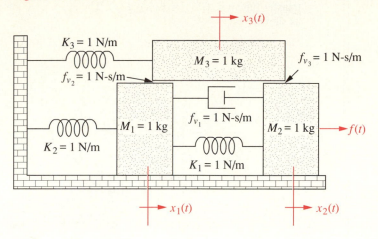

Figure P2.15

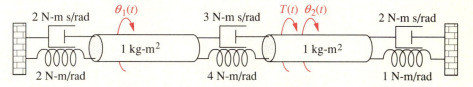

Figure P2.16

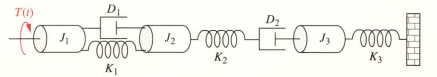

26. Find the transfer function, $G(s) = \theta_3(s)/T(s)$, in the rotational system shown in Figure P2.17.

Figure P2.17

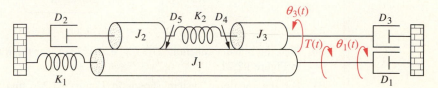

27. For the rotational mechanical system with gears shown in Figure P2.18, find the transfer function, $G(s) = \theta_3(s)/T(s)$. The gears have inertia and bearing friction as shown.

Figure P2.18

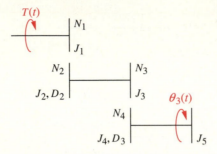

28. For the rotational system shown in Figure P2.19, find the transfer function, $G(s) = \theta_2(s)/T(s)$.

Figure P2.19

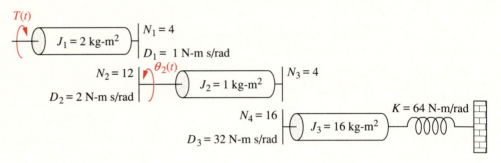

29. Find the transfer function, $G(s) = \theta_2(s)/T(s)$, for the rotational mechanical system shown in Figure P2.20.

Figure P2.20

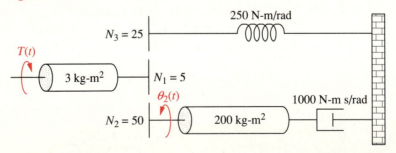

30. Find the transfer function, $G(s) = \theta_4(s)/T(s)$, for the rotational system shown in Figure P2.21.

31. For the rotational system shown in Figure P2.22, find the transfer function, $G(s) = \theta_4(s)/T(s)$.

Figure P2.21

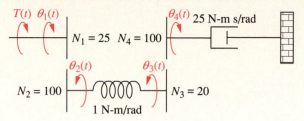

Figure P2.22

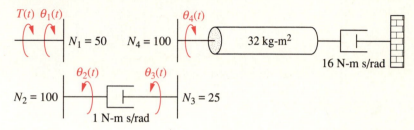

32. For the rotational system shown in Figure P2.23, find the transfer function, $G(s) = \theta_1(s)/T(s)$.

Figure P2.23

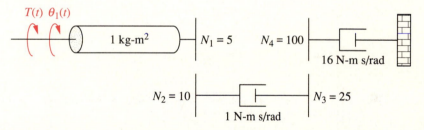

33. For the rotational system shown in Figure P2.24, write the equations of motion from which the transfer function, $G(s) = \theta_1(s)/T(s)$, can be found.

Figure P2.24

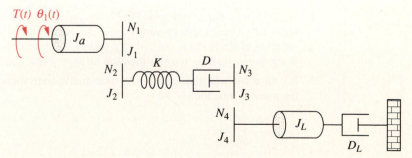

34. Find $G(s) = E_o(s)/T(s)$ for the system shown in Figure P2.25.

Figure P2.25

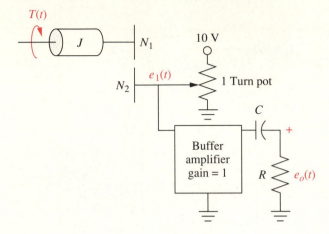

35. Given the rotational system shown in Figure P2.26, find the transfer function, $G(s) = \theta_6(s)/\theta_1(s)$.

Figure P2.26

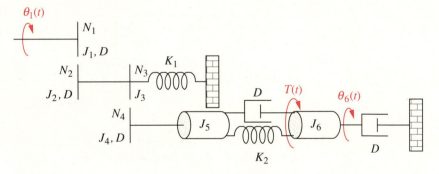

36. In the system shown in Figure P2.27, the inertia, J, of radius r is constrained to move only about the stationary axis A. A viscous damping force of translational value f_V exists between the bodies J and M. If an external force, $f(t)$, is applied to the mass, find the transfer function $G(s) = \theta(s)/F(s)$.

37. For the combined translational and rotational system shown in Figure P2.28, find the transfer function, $G(s) = X(s)/T(s)$.

Figure P2.27

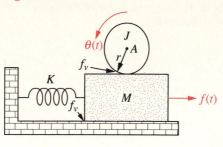

Figure P2.28

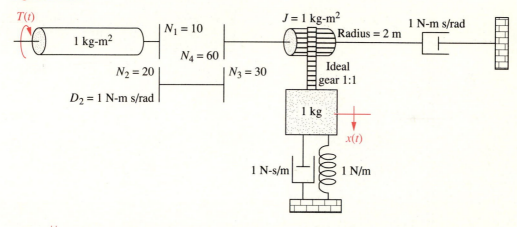

38. Given the combined translational and rotational system shown in Figure P2.29, find the transfer function, $G(s) = X(s)/T(s)$.

Figure P2.29

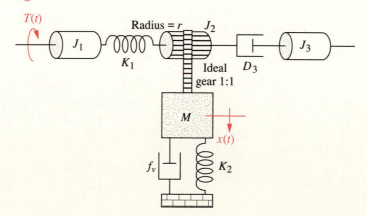

39. The motor whose torque-speed characteristics are shown in Figure P2.30 drives the load shown in the diagram. Some of the gears have inertia. Find the transfer function, $G(s) = \theta_2(s)/E_a(s)$.

Figure P2.30

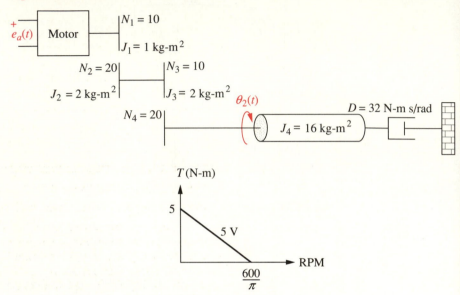

40. A dc motor develops 50 N-m of torque at a speed of 500 rad/s when 10 volts are applied. It stalls out at this voltage with 100 N-m of torque. If the inertia and damping of the armature are 5 kg-m^2 and 1 N-m s/rad respectively, find the transfer function, $G(s) = \theta_L(s)/E_a(s)$, of this motor if it drives an inertia load of 100 kg-m^2 through a gear train, as shown in Figure P2.31.

Figure P2.31

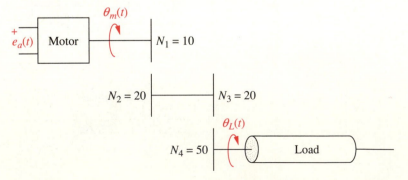

41. Find the transfer function, $G(s) = X(s)/E_a(s)$, for the system shown in Figure P2.32.

Figure P2.32

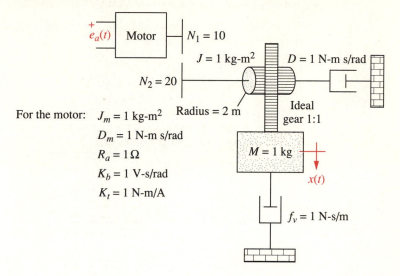

For the motor: $J_m = 1$ kg-m^2

$D_m = 1$ N-m s/rad

$R_a = 1\,\Omega$

$K_b = 1$ V-s/rad

$K_t = 1$ N-m/A

42. A system's output, c, is related to the system's input, r, by the straight-line relationship, $c = 5r + 7$. Is the system linear?

43. Consider the differential equation

$$\frac{d^2x}{dt^2} + 3\frac{dx}{dt} + 2x = f(x)$$

where $f(x)$ is the input and is a function of the output, x. If $f(x) = \sin x$, linearize the differential equation for small excursions near

a. $x = 0$

b. $x = \pi$

44. Consider the differential equation,

$$\frac{d^3x}{dt^3} + 10\frac{d^2x}{dt^2} + 31\frac{dx}{dt} + 30x = f(x)$$

where $f(x)$ is the input and is a function of the output, x. If $f(x) = e^{-x}$, linearize the differential equation for x near 0.

45. Many systems are *piecewise* linear. That is, over a *large* range of variable values, the system can be described linearly. A system with amplifier saturation is one such example. Given the following differential equation,

$$\frac{d^2x}{dt^2} + 15\frac{dx}{dt} + 50x = f(x)$$

assume that $f(x)$ is as shown in Figure P2.33. Write the differential equation for each of the following ranges of x:

a. $-\infty < x < -2$
b. $-2 < x < 2$
c. $2 < x < \infty$

Figure P2.33

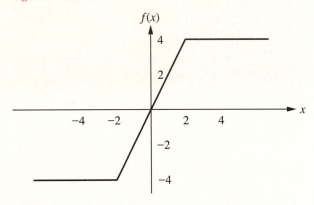

46. Given the nonlinear electrical network shown in Figure P2.34, find the transfer function relating the output nonlinear resistor voltage, $V_r(s)$, to the input source voltage, $V(s)$.

Figure P2.34

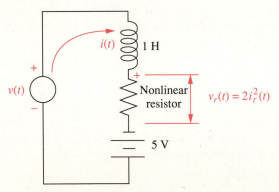

47. Consider the restaurant plate dispenser shown in Figure P2.35, which consists of a vertical stack of dishes supported by a compressed spring. As each plate is removed, the reduced weight on the dispenser causes the remaining plates to rise. Assume that the mass of the system minus the top plate is M, the viscous friction between the piston and the sides of the cylinder is f_v; the spring constant is K; and the weight of a single plate is W_D. Find the transfer function, $Y(s)/F(s)$, where $F(s)$ is the step reduction in force felt when the top plate is removed, and $Y(s)$ is the vertical displacement of the dispenser in an upward direction.

Figure P2.35 Plate Dispenser

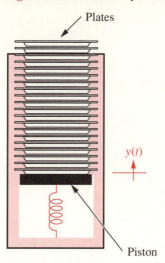

Figure P2.36 Antenna Azimuth Control System

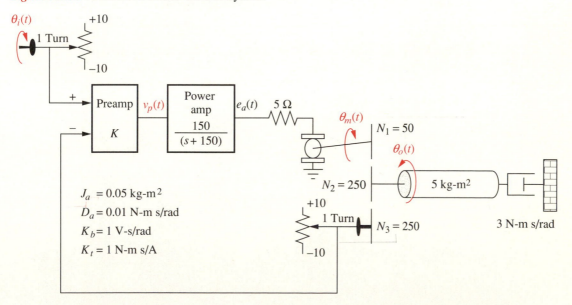

48. *Chapter-objective problem:* Given the schematic for the antenna azimuth position control system shown in Figure P2.36, evaluate the transfer function of each subsystem.

BIBLIOGRAPHY

Aggarwal, J. K. *Notes on Nonlinear Systems*. Van Nostrand Reinhold, New York, 1972.

Cannon, R. H., Jr. *Dynamics of Physical Systems*. McGraw-Hill, New York, 1967.

Cochin, I. *Analysis and Design of Dynamic Systems*. Harper and Row, New York, 1980.

Cook, P. A. *Nonlinear Dynamical Systems*. Prentice Hall, United Kingdom, 1986.

Davis, S. A., and Ledgerwood, B. K. *Electromechanical Components for Servomechanisms*. McGraw-Hill, New York, 1961.

Doebelin, E. O. *Measurement Systems Application and Design*. McGraw-Hill, New York, 1983.

Franklin, G. F.; Powell, J. D.; and Emami-Naeini, A. *Feedback Control of Dynamic Systems*. Addison-Wesley, Reading, Mass., 1986.

Hsu, J. C., and Meyer, A. U. *Modern Control Principles and Applications*. McGraw-Hill, New York, 1968.

Kuo, F. F. *Network Analysis and Synthesis*. Wiley, New York, 1966.

Lago, G., and Benningfield, L. M. *Control System Theory*. Ronald Press, New York, 1962.

Mablekos, V. E. *Electric Machine Theory for Power Engineers*. Harper & Row, Cambridge, Mass., 1980.

Milsum, J. H. *Biological Control Systems Analysis*. McGraw-Hill, New York, 1966.

Minorsky, N. *Theory of Nonlinear Control Systems*. McGraw-Hill, New York, 1969.

Nilsson, J. W. *Electric Circuits*. 2d ed. Addison-Wesley, Reading, Mass., 1986.

Van Valkenburg, M. E. *Network Analysis*. Prentice Hall, Englewood Cliffs, N.J., 1974.

Vidyasagar, M. *Nonlinear Systems Analysis*. Prentice Hall, Englewood Cliffs, N.J., 1978.

SYSTEM REPRESENTATION
IN THE TIME DOMAIN

3.1 Introduction

There are two approaches available for the analysis and design of feedback control systems. The first, which we began to study in Chapter 2, is known as the *classical*, or *frequency domain*, technique. This approach is based upon converting a system's differential equation to a transfer function, thus generating a mathematical model of the system that *algebraically* relates a representation of the output to a representation of the input. Replacing a differential equation with an algebraic equation not only simplifies the representation of individual subsystems, but also simplifies modeling interconnected subsystems.

The primary disadvantage of the classical approach is its limited applicability: it can only be applied to linear, time-invariant systems or systems that can be approximated as such.

A major advantage of frequency domain techniques is that our intuition is improved by mathematical tools that rapidly provide stability and transient response information, so that we can immediately see the effect of varying system parameters until an acceptable design is met.

With the arrival of space exploration, requirements for control systems increased in scope. Modeling systems by using linear, time-invariant differential equations and subsequent transfer functions became inadequate. The *state-space* approach (also referred to as the *modern*, or *time-domain*, approach) is a unified method for modeling, analyzing, and designing a wide range of systems. For example, the state-space approach can be used to represent nonlinear systems that have backlash, saturation, and dead zone. Also, the modern approach can handle, conveniently, systems with nonzero initial conditions. Time-varying systems (e.g., missiles with varying fuel levels; lift in an aircraft flying through a wide range of altitudes) can be represented in state space. Many systems do not have just a single input and a single output. Multiple input-output systems (e.g., a vehicle with input direction and input speed yielding an output direction and an output velocity) can be compactly represented in state space with a model similar in form and complexity to that used for single-input, single-output systems. The time-domain approach can be used to represent systems with a digital computer in the loop or to model systems for digital simulation. With a simulated system, system response can be obtained for changes in system parameters—an important design tool. The state-space approach is also attractive because of the availability of numerous state-space software packages for the personal computer.

The time-domain approach also can be used for the same class of systems modeled by the classical approach. This alternate model gives the control systems designer another perspective from which to create a design. While the state-space approach can be applied to a wide range of systems, it is not as intuitive as the classical approach. The designer has to engage in several calculations before the physical interpretation of the model is apparent, whereas in classical control, a few quick calculations or a graphic presentation of data rapidly yields the physical interpretation.

In this book the coverage of state-space techniques is to be regarded as an introduction to the subject, a springboard to advanced studies, and an alternate approach to frequency domain techniques. We will limit the state-space approach to linear, time-invariant systems or systems that can be linearized by the methods of Chapter 2. The study of other classes of systems is beyond the scope of this book. Since state-space analysis and design rely upon matrices and matrix operations, you may want to review this topic in Appendix B before continuing. We proceed now to establish the state-space approach as an alternate method for representing physical systems.

Chapter Objective

The student will be able to derive the state-space representation of each subsystem of the azimuth position control system of Figure 2.2.

3.2 Some Observations

This section sets the stage for the formal definition of the state-space representation by making some observations about systems and their variables. These observations will give us a physical feel and establish an impetus for the formal definitions and the mathematics of the next section. In the discussion that follows, some of the development has been placed in footnotes to avoid clouding the main issues with an excess of equations and to ensure that the concept is clear. Although we use two electrical networks as examples, we could just as easily have used a mechanical or any other physical system to illustrate the concepts.

We will now demonstrate that, for a system with many variables, such as inductor voltage, resistor voltage, capacitor charge, and so on, we need to use differential equations only to solve for a selected subset of system variables since all other remaining system variables can be evaluated algebraically from the variables in the subset. Our examples take the following approach:

1. We will select a particular *subset* of all possible system variables and call the variables in this subset *state variables*.

2. For an *n*th-order system, we will write *n simultaneous, first-order differential equations* in terms of the state variables. We will call this system of simultaneous differential equations *state equations*.

3. If we know the initial condition of all of the state variables at t_0 as well as the system input for $t \geq t_0$, we will be able to solve the simultaneous differential equations for the state variables for $t \geq t_0$.

4. We will *algebraically* combine the state variables with the system's input and find all of the other system variables for $t \geq 0$. We will call this algebraic equation the *output equation*.

5. We will consider the state equations and the output equations a viable representation of the system. We will call this representation of the system the *state-space representation*.

Let us now follow these steps through an example. Consider the *RL* network shown in Figure 3.1 with an initial current of $i(0)$.

   1. We select the current, $i(t)$, for which we will write and solve a differential equation using Laplace transforms.

   2. We write the loop equation,

$$L\frac{di}{dt} + Ri = v(t) \tag{3.1}$$

   3. Taking the Laplace transform, using Table 2.2, Item 7 and including the initial conditions, yields

$$L[sI(s) - i(0)] + RI(s) = V(s) \tag{3.2}$$

Figure 3.1 *RL* Network

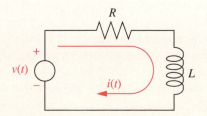

Assuming the input, $v(t)$, to be a unit step, $u(t)$, whose Laplace transform is $V(s) = 1/s$, we solve for $I(s)$ and get

$$I(s) = \frac{1}{R}\left(\frac{1}{s} - \frac{1}{s + \frac{R}{L}}\right) + \frac{i(0)}{s + \frac{R}{L}} \tag{3.3}$$

from which

$$i(t) = \frac{1}{R}\left(1 - e^{-(R/L)t}\right) + i(0)e^{-(R/L)t} \tag{3.4}$$

The function $i(t)$ is a subset of all possible network variables that we are able to find from Equation 3.4 if we know its initial condition, $i(0)$, and the input, $v(t)$. Thus, $i(t)$ is a state variable, and the differential equation (3.1) is a *state equation*.

4.   We can now solve for all of the other network variables *algebraically* in terms of $i(t)$ and the applied voltage, $v(t)$. For example, the voltage across the resistor is

$$v_R(t) = Ri(t) \tag{3.5}$$

The voltage across the inductor is

$$v_L(t) = v(t) - Ri(t) \tag{3.6}^1$$

The derivative of the current is

$$\frac{di}{dt} = \frac{1}{L}[v(t) - Ri(t)] \tag{3.7}^2$$

Thus, knowing the state variable, $i(t)$, and the input, $v(t)$, we can find the value, or *state,* of any network variable at any time, $t \geq t_0$. Hence, the algebraic equations, Equations 3.5 through 3.7, are *output equations*.

5. Since the variables of interest are completely described by Equation 3.1 and Equations 3.5 through 3.7, we say that the combined state equation (3.1) and the output equations (3.5 through 3.7) form a viable representation of the network, which we will call the *state-space representation*.

Equation 3.1, which describes the dynamics of the network, is not unique. This equation could be written in terms of any other network variable. For example, substituting $i = v_R/R$ into Equation 3.1 yields

$$\frac{L}{R}\frac{dv_R}{dt} + v_R = v(t) \tag{3.8}$$

which can be solved knowing that the initial condition $v_R(0) = Ri(0)$ and knowing $v(t)$. In this case, the state variable is $v_R(t)$. Similarly, all other network variables

---

[1] Since $v_L(t) = v(t) - v_R(t) = v(t) - Ri(t)$.
[2] Since $\dfrac{di}{dt} = \dfrac{1}{L}v_L(t) = \dfrac{1}{L}[v(t) - Ri(t)]$.

can now be written in terms of the state variable, $v_R(t)$, and the input, $v(t)$. Let us now extend our observations to a second-order system such as that shown in Figure 3.2.

1. Since the network is of second order, it will take two simultaneous, first-order differential equations to solve for two state variables. We select $i(t)$ and $q(t)$, the charge on the capacitor, as the two state variables.

2. Writing the loop equation yields

$$L\frac{di}{dt} + Ri + \frac{1}{C}\int i\,dt = v(t) \tag{3.9}$$

Converting to charge, using $i(t) = dq/dt$, we get

$$L\frac{d^2q}{dt^2} + R\frac{dq}{dt} + \frac{1}{C}q = v(t) \tag{3.10}$$

But an nth-order differential equation can be converted to n simultaneous first-order differential equations, with each equation of the form

$$\frac{dx_i}{dt} = a_{i1}x_1 + a_{i2}x_2 + \cdots + a_{in}x_n + b_i f(t) \tag{3.11}$$

where each x_i is a state variable, and the a_{ij}'s and b_i are constants for linear, time-invariant systems. We say that the right-hand side of Equation 3.11 is a *linear combination* of the state variables and the input, $f(t)$.

We can convert Equation 3.10 into two simultaneous, first-order differential equations in terms of $i(t)$ and $q(t)$. The first equation can be $dq/dt = i$. The second equation can be formed by substituting $\int i\,dt = q$ into Equation 3.9 and solving for di/dt. Summarizing the two resulting equations, we get

$$\frac{dq}{dt} = i \tag{3.12a}$$

$$\frac{di}{dt} = -\frac{1}{LC}q - \frac{R}{L}i + \frac{1}{L}v(t) \tag{3.12b}$$

3. These equations are the state equations and can be solved simultaneously for the state variables, $q(t)$ and $i(t)$, using the Laplace transform and the methods

Figure 3.2 *RLC* Network

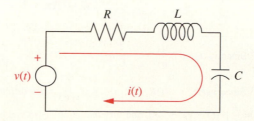

of Chapter 2, if we know the initial conditions for $q(t)$ and $i(t)$ and if we know $v(t)$, the input.

4. From these two state variables, we can solve for all other network variables. For example, the voltage across the inductor can be written in terms of the solved state variables and the input as

$$v_L(t) = -\frac{1}{C}q(t) - Ri(t) + v(t) \qquad (3.13)^3$$

Equation 3.13 is an *output equation;* we say that $v_L(t)$ is a *linear combination* of the state variables, $q(t)$ and $i(t)$, and the input, $v(t)$.

5. The combined state equations (3.12) and the output equation (3.13) form a viable representation of the network, which we call the *state-space representation.*

Another choice of two state variables can be made, for example, $v_R(t)$ and $v_L(t)$, the resistor and inductor voltage, respectively. The resulting set of simultaneous, first-order differential equations follows:

$$\frac{dv_R}{dt} = -\frac{R}{L}v_R - \frac{R}{L}v_C + \frac{R}{L}v(t) \qquad (3.14a)^4$$

$$\frac{dv_C}{dt} = \frac{1}{RC}v_R \qquad (3.14b)$$

Again, these state variables can be solved if we know their initial conditions along with $v(t)$, and all other network variables can be found as a linear combination of these state variables.

Is there a restriction on the choice of state variables? Yes! The restriction is that no state variable can be chosen if it can be expressed as a linear combination of the other state variables. For example, if $v_R(t)$ is chosen as a state variable, then $i(t)$ cannot be chosen, because $v_R(t)$ can be written as a linear combination of $i(t)$, namely, $v_R(t) = Ri(t)$. Under these circumstances we say that the state variables are *linearly dependent*. State variables must be *linearly independent;* that is, no state variable can be written as a linear combination of the other state variables, or else we would not have enough information to solve for all other system variables, and we could even have trouble writing the simultaneous equations themselves.

The state and output equations can be written in vector-matrix form if the system is linear. Thus, Equation 3.12, the state equation, can be written as

$$\dot{\mathbf{x}} = \mathbf{A}\mathbf{x} + \mathbf{B}u \qquad (3.15)$$

---

[3] Since $v_L(t) = L\dfrac{di}{dt} = -\dfrac{1}{C}q - Ri + v(t)$, where $\dfrac{di}{dt}$ can be found from Equation 3.9, and $\int i\,dt = q$.

[4] Since $v_R(t) = i(t)R$, and $v_C(t) = \dfrac{1}{C}\int i\,dt$, differentiating $v_R(t)$ yields $\dfrac{dv_R}{dt} = R\dfrac{di}{dt} = \dfrac{R}{L}v_L = \dfrac{R}{L}[v(t) - v_R - v_C]$, and differentiating $v_C(t)$ yields $\dfrac{dv_C}{dt} = \dfrac{1}{C}i = \dfrac{1}{RC}v_R$.

where

$$\dot{\mathbf{x}} = \begin{bmatrix} dq/dt \\ di/dt \end{bmatrix}; \quad \mathbf{A} = \begin{bmatrix} 0 & 1 \\ -1/LC & -R/L \end{bmatrix}$$

$$\mathbf{x} = \begin{bmatrix} q \\ i \end{bmatrix}; \quad \mathbf{B} = \begin{bmatrix} 0 \\ 1/L \end{bmatrix}; \quad u = v(t)$$

Equation 3.13, the output equation, can be written as

$$y = \mathbf{C}\mathbf{x} + Du \qquad (3.16)$$

where

$$y = v_L(t); \quad \mathbf{C} = [-1/C \quad -R]; \quad \mathbf{x} = \begin{bmatrix} q \\ i \end{bmatrix}; \quad D = 1; \quad u = v(t)$$

We call the combination of Equations 3.15 and 3.16 the *state-space representation* of the network of Figure 3.2. The state-space representation, therefore, consists of (1) the simultaneous, first-order differential equations from which the state variables can be solved and (2) the algebraic output equation from which all other system variables can be found. The state-space representation is not unique, since a different choice of state variables leads to a different representation of the same system.

In this section we used two electrical networks to demonstrate some principles that are the foundation of the state-space representation. The representations developed in this section were for single-input, single-output systems, where y, D, and u in Equations 3.15 and 3.16 are scalar quantities. In general, systems have multiple inputs and multiple outputs. For these cases, y and u become vector quantities, and D becomes a matrix. In Section 3.3, we will generalize the representation for multiple-input-output systems and summarize the concept of the state-space representation.

3.3 The General State-Space Representation

Now that we have represented a physical network in state space and have a good idea of the terminology and the concept, let us summarize and generalize the representation. First we will formalize some of the definitions that we came across in the last section.

Linear combination: A linear combination of n variables, x_i, for $i = 1$ to n, is given by the following sum, S:

$$S = K_n x_n + K_{n-1} x_{n-1} + \cdots + K_1 x_1 \qquad (3.17)$$

where each K_i is a constant.

Linear independence: A set of variables is said to be linearly independent if none of the variables can be written as a linear combination of the others. For example, given x_1, x_2, and x_3, if $x_2 = 5x_1 + 6x_3$, then the variables are not linearly independent, since one of them can be written as a linear combination of the other two. Now, what must be true so that one variable cannot be written as a linear combination of the other variables? Consider the example $K_2 x_2 = K_1 x_1 + K_3 x_3$. In this case one variable cannot be written as a linear combination of the other variables only if all $K_i = 0$, and no $x_i = 0$. Formally, then, variables x_i, for $i = 1$ to n, are said to be linearly independent if their linear combination, S, equals zero *only* if every $K_i = 0$ and *no* $x_i = 0$.

System variable: Any variable that responds to an input or initial conditions in a system.

State variables: The smallest set of linearly independent system variables such that the values of the members of the set at time t_0 along with known forcing functions completely determine the value of all system variables for all $t \geq t_0$.

State vector: A vector whose elements are the state variables.

State space: The n-dimensional space whose axes are the state variables. This is a new term and is illustrated in Figure 3.3, where the state variables are assumed to be a resistor voltage, v_R, and a capacitor voltage, v_C. These variables form the axes of the *state space*. A trajectory can be thought of as being mapped out by the state vector, $\mathbf{x}(t)$, for a range of t. Also shown is the state vector at the particular time $t = 4$.

State equations: A set of n simultaneous, first-order differential equations with n variables, where the n variables to be solved are the state variables.

Output equation: The algebraic equation that expresses the output variables of a system as linear combinations of the state variables.

Now that the definitions have been formally stated, we define the state-space representation of a system. A system is represented in state space by the following equations:

$$\dot{\mathbf{x}} = \mathbf{A}\mathbf{x} + \mathbf{B}\mathbf{u} \tag{3.18}$$

$$\mathbf{y} = \mathbf{C}\mathbf{x} + \mathbf{D}\mathbf{u} \tag{3.19}$$

Figure 3.3 Graphic Representation of State Space and a State Vector

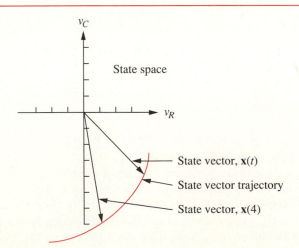

State space

State vector, $\mathbf{x}(t)$

State vector trajectory

State vector, $\mathbf{x}(4)$

where

    $\mathbf{x}$ = state vector

    $\dot{\mathbf{x}}$ = derivative of the state vector with respect to time

    $\mathbf{y}$ = output vector

    $\mathbf{u}$ = input or control vector

    $\mathbf{A}$ = system matrix

    $\mathbf{B}$ = input coupling matrix

    $\mathbf{C}$ = output matrix

    $\mathbf{D}$ = feedforward matrix

Equation 3.18 is called the *state equation* and the vector $\mathbf{x}$, the *state vector*, contains the state variables. Equation 3.18 can be solved for the state variables, which we demonstrate in Chapter 4. Equation 3.19 is called the *output equation*. This equation is used to calculate any other system variables.

As an example, for a linear, time-invariant, second-order system, with a single-input, $v(t)$, the state equations could take on the following form:

$$\frac{dx_1}{dt} = a_{11}x_1 + a_{12}x_2 + b_1 v(t) \tag{3.20a}$$

$$\frac{dx_2}{dt} = a_{21}x_1 + a_{22}x_2 + b_2 v(t) \tag{3.20b}$$

where x_1 and x_2 are the state variables. If there is a single output, the output equation could take on the following form:

$$y = c_1 x_1 + c_2 x_2 + d_1 v(t) \tag{3.21}$$

The choice of state variables for a given system is not unique. The requirement in choosing the state variables is that they be linearly independent and that a minimum number of them be chosen.

3.4 Applying the State-Space Representation

In this section we apply the state-space formulation to the representation of more complicated physical systems. The first step in representing a system is to select the state vector, which must be chosen according to the following considerations:

1. There is a minimum number of state variables that must be selected as components of the state vector.
2. The components of the state vector (i.e., this minimum number of state variables) must be linearly independent.

Let us review and clarify these statements.

Selecting Linearly Independent State Variables

The components of the state vector must be linearly independent. For example, following the definition of linear independence in Section 3.3, if x_1, x_2, and

x_3 are chosen as state variables, but $x_3 = 5x_1 + 4x_2$, then x_3 is not linearly independent of x_1 and x_2, since knowledge of the values of x_1 and x_2 will yield the value of x_3. Variables related by derivatives are linearly independent. For example, the voltage across an inductor, v_L, is linearly independent of the current through the inductor, i_L, since $v_L = L\,di_L/dt$. Thus, v_L cannot be evaluated as a linear combination of the current, i_L.

Determining the Minimum Number of State Variables

How do we know the minimum number of state variables to select? Typically, the minimum number required equals the order of the differential equation describing the system. For example, if a third-order differential equation describes the system, then three simultaneous, first-order differential equations are required along with three state variables. From the perspective of the transfer function, the order of the differential equation is the order of the denominator of the transfer function after canceling common factors in the numerator and denominator.

In most cases another way of determining the number of state variables is to count the number of independent energy-storage elements in the system.[5] The number of these energy-storage elements equals the order of the differential equation and the number of state variables. In Figure 3.2 there are two energy-storage elements, the capacitor and the inductor. Hence two state variables and two state equations are required for the system.

If too few state variables are selected, then it may be impossible to write particular output equations, since some system variables cannot be written as a linear combination of the reduced number of state variables. In many cases it may be impossible even to complete the writing of the state equations, since the derivatives of the state variables cannot be expressed as linear combinations of the reduced number of state variables.

If the minimum number of state variables is selected, but they are not linearly independent, then at best you may not be able to solve for all other system variables. At worst, you may not even be able to complete the writing of the state equations.

Often the state vector includes more than the minimum number of state variables required. Two possible cases exist. Many times state variables are chosen to be physical variables of a system, such as position and velocity in a mechanical system. Cases arise where these variables, although linearly independent, are also *decoupled*. That is, some linearly independent variables are not required in order to solve for any of the other linearly independent variables or any other dependent

---

[5]Sometimes it is not apparent in a schematic how many independent energy-storage elements there are. It is possible that more than the minimum number of energy-storage elements could be selected, leading to a state vector whose components number more than the minimum required and are not linearly independent. Selecting additional dependent energy-storage elements results in a system matrix of higher order and more complexity than required for the solution of the state equations.

system variable. Consider the case of a mass and viscous damper whose differential equation is $M \dfrac{dv}{dt} + Dv = f(t)$, where v is the velocity of the mass. Since this is a first-order equation, one state equation is all that is required to define this system in state space with velocity as the state variable. Also, since there is only one energy-storage element, mass, only one state variable is required to represent this system in state space. However, the mass also has an associated position, which is linearly independent of velocity. If we want to include position in the state vector along with velocity, then we add position as a state variable that is linearly independent of the other state variable, velocity. Many times, the writing of the state equations is simplified by including additional state variables.

Another case that increases the size of the state vector arises when the added variable is not linearly independent of the other members of the state vector. This usually occurs when a variable is selected as a state variable, but its dependence upon the other state variables is not immediately apparent. For example, energy-storage elements may be used to select the state variables, and the dependence of the variable associated with one energy-storage element upon the variables of other energy-storage elements may not be recognized. Thus, the dimension of the system matrix is increased unnecessarily, and the solution for the state vector, which we cover in Chapter 4, is more difficult. Also, adding dependent state variables affects the designer's ability to use state-space methods for design.[6]

We saw in Section 3.2 that the state-space representation is not unique. The following example demonstrates one technique for selecting state variables and representing a system in state-space. Our approach is to write the simple derivative equation for each energy-storage element and solve for each derivative term as a linear combination of any of the system variables and the input that are present in the equation. Next we select each differentiated variable as a state variable. Then we express all other system variables in the equations in terms of the state variables and the input. Finally, we write the output variables as linear combinations of the state variables and the input.

Example 3.1   Represent an electrical network in state space.

Problem   Given the electrical network of Figure 3.4, find a state-space representation if the output is the current through the resistor.

Solution   The following steps will yield a viable representation of the network in state space.

Step 1   Label all of the branch currents in the network. These include i_L, i_R, and i_C, as shown in Figure 3.4.

---

[6]See Chapter 12 for state-space design techniques.

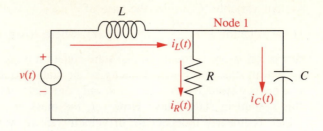

Figure 3.4 Electrical Network for Representation in State Space

Step 2 Select the state variables by writing the derivative equation for all energy-storage elements, that is, the inductor and the capacitor. Thus,

$$C\frac{dv_C}{dt} = i_C \tag{3.22}$$

$$L\frac{di_L}{dt} = v_L \tag{3.23}$$

From Equations 3.22 and 3.23, choose the state variables as the quantities that are differentiated, namely v_C and i_L. Using Equation 3.20 as a guide, we see that the state-space representation is complete if the right-hand sides of Equations 3.22 and 3.23 can be written as linear combinations of the state variables and the input.

Since i_C and v_L are not state variables, our next step is to express i_C and v_L as linear combinations of the state variables, v_C and i_L, and the input, $v(t)$.

Step 3 Apply network theory, such as Kirchhoff's voltage and current laws, to obtain i_C and v_L in terms of the state variables, v_C and i_L. At Node 1,

$$i_C = -i_R + i_L$$
$$= -\frac{1}{R}v_C + i_L \tag{3.24}$$

which yields i_C in terms of the state variables, v_C and i_L.

Around the outer loop,

$$v_L = -v_C + v(t) \tag{3.25}$$

which yields v_L in terms of the state variable, v_C, and the source, $v(t)$.

Step 4 Substitute the results of Equations 3.24 and 3.25 into Equations 3.22 and 3.23 to obtain the following state equations:

$$C\frac{dv_C}{dt} = -\frac{1}{R}v_C + i_L \tag{3.26a}$$

$$L\frac{di_L}{dt} = -v_C + v(t) \tag{3.26b}$$

or

$$\frac{dv_C}{dt} = -\frac{1}{RC}v_C + \frac{1}{C}i_L \tag{3.27a}$$

$$\frac{di_L}{dt} = -\frac{1}{L}v_C \qquad +\frac{1}{L}v(t) \tag{3.27b}$$

Step 5 Find the output equation. Since the output is $i_R(t)$,

$$i_R = \frac{1}{R}v_C \tag{3.28}$$

The final result for the state-space representation is found by representing Equations 3.27 and 3.28 in vector-matrix form as follows:

$$\begin{bmatrix} \dot{v}_C \\ \dot{i}_L \end{bmatrix} = \begin{bmatrix} -1/(RC) & 1/C \\ -1/L & 0 \end{bmatrix} \begin{bmatrix} v_C \\ i_L \end{bmatrix} + \begin{bmatrix} 0 \\ 1/L \end{bmatrix} v(t) \tag{3.29a}$$

$$i_R = \begin{bmatrix} 1/R & 0 \end{bmatrix} \begin{bmatrix} v_C \\ i_L \end{bmatrix} \tag{3.29b}$$

In order to clarify the representation of physical systems in state space, we will look at two more examples. The first is an electrical network with a dependent source. Although we will follow the same procedure as in the previous problem, this problem will yield increased complexity in applying network analysis to find the state equations.

Example 3.2 Find the state-space representation of an electrical network with a dependent source.

Problem Find the state and output equations for the electrical network shown in Figure 3.5 if the output vector is $\mathbf{y} = [v_{R_2} \quad i_{R_2}]^T$, where T means transpose.[7]

Solution Immediately notice that this network has a voltage-dependent current source.

Step 1 Label all of the branch currents on the network, as shown in Figure 3.5.

Step 2 Select the state variables by listing the voltage-current relationship for all of the energy-storage elements.

$$L\frac{di_L}{dt} = v_L \tag{3.30a}$$

$$C\frac{dv_C}{dt} = i_C \tag{3.30b}$$

From Equations 3.30, select the state variables to be the differentiated variables. Thus, the state variables, x_1 and x_2, are

$$x_1 = i_L; \qquad x_2 = v_C \tag{3.31}$$

---

[7] See Appendix B for a discussion of the transpose.

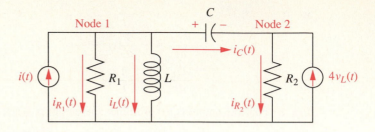

Figure 3.5 Electrical Network

Step 3   Remembering that the form of the state equation is

$$\dot{\mathbf{x}} = \mathbf{A}\mathbf{x} + \mathbf{B}u \qquad (3.32)$$

we see that the remaining task is to transform the right-hand side of Equations 3.30 into linear combinations of the state variables and input source current. Using Kirchhoff's voltage and current laws, we find v_L and i_C in terms of the state variables and the input current source.

Around the mesh containing L and C,

$$v_L = v_C + v_{R_2} = v_C + i_{R_2}R_2 \qquad (3.33)$$

But at Node 2, $i_{R_2} = i_C + 4v_L$. Substituting this relationship for i_{R_2} into Equation 3.33 yields

$$v_L = v_C + (i_C + 4v_L)R_2 \qquad (3.34)$$

Solving for v_L, we get

$$v_L = \frac{1}{1 - 4R_2}(v_C + i_C R_2) \qquad (3.35)$$

Notice that since v_C is a state variable, we only need to find i_C in terms of the state variables. We will then have obtained v_L in terms of the state variables.

Thus at Node 1 we can write the sum of the currents as

$$i_C = i(t) - i_{R_1} - i_L$$

$$= i(t) - \frac{v_{R_1}}{R_1} - i_L$$

$$= i(t) - \frac{v_L}{R_1} - i_L \qquad (3.36)$$

where $v_{R_1} = v_L$. Equations 3.35 and 3.36 are two equations relating v_L and i_C in terms of the state variables i_L and v_C. Rewriting Equations 3.35 and 3.36, we obtain two simultaneous equations yielding v_L and i_C as linear combinations of the state variables i_L and v_C:

$$(1 - 4R_2)v_L - R_2 i_C = v_C \qquad (3.37a)$$

$$-\frac{1}{R_1}v_L - i_C = i_L - i(t) \qquad (3.37b)$$

Solving Equations 3.37 simultaneously for v_L and i_C yields

$$v_L = \frac{1}{\Delta}[R_2 i_L - v_C - R_2 i(t)] \tag{3.38}$$

and

$$i_C = \frac{1}{\Delta}\left[(1 - 4R_2)i_L + \frac{1}{R_1}v_C - (1 - 4R_2)i(t)\right] \tag{3.39}$$

where

$$\Delta = -\left[(1 - 4R_2) + \frac{R_2}{R_1}\right] \tag{3.40}$$

Substituting Equations 3.38 and 3.39 into 3.30, simplifying, and writing the result in vector-matrix form renders the following state equation:

$$\begin{bmatrix} \dot{i}_l \\ \dot{v}_C \end{bmatrix} = \begin{bmatrix} R_2/(L\Delta) & -1/(L\Delta) \\ (1 - 4R_2)/(C\Delta) & 1/(R_1 C\Delta) \end{bmatrix}\begin{bmatrix} i_L \\ v_C \end{bmatrix}$$
$$+ \begin{bmatrix} -R_2/(L\Delta) \\ -(1 - 4R_2)/(C\Delta) \end{bmatrix}i(t) \tag{3.41}$$

Step 4 Derive the output equation. Since the specified output variables are v_{R_2} and i_{R_2}, we note that around the mesh containing C, L, and R_2,

$$v_{R_2} = -v_C + v_L \tag{3.42a}$$

$$i_{R_2} = i_C + 4v_L \tag{3.42b}$$

Substituting Equations 3.38 and 3.39 into 3.42, v_{R_2} and i_{R_2} are obtained as linear combinations of the state variables, i_L and v_C. In vector-matrix form, the output equation is

$$\begin{bmatrix} v_{R_2} \\ i_{R_2} \end{bmatrix} = \begin{bmatrix} R_2/\Delta & -(1 + 1/\Delta) \\ 1/\Delta & (1 - 4R_1)/(\Delta R_1) \end{bmatrix}\begin{bmatrix} i_L \\ v_C \end{bmatrix} + \begin{bmatrix} -R_2/\Delta \\ -1/\Delta \end{bmatrix}i(t) \tag{3.43}$$

In the next example we find the state-space representation for a mechanical system. It is more convenient when working with mechanical systems to obtain the state equations directly from the equations of motion rather than from the energy-storage elements. Consider, for example, the energy-storage element of a spring, where $F = Kx$. This relationship does not contain the derivative of a physical variable as in the case of electrical networks, where $i = C\dfrac{dv}{dt}$ for capacitors, and $v = L\dfrac{di}{dt}$ for inductors. Thus, in mechanical systems, we change our selection of state variables to be the position and velocity of each point of

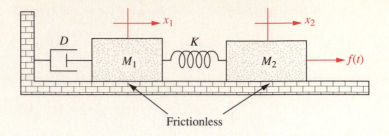

Figure 3.6 A Translational Mechanical System

linearly independent motion. In the example we will see that although there are three energy-storage elements, there will be four state variables; an additional linearly independent state variable is included for the convenience of writing the state equations.

Example 3.3

Find the state-space representation of a translational mechanical system.

Problem Find the state equations for the translational mechanical system shown in Figure 3.6.

Solution First, write the differential equations for the network in Figure 3.6, using the methods of Chapter 2 to find the Laplace transformed equations of motion. Next, take the inverse Laplace transform of these equations, assuming zero initial conditions, and obtain

$$M_1 \frac{d^2x_1}{dt^2} + D\frac{dx_1}{dt} + Kx_1 - Kx_2 = 0 \tag{3.44}$$

$$-Kx_1 + M_2 \frac{d^2x_2}{dt^2} + Kx_2 = f(t) \tag{3.45}$$

Now, let $d^2x_1/dt^2 = dv_1/dt$, and $d^2x_2/dt^2 = dv_2/dt$, and then select x_1, v_1, x_2, and v_2 as state variables. Next, form two of the state equations by solving Equation 3.44 for dv_1/dt and Equation 3.45 for dv_2/dt. Finally, add $dx_1/dt = v_1$, and $dx_2/dt = v_2$ to complete the set of state equations. Hence,

$$\frac{dx_1}{dt} = \qquad\qquad + v_1 \tag{3.46a}$$

$$\frac{dv_1}{dt} = -\frac{K}{M_1}x_1 - \frac{D}{M_1}v_1 + \frac{K}{M_1}x_2 \tag{3.46b}$$

$$\frac{dx_2}{dt} = \qquad\qquad\qquad + v_2 \tag{3.46c}$$

$$\frac{dv_2}{dt} = +\frac{K}{M_2}x_1 \qquad\quad - \frac{K}{M_2}x_2 \qquad + \frac{1}{M_2}f(t) \tag{3.46d}$$

In vector-matrix form,

$$
\begin{bmatrix} \dot{x}_1 \\ \dot{v}_1 \\ \dot{x}_2 \\ \dot{v}_2 \end{bmatrix} = \begin{bmatrix} 0 & 1 & 0 & 0 \\ -K/M_1 & -D/M_1 & K/M_1 & 0 \\ 0 & 0 & 0 & 1 \\ K/M_2 & 0 & -K/M_2 & 0 \end{bmatrix} \begin{bmatrix} x_1 \\ v_1 \\ x_2 \\ v_2 \end{bmatrix} + \begin{bmatrix} 0 \\ 0 \\ 0 \\ 1/M_2 \end{bmatrix} f(t) \qquad (3.47)
$$

where the dot indicates differentiation with respect to time. What is the output equation if the output is $x_2(t)$?

In this section we applied the state-space representation to electrical and mechanical systems. In the next section we see how to convert a transfer function description to a state-space representation.

3.5 Converting Transfer Functions to the State-Space, Phase-Variable Form

In this section we learn how to convert a transfer function representation to a state-space representation. One advantage of the state-space representation is that it can be used for the simulation of physical systems on the digital computer. Thus, if we want to simulate a system that is represented by a transfer function, we must first convert the transfer function representation to state space.

At first, we select a set of state variables, called *phase variables*, where each subsequent state variable is defined to be the derivative of the previous state variable. In Chapter 5, we show how to make other choices for the state variables.

Let us begin by showing how to represent a general, nth-order, linear differential equation with constant coefficients in state space in phase-variable form. We will then show how to apply this representation to transfer functions.

Consider the differential equation

$$
\frac{d^n y}{dt^n} + a_{n-1} \frac{d^{n-1} y}{dt^{n-1}} + \cdots + a_1 \frac{dy}{dt} + a_0 y = u(t) \qquad (3.48)
$$

A convenient way to choose state variables is to choose the output, $y(t)$, and its $(n-1)$ derivatives as the state variables. This choice is called the *phase-variable* choice. Choosing the state variables x_i, we get

$$
x_1 = y \qquad (3.49a)
$$

$$
x_2 = \frac{dy}{dt} \qquad (3.49b)
$$

$$
x_3 = \frac{d^2 y}{dt^2} \qquad (3.49c)
$$

$$
\vdots
$$

$$
x_n = \frac{d^{n-1} y}{dt^{n-1}} \qquad (3.49d)
$$

and differentiating both sides yields

$$\dot{x}_1 = \frac{dy}{dt} \tag{3.50a}$$

$$\dot{x}_2 = \frac{d^2y}{dt^2} \tag{3.50b}$$

$$\dot{x}_3 = \frac{d^3y}{dt^3} \tag{3.50c}$$

$$\vdots$$

$$\dot{x}_n = \frac{d^ny}{dt^n} \tag{3.50d}$$

where the dot above the x signifies differentiation with respect to time.

Substituting the definitions of Equations 3.49 into Equations 3.50, the state equations are evaluated as

$$\dot{x}_1 = x_2 \tag{3.51a}$$

$$\dot{x}_2 = x_3 \tag{3.51b}$$

$$\vdots$$

$$\dot{x}_{n-1} = x_n \tag{3.51c}$$

$$\dot{x}_n = -a_0 x_1 - a_1 x_2 \cdots - a_{n-1}x_n + u(t) \tag{3.51d}$$

where Equation 3.51 was obtained from Equation 3.48 by solving for d^ny/dt^n and using Equations 3.49. In vector-matrix form, Equation 3.51 becomes

$$
\begin{bmatrix} \dot{x}_1 \\ \dot{x}_2 \\ \dot{x}_3 \\ \vdots \\ \dot{x}_{n-1} \\ \dot{x}_n \end{bmatrix} =
\begin{bmatrix}
0 & 1 & 0 & 0 & 0 & 0 & \cdots & 0 \\
0 & 0 & 1 & 0 & 0 & 0 & \cdots & 0 \\
0 & 0 & 0 & 1 & 0 & 0 & \cdots & 0 \\
\vdots & & & & & & & \\
0 & 0 & 0 & 0 & 0 & 0 & \cdots & 1 \\
-a_0 & -a_1 & & & & & \cdots & -a_{n-1}
\end{bmatrix}
\begin{bmatrix} x_1 \\ x_2 \\ x_3 \\ \vdots \\ x_{n-1} \\ x_n \end{bmatrix} +
\begin{bmatrix} 0 \\ 0 \\ 0 \\ \vdots \\ 0 \\ 1 \end{bmatrix} u(t)
\tag{3.52}
$$

Equation 3.52 is the phase-variable form of the state equations. This form is easily recognized by the unique pattern of 1's and 0's and the negative of the coefficients of the differential equation written in reverse order in the last row of the system matrix.

Finally, since the solution to the differential equation is $y(t)$, or x_1, the output equation is

$$
y = \begin{bmatrix} 1 & 0 & 0 & \cdots & 0 \end{bmatrix}
\begin{bmatrix} x_1 \\ x_2 \\ x_3 \\ \vdots \\ x_{n-1} \\ x_n \end{bmatrix}
\tag{3.53}
$$

In summary, then, to convert a transfer function into state equations in phase-variable form, we first convert the transfer function to a differential equation by cross-multiplying and taking the inverse Laplace transform, assuming zero initial conditions. Then we represent the differential equation in state space in phase-variable form. An example illustrates the process.

Example 3.4

Represent a transfer function with a constant term in the numerator in state space in phase-variable form.

Problem   Find the state-space representation in phase-variable form for the transfer function shown in Figure 3.7.

Solution

Step 1   Find the associated differential equation. Since

$$\frac{C(s)}{R(s)} = \frac{24}{(s^3 + 9s^2 + 26s + 24)} \tag{3.54}$$

cross-multiplying yields

$$(s^3 + 9s^2 + 26s + 24)C(s) = 24R(s) \tag{3.55}$$

The corresponding differential equation is found by taking the inverse Laplace transform, assuming zero initial conditions:

$$\dddot{c} + 9\ddot{c} + 26\dot{c} + 24c = 24r(t) \tag{3.56}$$

Step 2   Select the state variables.

Choosing the state variables as successive derivatives, we get

$$x_1 = c \tag{3.57a}$$

$$x_2 = \dot{c} \tag{3.57b}$$

$$x_3 = \ddot{c} \tag{3.57c}$$

Differentiating both sides and making use of Equations 3.57 to find $\dot{x}_1$ and $\dot{x}_2$, and Equation 3.56 to find $\dddot{c} = \dot{x}_3$, we obtain the state equations. Since the output is $c = x_1$, the combined state and output equations are

$$\dot{x}_1 = \qquad\qquad x_2 \tag{3.58a}$$

$$\dot{x}_2 = \qquad\qquad x_3 \tag{3.58b}$$

$$\dot{x}_3 = -24x_1 - 26x_2 - 9x_3 + 24r(t) \tag{3.58c}$$

$$y = c = x_1 \tag{3.58d}$$

Figure 3.7 Transfer Function

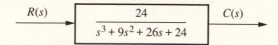

$R(s)$ → $\dfrac{24}{s^3 + 9s^2 + 26s + 24}$ → $C(s)$

In vector-matrix form,

$$\begin{bmatrix} \dot{x}_1 \\ \dot{x}_2 \\ \dot{x}_3 \end{bmatrix} = \begin{bmatrix} 0 & 1 & 0 \\ 0 & 0 & 1 \\ -24 & -26 & -9 \end{bmatrix} \begin{bmatrix} x_1 \\ x_2 \\ x_3 \end{bmatrix} + \begin{bmatrix} 0 \\ 0 \\ 24 \end{bmatrix} r(t) \qquad (3.59a)$$

$$y = \begin{bmatrix} 1 & 0 & 0 \end{bmatrix} \begin{bmatrix} x_1 \\ x_2 \\ x_3 \end{bmatrix} \qquad (3.59b)$$

Notice that the third row of the system matrix has the same coefficients as the denominator of the transfer function but negative and in reverse order.

The transfer function of Example 3.4 has a constant term in the numerator. If a transfer function has a polynomial in s in the numerator that is of order less than the polynomial in the denominator, as shown in Figure 3.8(a), then the numerator and denominator can be handled separately. First, separate the transfer function into two cascaded transfer functions, as shown in Figure 3.8(b); the first is the denominator, and the second is just the numerator. The first transfer function with just the denominator is converted to the phase-variable representation in state space just as we demonstrated in the last example. Hence, the phase variable x_1 is the output, and the rest of the phase variables are internal to Block 1, as shown in Figure 3.8(b). The second transfer function with just the numerator yields

$$Y(s) = C(s) = (a_2 s^2 + a_1 s + a_0) X_1(s) \qquad (3.60)$$

where, after taking the inverse Laplace transform with zero initial conditions,

$$y(t) = a_2 \frac{d^2 x_1}{dt^2} + a_1 \frac{dx_1}{dt} + a_0 x_1 \qquad (3.61)$$

Figure 3.8 Decomposing a Transfer Function

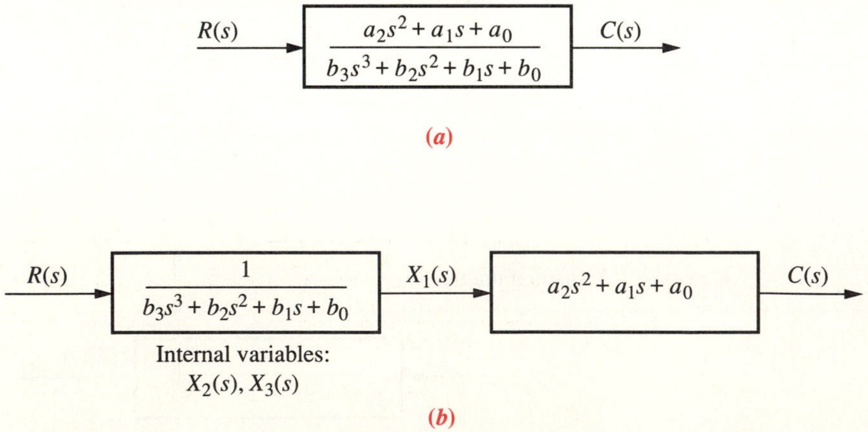

$$R(s) \rightarrow \boxed{\frac{a_2 s^2 + a_1 s + a_0}{b_3 s^3 + b_2 s^2 + b_1 s + b_0}} \rightarrow C(s)$$

(a)

$$R(s) \rightarrow \boxed{\frac{1}{b_3 s^3 + b_2 s^2 + b_1 s + b_0}} \xrightarrow{X_1(s)} \boxed{a_2 s^2 + a_1 s + a_0} \rightarrow C(s)$$

Internal variables:
$X_2(s), X_3(s)$

(b)

But the derivative terms are the definitions of the phase variables obtained in the first block. Thus, writing the terms in reverse order to conform to an output equation,

$$y(t) = a_0 x_1 + a_1 x_2 + a_2 x_3 \tag{3.62}$$

Hence, the second block simply forms a specified linear combination of the state variables developed in the first block.

From another perspective, the denominator of the transfer function yields the state equations while the numerator yields the output equation. The next example demonstrates the process.

Example 3.5 Find the state-space representation of a transfer function with a polynomial in s in the numerator.

Problem Find the state-space representation of the transfer function shown in Figure 3.9(a).

Solution This problem differs from Example 3.4 since the numerator has a polynomial in s instead of just a constant term.

Step 1 Separate the system into two cascaded blocks, as shown in Figure 3.9(b). The first block contains the denominator, and the second block contains the numerator.

Step 2 Find the state equations for the block containing the denominator. We notice that the first block's numerator is $1/24$ that of Example 3.4. Thus, the state equations are the same except that this system's input coupling matrix is $1/24$ that of Example 3.4. Hence, the state equation is

$$\begin{bmatrix} \dot{x}_1 \\ \dot{x}_2 \\ \dot{x}_3 \end{bmatrix} = \begin{bmatrix} 0 & 1 & 0 \\ 0 & 0 & 1 \\ -24 & -26 & -9 \end{bmatrix} \begin{bmatrix} x_1 \\ x_2 \\ x_3 \end{bmatrix} + \begin{bmatrix} 0 \\ 0 \\ 1 \end{bmatrix} r(t) \tag{3.63}$$

Figure 3.9 (a) Transfer Function; (b) Decomposed Transfer Function

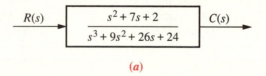

$$R(s) \longrightarrow \boxed{\dfrac{s^2 + 7s + 2}{s^3 + 9s^2 + 26s + 24}} \longrightarrow C(s)$$

(a)

$$R(s) \longrightarrow \boxed{\dfrac{1}{s^3 + 9s^2 + 26s + 24}} \xrightarrow{X_1(s)} \boxed{s^2 + 7s + 2} \longrightarrow C(s)$$

Internal variables:
$X_2(s), X_3(s)$

(b)

Step 3 Introduce the effect of the block with the numerator. The second block of Figure 3.9(*b*) states that

$$C(s) = (s^2 + 7s + 2)X_1(s) \tag{3.64}$$

Taking the inverse Laplace transform with zero initial conditions, we get

$$c = \ddot{x}_1 + 7\dot{x}_1 + 2x_1 \tag{3.65}$$

But

$$x_1 = x_1$$
$$\dot{x}_1 = x_2$$
$$\ddot{x}_1 = x_3$$

Hence,

$$y(t) = c(t) = x_3 + 7x_2 + 2x_1 \tag{3.66}$$

Thus, the last box of Figure 3.9(*b*) "collects" the states and generates the output equation. From Equation 3.66,

$$y = \begin{bmatrix} 2 & 7 & 1 \end{bmatrix} \begin{bmatrix} x_1 \\ x_2 \\ x_3 \end{bmatrix} \tag{3.67}$$

Although the second block of Figure 3.9(*b*) shows differentiation, this block was implemented without differentiation because of the partitioning that was applied to the transfer function. The last block simply collected derivatives that were already formed by the first block.

In this section we converted a transfer function into state equations by first separating the numerator and denominator terms into two cascaded transfer functions. The first transfer function containing the denominator generated the state equation, whereas the second transfer function containing the numerator generated the output equation. Now that we have converted transfer functions to the state-space representation, let us see how to go in the other direction and convert state equations into transfer functions.

3.6 Converting the State Equations to a Transfer Function

In Chapters 2 and 3, we have explored two methods of representing systems: (1) the transfer function representation and (2) the state-space representation. In the last section we united the two representations by converting transfer functions into state equations; now we move in the opposite direction and convert the state-space representation into a transfer function.

Given the state and output equations

$$\dot{\mathbf{x}} = \mathbf{A}\mathbf{x} + \mathbf{B}\mathbf{u} \qquad (3.68a)$$

$$\mathbf{y} = \mathbf{C}\mathbf{x} + \mathbf{D}\mathbf{u} \qquad (3.68b)$$

Take the Laplace transform assuming zero initial conditions:[8]

$$s\mathbf{X}(s) = \mathbf{A}\mathbf{X}(s) + \mathbf{B}\mathbf{U}(s) \qquad (3.69a)$$

$$\mathbf{Y}(s) = \mathbf{C}\mathbf{X}(s) + \mathbf{D}\mathbf{U}(s) \qquad (3.69b)$$

Solving for $\mathbf{X}(s)$ in Equation 3.69a,

$$(s\mathbf{I} - \mathbf{A})\mathbf{X}(s) = \mathbf{B}\mathbf{U}(s) \qquad (3.70)$$

or

$$\mathbf{X}(s) = (s\mathbf{I} - \mathbf{A})^{-1}\mathbf{B}\mathbf{U}(s) \qquad (3.71)$$

where $\mathbf{I}$ is the identity matrix.

Substituting Equation 3.71 into Equation 3.69b yields

$$\mathbf{Y}(s) = \mathbf{C}(s\mathbf{I} - \mathbf{A})^{-1}\mathbf{B}\mathbf{U}(s) + \mathbf{D}\mathbf{U}(s)$$

$$= [\mathbf{C}(s\mathbf{I} - \mathbf{A})^{-1}\mathbf{B} + \mathbf{D}]\mathbf{U}(s) \qquad (3.72)$$

We call the matrix $[\mathbf{C}(s\mathbf{I}-\mathbf{A})^{-1}\mathbf{B}+\mathbf{D}]$ the transfer function matrix, since it relates the output vector, $\mathbf{Y}(s)$, to the input vector, $\mathbf{U}(s)$. However, if $\mathbf{U}(s) = U(s)$, and $\mathbf{Y}(s) = Y(s)$, are scalars, then we can find the transfer function

$$T(s) = \frac{Y(s)}{U(s)} = \mathbf{C}(s\mathbf{I} - \mathbf{A})^{-1}\mathbf{B} + \mathbf{D} \qquad (3.73)$$

Let us look at an example.

Example 3.6   Find a system's transfer function if the system is represented in state space.

Problem   Given the system defined by Equations 3.74, find the transfer function, $T(s) = Y(s)/R(s)$, where $R(s)$ is the input and $Y(s)$ is the output.

$$\dot{\mathbf{x}} = \begin{bmatrix} 0 & 1 & 0 \\ 0 & 0 & 1 \\ -1 & -2 & -3 \end{bmatrix}\mathbf{x} + \begin{bmatrix} 10 \\ 0 \\ 0 \end{bmatrix}u(t) \qquad (3.74a)$$

$$y = \begin{bmatrix} 1 & 0 & 0 \end{bmatrix}\mathbf{x} \qquad (3.74b)$$

---

[8]The Laplace transform of a vector is found by taking the Laplace transform of each component. Since $\dot{\mathbf{x}}$ consists of the derivatives of the state variables, the Laplace transform of $\dot{\mathbf{x}}$ with zero initial conditions yields each component with the form $sX_i(s)$, where $X_i(s)$ is the Laplace transform of the state variable. Factoring out the complex variable, s, in each component yields the Laplace transform of $\dot{\mathbf{x}}$ as $s\mathbf{X}(s)$ where $\mathbf{X}(s)$ is a column vector with components $X_i(s)$.

Solution   The solution revolves around finding the term $(s\mathbf{I} - \mathbf{A})^{-1}$ in Equation 3.73.[9] All other terms are already defined. Hence, first find $(s\mathbf{I} - \mathbf{A})$:

$$(s\mathbf{I} - \mathbf{A}) = \begin{bmatrix} s & 0 & 0 \\ 0 & s & 0 \\ 0 & 0 & s \end{bmatrix} - \begin{bmatrix} 0 & 1 & 0 \\ 0 & 0 & 1 \\ -1 & -2 & -3 \end{bmatrix} = \begin{bmatrix} s & -1 & 0 \\ 0 & s & -1 \\ 1 & 2 & s+3 \end{bmatrix} \quad (3.75)$$

Now form $(s\mathbf{I} - \mathbf{A})^{-1}$:

$$(s\mathbf{I} - \mathbf{A})^{-1} = \frac{\text{adj}(s\mathbf{I} - \mathbf{A})}{\det(s\mathbf{I} - \mathbf{A})} = \frac{\begin{bmatrix} (s^2 + 3s + 2) & s+3 & 1 \\ -1 & s(s+3) & s \\ -s & -(2s+1) & s^2 \end{bmatrix}}{s^3 + 3s^2 + 2s + 1} \quad (3.76)$$

Substituting $(s\mathbf{I} - \mathbf{A})^{-1}$, $\mathbf{B}$, $\mathbf{C}$, and $\mathbf{D}$ into Equation 3.73, where

$$\mathbf{B} = \begin{bmatrix} 10 \\ 0 \\ 0 \end{bmatrix}$$

$$\mathbf{C} = \begin{bmatrix} 1 & 0 & 0 \end{bmatrix}$$

$$\mathbf{D} = \mathbf{0}$$

we obtain the final result for the transfer function:

$$T(s) = \frac{10(s^2 + 3s + 2)}{s^3 + 3s^2 + 2s + 1} \quad (3.77)$$

In Example 3.6, the state equations in phase-variable form were converted to transfer functions. In Chapter 5 we will see that other forms besides the phase-variable form can be used to represent a system in state space. The method of finding the transfer function representation for these other forms is the same as that presented in this section.

3.7 Linearization

A prime advantage of the state-space representation over the transfer function representation is the ability to represent systems with nonlinearities, such as the one shown in Figure 3.10. The ability to represent nonlinear systems does not imply the ability to solve their state equations for the state variables and the output. Techniques do exist for the solution of some nonlinear state equations, but this study is beyond the scope of this course. In Chapter 4, however, we will see how to use the digital computer to solve state equations. This method can also be used for nonlinear state equations.

If we are interested in small perturbations about an equilibrium point, as

---

[9]See Appendix B, which discusses the evaluation of the matrix inverse.

Figure 3.10 Odex I, a Technology Demonstrator, Walks the
Steps (Courtesy of Odetics, Inc.)

we were when we studied linearization in Chapter 2, then we can also linearize
the state equations about the equilibrium point. The key to linearization about
an equilibrium point is, once again, the Taylor series. In the following example
we write the state equations for a simple pendulum. First, we show that we can
represent a nonlinear system in state space; then, we linearize the pendulum about
its equilibrium point, the vertical position with zero velocity.

Example 3.7   Represent and linearize a nonlinear system in state space.

Problem   First, represent the simple pendulum shown in Figure 3.11(*a*) in state
space: Mg is the weight, T is an applied torque in the θ direction, and L is the length
of the pendulum. Assume the mass is evenly distributed with the center of mass at
$L/2$. Then linearize the state equations about the pendulum's equilibrium point—the
vertical position with zero angular velocity.

Solution   First, draw a free-body diagram as shown in Figure 3.11(*c*). Summing the
torques, we get

$$J\frac{d^2\theta}{dt^2} + \frac{MgL}{2}\sin\theta = T \tag{3.78}$$

where J is the moment of inertia of the pendulum around the point of rotation. Select
the state variables x_1 and x_2 as phase variables. Letting $x_1 = \theta$, and $x_2 = d\theta/dt$,

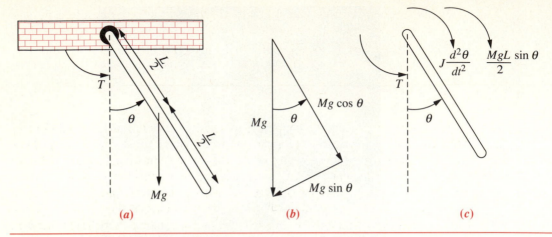

Figure 3.11 (**a**) Simple Pendulum; (**b**) Component Forces of Mg; (**c**) Free-Body Diagram

we write the state equations as

$$\dot{x}_1 = x_2 \tag{3.79a}$$

$$\dot{x}_2 = -\frac{MgL}{2J}\sin x_1 + \frac{T}{J} \tag{3.79b}$$

where $\dot{x}_2 = d^2\theta/dt^2$ is evaluated from Equation 3.78.

Thus, we have represented a nonlinear system in state space. Now we proceed to linearize the equation about the equilibrium point, $x_1 = 0$, $x_2 = 0$, that is, $\theta = 0$ and $d\theta/dt = 0$.

Let x_1 and x_2 be perturbed about the equilibrium point, or

$$x_1 = 0 + \delta x_1 \tag{3.80a}$$

$$x_2 = 0 + \delta x_2 \tag{3.80b}$$

Using Equation 2.156, we obtain

$$\sin x_1 - \sin 0 = \left.\frac{d(\sin x_1)}{dx_1}\right|_{x_1=0} \delta x_1 = \delta x_1 \tag{3.81}$$

from which,

$$\sin x_1 = \delta x_1 \tag{3.82}$$

Substituting Equations 3.80 and 3.82 into 3.79 yields the following state equations:

$$\dot{\delta x}_1 = \delta x_2 \tag{3.83a}$$

$$\dot{\delta x}_2 = -\frac{MgL}{2J}\delta x_1 + \frac{T}{J} \tag{3.83b}$$

which are linear and a good approximation to Equations 3.79 for small excursions away from the equilibrium point. What is the output equation?

In this section we saw that a nonlinear system can be represented in state space. Further, we saw that the nonlinear equations can be linearized. Let us look at more examples that demonstrate the chapter's objectives.

3.8 Chapter-Objective Demonstration Problems

We have covered the state-space representation of individual physical subsystems in this chapter. In Chapter 5, we will assemble individual subsystems into feedback control systems and represent the entire feedback system in state space. Chapter 5 also shows how the state-space representation, via signal-flow diagrams, can be used to interconnect these subsystems and permit the state-space representation of the whole closed-loop system. In this section we look at the antenna azimuth control system and demonstrate the concepts of this chapter by representing each subsystem in state space.

Example 3.8   Find the state-space representation of each subsystem composing an antenna azimuth control system.

Problem   Find the state-space representation in phase-variable form for each dynamic subsystem in the antenna azimuth control system shown in Figure 2.36. By *dynamic*, we mean that the system does not reach the steady state instantaneously. For example, a system described by a differential equation of first order or higher is a dynamic system. A pure gain, on the other hand, is an example of a nondynamic system, since the steady state is reached instantaneously.

Solution   In the chapter-objective demonstration problem of Chapter 2, each subsystem was identified. We found that the power amplifier and the motor and load were dynamic systems. The preamplifier and the potentiometers are pure gains and hence respond instantaneously. Hence, we will find the state-space representation only of the power amplifier and the motor and load.

Power Amplifier   The transfer function of the power amplifier is given in Figure 2.36 as $G(s) = 100/(s + 100)$. We will convert this transfer function to its state-space representation. Letting $v_p(t)$ represent the power amplifier input and $e_a(t)$ represent the power amplifier output,

$$G(s) = \frac{E_a(s)}{V_p(s)} = \frac{100}{(s + 100)} \tag{3.84}$$

Cross multiplying, $(s + 100)E_a(s) = 100V_p(s)$, from which the differential equation can be written as

$$\frac{de_a}{dt} + 100e_a = 100v_p(t) \tag{3.85}$$

Rearranging Equation 3.85 leads to the state equation with e_a as the state variable:

$$\frac{de_a}{dt} = -100e_a + 100v_p(t) \tag{3.86}$$

Since the output of the power amplifier is $e_a(t)$, the output equation is

$$y = e_a \qquad (3.87)$$

Motor and Load   We now find the state-space representation for the motor and load. The elements of the derivation were covered in Section 2.8 but are repeated here for continuity. Starting with Kirchhoff's voltage equation around the armature circuit, we find

$$e_a(t) = i_a(t)R_a + K_b \frac{d\theta_m}{dt} \qquad (3.88)$$

where $e_a(t)$ is the armature input voltage, $i_a(t)$ is the armature current, R_a is the armature resistance, K_b is the armature constant, and θ_m is the angular displacement of the armature.

The torque, $T_m(t)$, delivered by the motor is related separately to the armature current and the load seen by the armature. From Section 2.8,

$$T_m(t) = K_t i_a(t) = J_m \frac{d^2\theta_m}{dt^2} + D_m \frac{d\theta_m}{dt} \qquad (3.89)$$

where J_m is the equivalent inertia as seen by the armature, and D_m is the equivalent viscous damping as seen by the armature.

Solving Equation 3.89 for $i_a(t)$ and substituting the result into Equation 3.88 yields

$$e_a(t) = \left(\frac{R_a J_m}{K_t}\right) \frac{d^2\theta_m}{dt^2} + \left(\frac{D_m R_a}{K_t} + K_b\right) \frac{d\theta_m}{dt} \qquad (3.90)$$

Defining the state variables x_1 and x_2 as

$$x_1 = \theta_m \qquad (3.91a)$$

$$x_2 = \frac{d\theta_m}{dt} \qquad (3.91b)$$

and substituting into Equation 3.90, we get

$$e_a(t) = \left(\frac{R_a J_m}{K_t}\right) \frac{dx_2}{dt} + \left(\frac{D_m R_a}{K_t} + K_b\right) x_2 \qquad (3.92)$$

Solving for dx_2/dt yields

$$\frac{dx_2}{dt} = -\frac{1}{J_m}\left(D_m + \frac{K_t K_b}{R_a}\right) x_2 + \left(\frac{K_t}{R_a J_m}\right) e_a(t) \qquad (3.93)$$

Using Equations 3.91 and 3.93, the state equations are written as

$$\frac{dx_1}{dt} = x_2 \qquad (3.94a)$$

$$\frac{dx_2}{dt} = -\frac{1}{J_m}\left(D_m + \frac{K_t K_b}{R_a}\right) x_2 + \left(\frac{K_t}{R_a J_m}\right) e_a(t) \qquad (3.94b)$$

The output, θ_o, is $1/10$ the displacement of the armature, which is x_1. Hence the output equation is

$$y = 0.1 x_1 \qquad (3.95)$$

In vector-matrix form,

$$\dot{\mathbf{x}} = \begin{bmatrix} 0 & 1 \\ 0 & -\dfrac{1}{J_m}\left(D_m + \dfrac{K_t K_b}{R_a}\right) \end{bmatrix}\mathbf{x} + \begin{bmatrix} 0 \\ \dfrac{K_t}{R_a J_m} \end{bmatrix} e_a(t) \qquad (3.96a)$$

$$y = [0.1 \quad 0\,]\mathbf{x} \qquad (3.96b)$$

But from the chapter-objective demonstration problem in Chapter 2, $J_m = 0.03$, and $D_m = 0.02$. Also $K_t/R_a = 0.0625$ and $K_b = 0.5$. Substituting the values into Equations 3.96, we obtain the final state-space representation:

$$\dot{\mathbf{x}} = \begin{bmatrix} 0 & 1 \\ 0 & -1.71 \end{bmatrix}\mathbf{x} + \begin{bmatrix} 0 \\ 2.083 \end{bmatrix} e_a(t) \qquad (3.97a)$$

$$y = [0.1 \quad 0\,]\,\mathbf{x} \qquad (3.97b)$$

An advantage of state-space representation over the transfer function representation is the ability to focus on component parts of a system and write n simultaneous, first-order differential equations rather than attempting to represent the system as a single, nth-order differential equation as we have done with the transfer function. Also, multiple-input-output systems can be conveniently represented in state space. The following example demonstrates both of these concepts.

Example 3.9   Represent a multiple-output system in state space.

Problem   In the pharmaceutical industry we want to describe the distribution of a drug in the body. A simple model divides the process into compartments: the dosage, the absorption site, the blood, the peripheral compartment, and the urine. The rate of change of the amount of a drug in a compartment is proportional to the amount of the drug in the previous compartment diminished by an amount of the drug proportional to the quantity in the given compartment. Figure 3.12 summarizes the system. Here, each x_i is the amount of drug in that particular compartment (Lordi, 1972). Represent the system in state space, where the outputs are the amounts of the drug in each compartment.

Solution   The flow rate of the drug into any given compartment is proportional to the concentration of the drug in the previous compartment, and the flow rate out of a given compartment is proportional to the concentration of the drug within its own compartment. We now write the flow rate for each compartment.
   The dosage is released to the absorption site at a rate proportional to the dosage concentration, or

$$\frac{dx_1}{dt} = -K_1 x_1 \qquad (3.98)$$

The flow into the absorption site is proportional to the concentration of the drug at the dosage site. The flow from the absorption site into the blood is proportional to the

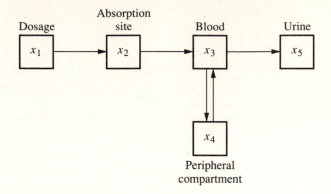

Figure 3.12 Drug-Level Concentrations in a Human

concentration of the drug at the absorption site. Hence,

$$\frac{dx_2}{dt} = K_1 x_1 - K_2 x_2 \tag{3.99}$$

Similarly, the net flow rate into the blood and peripheral compartment is

$$\frac{dx_3}{dt} = K_2 x_2 - K_3 x_3 + K_4 x_4 - K_5 x_3 \tag{3.100}$$

$$\frac{dx_4}{dt} = K_5 x_3 - K_4 x_4 \tag{3.101}$$

where $(K_4 x_4 - K_5 x_3)$ is the net flow rate into the blood from the peripheral compartment. Finally, the amount of the drug in the urine is increased as the blood releases the drug to the urine at a rate proportional to the concentration of the drug in the blood. Thus,

$$\frac{dx_5}{dt} = K_3 x_3 \tag{3.102}$$

Equations 3.98 through 3.102 are the state equations. The output equation is a vector that contains each of the amounts, x_i. Thus, in vector-matrix form,

$$\dot{\mathbf{x}} = \begin{bmatrix} -K_1 & 0 & 0 & 0 & 0 \\ K_1 & -K_2 & 0 & 0 & 0 \\ 0 & K_2 & -(K_3 + K_5) & K_4 & 0 \\ 0 & 0 & K_5 & -K_4 & 0 \\ 0 & 0 & K_3 & 0 & 0 \end{bmatrix} \mathbf{x} \tag{3.103a}$$

$$\mathbf{y} = \begin{bmatrix} 1 & 0 & 0 & 0 & 0 \\ 0 & 1 & 0 & 0 & 0 \\ 0 & 0 & 1 & 0 & 0 \\ 0 & 0 & 0 & 1 & 0 \\ 0 & 0 & 0 & 0 & 1 \end{bmatrix} \mathbf{x} \tag{3.103b}$$

You may wonder how there can be a solution to these equations if there is no input. In Chapter 4, when we study how to solve the state equations, we will see that initial conditions will yield solutions without forcing functions. For this problem, an initial condition on the amount of dosage, x_1, will generate drug quantities in all other compartments.

3.9 Summary

This chapter dealt with the state-space representation of physical systems, which took the form of a state equation,

$$\dot{\mathbf{x}} = \mathbf{A}\mathbf{x} + \mathbf{B}\mathbf{u} \tag{3.104}$$

and an output equation,

$$\mathbf{y} = \mathbf{C}\mathbf{x} + \mathbf{D}\mathbf{u} \tag{3.105}$$

Vector **x** is called the *state vector* and contains variables, called *state variables*. The state variables along with the input can be combined algebraically to form the output equation, Equation 3.105, from which any other system variables can be found. State variables, which can represent physical quantities such as current or voltage, are chosen to be linearly independent. The choice of state variables is not unique and affects how the matrices **A, B, C,** and **D** look. We will solve the state and output equations for **x** and **y** in Chapter 4.

In this chapter transfer functions were represented in state space. The form selected was the phase-variable form, which consists of state variables that are successive derivatives of each other. In three-dimensional state space, the resulting system matrix, **A,** for the phase-variable representation is of the form

$$\begin{bmatrix} 0 & 1 & 0 \\ 0 & 0 & 1 \\ -a_3 & -a_2 & -a_1 \end{bmatrix} \tag{3.106}$$

where the a_i's are the coefficients of the characteristic polynomial or denominator of the system transfer function. We concluded the chapter with a discussion of how to convert the state equations to a transfer function.

In conclusion then, for linear, time-invariant systems, the state-space representation is simply another way of mathematically modeling the system. One major advantage of applying the state-space representation to these linear systems is that it allows computer simulation. Programming the system on the digital computer and watching the system's response is an invaluable analysis and design tool. We cover simulation in the next chapter.

REVIEW QUESTIONS

1. Give two reasons for modeling systems in state space.
2. State an advantage of the transfer function approach over the state-space approach.
3. Define *state variables*.
4. Define *state*.
5. Define *state vector*.
6. Define *state space*.
7. What is required to represent a system in state space?
8. An eighth-order system would be represented in state space with how many state equations?
9. If the state equations are a system of first-order differential equations whose solution yields the state variables, then the output equation performs what function?
10. What is meant by *linear independence*?
11. What factors influence the choice of state variables in any system?
12. What is a convenient choice of state variables for electrical networks?
13. If an electrical network has three energy-storage elements, is it possible to have a state-space representation with more than three state variables? Explain.
14. What is meant by the phase-variable form of the state-equation?
15. When converting state equations to a transfer function, where do the poles of the transfer function come from?

PROBLEMS

1. Represent the electrical network shown in Figure P3.1 in state space, where $v_o(t)$ is the output.

Figure P3.1

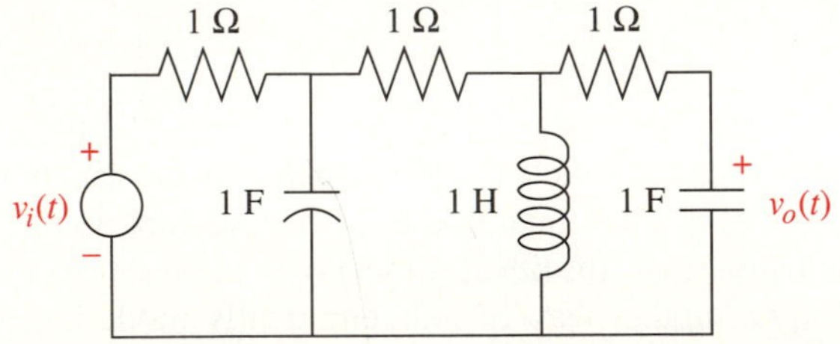

2. Represent the electrical network shown in Figure P3.2 in state space, where $i_R(t)$ is the output.
3. Represent the translational mechanical system shown in Figure P3.3 in state space, where $x_3(t)$ is the output.

Figure P3.2

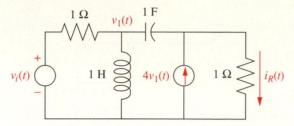

Figure P3.3

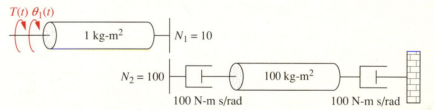

4. Represent the rotational mechanical system shown in Figure P3.4 in state space, where $\theta_1(t)$ is the output.

Figure P3.4

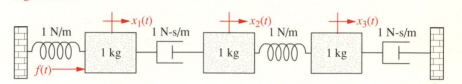

5. Find the state-space representation of the network shown in Figure P3.5 if the output is $v_o(t)$, and the mesh currents $i_1(t)$, $i_2(t)$, and $i_3(t)$ are among the state variables.

Figure P3.5

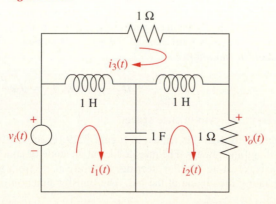

6. Find the state-space representation in phase-variable form for the system shown in Figure P3.6.

   Figure P3.6

   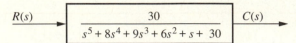

   $$R(s) \longrightarrow \boxed{\dfrac{30}{s^5 + 8s^4 + 9s^3 + 6s^2 + s + 30}} \longrightarrow C(s)$$

7. For the system shown in Figure P3.7, write the state equations and the output equation for the phase-variable representation.

   Figure P3.7

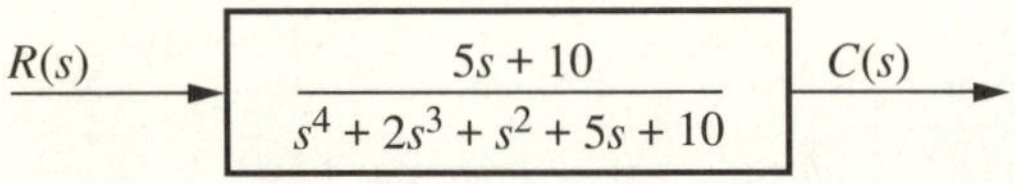

   $$R(s) \longrightarrow \boxed{\dfrac{5s + 10}{s^4 + 2s^3 + s^2 + 5s + 10}} \longrightarrow C(s)$$

8. Find the transfer function, $G(s) = Y(s)/R(s)$, for the following system represented in state space.

$$\dot{\mathbf{x}} = \begin{bmatrix} 0 & 1 & 0 \\ 0 & 0 & 1 \\ -3 & -2 & -5 \end{bmatrix} \mathbf{x} + \begin{bmatrix} 0 \\ 0 \\ 10 \end{bmatrix} r(t)$$

$$y = \begin{bmatrix} 1 & 0 & 0 \end{bmatrix} \mathbf{x}$$

9. Find the transfer function and poles of the system represented in state space below:

$$\dot{\mathbf{x}} = \begin{bmatrix} -2 & 1 & 0 \\ 0 & 0 & 1 \\ 0 & -2 & -4 \end{bmatrix} \mathbf{x} + \begin{bmatrix} 0 \\ 0 \\ 1 \end{bmatrix} u(t)$$

$$y = \begin{bmatrix} 1 & 1 & 0 \end{bmatrix} \mathbf{x}$$

10. Gyros are used on space vehicles, aircraft, and ships for inertial navigation. The gyro shown in Figure P3.8 is a rate gyro restrained by springs connected between the inner gimbal and the outer gimbal (frame) as shown. A rotational rate about the z-axis causes the rotating disk to precess about the x-axis. Hence the input is a rotational rate about the z-axis, and the output is an angular displacement about the x-axis. Since the outer gimbal is secured to the vehicle, the displacement about the x-axis is a measure of the vehicle's angular rate about the z-axis. The equation of motion is

$$J_x \frac{d^2\theta_x}{dt^2} + D_x \frac{d\theta_x}{dt} + K_x\theta_x = J\omega \frac{d\theta_z}{dt}$$

   Represent the gyro in state space.

11. A missile in flight, as shown in Figure P3.9, is subject to several forces: thrust, lift, drag, and gravity. The missile flies at an angle of attack, α, from its longitudinal axis, creating lift. For steering, the body angle from vertical, ϕ, is controlled by rotating the engine at the tail. The transfer function relating the body angle,

ROBLEMS **145**

ϕ, to the angular displacement, δ, of the engine is of the form

$$\frac{\Phi(s)}{\delta(s)} = \frac{K_a s + K_b}{K_3 s^3 + K_2 s^2 + K_1 s + K_0}$$

Represent the missile steering control in state space.

Figure P3.8 Gyro System

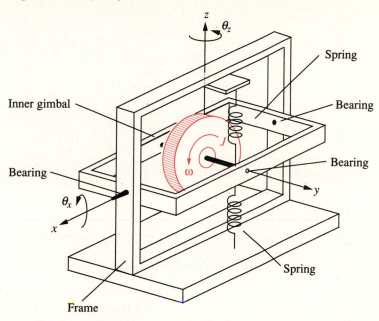

Figure P3.9 Missile

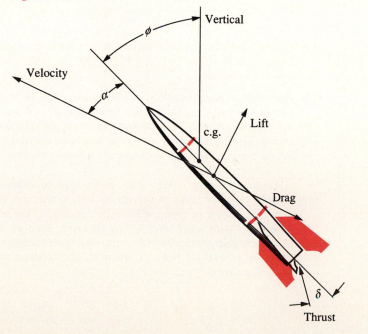

12. Given the dc servomotor and load shown in Figure P3.10, represent the system in state space, where the state variables are the armature current, i_a, load displacement, θ_L, and load angular velocity, ω_L. Assume that the output is the angular displacement of the armature. Do not neglect armature inductance.

Figure P3.10 Positional Control System

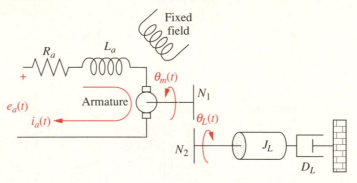

13. Consider the mechanical system of Figure P3.11. If the spring is nonlinear, and the force, F_s, required to stretch the spring is $F_s = 2x_1^2$, represent the system in state space linearized about $x_1 = 1$ if the output is x_2.

Figure P3.11 Nonlinear Mechanical System

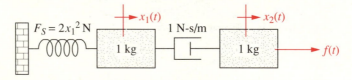

14. *Chapter-objective problem:* Given the schematic for the antenna azimuth position control system shown in Figure P2.36, find the state-space representation of each dynamic subsystem.

15. Underground water supplies, called *aquifers*, can be used in arid areas for agriculture and industry. The aquifer system consists of a number of interconnected natural storage tanks. Natural water flows through the sand and sandstone of the aquifer system, changing the water levels in the tanks on its way to the sea. People can use the water in these natural underground storage tanks for irrigation and industry.

Further, a water conservation policy can be established whereby water is pumped between tanks to prevent its loss to the sea. A model for the aquifer system is shown in Figure P3.12. In this model, the aquifer is represented by three tanks with water level h_i, called *head*. Each g_i is the natural water flow to the sea and is proportional to the difference in head between two adjoining tanks, or $g_i = C_i(h_i - h_{i-1})$, where C_i is a constant of proportionality and the units of g_i are m^3/yr.

Figure P3.12 Aquifer System Model

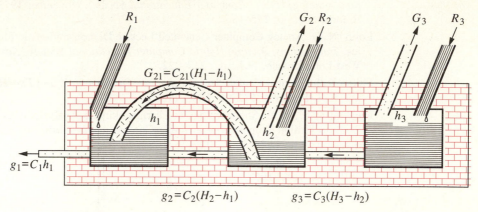

The engineered flow consists of three components also measured in m^3/yr: (1) flow from the tanks for irrigation and industry, G_i; (2) replenishing of the tanks from wells, R_i; and (3) flow, G_{21}, created by the water conservation policy to prevent loss to the sea. In this model, water for irrigation and industry will be taken only from Tank 2 and Tank 3. Thus, discharge of water from wells will only take place in Tank 2 and Tank 3. Water conservation will take place only between Tank 1 and Tank 2, as follows. Let H_1 be a reference head for Tank 1. If the water level in Tank 1 falls below H_1, water will be pumped from Tank 2 to Tank 1 to replenish the head. If h_1 is higher than H_1, water will be pumped back to Tank 2 to prevent loss to the sea. Calling this *flow for conservation*, G_{21}, we can say this flow is proportional to the difference between the head of Tank 1, h_1, and the reference head, H_1, or $G_{21} = C_{21}(H_1 - h_1)$.

The net flow into the tank is proportional to the rate of change of head in each tank. Thus, $K_i \dfrac{dh_i}{dt} = \sum R_i + \sum g_i + \sum G_{ij} - \sum G_i$ (Kandel, 1973). Represent the aquifer system in state space, where the state variables and the outputs are the heads of each tank.

BIBLIOGRAPHY

Cereijo, M. R. State Variable Formulations. *Instruments and Control Systems*, December 1969, pp. 87–88.

Cochin, I. *Analysis and Design of Dynamic Systems*. Harper & Row, New York, 1980.

Franklin, G. F.; Powell, J. D.; and Emami-Naeini, A. *Feedback Control of Dynamic Systems*. Addison-Wesley, Reading, Mass., 1986.

Inigo, R. M. Observer and Controller Design for D.C. Positional Control Systems Using State Variables. *Transactions, Analog/Hybrid Computer Educational Society*, December 1974. West Long Branch, N.J., pp. 177–189.

Kailath, T. *Linear Systems*. Prentice Hall, Englewood Cliffs, N.J., 1980.

Kandel, A. Analog Simulation of Groundwater Mining in Coastal Aquifers. *Transactions, Analog/Hybrid Computer Educational Society*, November 1973. West Long Branch, N.J, pp. 175–183.

Lordi, N. G. Analog Computer Generated Lecture Demonstrations in Pharmacokinetics. *Transactions, Analog/Hybrid Computer Educational Society*, November 1972. West Long Branch, N.J., pp. 217–222.

Philco Technological Center. *Servomechanism Fundamentals and Experiments*. Prentice Hall, Englewood Cliffs, N.J., 1980.

Riegelman, S., et al. Shortcomings in Pharmacokinetic Analysis by Conceiving the Body to Exhibit Properties of a Single Compartment. *Journal of Pharmaceutical Sciences* 57 no.1, 1968, pp. 117–123.

Timothy, L. K., and Bona, B. E. *State Space Analysis: An Introduction*. McGraw-Hill, New York, 1968.

4

TIME RESPONSE
FOR THE SYSTEM MODEL

4.1 Introduction

In Chapter 2 we showed how transfer functions can represent linear, time-invariant systems. In Chapter 3 systems were represented directly in the time domain via the state and output equations. After the engineer obtains a mathematical representation of a subsystem, the subsystem is analyzed for its transient and steady-state responses to see if these characteristics yield the desired behavior. This chapter is devoted to analysis of system transient response.

It may appear more logical to continue with Chapter 5, which covers the modeling of closed-loop systems, rather than to break the modeling sequence with the analysis presented here in Chapter 4. However, the student should not continue too far into system representation without knowing the application for the effort expended. Thus, this chapter demonstrates applications of the system representation by evaluating the transient response from the system model. Logically, this approach is not far from reality, since the engineer may indeed want to evaluate the response of a subsystem prior to inserting it into the closed-loop system.

After describing a valuable analysis and design tool, poles and zeros, we will begin analyzing our models to find the step response of first- and second-order systems, where the order refers to the order of the equivalent differential equation representing the system—the order of the denominator of the transfer function after cancellation of common factors in the numerator or the number of simultaneous first-order equations required for the state-space representation.

Chapter Objective

Given the azimuth position control system of Figure 2.2 and assuming an open-loop system (feedback path disconnected), the student will be able to

1. Predict, by inspection, the form of the open-loop angular velocity response of the load to a step-voltage input to the power amplifier
2. Find the damping ratio and natural frequency of the open-loop system[1]

---

[1] These terms will be defined later in this chapter.

3. Derive the complete analytical expression for the open-loop angular velocity response of the load to a step-voltage input to the power amplifier, using transfer functions

*4. Obtain the open-loop state equations and simulate the open-loop system on a digital computer.[2]

Also, given any underdamped, two-pole system, the student will be able to evaluate the following transient response characteristics: percent overshoot, settling time, peak time, and rise time.

4.2 The Concept of Poles and Zeros and System Response

The output response of a system is the sum of two responses: the *forced* response and the *natural* response.[3] Although many techniques, such as solving a differential equation or taking the inverse Laplace transform, enable us to evaluate this output response, these techniques are laborious and time-consuming. Productivity is aided by analysis and design techniques that yield results in a minimum of time. If the technique is so rapid that we feel we derive the desired result by inspection, we sometimes use the attribute *qualitative* to describe the method. A knowledge of poles and zeros and their relationship to the time response of a system is such a technique. Learning this relationship gives us a qualitative "handle" on problems. The concept of poles and zeros, fundamental to the design and analysis of control systems, simplifies the evaluation of a system's response. The reader is encouraged to fully master the concepts of poles and zeros and their application to problems throughout this book. Let us begin with two definitions.

Poles of a Transfer Function

The *poles* of a transfer function are (1) those values of the Laplace transform variable, s, that cause the transfer function to become infinite or (2) any roots of the denominator that are common to roots of the numerator of the transfer function.

Strictly speaking, the poles of a transfer function satisfy (1) of the definition. For example, the roots of the characteristic polynomial in the denominator are values of s that make the transfer function infinite and are thus poles. However, if a factor of the denominator can be canceled by the same factor in the numerator, the root of this factor no longer causes the transfer function to become infinite. In control systems we often refer to the root of the canceled factor in the denominator as a pole even though the transfer function will not be infinite at this value. Hence we include (2) of the definition.

---

[2]Objective 4 is included for those students pursuing the state-space approach.
[3]The forced response is also called the *steady-state response,* or *particular solution.* The natural response is also called the *homogeneous solution.*

Zeros of a Transfer Function

The *zeros* of a transfer function are (1) those values of the Laplace transform variable, s, that cause the transfer function to become zero or (2) any roots of the numerator that are common to roots of the denominator of the transfer function.

Strictly speaking, the zeros of a transfer function satisfy (1) of the definition. For example, the roots of the numerator are values of s that make the transfer function zero and are thus zeros. However, if a factor of the numerator can be canceled by the same factor in the denominator, the root of this factor no longer causes the transfer function to become zero. In control systems we often refer to the root of the canceled factor in the numerator as a zero even though the transfer function will not be zero at this value. Hence we include (2) of the definition.

Poles and Zeros of a First-Order System: An Example

Given the transfer function $G(s)$ in Figure 4.1(a), a pole exists at $s = -5$ and a zero exists at -2. These values are plotted on the complex s-plane in Figure 4.1(b) using an $\times$ for the pole and a $\bigcirc$ for the zero. To show the properties of the poles and zeros, let us find the unit step response of the system. From Figure 4.1(a),

$$C(s) = \frac{(s + 2)}{s(s + 5)} = \frac{A}{s} + \frac{B}{s + 5} = \frac{2/5}{s} + \frac{3/5}{s + 5} \tag{4.1}$$

where

$$A = \frac{(s + 2)}{(s + 5)}\Bigg|_{s \to 0} = \frac{2}{5}$$

$$B = \frac{(s + 2)}{s}\Bigg|_{s \to -5} = \frac{3}{5}$$

Thus,

$$c(t) = \frac{2}{5} + \frac{3}{5}e^{-5t} \tag{4.2}$$

From the development summarized in Figure 4.1(c), we draw the following conclusions:

1. A pole of the input function generates the form of the *forced* response (i.e., the pole at the origin generated a step function at the output).
2. A pole of the transfer function generates the form of the *natural* response (i.e., the pole at -5 generated e^{-5t}).
3. A pole on the real axis generates an *exponential* response of the form $e^{-\alpha t}$, where $-\alpha$ is the pole location on the real axis. Thus, the further to the left a pole is on the negative real axis, the faster the exponential

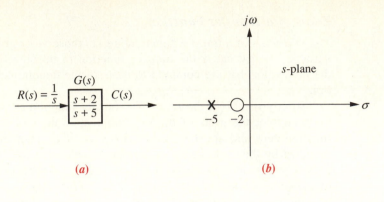

(a) (b)

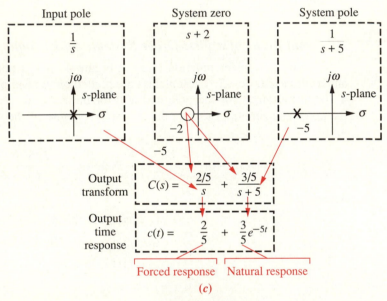

(c)

Figure 4.1 (*a*) A System Showing Input and Output;
(*b*) Pole-Zero Plot of the System; (*c*) Evolution of a System
Response

transient response will decay to zero (i.e., again the pole at -5 generated
e^{-5t}; see Figure 4.2 for the general case).

4. The zero helps generate the *amplitudes* for both the steady-state and transient responses (this can be seen from the calculation of A and B in Equation 4.1).

Let us now look at an example that demonstrates the technique of using poles
to obtain the form of the system response. We will learn to write the form of
the response by inspection. Each pole of the system transfer function that is on
the real axis generates an exponential response that is a component of the natural
response. The input pole generates the forced response.

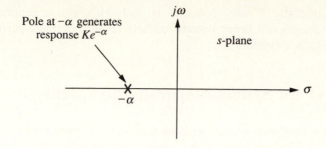

figure

Pole at $-\alpha$ generates response $Ke^{-\alpha}$

$j\omega$

s-plane

$-\alpha$

σ

Figure 4.2 Effect of a Real-Axis Pole upon Transient Response

Example 4.1   Use poles to evaluate system response.

Problem   Given the system of Figure 4.3, write the output, $c(t)$, in general terms. Specify the forced and natural parts of the solution.

Solution   By inspection, each system pole generates an exponential as part of the natural response. The input's pole generates the forced response. Thus,

$$C(s) \equiv \frac{K_1}{s} + \frac{K_2}{(s + 2)} + \frac{K_3}{(s + 4)} + \frac{K_4}{(s + 5)} \qquad (4.3)$$

Forced response   Natural response

Taking the inverse Laplace transform, we get

$$c(t) \equiv K_1 + K_2 e^{-2t} + K_3 e^{-4t} + K_4 e^{-5t} \qquad (4.4)$$

Forced response   Natural response

In this section we learned that poles determine the nature of the time response: poles of the input function determine the form of the forced response, and poles of the transfer function determine the form of the natural response. Zeros of the input or transfer function contribute to the amplitudes of the component parts of the total

Figure 4.3 System for Example 4.1

$R(s) = \frac{1}{s}$ $\dfrac{(s + 3)}{(s + 2)(s + 4)(s + 5)}$ $C(s)$

response. Finally, poles on the real axis generate exponential responses. We now discuss first-order systems without zeros to define a performance specification for such a system.

4.3 First-Order System Responses and Specifications

A first-order system without zeros can be described by the transfer function shown in Figure 4.4(a). If the input is a unit step, where $R(s) = 1/s$, the Laplace transform of the step response is $C(s)$, where

$$C(s) = R(s)G(s) = \frac{a}{s(s + a)} \tag{4.5}$$

Taking the inverse transform, the step response is given by

$$c(t) = c_f(t) + c_n(t) = 1 - e^{-at} \tag{4.6}$$

where the input pole at the origin generated the forced response $c_f(t) = 1$, and the system pole at $-a$ generated the natural response $c_n(t) = -e^{-at}$. Equation 4.6 is plotted in Figure 4.5.

Let us examine the significance of parameter a, the only parameter needed to describe the transient response. When $t = 1/a$,

$$e^{-at}\Big|_{t=1/a} = e^{-1} = 0.37 \tag{4.7}$$

or

$$x(t) = 1 - e^{-at}\Big|_{t=1/a} = 1 - 0.37 = 0.63 \tag{4.8}$$

Time Constant

We call $1/a$ the *time constant* of the response. From Equation 4.7, the time constant can be described as the time for e^{-at} to decay to 37% of its initial value.

Figure 4.4 (*a*) First-Order System; (*b*) Pole Plot

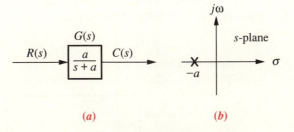

(*a*)   (*b*)

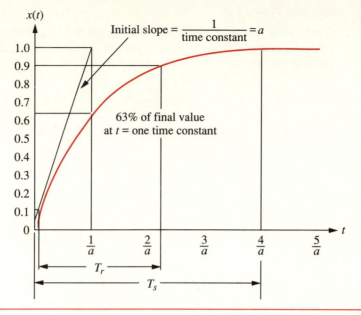

Figure 4.5 First-Order System Response to a Unit Step

Alternately, from Equation 4.8, the time constant is the time it takes for the step response to rise to 63% of its final value (see Figure 4.5).

The reciprocal of the time constant has the units (1/seconds), or frequency. Thus, we can call the parameter a the *exponential frequency*. Since the derivative of e^{-at} is $-a$ when $t = 0$, a is the initial rate of change of the exponential at $t = 0$. Thus, the time constant can be considered a transient response specification for a first-order system, since it is related to the speed at which the system responds to a step input.

The time constant can also be evaluated from the pole plot (see Figure 4.4(*b*)). Since the pole of the transfer function is at $-a$, we can say the pole is located at the *reciprocal* of the time constant, and the further the pole from the imaginary axis, the faster the transient response.

Let us look at other transient response specifications such as *rise time, T_r*, and *settling time, T_s*, as shown in Figure 4.5.

Rise Time, T_r

Rise time is defined as the time for the waveform to go from 0.1 to 0.9 of its final value. Rise time is found by solving Equation 4.6 for the difference in time at $c(t) = 0.9$ and $c(t) = 0.1$. Hence,

$$T_r = \frac{2.31}{a} - \frac{0.11}{a} = \frac{2.2}{a} \tag{4.9}$$

Settling Time, T_s

Settling time is defined as the time for the response to reach 2% of its final value.[4] Letting $c(t) = 0.98$ in Equation 4.6 and solving for time, t, we find the settling time to be

$$T_s = \frac{4}{a} \tag{4.10}$$

Let us now extend the concepts of poles and zeros and transient response to second-order systems.

4.4 Second-Order System Responses: Some Observations

Compared to the simplicity of a first-order system, a second-order system exhibits a wide range of responses that must be analyzed and described. Whereas varying a first-order system's parameter simply changes the speed of the response, changes in the parameters of a second-order system can change the *form* of the response. For example, a second-order system can display characteristics much like a first-order system or, depending upon component values, display damped or pure oscillations for its transient response.

To become familiar with the wide range of responses before formalizing our discussion in the next section, we will take a look at the numerical examples of second-order system responses shown in Figure 4.6. All examples are derived from Figure 4.6(*a*), the general case, which has two finite poles and no zeros. The term in the numerator is simply a scale or input multiplying factor that can take on any value without affecting the form of the derived results. By assigning appropriate values to parameters a and b, we can show all possible second-order transient responses. The unit step response can then be found using $C(s) = R(s)G(s)$, where $R(s) = 1/s$, followed by a partial fraction expansion and the inverse Laplace transform. Details are left to the student as an end-of-chapter problem, for which you may want to review Section 2.2.

We now explain each response and show how we can use the poles to determine the nature of the response without going through the previously described procedure of a partial fraction expansion followed by the inverse Laplace transform.

Overdamped Response, Figure 4.6(b)

For this response, $C(s) = \dfrac{9}{s(s^2 + 9s + 9)} = \dfrac{9}{s(s + 7.854)(1.146)}$. This function has a pole at the origin that comes from the unit step input and two real poles that come from the system. The input pole at the origin generates the constant forced response; each of the two system poles on the real axis generates an exponential natural response whose exponential frequency is equal to the pole

---

[4]Strictly speaking, this is the definition of the *2% settling time*. Other percentages, for example 5%, can also be used. We will use *settling time* throughout the book to mean 2% settling time.

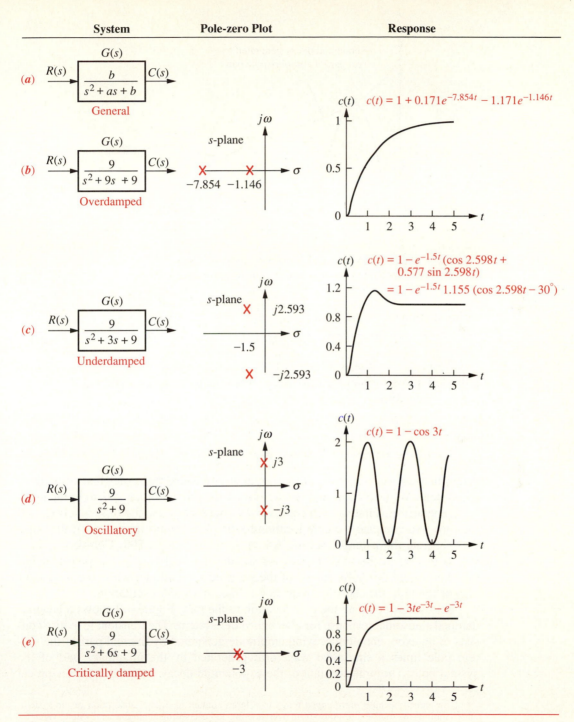

Figure 4.6 Second-Order Systems, Pole Plots, and Responses

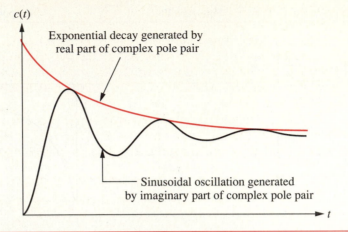

$c(t)$

Exponential decay generated by real part of complex pole pair

Sinusoidal oscillation generated by imaginary part of complex pole pair

t

Figure 4.7 Function of Real and Imaginary Parts of Complex Poles in Generating Second-Order Step Response

location. Hence, the output initially could have been written as

$$c(t) = K_1 + K_2 e^{-7.854t} + K_3 e^{-1.146t} \tag{4.11}$$

This response, shown in Figure 4.6(b), is called *overdamped*.[5] We see that the poles tell us the form of the response without the tedious calculation of the inverse Laplace transform.

Underdamped Response, Figure 4.6(c)

For this response, $C(s) = \dfrac{9}{s(s^2 + 3s + 9)}$. This function has a pole at the origin that comes from the unit step input and two complex poles that come from the system. We now compare the response of the second-order system to the poles that generated it. First we will compare the pole location to the time function, and then we will compare the pole location to the plot. From Figure 4.6(c), the poles that generate the natural response Are at $s = -1.5 \pm j2.598$. Comparing these values to $c(t)$ in the same figure, we see that the real part of the pole matches the exponential decay frequency of the sinusoid's amplitude, while the imaginary part of the pole matches the frequency of the sinusoidal oscillation.

Let us now compare the pole location to the plot. Figure 4.7 shows a general, damped sinusoidal response for a second-order system. The transient response consists of an exponentially decaying amplitude generated by the real part of the system pole times a sinusoidal waveform generated by the imaginary part of the system pole. The time constant of the exponential decay is equal to the reciprocal

---

[5]So named because *overdamped* refers to a large amount of energy absorption in the system, which inhibits the transient response from overshooting and oscillating about the steady-state value for a step input. As the energy absorption is reduced, an overdamped system will become underdamped and exhibit overshoot.

of the real part of the system pole. The value of the imaginary part is the actual frequency of the sinusoid, as depicted in Figure 4.7. This sinusoidal frequency is given the name *damped frequency of oscillation,* ω_d. Finally, the steady-state response (unit step) was generated by the input pole located at the origin. We call the type of response shown in Figure 4.7 an *underdamped response,* one which approaches a steady-state value via a transient response that is a damped oscillation.

The following example demonstrates how a knowledge of the relationship between the pole location and the transient response can lead rapidly to the response form without calculating the inverse Laplace transform.

Example 4.2

Use the poles of a transfer function to evaluate the form of the step response for an underdamped second-order system.

Problem   By inspection, write the form of the step response of the system in Figure 4.8.

Solution   First, we determine that the form of the forced response is a step. Next, we find the form of the natural response. Factoring the denominator of the transfer function in Figure 4.8, we find the poles to be $s = -5 \pm j13.23$. The real part, -5, is the exponential frequency for the damping. It is also the reciprocal of the time constant of the decay of the oscillations. The imaginary part, 13.23, is the radian frequency for the sinusoidal oscillations. Using our previous discussion and Figure 4.6(c) as a guide, we obtain

$$c(t) = K_1 + e^{-5t}(K_2 \cos 13.23t + K_3 \sin 13.23t)$$
$$= K_1 + K_4 e^{-5t}(\cos 13.23t - \phi) \tag{4.12}$$

where $\phi = \tan^{-1}\dfrac{K_3}{K_2}$, $K_4 = \sqrt{K_2^2 + K_3^2}$, and $c(t)$ is a constant plus an exponentially damped sinusoid.

We will revisit the second-order underdamped response in Sections 4.5 and 4.6, where we generalize the discussion and derive some results that relate the pole position to other parameters of the response.

Oscillatory Response, Figure 4.6(d)

For this response, $C(s) = \dfrac{9}{s(s^2 + 9)}$. This function has a pole at the origin that comes from the unit step input and two imaginary poles that come from the

Figure 4.8 System for Example 4.2

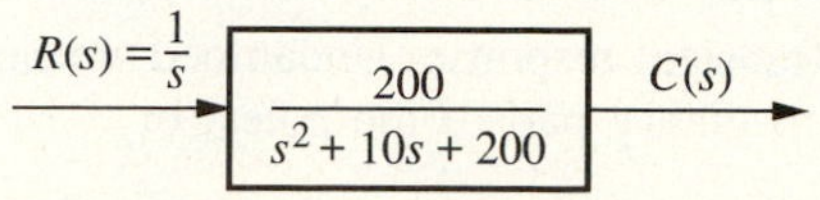

$$R(s) = \frac{1}{s} \longrightarrow \boxed{\dfrac{200}{s^2 + 10s + 200}} \longrightarrow C(s)$$

system. The input pole at the origin generates the constant forced response, and the two system poles on the imaginary axis at $\pm j3$ generate a sinusoidal natural response whose frequency is equal to the location of the imaginary poles. Hence, the output can be estimated as

$$c(t) = K_1 + K_4 \cos(3t - \phi) \tag{4.13}$$

This type of response, shown in Figure 4.6(d), is called *oscillatory*.

Critically Damped Response, Figure 4.6(e)

For this response, $C(s) = \dfrac{9}{s(s^2 + 6s + 9)} = \dfrac{9}{s(s + 3)^2}$. This function has a pole at the origin that comes from the unit step input and two multiple real poles that come from the system. The input pole at the origin generates the constant forced response, and the two poles on the real axis at -3 generate a natural response consisting of an exponential and an exponential multiplied by time, where the exponential frequency is equal to the location of the real poles. Hence, the output can be estimated as

$$c(t) = K_1 + K_2 e^{-3t} + K_3 t e^{-3t} \tag{4.14}$$

This type of response, shown in Figure 4.6(e), is called *critically damped*. Critically damped responses are the fastest possible without the overshoot that is characteristic of the underdamped response.

We now summarize our observations. In this section we defined the following responses and found their characteristics:

1. *Overdamped responses:*

   Poles: Two real at $-\sigma_1$, $-\sigma_2$

   Transient response: Two exponentials with time constants equal to the reciprocal of the pole locations, or

   $$c(t) = K_1 e^{-\sigma_1 t} + K_2 e^{-\sigma_2 t}$$

2. *Underdamped responses:*

   Poles: Two complex at $-\sigma_d \pm j\omega_d$

   Transient response: Damped sinusoid with an exponential envelope whose time constant is equal to the reciprocal of the pole's real part. The radian frequency of the sinusoid, the damped frequency of oscillation, is equal to the imaginary part of the poles, or

   $$c(t) = A e^{-\sigma_d t} \cos(\omega_d t - \phi)$$

3. *Oscillatory responses:*

   Poles: Two imaginary at $\pm j\omega_1$

   Transient response: Undamped sinusoid with radian frequency equal to the imaginary part of the poles, or

   $$c(t) = A \cos(\omega_1 t - \phi)$$

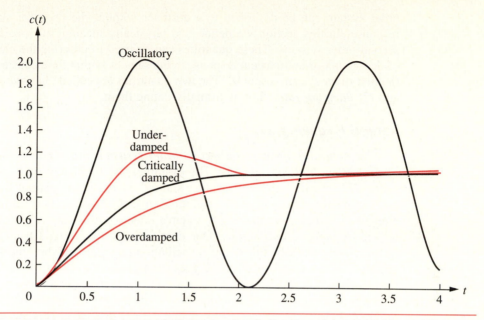

Figure 4.9 Step Responses for the Four Damping Cases of a Second-Order System

4. *Critically damped responses:*

Poles: Two real at $-\sigma_1$

Transient response: One term is an exponential whose time constant is equal to the reciprocal of the pole location. Another term is the product of time, t, and an exponential with time constant equal to the reciprocal of the pole location, or

$$c(t) = K_1 e^{-\sigma_1 t} + K_2 t e^{-\sigma_1 t}$$

The four cases are graphically summarized in Figure 4.9. Notice that the critically damped case is the division between the overdamped cases and the underdamped cases and is the fastest response without overshoot.

In the next section we formalize and generalize our discussion of second-order responses and define two specifications that are used for the design and analysis of second-order systems. In Section 4.6, we focus on the *underdamped* case and derive some specifications unique to this response that we use later for analysis and design.

4.5 The General Second-Order System and Its Specifications

Now that we have become familiar with second-order systems and their responses through numerical examples, we will generalize the discussion and establish a set of quantitative specifications defined in such a way that the response of a second-order system can be described to a designer without the need for sketching the response. In this section we define two physically meaningful specifications for second-order systems. These quantities can be used to describe the characteristics of the second-order transient response in the same way that time constants describe the first-order system response. The two quantities are called (1) *natural frequency* and (2) *damping ratio*. Let us formally define them.

Natural Frequency, ω_n

The natural frequency of a second-order system is the frequency of oscillation of the system without damping. For example, the frequency of oscillation of a series *RLC* circuit with the resistance shorted would be the natural frequency.

Before we state our next definition, some explanation is in order. We have already seen that a second-order system's underdamped step response is characterized by damped oscillations. Our next definition is derived from the need to quantitatively describe this damped oscillation regardless of the time scale. Thus, a system whose transient response goes through three cycles in a millisecond before reaching the steady state would have the same measure as a system that went through three cycles in a millennium. For example, the underdamped curve in Figure 4.9 has an associated measure that defines its shape. This measure remains the same even if we change the time base from seconds to microseconds or to millennia.

A viable definition for this quantity is one that compares the exponential decay frequency of the envelope to the natural frequency. This ratio is constant regardless of the time scale of the response. Also, the reciprocal, which is proportional to the ratio of the natural period to the exponential time constant, also remains the same regardless of the time base.

Damping Ratio, ζ

We define the damping ratio, ζ, to be

$$\zeta = \frac{\text{Exponential decay frequency}}{\text{Natural frequency (rad/second)}} = \frac{1}{2\pi} \frac{\text{Natural period (seconds)}}{\text{Exponential time constant}}$$

Let us now revise our description of the second-order system to reflect the new definitions. The general second-order system shown in Figure 4.6(*a*) can be transformed to show the quantities ζ and ω_n. Consider the general system

$$G(s) = \frac{b}{s^2 + as + b} \tag{4.15}$$

Without damping, the poles would be on the $j\omega$ axis, and the response would be an undamped sinusoid. For the poles to be purely imaginary, $a = 0$. Hence,

$$G(s) = \frac{b}{s^2 + b} \tag{4.16}$$

By definition, the natural frequency, ω_n, is the frequency of oscillation of this system. Since the poles of this system are on the $j\omega$ axis at $\pm j \sqrt{b}$,

$$\omega_n = \sqrt{b} \tag{4.17}$$

Hence,

$$b = \omega_n{}^2 \tag{4.18}$$

Now what is the term a in Equation 4.15? Assuming an underdamped system, the complex poles have a real part equal to $-a/2$. The magnitude of this value is then the exponential decay frequency described in Section 4.4. Hence,

$$\zeta = \frac{\text{Exponential decay frequency}}{\text{Natural frequency (rad/second)}} = \frac{a/2}{\omega_n} \tag{4.19}$$

from which,

$$a = 2\zeta\omega_n \tag{4.20}$$

Our general second-order transfer function finally looks like this:

$$G(s) = \frac{\omega_n{}^2}{s^2 + 2\zeta\omega_n s + \omega_n{}^2} \tag{4.21}$$

In the following example we find numerical values for ζ and ω_n by matching the transfer function to Equation 4.21.

Example 4.3   Find ζ and ω_n for a second-order system.

Problem   Given the transfer function of Equation 4.22, find ζ and ω_n.

$$G(s) = \frac{36}{s^2 + 4.2s + 36} \tag{4.22}$$

Solution   Comparing Equation 4.22 to 4.21, $\omega_n{}^2 = 36$, from which $\omega_n = 6$. Also, $2\zeta\omega_n = 4.2$. Substituting the value of ω_n, $\zeta = 0.35$.

Now that we have defined ζ and ω_n, let us relate these quantities to the pole location. Solving for the poles of the transfer function in Equation 4.21 yields

$$s_{1,2} = -\zeta\omega_n \pm \omega_n \sqrt{\zeta^2 - 1} \tag{4.23}$$

From Equation 4.23, we see that the various cases of second-order response are a function of ζ and are summarized in Figure 4.10.[6]

In the following example we find the numerical value of ζ and determine the nature of the transient response.

---

[6]The student should verify Figure 4.10 as an exercise.

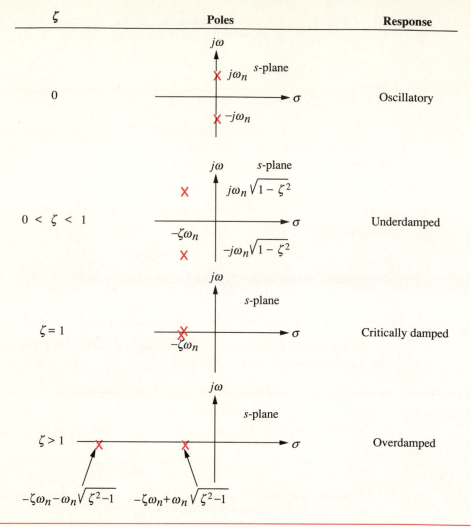

ζ	Poles	Response

Figure 4.10 Variation of Second-Order Response as a Function of Damping Ratio

---

Example 4.4 Find the nature of the second-order system response via the value of the damping ratio.

Problem For each of the systems shown in Figure 4.11, find the value of ζ and report the kind of response expected.

Solution First, match the form of these systems to the forms shown in Equations 4.15 and 4.21. Since $a = 2\zeta\omega_n$ and $\omega_n = \sqrt{b}$,

$$\zeta = \frac{a}{2\sqrt{b}} \tag{4.24}$$

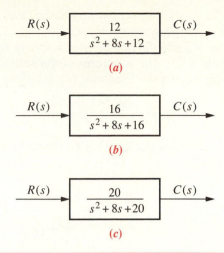

Figure 4.11 Systems for Example 4.4

Using the values of a and b from each of the systems of Figure 4.11, we find $\zeta = 1.155$ for system (a), which is thus overdamped, since $\zeta > 1$; $\zeta = 1$ for system (b), which is thus critically damped and $\zeta = 0.894$ for system (c), which is thus underdamped, since $\zeta < 1$.

In this section we defined two specifications or parameters of second-order systems: natural frequency, ω_n, and damping ratio, ζ. We saw that the nature of the response obtained was related to the value of ζ. Variations of damping ratio alone yield the complete range of overdamped, critically damped, underdamped, and oscillatory responses.

Now that we have generalized the second-order transfer function in terms of ζ and ω_n, let us analyze the step response of an *underdamped* second-order system. Not only will this response be found in terms of ζ and ω_n, but more specifications indigenous to the underdamped case will be defined.

4.6 Further Analysis for Underdamped Second-Order Systems

The underdamped second-order system, a common model for physical problems, displays unique behavior that must be itemized; a detailed description of the underdamped response is necessary both for analysis and design. Our first objective is to define transient specifications associated with underdamped responses. Next, we relate these specifications to the pole location, drawing an association between pole location and the form of the underdamped second-order response. Finally, we tie the pole location to system parameters, thus closing the loop: desired response generates required system components.

Let us begin by finding the step response for the general second-order system of Equation 4.21. The transform of the response, $C(s)$, is the transform of the input times the transfer function, or

$$C(s) = \frac{\omega_n{}^2}{s(s^2 + 2\zeta\omega_n s + \omega_n{}^2)} = \frac{K_1}{s} + \frac{K_2 s + K_3}{s^2 + 2\zeta\omega_n s + \omega_n{}^2} \tag{4.25}$$

where it is assumed that $\zeta < 1$ (i.e., the underdamped case). Expanding by partial fractions yields

$$C(s) = \frac{1}{s} - \frac{(s + \zeta\omega_n) + \dfrac{\zeta}{\sqrt{1 - \zeta^2}}\omega_n \sqrt{1 - \zeta^2}}{(s + \zeta\omega_n)^2 + \omega_n{}^2(1 - \zeta^2)} \tag{4.26}$$

from which

$$c(t) = 1 - e^{-\zeta\omega_n t}\left(\cos\omega_n\sqrt{1 - \zeta^2}t + \frac{\zeta}{\sqrt{1 - \zeta^2}}\sin\omega_n\sqrt{1 - \zeta^2}t\right)$$

$$= 1 - \frac{1}{\sqrt{1 - \zeta^2}}e^{-\zeta\omega_n t}\cos\left(\omega_n\sqrt{1 - \zeta^2}t - \phi\right) \tag{4.27}$$

where $\phi = \tan^{-1}\dfrac{\zeta}{\sqrt{1 - \zeta^2}}$.

A plot of this response appears in Figure 4.12 for various values of ζ, plotted along a time axis normalized to the natural frequency. We now see the relationship between the value of ζ and the type of response obtained: The lower the value of ζ, the more oscillatory the response. The natural frequency is a time-axis scale factor and does not affect the nature of the response other than to scale it in time.

We have defined two parameters associated with second-order systems, ζ and ω_n. Other parameters associated with the underdamped response are percent overshoot, peak time, settling time, and rise time. These specifications are defined as follows (see also Figure 4.13):

1. *Peak time, T_p:* The time required to reach the first, or maximum, peak.

2. *Percent overshoot, %OS:* The amount that the waveform overshoots the steady-state, or final, value at the peak time, expressed as a percentage of the steady-state value.

3. *Settling time, T_s:* The amount of time required for the transient's damped oscillations to stay within $\pm 2\%$ of the steady-state value.

4. *Rise time, T_r:* The time required for the waveform to go from 0.1 of the final value to 0.9 of the final value.

Notice that the definitions for settling time and rise time are basically the same definitions we made for the first-order response. We should mention that all definitions

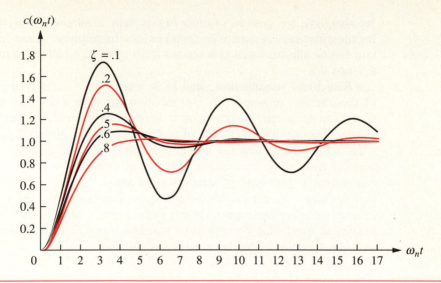

Figure 4.12 Second-Order Underdamped Responses for Various Values of Damping Ratio

Figure 4.13 Second-Order Underdamped Response Specifications

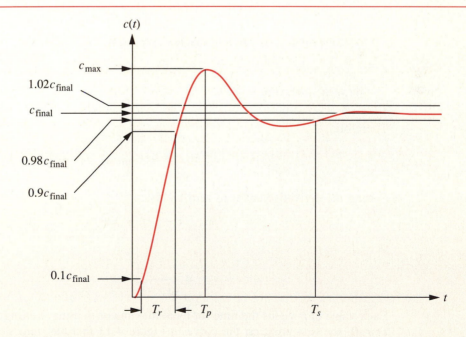

are also valid for systems of order higher than 2, although analytical expressions for these parameters cannot be found unless the response of the higher-order system can be approximated as a second-order system, which we will do in Sections 4.7 and 4.8.

Rise time, settling time, and peak time yield information about the speed of the transient response. This information can help a designer determine if the speed and the nature of the response do or do not degrade the performance of the system. For example, the speed of an entire computer system depends on the time it takes for a floppy disk drive head to reach steady state and read data; passenger comfort depends in part on the suspension system of a car and the number of oscillations it goes through after hitting a bump.

We now evaluate T_p, $\%OS$, and T_s as functions of ζ and ω_n. Later in this chapter we relate these specifications to the location of the system poles. A precise analytical expression for rise time cannot be obtained, thus we present a plot showing the relationship between ζ and rise time.

Evaluation of T_p

T_p is found by differentiating $c(t)$ in Equation 4.27 and finding the first zero crossing after $t = 0$. This task is simplified by "differentiating" in the frequency domain using Item 7 of Table 2.2. Assuming zero initial conditions and using Equation 4.25, we get

$$\mathcal{L}[\dot{c}(t)] = sC(s) = \frac{\omega_n^2}{s^2 + 2\zeta\omega_n s + \omega_n^2} \tag{4.28}$$

Completing squares in the denominator, we have

$$\mathcal{L}[\dot{c}(t)] = \frac{\omega_n^2}{(s + \zeta\omega_n)^2 + \omega_n^2(1 - \zeta^2)} = \frac{\dfrac{\omega_n}{\sqrt{1 - \zeta^2}}\omega_n\sqrt{1 - \zeta^2}}{(s + \zeta\omega_n)^2 + \omega_n^2(1 - \zeta^2)} \tag{4.29}$$

Therefore,

$$\dot{c}(t) = \frac{\omega_n}{\sqrt{1 - \zeta^2}}e^{-\zeta\omega_n t}\sin\omega_n\sqrt{1 - \zeta^2}t \tag{4.30}$$

Setting the derivative equal to zero yields

$$\omega_n\sqrt{1 - \zeta^2}t = n\pi \tag{4.31}$$

or

$$t = \frac{n\pi}{\omega_n\sqrt{1 - \zeta^2}} \tag{4.32}$$

Each value of n yields the time for local maxima or minima. Letting $n = 0$ yields $t = 0$, the first point on the curve in Figure 4.13 that has zero slope. The first peak, which occurs at the peak time, T_p, is found by letting $n = 1$ in Equation 4.32. Thus,

$$T_p = \frac{\pi}{\omega_n \sqrt{1 - \zeta^2}} \tag{4.33}$$

Evaluation of %OS

From Figure 4.13, the percent overshoot, %OS, is given by

$$\%OS = \frac{c_{max} - c_{final}}{c_{final}} \times 100 \tag{4.34}$$

c_{max} is found by evaluating $c(t)$ at the peak time, $c(T_p)$. Using Equation 4.33 for T_p and substituting into Equation 4.27 yields

$$c_{max} = c(T_p) = 1 - e^{-(\zeta\pi/\sqrt{1-\zeta^2})}\left(\cos \pi + \frac{\zeta}{\sqrt{1-\zeta^2}}\sin \pi\right)$$

$$= 1 + e^{-(\zeta\pi/\sqrt{1-\zeta^2})} \tag{4.35}$$

For the unit step used for Equation 4.27,

$$c_{final} = 1 \tag{4.36}$$

Substituting Equations 4.35 and 4.36 into Equation 4.34, we finally obtain

$$\%OS = e^{-(\zeta\pi/\sqrt{1-\zeta^2})} \times 100 \tag{4.37}$$

Notice that the percent overshoot is a function only of the damping ratio, ζ.

Whereas Equation 4.37 allows one to find %OS given ζ, the inverse of the equation allows one to solve for ζ given %OS. The inverse is given by

$$\zeta = \frac{-\ln(\%OS/100)}{\sqrt{\pi^2 + \ln^2(\%OS/100)}} \tag{4.38}$$

The derivation of Equation 4.38 is left as an exercise for the student. Equation 4.37 (or, equivalently, 4.38) is plotted in Figure 4.14.

Evaluation of T_s

In order to find the settling time, we must find the time for which $c(t)$ in Equation 4.27 reaches and stays within ±2% of the steady-state value, c_{final}. Using our definition, the settling time is the time it takes for the amplitude of the decaying sinusoid in Equation 4.27 to reach 0.02, or

$$e^{-\zeta\omega_n t}\frac{1}{\sqrt{1-\zeta^2}} = 0.02 \tag{4.39}$$

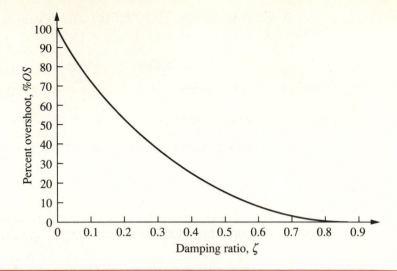

Figure 4.14 The Relationship between Percent Overshoot and Damping Ratio

This equation is a conservative estimate, since we are assuming that $\cos(\omega_n \sqrt{1 - \zeta^2}t - \phi) = 1$ at the settling time. Solving for t, the settling time is

$$T_s = \frac{-\ln(0.02\sqrt{1 - \zeta^2})}{\zeta\omega_n} \tag{4.40}$$

The reader can verify that the numerator of Equation 4.40 varies from 3.91 to 4.74 as ζ varies from 0 to 0.9. Let us agree upon an approximation for the settling time that will be used for all values of ζ; let it be

$$T_s = \frac{4}{\zeta\omega_n} \tag{4.41}$$

Evaluation of T_r

A precise analytical relationship between rise time and damping ratio, ζ, cannot be found. However, using a computer and Equation 4.27, the rise time can be found. The results are shown in Figure 4.15. Let us look at an example.

Example 4.5   Find T_p, %OS, T_r, and T_s from a transfer function.

Problem   Given the transfer function

$$G(s) = \frac{100}{s^2 + 15s + 100} \tag{4.42}$$

find T_p, %OS, and T_s.

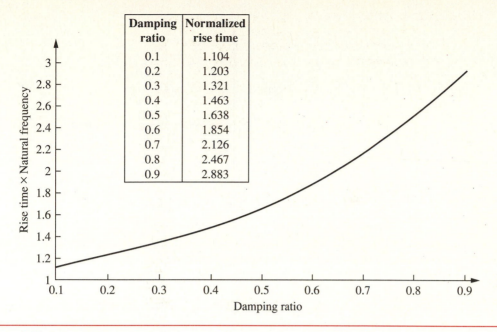

Damping ratio	Normalized rise time
0.1	1.104
0.2	1.203
0.3	1.321
0.4	1.463
0.5	1.638
0.6	1.854
0.7	2.126
0.8	2.467
0.9	2.883

Figure 4.15 Damping Ratio as a Function of Normalized Rise Time for a Second-Order Underdamped Response

Solution ω_n and ζ are calculated to be 10 and 0.75 respectively. Now, substitute ζ and ω_n into Equations 4.33, 4.37, and 4.41 and find respectively that $T_p = 0.475$ seconds, $\%OS = 2.838$, and $T_s = 0.533$ seconds. Using the table in Figure 4.15, we find that the normalized rise time is approximately 2.3 seconds. Dividing by ω_n yields $T_r = 0.23$ seconds. Notice how this problem demonstrates that we can find T_p, $\%OS$, and T_s without the tedious task of taking an inverse Laplace transform, plotting the output response, and taking measurements from the plot.

We now have expressions that relate peak time, percent overshoot, and settling time to the natural frequency and the damping ratio. Now let us relate these quantities to the location of the poles that generate these characteristics.

The pole plot for a general, underdamped second-order system, previously shown in Figure 4.10, is reproduced and expanded in Figure 4.16 for focus. We see from the Pythagorean theorem that the radial distance from the origin to the pole is the natural frequency, ω_n, and the $\cos\theta = \zeta$.

Now, comparing Equations 4.33 and 4.41 with the pole location, we evaluate peak time and settling time in terms of the pole location. Thus

$$T_p = \frac{\pi}{\omega_n \sqrt{1 - \zeta^2}} = \frac{\pi}{\omega_d} \qquad (4.43)$$

$$T_s = \frac{4}{\zeta\omega_n} = \frac{4}{\sigma_d} \qquad (4.44)$$

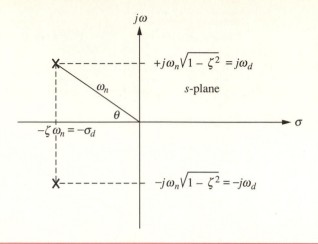

Figure 4.16 Pole Plot for an Underdamped Second-Order System

where ω_d is the imaginary part of the pole and is called the *damped frequency of oscillation*, and σ_d is the real part of the pole and is the *exponential damping frequency*.

Equation 4.43 shows that T_p is inversely proportional to the imaginary part of the pole. Since horizontal lines on the s-plane are lines of constant imaginary value, horizontal lines are also lines of constant peak time. Similarly, Equation 4.44 tells us that settling time is inversely proportional to the real part of the pole. Since vertical lines on the s-plane are lines of constant real value, vertical lines are also lines of constant settling time. Finally, since $\zeta = \cos\theta$, radial lines are lines of constant ζ. Since percent overshoot is only a function of ζ, radial lines are thus lines of constant percent overshoot, $\%OS$. These concepts are depicted in Figure 4.17, where lines of constant T_p, T_s, and $\%OS$ are labeled on the s-plane.

At this point, we can understand the significance of Figure 4.17 by examining the actual step response of comparative systems. Depicted in Figure 4.18(a) are the step responses as the poles are moved in a vertical direction, keeping the real part the same. As the poles move in a vertical direction, the frequency increases, but the envelope remains the same since the real part of the pole is not changing. The figure shows a constant exponential envelope, even though the sinusoidal response is changing frequency. Since all curves fit under the same exponential decay curve, the settling time is virtually the same for all waveforms. Note that as overshoot increases, the rise time decreases.

Let us move the poles to the right or left. Since the imaginary part is now constant, movement of the poles yields the responses of Figure 4.18(b). Here the frequency is constant over the range of variation of the real part. As the pole moves to the left, the response damps out more rapidly, while the frequency remains the same. Notice that the peak time is the same for all waveforms because the imaginary part remains the same.

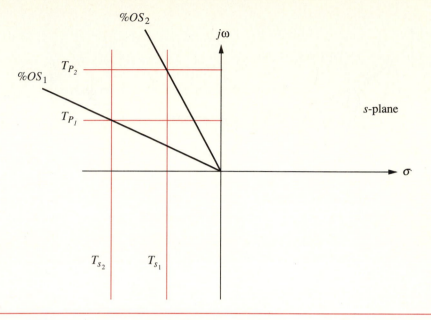

Figure 4.17 Lines of Constant Peak Time, T_p, Settling Time, T_s, and Percent Overshoot, %OS

Moving the poles along a constant radial line yields the responses shown in Figure 4.18(c). Here the percent overshoot remains the same. Notice also that the responses look exactly alike, except for their speed. The further the poles are from the origin, the more rapid the response. We conclude this section with some examples that demonstrate the relationship between the pole location and the specifications of the second-order underdamped response.

Example 4.6

Find T_p, %OS, and T_s from the pole location.

Problem Given the pole plot shown in Figure 4.19, find ζ, ω_n, T_p, %OS, and T_s.

Solution The damping ratio is given by $\zeta = \cos\theta = \cos[\arctan(7/3)] = 0.394$. The natural frequency, ω_n, is the radial distance from the origin to the pole, or $\omega_n = \sqrt{7^2 + 3^2} = 7.616$. The peak time is

$$T_p = \frac{\pi}{\omega_d} = \frac{\pi}{7} = 0.449 \text{ seconds} \tag{4.45}$$

The percent overshoot is

$$\%OS = e^{-(\zeta\pi/\sqrt{1-\zeta^2})} \times 100 = 26.018\% \tag{4.46}$$

The approximate settling time is

$$T_s = \frac{4}{\sigma_d} = \frac{4}{3} = 1.333 \text{ seconds} \tag{4.47}$$

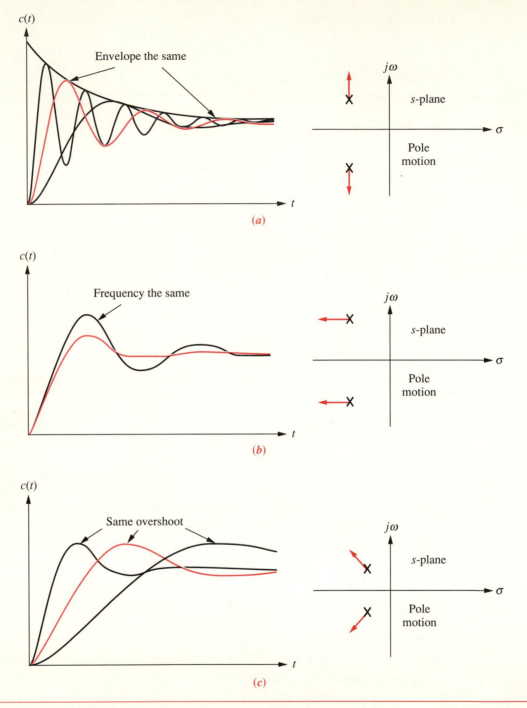

Figure 4.18 Step Responses of Second-Order Underdamped Systems:
(a) As Poles Move with Constant Real Part; **(b)** As Poles Move with Constant
Imaginary Part; **(c)** As Poles Move with Constant Damping Ratio

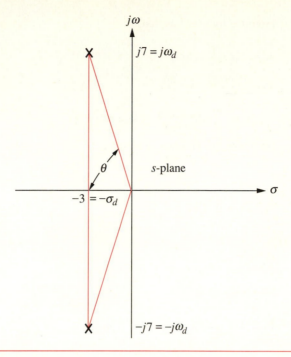

Figure 4.19 Pole Plot for Example 4.6

Let us apply the concepts to a simple design problem consisting of a physical network whose component values we want to design to meet a response specification.

Example 4.7

Design system components to meet transient response specifications.

Problem Given the system shown in Figure 4.20, find J and D to yield 20% overshoot and a settling time of 2 seconds for a step input of torque $T(t)$.

Solution First, the transfer function for the system is

$$G(s) = \frac{1/J}{s^2 + \dfrac{D}{J}s + \dfrac{K}{J}} \tag{4.48}$$

Figure 4.20 Rotational Mechanical System for Example 4.7

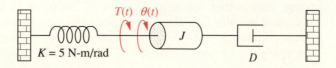

From the transfer function,

$$\omega_n = \sqrt{\frac{K}{J}} \tag{4.49}$$

and

$$2\zeta\omega_n = \frac{D}{J} \tag{4.50}$$

But, from the problem statement,

$$T_s = 2 = \frac{4}{\zeta\omega_n} \tag{4.51}$$

or $\zeta\omega_n = 2$. Hence,

$$2\zeta\omega_n = 4 = \frac{D}{J} \tag{4.52}$$

Also, from Equations 4.49 and 4.51,

$$\zeta = \frac{4}{2\omega_n} = 2\sqrt{\frac{J}{K}} \tag{4.53}$$

From Equation 4.38, a 20% overshoot implies a $\zeta = 0.456$. Therefore, from Equation 4.53,

$$\zeta = 2\sqrt{\frac{J}{K}} = 0.456 \tag{4.54}$$

Hence,

$$\frac{J}{K} = 0.052 \tag{4.55}$$

From the problem statement, $K = 5$ N-m/rad. Combining this value with Equations 4.52 and 4.55, $D = 1.04$ N-m s/rad, and $J = 0.26$ kg-m^2.

Now that we have analyzed systems with two poles, how does the addition of another pole affect the response? We answer this question in the next section.

4.7 System Responses with Additional Poles

In the last section we analyzed systems with one or two poles. It must be emphasized that the formulae describing percent overshoot, settling time, and peak time were derived only for a system with two complex poles and no zeros. If a system such as that shown in Figure 4.21 has more than two poles or has zeros, we cannot use the formulae to calculate the performance specifications that we derived. However, under certain conditions, a system with more than two poles or with zeros can be approximated as a second-order system that has just two complex *dominant poles*. Once we justify this approximation, the formulae for percent overshoot, settling time, and peak time can be applied to these higher-order sys-

Figure 4.21 A Robot Applying an Average of Twelve Body Welds to Buick LeSabre and Oldsmobile 88 Model Automobiles (Courtesy of Flint Automotive Division; General Motors Corporation)

tems using the location of the dominant poles. In this section we investigate the effect of an additional pole on the second-order response. In the next section we analyze the effect of adding a zero to a two-pole system.

Let us now look at the conditions that would have to exist in order to approximate the behavior of a three-pole system as that of a two-pole system. Consider a three-pole system with complex poles and a third pole on the real axis. Assuming that the complex poles are at $-\zeta\omega_n \pm j\omega_n \sqrt{1 - \zeta^2}$, and the real pole is at $-\alpha_r$, the step response of the system can be determined from a partial fraction expansion. Thus the output transform is

$$C(s) = \frac{A}{s} + \frac{B(s + \zeta\omega_n) + C\omega_d}{(s + \zeta\omega_n)^2 + \omega_d^2} + \frac{D}{s + \alpha_r} \tag{4.56}$$

or, in the time domain,

$$c(t) = Au(t) + e^{-\zeta\omega_n t}(B\cos\omega_d t + C\sin\omega_d t) + De^{-\alpha_r t} \tag{4.57}$$

The component parts of $c(t)$ are shown in Figure 4.22 for three cases of α_r. For Case I, $\alpha_r = \alpha_{r_1}$ and is not much larger than $\zeta\omega_n$; for Case II, $\alpha_r = \alpha_{r_2}$ and is much larger than $\zeta\omega_n$; and for Case III, $\alpha_r = \infty$. Let us direct our attention to Equation 4.57 and Figure 4.22.

If $\alpha_r \gg \zeta\omega_n$ (Case II), the pure exponential will die out much more rapidly

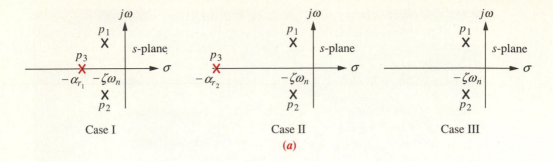

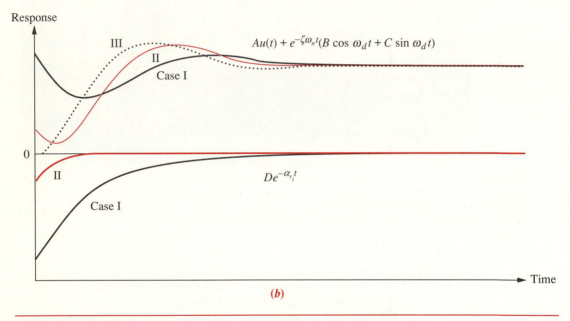

Figure 4.22 Component Responses of a Three-Pole System: (**a**) Pole Plot; (**b**) Component Responses: Nondominant Pole Is Near the Dominant Second-Order Pair (Case I), Far from the Dominant Second-Order Pair (Case II), and at Infinity (Case III)

than the second-order underdamped step response. If the pure exponential term decays to an insignificant value at the time of the first overshoot, such parameters as percent overshoot, settling time, and peak time will be generated by the second-order underdamped step response component which approaches the response of a pure second-order system (Case III).

If α_r is not much greater than $\zeta\omega_n$ (Case I), the real pole's transient response will not decay to insignificance at the peak time or settling time generated by the second-order pair. In this case, the exponential decay is significant, and the system cannot be represented as a second-order system.

The next question is how much further from the dominant poles does the third pole have to be in order for its effect to be negligible on the second-order response? The answer, of course, depends upon the accuracy that you are looking for. However, in this book, we will assume that the exponential decay is negligible after five time constants. Thus, if the real pole is five times further to the left than the dominant poles, we will assume that the system is represented by its dominant second-order pair of poles.

What about the magnitude of the exponential decay? Can its magnitude be so large so that its contribution at the peak time is not negligible? We can show, through a partial fraction expansion, that the residue of the third pole, in a three-pole system with dominant second-order poles and no zeros, will actually reduce in magnitude as the third pole is moved further into the left half-plane. Assume a step response, $C(s)$, of a three-pole system:

$$C(s) = \frac{bc}{s(s^2 + as + b)(s + c)} = \frac{A}{s} + \frac{Bs + C}{s^2 + as + b} + \frac{D}{s + c} \tag{4.58}$$

where we assume that the nondominant pole is located at $-c$ on the real axis, and that the steady-state response approaches unity. Evaluating the constants in the numerator of each term,

$$A = 1; \qquad B = \frac{ca - c^2}{c^2 + b - ca} \tag{4.59a}$$

$$C = \frac{ca^2 - c^2a - bc}{c^2 + b - ca}; \qquad D = \frac{-b}{c^2 + b - ca} \tag{4.59b}$$

As the nondominant pole approaches ∞, or $c \to \infty$,

$$A = 1; \qquad B = -1; \qquad C = -a; \qquad D = 0 \tag{4.60}$$

Thus, for this example, D, the residue of the nondominant pole and its response, becomes zero as the nondominant pole approaches infinity.

The designer can also choose to forego extensive residue analysis, since all system designs should be simulated to determine final acceptance. In this case, the control systems engineer can use the "five times" rule-of-thumb as a necessary but not sufficient condition to increase the confidence in the second-order approximation during design, but then simulate the completed design.

Let us look at an example that compares the responses of two different three-pole systems with that of a second-order system.

Example 4.8

Compare the responses of three-pole systems.

Problem Find the step response of each of the transfer functions shown in Equations 4.61 through 4.63 and compare them.

$$T_1(s) = \frac{24.542}{s^2 + 4s + 24.542} \tag{4.61}$$

$$T_2(s) = \frac{245.42}{(s + 10)(s^2 + 4s + 24.542)} \tag{4.62}$$

$$T_3(s) = \frac{73.626}{(s + 3)(s^2 + 4s + 24.542)} \tag{4.63}$$

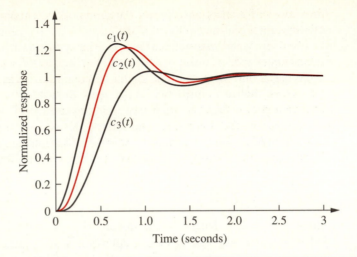

Figure 4.23 Step Responses of System $T_1(s)$, System $T_2(s)$, and System $T_3(s)$

Solution   The step response, $C_i(s)$, for the transfer function, $T_i(s)$, can be found by multiplying the transfer function by $1/s$, a step input, and using partial fraction expansion followed by the inverse Laplace transform to find the response, $c_i(t)$. With the details left as an exercise for the student, the results are

$$c_1(t) = 1 - 1.09e^{-2t}\cos(4.532t - 23.8°) \tag{4.64}$$

$$c_2(t) = 1 - 0.29e^{-10t} - 1.189e^{-2t}\cos(4.532t - 53.34°) \tag{4.65}$$

$$c_3(t) = 1 - 1.14e^{-3t} + 0.707e^{-2t}\cos(4.532t + 78.63°) \tag{4.66}$$

The three responses are plotted in Figure 4.23. Notice that $c_2(t)$, with its third pole at -10 and furthest from the dominant poles, is the better approximation of $c_1(t)$, the pure second-order system response; $c_3(t)$, with a third pole close to the dominant poles, yields the most error.

Now that we have seen the effect of an additional pole, let us add a zero to the second-order system.

4.8 System Responses with Zeros

In Section 4.2 we saw that the zeros of a response affect the residue, or amplitude, of a response component but do not affect the nature of the response, that is, exponential, damped sinusoid, and so on. In this section we add a real zero to a two-pole system. The zero will be added first in the left half-plane and then in the right half-plane and its effect noted and analyzed. We conclude the section by talking about pole-zero cancellation.

Starting with a two-pole system with poles at $(-1 \pm j2.828)$, we consecutively add zeros at -3, -5, and -10. The normalized results are plotted in Figure 4.24. We can see that the closer the zero is to the dominant poles, the greater its effect on the transient response. As the zero moves away from the dominant poles, the response approaches that of the two-pole system. This analysis can be reasoned via the partial fraction expansion. If we assume a group of poles and a zero far from the poles, the residue of each pole will be affected the same by the zero. Hence the relative amplitudes remain appreciably the same. For example, assume the partial fraction shown in Equation 4.67:

$$T(s) = \frac{(s + a)}{(s + b)(s + c)} = \frac{A}{s + b} + \frac{B}{s + c}$$

$$= \frac{(-b + a)/(-b + c)}{s + b} + \frac{(-c + a)/(-c + b)}{s + c} \quad (4.67)$$

If the zero is located far from the poles, then a is large compared to b and c. Therefore,

$$T(s) = a\left[\frac{1/(-b + c)}{s + b} + \frac{1/(-c + b)}{s + c}\right] = \frac{a}{(s + b)(s + c)} \quad (4.68)$$

Hence, the zero looks like a simple gain factor and does not change the relative amplitudes of the components of the response.

Figure 4.24 The Effect of Adding a Zero to a Two-Pole System

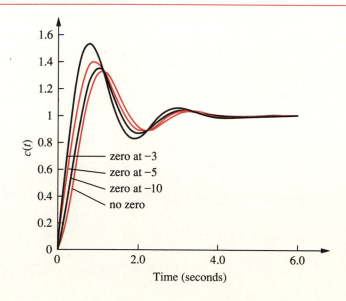

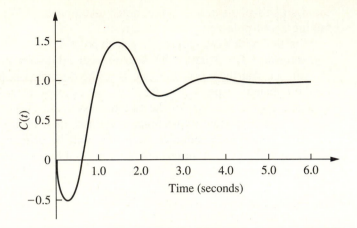

Figure 4.25 Step Response of a Nonminimum-Phase System

A zero in the right half-plane added to a two-pole system produces the interesting phenomenon shown in Figure 4.25. Notice how the response begins to turn toward the negative direction even though the final value is positive. A system that exhibits this phenomenon is known as a *nonminimum-phase* system. If a motorcycle or airplane is a nonminimum-phase system, it would initially veer left when commanded to steer right.

We conclude this section by talking about pole-zero cancellation and its effect upon our ability to make second-order approximations to a system. Assume a three-pole system with a zero as shown in Equation 4.69. If the pole $(s + p_3)$ and the zero $(s + z)$ cancel out, we are left with

$$T(s) = \frac{K(s + z)}{(s + p_3)(s^2 + as + b)} \tag{4.69}$$

as a second-order transfer function. From another perspective, if the zero at $-z$ is very close to the pole at $-p_3$ then a partial fraction expansion of Equation 4.69 will show that the residue of the exponential decay is much smaller than the amplitude of the second-order response. Let us look at an example.

Example 4.9   Evaluate the residues to determine pole-zero cancellation.

Problem   For each of the response functions in Equations 4.70 and 4.71, determine whether there is cancellation between the zero and the pole closest to the zero. For any function for which pole-zero cancellation is valid, find the approximate response.

$$C_1(s) = \frac{26.25(s + 4)}{s(s + 3.5)(s + 5)(s + 6)} \tag{4.70}$$

$$C_2(s) = \frac{26.25(s + 4)}{s(s + 4.01)(s + 5)(s + 6)} \tag{4.71}$$

Solution   The partial fraction expansion of Equation 4.70 is

$$C_1(s) = \frac{1}{s} - \frac{3.5}{s+5} + \frac{3.5}{s+6} - \frac{1}{s+3.5} \tag{4.72}$$

The residue of the pole $(s + 3.5)$ closest to the zero $(s + 4)$ is equal to 1 and is not negligible compared to the other residues. Thus a second-order step response approximation cannot be made for $C_1(s)$. The partial fraction expansion for $C_2(s)$ is

$$C_2(s) = \frac{0.87}{s} - \frac{5.3}{s+5} + \frac{4.4}{s+6} + \frac{0.033}{s+4.01} \tag{4.73}$$

The residue of the pole $(s + 4.01)$ closest to the zero is equal to 0.034, which is almost two orders of magnitude below any of the other residues. Hence, we make a second-order approximation by neglecting the response generated by the pole at -4.01. Thus,

$$c_2(s) \approx \frac{0.87}{s} - \frac{5.3}{s+5} + \frac{4.4}{s+6} \tag{4.74}$$

and the response $c_2(t)$ is approximately

$$.c_2(t) \approx 0.87 - 5.3e^{-5t} + 4.4e^{-6t} \tag{4.75}$$

Our coverage so far has dealt with finding the time response using the Laplace transform in the frequency domain. Another way to solve for the response is to use state-space techniques in the time domain. This topic is the subject of the next three sections.

* 4.9 Solving the State Equations via the Laplace Transform

In Chapter 3, systems were modeled in state space where the state-space representation consisted of a state equation and an output equation. In this section, we will use the Laplace transform to solve the state equations for the state and output vectors.

Consider the state equation

$$\dot{\mathbf{x}} = \mathbf{A}\mathbf{x} + \mathbf{B}\mathbf{u} \tag{4.76}$$

and the output equation

$$\mathbf{y} = \mathbf{C}\mathbf{x} + \mathbf{D}\mathbf{u} \tag{4.77}$$

Taking the Laplace transform of both sides of the state equation yields

$$s\mathbf{X}(s) - \mathbf{x}(0) = \mathbf{A}\mathbf{X}(s) + \mathbf{B}\mathbf{U}(s) \tag{4.78}$$

In order to separate $\mathbf{X}(s)$, replace $s\mathbf{X}(s)$ with $s\mathbf{IX}(s)$, where $\mathbf{I}$ is an $n \times n$ identity matrix, and n is the order of the system. Combining all of the $\mathbf{X}(s)$ terms, we get

$$(s\mathbf{I} - \mathbf{A})\mathbf{X}(s) = \mathbf{x}(0) + \mathbf{BU}(s) \qquad (4.79)$$

Solving for $\mathbf{X}(s)$ by premultiplying both sides of Equation 4.79 by $(s\mathbf{I} - \mathbf{A})^{-1}$, the final solution for $\mathbf{X}(s)$ is

$$\mathbf{X}(s) = (s\mathbf{I} - \mathbf{A})^{-1}\mathbf{x}(0) + (s\mathbf{I} - \mathbf{A})^{-1}\mathbf{BU}(s)$$

$$= \frac{\text{adj}(s\mathbf{I} - \mathbf{A})}{\det(s\mathbf{I} - \mathbf{A})}[\mathbf{x}(0) + \mathbf{BU}(s)] \qquad (4.80)$$

Taking the Laplace transform of the output equation yields

$$\mathbf{Y}(s) = \mathbf{CX}(s) + \mathbf{DU}(s) \qquad (4.81)$$

Eigenvalues and Transfer Function Poles

We saw that the poles of the transfer function determine the nature of the transient response of the system. Is there an equivalent quantity in the state-space representation that yields the same information? In Section 5.8 we will formally define the roots of $\det(s\mathbf{I} - \mathbf{A}) = 0$ (see the denominator of Equation 4.80) to be *eigenvalues* of the system matrix, $\mathbf{A}$.[7] Let us show that the eigenvalues are equal to the poles of the system's transfer function. Let the output, $\mathbf{Y}(s)$, and the input, $\mathbf{U}(s)$, be scalar quantities, $Y(s)$ and $U(s)$ respectively. Further, to conform to the definition of a transfer function, let $\mathbf{x}(0)$, the initial state vector, equal $\mathbf{0}$, the null vector. Substituting Equation 4.80 into Equation 4.81 and solving for the transfer function, $Y(s)/U(s)$ yields

$$\frac{Y(s)}{U(s)} = \mathbf{C}\left[\frac{\text{adj}(s\mathbf{I} - \mathbf{A})}{\det(s\mathbf{I} - \mathbf{A})}\right]\mathbf{B} + \mathbf{D}$$

$$= \frac{\mathbf{C}\,\text{adj}(s\mathbf{I} - \mathbf{A})\mathbf{B} + \mathbf{D}\det(s\mathbf{I} - \mathbf{A})}{\det(s\mathbf{I} - \mathbf{A})} \qquad (4.82)$$

The roots of the denominator of Equation 4.82 are the poles of the system. Since the denominators of Equations 4.80 and 4.82 are identical, the system poles equal the eigenvalues. Hence, if a system is represented in state-space. we can find the poles from $\det(s\mathbf{I} - \mathbf{A}) = 0$. We will be more formal about these facts when we discuss stability in Chapter 6.

The following example demonstrates the method for solving the state equations using the Laplace transform as well as finding the eigenvalues and system poles.

---

[7]Sometimes the symbol λ is used in place of the complex variable s when solving the state equations without the use of the Laplace transform. Thus it is common to see the characteristic equation also written as $\det(\lambda\mathbf{I} - \mathbf{A}) = 0$.

Example 4.10

Solve the state equations via the Laplace transform, and find the eigenvalues and poles of a system.

Problem

a. Solve the following state equation and obtain the output, $y(t) = x_1(t) + x_2(t)$.
b. Find the eigenvalues and the system poles.
Given:

$$\dot{\mathbf{x}} = \begin{bmatrix} 0 & 1 & 0 \\ 0 & 0 & 1 \\ -24 & -26 & -9 \end{bmatrix} \mathbf{x} + \begin{bmatrix} 0 \\ 0 \\ 1 \end{bmatrix} e^{-t} \tag{4.83}$$

and

$$\mathbf{x}(0) = \begin{bmatrix} 1 \\ 0 \\ 2 \end{bmatrix} \tag{4.84}$$

Solution

a. Equation 4.83 is in the same form as Equation 4.76. We will solve the problem by finding the component parts of Equation 4.80. Since

$$s\mathbf{I} = \begin{bmatrix} s & 0 & 0 \\ 0 & s & 0 \\ 0 & 0 & s \end{bmatrix} \tag{4.85}$$

then

$$(s\mathbf{I} - \mathbf{A}) = \begin{bmatrix} s & -1 & 0 \\ 0 & s & -1 \\ 24 & 26 & s+9 \end{bmatrix} \tag{4.86}$$

and

$$(s\mathbf{I} - \mathbf{A})^{-1} = \frac{\begin{bmatrix} (s^2 + 9s + 26) & (s + 9) & 1 \\ -24 & s^2 + 9s & s \\ -24s & -(26s + 24) & s^2 \end{bmatrix}}{s^3 + 9s^2 + 26s + 24} \tag{4.87}$$

Since $\mathbf{U}(s)$, the Laplace transform for e^{-t}, is $1/(s + 1)$, $\mathbf{X}(s)$ can be calculated. Rewriting Equation 4.80 as

$$\mathbf{X}(s) = (s\mathbf{I} - \mathbf{A})^{-1}[\mathbf{x}(0) + \mathbf{B}\mathbf{U}(s)] \tag{4.88}$$

and using $\mathbf{B}$ and $\mathbf{x}(0)$ from Equations 4.83 and 4.84, we get

$$X_1(s) = \frac{(s^3 + 10s^2 + 37s + 29)}{(s + 1)(s + 2)(s + 3)(s + 4)} \tag{4.89a}$$

$$X_2(s) = \frac{(2s^2 - 21s - 24)}{(s + 1)(s + 2)(s + 3)(s + 4)} \tag{4.89b}$$

$$X_3(s) = \frac{s(2s^2 - 21s - 24)}{(s + 1)(s + 2)(s + 3)(s + 4)} \tag{4.89c}$$

The output equation is found directly from the problem statement for this example. Performing the indicated addition yields

$$Y(s) = X_1(s) + X_2(s) = \begin{bmatrix} 1 & 1 & 0 \end{bmatrix} \begin{bmatrix} X_1(s) \\ X_2(s) \\ X_3(s) \end{bmatrix} \tag{4.90}$$

or

$$
\begin{aligned}
Y(s) &= \frac{(s^3 + 12s^2 + 16s + 5)}{(s + 1)(s + 2)(s + 3)(s + 4)} \\
&= \frac{-6.5}{s + 2} + \frac{19}{s + 3} - \frac{11.5}{s + 4}
\end{aligned}
\tag{4.91}
$$

where the pole at -1 canceled a zero at -1. Taking the inverse Laplace transform, we get

$$y(t) = -6.5e^{-2t} + 19e^{-3t} - 11.5e^{-4t} \tag{4.92}$$

b. The denominator of Equation 4.87, which is $\det(s\mathbf{I} - \mathbf{A})$, is the denominator of the system shown in Figure 3.7. Thus $\det(s\mathbf{I} - \mathbf{A}) = 0$ furnishes both the poles of the system and the eigenvalues $-2, -3, -4$.

We now look at another technique for solving the state equations. Rather than using the Laplace transform, we solve the equations directly in the time domain. We will find that the final solution consists of two parts that are different from the forced and natural responses found during the Laplace transform solution.

* 4.10 Solving the State Equations Directly in the Time Domain

Another method for solving the state equations is closely allied to the classical solution of differential equations in the time domain. First, assume a homogeneous state equation of the form

$$\dot{\mathbf{x}}(t) = \mathbf{A}\mathbf{x}(t) \tag{4.93}$$

Since we want to solve for $\mathbf{x}$, we will assume a series solution, just as we did in elementary scalar differential equations. Thus,

$$\mathbf{x}(t) = \mathbf{b_0} + \mathbf{b_1}t + \mathbf{b_2}t^2 + \cdots + \mathbf{b_k}t^k + \mathbf{b_{k+1}}t^{k+1} + \cdots \tag{4.94}$$

Substituting Equation 4.94 into 4.93 we get

$$
\begin{aligned}
\mathbf{b_1} &+ 2\mathbf{b_2}t + \cdots + k\mathbf{b_k}t^{k-1} + (k + 1)\mathbf{b_{k+1}}t^k + \cdots \\
&= \mathbf{A}(\mathbf{b_0} + \mathbf{b_1}t + \mathbf{b_2}t^2 + \cdots + \mathbf{b_k}t^k + \mathbf{b_{k+1}}t^{k+1} + \cdots)
\end{aligned}
\tag{4.95}
$$

Equating like coefficients yields

$$\mathbf{b_1} = \mathbf{Ab_0} \tag{4.96a}$$

$$\mathbf{b_2} = \frac{1}{2}\mathbf{Ab_1} = \frac{1}{2}\mathbf{A^2b_0} \tag{4.96b}$$

$$\vdots$$

$$\mathbf{b_k} = \frac{1}{k!}\mathbf{A^k b_0} \tag{4.96c}$$

$$\mathbf{b_{k+1}} = \frac{1}{(k+1)!}\mathbf{A^{k+1}b_0} \tag{4.96d}$$

$$\vdots$$

Substituting these values into Equation 4.94 yields

$$\mathbf{x}(t) = \mathbf{b_0} + \mathbf{Ab_0}t + \frac{1}{2}\mathbf{A^2b_0}t^2 + \cdots + \frac{1}{k!}\mathbf{A^k b_0}t^k$$

$$+ \frac{1}{(k+1)!}\mathbf{A^{k+1}b_0}t^{k+1} + \cdots$$

$$= \left(\mathbf{I} + \mathbf{A}t + \frac{1}{2}\mathbf{A^2}t^2 + \cdots + \frac{1}{k!}\mathbf{A^k}t^k + \frac{1}{(k+1)!}\mathbf{A^{k+1}}t^{k+1} + \cdots\right)\mathbf{b_0} \tag{4.97}$$

But, from Equation 4.94,

$$\mathbf{x}(0) = \mathbf{b_0} \tag{4.98}$$

Therefore,

$$\mathbf{x}(t) = \left(\mathbf{I} + \mathbf{A}t + \frac{1}{2}\mathbf{A^2}t^2 + \cdots + \frac{1}{k!}\mathbf{A^k}t^k + \frac{1}{(k+1)!}\mathbf{A^{k+1}}t^{k+1} + \cdots\right)\mathbf{x}(0) \tag{4.99}$$

Let

$$e^{\mathbf{A}t} = \left(\mathbf{I} + \mathbf{A}t + \frac{1}{2}\mathbf{A^2}t^2 + \cdots + \frac{1}{k!}\mathbf{A^k}t^k + \frac{1}{(k+1)!}\mathbf{A^{k+1}}t^{k+1} + \cdots\right) \tag{4.100}$$

where $e^{\mathbf{A}t}$ is simply a notation for the matrix formed by the right-hand side of Equation 4.100. We use this definition because the right-hand side of Equation 4.100 resembles a power series expansion of e^{at}, or

$$e^{at} = \left(1 + at + \frac{1}{2}a^2t^2 + \cdots + \frac{1}{k!}a^k t^k + \frac{1}{(k+1)!}a^{k+1}t^{k+1} + \cdots\right) \tag{4.101}$$

Using Equation 4.99, we have

$$\mathbf{x}(t) = e^{\mathbf{A}t}\mathbf{x}(0) \tag{4.102}$$

We give a special name to $e^{\mathbf{A}t}$: it is called the *state-transition matrix*, since it performs a transformation on $\mathbf{x}(0)$, taking $\mathbf{x}$ from the initial state, $\mathbf{x}(0)$, to the state $\mathbf{x}(t)$ at any time, t. The symbol, $\boldsymbol{\Phi}(t)$, is used to denote $e^{\mathbf{A}t}$. Thus,

$$\boldsymbol{\Phi}(t) = e^{\mathbf{A}t} \tag{4.103}$$

and

$$\mathbf{x}(t) = \boldsymbol{\Phi}(t)\mathbf{x}(0) \tag{4.104}$$

There are some properties of $\boldsymbol{\Phi}(t)$ that we will use later when we actually solve for $\mathbf{x}(t)$. From Equation 4.104,

$$\mathbf{x}(0) = \boldsymbol{\Phi}(0)\mathbf{x}(0) \tag{4.105}$$

Hence, the first property of $\boldsymbol{\Phi}(t)$ is

$$\boldsymbol{\Phi}(0) = \mathbf{I} \tag{4.106}$$

where $\mathbf{I}$ is the identity matrix. Also, differentiating Equation 4.104 and setting this equal to Equation 4.93 yields

$$\dot{\mathbf{x}}(t) = \dot{\boldsymbol{\Phi}}(t)\mathbf{x}(0) = \mathbf{A}\mathbf{x}(t) \tag{4.107}$$

which, at $t = 0$, yields

$$\dot{\boldsymbol{\Phi}}(0)\mathbf{x}(0) = \mathbf{A}\mathbf{x}(0). \tag{4.108}$$

Thus, the second property of $\boldsymbol{\Phi}(t)$ follows from Equation 4.108:

$$\dot{\boldsymbol{\Phi}}(0) = \mathbf{A} \tag{4.109}$$

In summary, then, the solution to the homogeneous, or unforced, system is

$$\mathbf{x}(t) = \boldsymbol{\Phi}(t)\mathbf{x}(0) \tag{4.110}$$

where

$$\boldsymbol{\Phi}(0) = \mathbf{I} \tag{4.111}$$

and

$$\dot{\boldsymbol{\Phi}}(0) = \mathbf{A} \tag{4.112}$$

Let us now solve the forced, or nonhomogeneous, problem. Given the forced state-equation

$$\dot{\mathbf{x}}(t) = \mathbf{A}\mathbf{x}(t) + \mathbf{B}\mathbf{u}(t) \tag{4.113}$$

rearrange and multiply both sides by $e^{-\mathbf{A}t}$:

$$e^{-\mathbf{A}t}[\dot{\mathbf{x}}(t) - \mathbf{A}\mathbf{x}(t)] = e^{-\mathbf{A}t}\mathbf{B}\mathbf{u}(t) \tag{4.114}$$

Realizing that the left-hand side is equal to the derivative of the product $e^{-\mathbf{A}t}\mathbf{x}(t)$, we obtain

$$\frac{d}{dt}[e^{-\mathbf{A}t}\mathbf{x}(t)] = e^{-\mathbf{A}t}\mathbf{B}\mathbf{u}(t) \tag{4.115}$$

Integrating both sides yields

$$[e^{-\mathbf{A}t}\mathbf{x}(t)]\Big|_0^t = e^{-\mathbf{A}t}\mathbf{x}(t) - \mathbf{x}(0) = \int_0^t e^{-\mathbf{A}\tau}\mathbf{B}\mathbf{u}(\tau)\,d\tau \tag{4.116}$$

since $e^{-\mathbf{A}t}$ evaluated at $t = 0$ is the identity matrix (from Equation 4.100). Solving for $\mathbf{x}(t)$ in Equation 4.116 we obtain

$$\mathbf{x}(t) = e^{\mathbf{A}t}\mathbf{x}(0) + \int_0^t e^{\mathbf{A}(t-\tau)}\mathbf{B}\mathbf{u}(\tau)\,d\tau$$

$$= \mathbf{\Phi}(t)\mathbf{x}(0) + \int_0^t \mathbf{\Phi}(t-\tau)\mathbf{B}\mathbf{u}(\tau)\,d\tau \qquad (4.117)$$

where $\mathbf{\Phi}(t) = e^{\mathbf{A}t}$ by definition.

The integral shown in Equation 4.117 is called the *convolution integral*. Notice that the first term on the right-hand side of the equation is the response due to the initial state vector. Notice also that this term is the only term dependent upon the initial state vector and not the input. We call this part of the response the *zero-input response*, since this is the total response if the input is zero. The second term, the convolution integral, is only dependent upon the input, $\mathbf{u}$, and the input coupling matrix $\mathbf{B}$, but not the initial state vector. We call this part of the response the *zero-state response*, since this part of the response is the total response if the initial state vector is zero. Thus, there is a partitioning of the response different from the forced/natural response we have seen when solving differential equations where the amplitudes of both the forced and natural responses depend upon the initial conditions.

Before proceeding with an example, let us examine the form that the elements of $\mathbf{\Phi}(t)$ take for linear, time-invariant systems. The first term of Equation 4.80, the Laplace transform of the response for unforced systems, is the transform of Equation 4.110. Thus,

$$\mathcal{L}[\mathbf{x}(t)] = \mathcal{L}[\mathbf{\Phi}(t)\mathbf{x}(0)] = (s\mathbf{I} - \mathbf{A})^{-1}\mathbf{x}(0) \qquad (4.118)$$

from which we can see that $(s\mathbf{I} - \mathbf{A})^{-1}$ is the Laplace transform of the state-transition matrix, $\mathbf{\Phi}(t)$. We have already seen that the denominator of $(s\mathbf{I} - \mathbf{A})^{-1}$ is a polynomial in s whose roots are the system poles. This polynomial is found from the equation $\det(s\mathbf{I} - \mathbf{A}) = 0$. Since $\mathcal{L}^{-1}\left[(s\mathbf{I} - \mathbf{A})^{-1}\right] = \mathcal{L}^{-1}\left[\dfrac{\text{adj}(s\mathbf{I} - \mathbf{A})}{\det(s\mathbf{I} - \mathbf{A})}\right] = $ $\mathbf{\Phi}(t)$, each term of $\mathbf{\Phi}(t)$ would be the sum of exponentials generated by the system's poles.

Let us summarize the concepts with a numerical example.

Example 4.11   Solve the state equations directly in the time domain.

Problem   For the state equation and initial state vector shown in Equations 4.119, find the state-transition matrix and then solve for $\mathbf{x}(t)$.

$$\dot{\mathbf{x}}(t) = \begin{bmatrix} 0 & 1 \\ -8 & -6 \end{bmatrix}\mathbf{x}(t) + \begin{bmatrix} 0 \\ 1 \end{bmatrix}u(t) \qquad (4.119a)$$

$$\mathbf{x}(0) = \begin{bmatrix} 1 \\ 0 \end{bmatrix} \qquad (4.119b)$$

Solution   Since the state equation is in the form

$$\dot{\mathbf{x}}(t) = \mathbf{A}\mathbf{x}(t) + \mathbf{B}u(t) \tag{4.120}$$

find the eigenvalues using det $(s\mathbf{I} - \mathbf{A}) = 0$. Hence,

$$s^2 + 6s + 8 = 0 \tag{4.121}$$

from which $s_1 = -2$ and $s_2 = -4$. Since each term of the state-transition matrix is the sum of responses generated by the poles (eigenvalues), we assume a state-transition matrix of the form

$$\mathbf{\Phi}(t) = \begin{bmatrix} (K_1 e^{-2t} + K_2 e^{-4t}) & (K_3 e^{-2t} + K_4 e^{-4t}) \\ (K_5 e^{-2t} + K_6 e^{-4t}) & (K_7 e^{-2t} + K_8 e^{-4t}) \end{bmatrix} \tag{4.122}$$

In order to find the values of the constants, we make use of the properties of the state-transition matrix. Since

$$\mathbf{\Phi}(0) = \mathbf{I} \tag{4.123}$$

$$K_1 + K_2 = 1 \tag{4.124a}$$

$$K_3 + K_4 = 0 \tag{4.124b}$$

$$K_5 + K_6 = 0 \tag{4.124c}$$

$$K_7 + K_8 = 1 \tag{4.124d}$$

and since

$$\dot{\mathbf{\Phi}}(0) = \mathbf{A} \tag{4.125}$$

then

$$-2K_1 - 4K_2 = 0 \tag{4.126a}$$

$$-2K_3 - 4K_4 = 1 \tag{4.126b}$$

$$-2K_5 - 4K_6 = -8 \tag{4.126c}$$

$$-2K_7 - 4K_8 = -6 \tag{4.126d}$$

The constants are solved by taking two simultaneous equations four times. For example, Equation 4.124a can be solved simultaneously with Equation 4.126a to yield the values of K_1 and K_2. Proceeding in a similar manner, all of the constants can be found. Therefore,

$$\mathbf{\Phi}(t) = \begin{bmatrix} (2e^{-2t} - e^{-4t}) & \left(\frac{1}{2}e^{-2t} - \frac{1}{2}e^{-4t}\right) \\ (-4e^{-2t} + 4e^{-4t}) & (-e^{-2t} + 2e^{-4t}) \end{bmatrix} \tag{4.127}$$

Also,

$$\mathbf{\Phi}(t - \tau)\mathbf{B} = \begin{bmatrix} \left(\frac{1}{2}e^{-2(t-\tau)} - \frac{1}{2}e^{-4(t-\tau)}\right) \\ (-e^{-2(t-\tau)} + 2e^{-4(t-\tau)}) \end{bmatrix} \tag{4.128}$$

Hence the first term of Equation 4.117 is

$$\Phi(t)\mathbf{x}(0) = \begin{bmatrix} (2e^{-2t} - e^{-4t}) \\ (-4e^{-2t} + 4e^{-4t}) \end{bmatrix} \tag{4.129}$$

The last term of Equation 4.117 is

$$\int_0^t \Phi(t - \tau)\mathbf{B}\mathbf{u}(\tau)\,d\tau = \begin{bmatrix} \dfrac{1}{2}e^{-2t}\int_0^t e^{2\tau}d\tau - \dfrac{1}{2}e^{-4t}\int_0^t e^{4\tau}d\tau \\ -e^{-2t}\int_0^t e^{2\tau}d\tau + 2e^{-4t}\int_0^t e^{4\tau}d\tau \end{bmatrix}$$

$$= \begin{bmatrix} \dfrac{1}{8} - \dfrac{1}{4}e^{-2t} + \dfrac{1}{8}e^{-4t} \\ \dfrac{1}{2}e^{-2t} - \dfrac{1}{2}e^{-4t} \end{bmatrix} \tag{4.130}$$

The final result is found by adding Equations 4.129 and 4.130. Hence,

$$\mathbf{x}(t) = \Phi(t)\mathbf{x}(0) + \int_0^t \Phi(t - \tau)\mathbf{B}\mathbf{u}(\tau)\,d\tau = \begin{bmatrix} \dfrac{1}{8} + \dfrac{7}{4}e^{-2t} - \dfrac{7}{8}e^{-4t} \\ -\dfrac{7}{2}e^{-2t} + \dfrac{7}{2}e^{-4t} \end{bmatrix} \tag{4.131}$$

Another method can be used to find the state-transition matrix in the previous example. It was pointed out that the state-transition matrix is the inverse Laplace transform of $(s\mathbf{I} - \mathbf{A})^{-1}$. We make use of this fact to find the state-transition matrix in the next example.

Example 4.12   Find the state-transition matrix from the inverse Laplace transform of $(s\mathbf{I} - \mathbf{A})^{-1}$.

Problem   Find the state-transition matrix of Example 4.11, using $(s\mathbf{I} - \mathbf{A})^{-1}$.

Solution   We use the fact that $\Phi(t)$ is the inverse Laplace transform of $(s\mathbf{I} - \mathbf{A})^{-1}$. Thus, first find $(s\mathbf{I} - \mathbf{A})$ as

$$(s\mathbf{I} - \mathbf{A}) = \begin{bmatrix} s & -1 \\ 8 & (s + 6) \end{bmatrix} \tag{4.132}$$

from which

$$(s\mathbf{I} - \mathbf{A})^{-1} = \frac{\begin{bmatrix} s + 6 & 1 \\ -8 & s \end{bmatrix}}{s^2 + 6s + 8} = \begin{bmatrix} \dfrac{s + 6}{s^2 + 6s + 8} & \dfrac{1}{s^2 + 6s + 8} \\ \dfrac{-8}{s^2 + 6s + 8} & \dfrac{s}{s^2 + 6s + 8} \end{bmatrix} \tag{4.133}$$

Expanding each term in the matrix on the right by partial fractions yields

$$(s\mathbf{I} - \mathbf{A})^{-1} = \begin{bmatrix} \left(\dfrac{2}{s+2} - \dfrac{1}{s+4}\right) & \left(\dfrac{1/2}{s+2} - \dfrac{1/2}{s+4}\right) \\ \left(\dfrac{-4}{s+2} + \dfrac{4}{s+4}\right) & \left(\dfrac{-1}{s+2} + \dfrac{2}{s+4}\right) \end{bmatrix} \tag{4.134}$$

Finally, taking the inverse Laplace transform of each term, we obtain

$$\mathbf{\Phi}(t) = \begin{bmatrix} (2e^{-2t} - e^{-4t}) & \left(\dfrac{1}{2}e^{-2t} - \dfrac{1}{2}e^{-4t}\right) \\ (-4e^{-2t} + 4e^{-4t}) & (-e^{-2t} + 2e^{-4t}) \end{bmatrix} \tag{4.135}$$

In the next section we show how to use the digital computer to simulate a system represented in state space and produce a plot of the output.

* 4.11 Numerical Solution of the State Equations for Computer Simulation

One advantage of state equations is that we can use this representation to simulate control systems on the digital computer. This section is devoted to demonstrating this concept. Consider the system represented in state space by Equations 4.136.

$$\begin{bmatrix} \dot{x}_1 \\ \dot{x}_2 \end{bmatrix} = \begin{bmatrix} 0 & 1 \\ -2 & -3 \end{bmatrix} \begin{bmatrix} x_1 \\ x_2 \end{bmatrix} + \begin{bmatrix} 0 \\ 1 \end{bmatrix} u(t) \tag{4.136a}$$

$$y(t) = \begin{bmatrix} 2 & 3 \end{bmatrix} \begin{bmatrix} x_1 \\ x_2 \end{bmatrix} \tag{4.136b}$$

$$\begin{bmatrix} x_1(0) \\ x_2(0) \end{bmatrix} = \begin{bmatrix} 1 \\ -2 \end{bmatrix} \tag{4.136c}$$

This system is represented in phase-variable form and has a unit step input, $u(t)$. We are about to formulate a solution for the system output, $y(t)$, by numerically integrating the differential equation on the digital computer. We will use a method called *Euler's approximation*, where the area to be integrated is approximated as a rectangle. The solution obtained on the computer is an actual time waveform plot rather than the closed-form expression we arrived at via the Laplace transform.

Writing the state equations explicitly, we have

$$\frac{dx_1}{dt} = x_2 \tag{4.137a}$$

$$\frac{dx_2}{dt} = -2x_1 - 3x_2 + 1 \tag{4.137b}$$

If we approximate dx by Δx and t by Δt and multiply through by Δt,[8]

$$\Delta x_1 = x_2 \Delta t \tag{4.138a}$$

$$\Delta x_2 = (-2x_1 - 3x_2 + 1)\Delta t \tag{4.138b}$$

We can say that the value at the next interval for either state variable is approximately the current value plus the change. Thus,

$$x_1(t + \Delta t) = x_1(t) + \Delta x_1 \tag{4.139a}$$

$$x_2(t + \Delta t) = x_2(t) + \Delta x_2 \tag{4.139b}$$

Finally, from the output equation, Equation 4.136b, $y(t)$ at the next time interval, $y(t + \Delta t)$, is

$$y(t + \Delta t) = 2x_1(t + \Delta t) + 3x_2(t + \Delta t) \tag{4.140}$$

Let us see how this would work on the digital computer. From the problem statement, x_1 and x_2 begin at $t = 0$ with values 1 and -2 respectively. If we assume a Δt interval of 0.1 seconds, the change in x_1 and x_2 from 0 to 0.1 seconds is found from Equations 4.138 to be

$$\Delta x_1 = x_2(0)\Delta t = -0.2 \tag{4.141a}$$

$$\Delta x_2 = [-2x_1(0) - 3x_2(0) + 1]\Delta t = 0.5 \tag{4.141b}$$

from which the state variables at $t = 0.1$ are found from Equations 4.139 to be

$$x_1(0.1) = x_1(0) + \Delta x_1 = 0.8 \tag{4.142a}$$

$$x_2(0.1) = x_2(0) + \Delta x_2 = -1.5 \tag{4.142b}$$

Finally, the output at $t = 0.1$ is calculated from Equation 4.140 to be

$$y(0.1) = 2x_1(0.1) + 3x_2(0.1) = -2.9 \tag{4.143}$$

The values of the state variables at $t = 0.1$ seconds are used to calculate the values of the state variables and the output at the next interval of time, $t = 0.2$ seconds. Once again, the changes in x_1 and x_2 are

$$\Delta x_1 = x_2(0.1)\Delta t = -0.15 \tag{4.144a}$$

$$\Delta x_2 = [-2x_1(0.1) - 3x_2(0.1) + 1]\Delta t = 0.39 \tag{4.144b}$$

from which the state variables at $t = 0.2$ are found to be

$$x_1(0.2) = x_1(0.1) + \Delta x_1 = 0.65 \tag{4.145a}$$

$$x_2(0.2) = x_2(0.1) + \Delta x_2 = -1.11 \tag{4.145b}$$

---

[8]Δt is selected to be small and, initially, at least an order of magnitude less than the system's time constants. In order to determine empirically how small Δt should be, the value of Δt can be successively reduced after each response has been calculated by the computer until the difference between the current response and the previous response is negligible. If Δt is too large, then error results from inaccurately representing the area under the state variable curve. If Δt is too small, then round-off error will accumulate during the computation because of the numerous calculations.

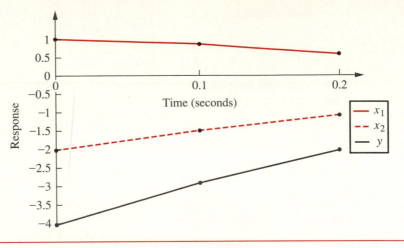

Figure 4.26 Plot of the State-Variables and the Output for the System of Equations 4.136

Finally, the output at $t = 0.2$ is calculated as

$$y(0.2) = 2x_1(0.2) + 3x_2(0.2) = -2.03 \qquad (4.146)$$

The results are summarized in Figure 4.26. Continuing in like manner until $t = t_f$, the maximum desired time, the response for $0 \leq t \leq t_f$ can be obtained.

The Microsoft® QuickBASIC program listing that implements the example just presented can be found in Appendix C. This program will allow the user to input any linear system of state equations and get the response for a step input.

4.12 Chapter-Objective Demonstration Problems

In this chapter we made use of the transfer functions derived in Chapter 2 and the state equations derived in Chapter 3 to obtain the output response of an open-loop system. We also showed the importance of the poles of a system in determining the transient response. This chapter-objective problem uses these concepts to analyze the antenna azimuth control system. The open-loop function that we will deal with consists of a power amplifier and motor with load. In Chapter 5 we will see that the equivalent single transfer function is simply the product of individual transfer functions of the power amplifier and the motor with its load. We make use of this fact now in order to make the problem meaningful.

Example 4.13   For an open-loop system, predict the open-loop response; find the damping ratio and natural frequency of the open-loop system; derive the open-loop response, using transfer functions; and simulate the open-loop system.

Problem   For the schematic of the azimuth position control system shown in Figure 2.36, assume an open-loop system (feedback path disconnected).

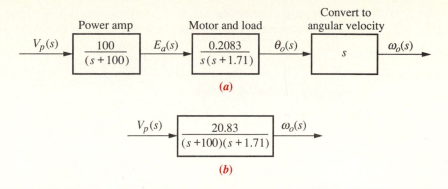

Figure 4.27 Antenna Azimuth Position Control System for Angular Velocity: (**a**) Forward Path; (**b**) Equivalent Forward Path

> **a.** Predict, by inspection, the form of the open-loop angular velocity response of the load to a step-voltage input to the power amplifier.
> **b.** Find the damping ratio and natural frequency of the open-loop system.
> **c.** Derive the complete analytical expression for the open-loop angular velocity response of the load to a step-voltage input to the power amplifier, using transfer functions.
> ***d.** Obtain the open-loop state equations and simulate the open-loop system on a digital computer.

Solution The transfer function of the power amplifier is given in Figure 2.36 and the transfer function of the motor and its load was derived in Section 2.11. The two subsystems are shown interconnected in Figure 4.27(a). Differentiating the angular position of the motor and load output by multiplying by s, we obtain the output angular velocity, ω_o, as shown in Figure 4.27(a). The equivalent transfer function representing the three blocks in Figure 4.27(a) is the product of the individual transfer functions and is shown in Figure 4.27(b).[9]

   a. Using the transfer function shown in Figure 4.27(b), we can predict the nature of the step response. The step response consists of the steady-state response generated by the step input and the transient response, which is the sum of two exponentials generated by each pole of the transfer function. Hence the form of the response is

$$\omega_o(t) = A + Be^{-100t} + Ce^{-1.71t} \qquad (4.147)$$

   b. The damping ratio and natural frequency of the open-loop system can be found by expanding the denominator of the transfer function. Since the open-loop transfer function is

$$G(s) = \frac{20.83}{s^2 + 101.71s + 171} \qquad (4.148)$$

$\omega_n = \sqrt{171} = 13.08$ and $\zeta = 3.89$.

---

[9]This relationship will be derived in Chapter 5.

c. In order to derive the angular velocity response to a step input, we multiply the transfer function of Equation 4.148 by a step input, $1/s$, and obtain

$$\omega_o(s) = \frac{20.83}{s(s + 100)(s + 1.71)} \qquad (4.149)$$

Expanding into partial fractions, we get

$$\omega_o(s) = \frac{0.122}{s} + \frac{2.12 \times 10^{-3}}{s + 100} - \frac{0.124}{s + 1.71} \qquad (4.150)$$

Transforming to the time domain yields

$$\omega_o(t) = 0.122 + (2.12 \times 10^{-3})e^{-100t} - 0.124e^{-1.71t} \qquad (4.151)$$

*d. In order to simulate the open-loop system on the digital computer, we first convert the transfer function into the state-space representation. Using Equation 4.148, we have

$$\frac{\omega_o(s)}{V_p(s)} = \frac{20.83}{s^2 + 101.71s + 171} \qquad (4.152)$$

Figure 4.28 Simulation of Open-Loop System for the Antenna Azimuth Control System: (*a*) Program Input; (*b*) Angular Velocity Step Response

```
Do you want a hard copy of parameters?(y,n) > n
Enter System Parameters
Enter System Order > 2
Enter System Matrix
Enter a( 1 , 1 ) > 0
Enter a( 1 , 2 ) > 1
Enter a( 2 , 1 ) > -171
Enter a( 2 , 2 ) > -101.71
Enter Input Coupling Matrix
Enter b( 1 ) > 0
Enter b( 2 ) > 20.83                          *--------------------Y(0)= 0
Enter Output Matrix                           :        *          Y(.5 )= 6.949493E-02
Enter c( 1 ) > 1                              :           *       Y( 1 )= 9.972681E-02
Enter c( 2 ) > 0                              :            *      Y(1.5 )= .1124892
Do you want a hard copy of plot?(y,n) > n :               *      Y( 2 )= .1178769
Enter Initial Conditions                      :             *     Y( 2.5 )= .1201513
Enter X( 1 ) > 0                              :             *     Y( 3 )= .1211114
Enter X( 2 ) > 0                              :             *     Y( 3.5 )= .1215167
Enter Plot Parameters                         :             *     Y( 4 )= .1216878
Iteration Interval > .01                      :             *     Y( 4.5 )= .1217601
Plot Interval > .5                            :             *     Y( 5 )= .1217906
Maximum Time for Analysis >  5        t axis
Maximum Amplitude for Plot > .2
```

        (*a*)   (*b*)

Cross-multiplying and taking the inverse Laplace transform with zero initial conditions, we have

$$\ddot{\omega}_o + 101.71\dot{\omega}_o + 171\omega_o = 20.83v_p \qquad (4.153)$$

Defining the phase-variables as

$$x_1 = \omega_o \qquad (4.154a)$$

$$x_2 = \dot{\omega}_o \qquad (4.154b)$$

and using Equation 4.153, the state equations are written as

$$\dot{x}_1 = x_2 \qquad (4.155a)$$

$$\dot{x}_2 = -171x_1 - 101.71x_2 + 20.83v_p \qquad (4.155b)$$

where $v_p = 1$, a unit step. Since $x_1 = \omega_o$ is the output, the output equation is

$$y = x_1 \qquad (4.156)$$

Equations 4.155 and 4.156 can be programmed using the methods of Section 4.11. The parameters used for the program in Appendix C are shown in Figure 4.28(*a*), and the output response is shown in Figure 4.28(*b*).

Let us look at another example that uses a second-order underdamped system.

Example 4.14 Find the roll response of a ship at sea.

Problem Consider the ship shown in Figure 4.29. A disturbance can cause the ship to roll about the axis shown. If the transfer function relating the roll-angle output to the disturbance-torque input is

$$\frac{\theta(s)}{T_D(s)} = \frac{2.25}{s^2 + 0.45s + 2.25} \qquad (4.157)$$

Figure 4.29 A Ship at Sea, Showing Roll Axis

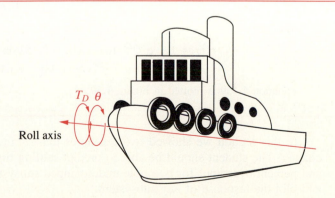

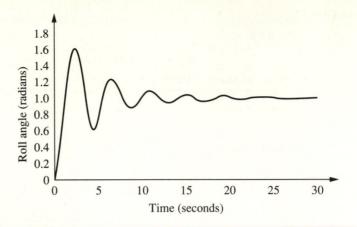

Figure 4.30 Roll-Angle Response of a Ship at Sea to a
Step Input in Torque

find the following:
  a. The natural frequency, damping ratio, peak time, settling time, rise time, and
  percent overshoot
  b. The output response to a step disturbance in torque

Solution
  a. From Equation 4.21, the natural frequency $\omega_n = \sqrt{2.25} = 1.5 \text{rad/s}$, and
$2\zeta\omega_n = 0.45$; hence, $\zeta = 0.15$. The poles of the transfer function, s_1 and s_2, are
$-0.225 \pm j1.483$. Hence, the peak time $T_p = \pi/\omega_d = \pi/1.483 = 2.12$ seconds.
The settling time $T_s = 4/\sigma_d = 4/0.225 = 17.78$ seconds. Using Figure 4.15 and
ω_n, the rise time is approximately 0.77 seconds. The percent overshoot is calculated
as $\%OS = e^{-(\zeta\pi/\sqrt{1-\zeta^2})} \times 100 = 62.1\%$.
  b. To find the step response to a unit step disturbance in torque, we find the inverse
Laplace transform of $C(s) = \dfrac{2.25}{s(s^2 + 0.45s + 2.25)}$, which is the transfer function
times the input transform, $1/s$. Expanding by partial fractions,

$$C(s) = \frac{2.25}{s(s^2 + 0.45s + 2.25)} = \frac{1}{s} - \frac{(s + 0.225) + 0.152(1.48)}{(s + 0.225)^2 + 1.48^2} \qquad (4.158)$$

from which

$$c(t) = 1 - e^{-0.225t}(\cos 1.48t + 0.152 \sin 1.48t)$$
$$= 1 - 1.01e^{-0.225t} \cos (1.48t - 8.64°) \qquad (4.159)$$

Equation 4.159 is plotted in Figure 4.30.

In this section we showed two examples that summarize the essence of this
chapter. The student should be able to predict settling time, peak time, rise time,
and percent overshoot for two-pole underdamped subsystems as well as calculate
and plot the response of any subsystem.

4.13 Summary

In this chapter, we took the system models developed in Chapters 2 and 3 and found the output response for a given input, usually a step. The step response yields a clear picture of the system's transient response. We performed this analysis for two types of systems, first-order and second-order, which are representative of many physical systems. We then formalized our findings and arrived at numerical specifications describing the responses.

For first-order systems having a single pole on the real axis, the specification of transient response that we derived was the time constant. The time constant is the reciprocal of the real-axis pole location. This specification gives us an indication of the speed of the transient response. In particular, the time constant is the time for the step response to reach 63% of its final value.

Second-order systems are more complex. Depending upon the values of system components, a second-order system can exhibit four kinds of behavior: (1) overdamped, (2) underdamped, (3) oscillatory, and (4) critically damped. We found that the poles of the input generate the forced response, whereas the system poles generate the transient response. If the system poles are real, then the system exhibits overdamped behavior. These exponential responses have time constants equal to the reciprocals of the pole locations. Purely imaginary poles yield undamped sinusoidal oscillations whose radian frequency is equal to the magnitude of the imaginary pole. Systems with complex poles display underdamped responses. The real part of the complex pole dictates the exponential decay envelope, and the imaginary part of the pole dictates the sinusoidal radian frequency. The exponential decay envelope has a time constant equal to the reciprocal of the real part of the pole, and the sinusoid has a radian frequency equal to the imaginary part of the pole.

Specifications were developed for all second-order cases. These specifications are called the *damping ratio*, ζ, and *natural frequency*, ω_n. The damping ratio gives us an idea about the nature of the transient response. It gives us a feel for the amount of overshoot and oscillation that the response undergoes, regardless of time scaling. We found that the value of ζ determines the form of the second-order natural response. In particular: if $\zeta = 0$, the response is oscillatory; if $\zeta < 1$, the response is underdamped; if $\zeta = 1$, the response is critically damped; if $\zeta > 1$, the response is overdamped. The natural frequency is the frequency of oscillation if all of the damping is removed. This quantity acts as a scaling factor for the response, and its value gives us an indication of the speed of the response.

For the underdamped case, we defined several transient response specifications. These specifications are percent overshoot, $\%OS$, peak time, T_p, settling time, T_s, and rise time, T_r. The peak time is inversely proportional to the imaginary part of the complex pole. Thus, horizontal lines on the s-plane are lines of constant peak time. We found that percent overshoot is only a function of damping ratio: radial lines are lines of constant percent overshoot. Finally, we found settling time to be inversely proportional to the real part of the complex pole. Hence, vertical lines on the s-plane are lines of constant settling time.

We found that peak time, percent overshoot, and settling time are related to pole location. Thus, we can design transient responses by relating a desired response to a pole location and then relating that pole location to a transfer function and the system's components.

In this chapter we also evaluated the output response using the state-space approach. The response found in this way was separated into the zero-input response, and the zero-state response, whereas the frequency response method yielded a total response divided into the natural response and the forced response. Finally, state-space methods were used to formulate digital computer simulations for physical systems.

REVIEW QUESTIONS

1. Name the performance specification for first-order systems.
2. What does the performance specification for a first-order system tell us?
3. In a system with an input and an output, what poles generate the steady-state response?
4. In a system with an input and an output, what poles generate the transient response?
5. The imaginary part of a pole generates what part of a response?
6. The real part of a pole generates what part of a response?
7. What is the difference between the natural frequency and the damped frequency of oscillation?
8. If a pole is moved with a constant imaginary part, what will the responses have in common?
9. If a pole is moved with a constant real part, what will the responses have in common?
10. If a pole is moved along a radial line extending from the origin, what will the responses have in common?
11. List five specifications for a second-order underdamped system.
12. For Question 11, how many specifications completely determine the response?
13. What pole locations characterize (1) the underdamped system, (2) the overdamped system, (3) the critically damped system?
14. Name two conditions under which the response generated by a pole can be neglected.
15. How can you justify pole-zero cancellation?
*16. Does the solution of the state equation yield the output response of the system? Explain.
*17. What is the relationship between $(s\mathbf{I} - \mathbf{A})$, which appeared during the Laplace transformation solution of the state equations, and the state-transition matrix, which appeared during the classical solution of the state equation?

*18. Name a major advantage of using time-domain techniques for the solution of the response.

*19. Name a major advantage of using frequency-domain techniques for the solution of the response.

*20. What three pieces of information must be given in order to solve for the output response of a system using state-space techniques?

*21. How can the poles of a system be found from the state equations?

PROBLEMS

1. Derive the output responses for all parts of Figure 4.6.

2. Find the output response, $c(t)$, for each of the systems shown in Figure P4.1. Also find the time constant, rise time, and settling time for each case.

Figure P4.1

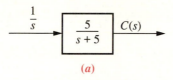

(a)

(b)

3. Find the capacitor voltage in the network shown in Figure P4.2 if the switch closes at $t = 0$. Assume zero initial conditions. Also find the time constant, rise time, and settling time for the capacitor voltage.

Figure P4.2

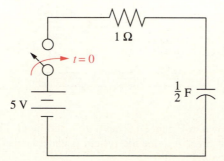

4. For each of the transfer functions shown below, find the location of the poles and zeros, plot them on the s-plane, and then write an expression for the general form of the step response without solving for the inverse Laplace transform. State the nature of each response (i.e., overdamped, underdamped, etc.).

a. $T(s) = \dfrac{2}{s + 2}$

b. $T(s) = \dfrac{5}{(s + 3)(s + 6)}$

c. $T(s) = \dfrac{10(s + 7)}{(s + 10)(s + 20)}$

d. $T(s) = \dfrac{20}{s^2 + 6s + 144}$

e. $T(s) = \dfrac{s + 2}{s^2 + 9}$

f. $T(s) = \dfrac{(s + 5)}{(s + 10)^2}$

5. Write the general form of the capacitor voltage for the electrical network shown in Figure P4.3.

Figure P4.3

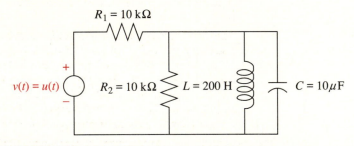

$R_1 = 10\ \text{k}\Omega$

$v(t) = u(t)$ $R_2 = 10\ \text{k}\Omega$ $L = 200\ \text{H}$ $C = 10\mu\text{F}$

6. Solve for $x(t)$ in the system shown in Figure P4.4 if $f(t)$ is a unit step.

Figure P4.4

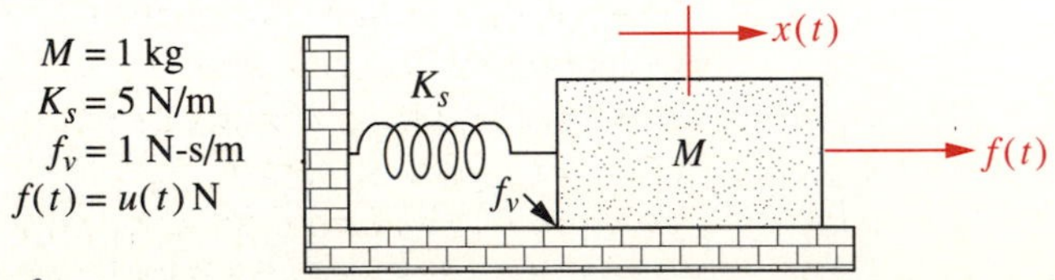

$M = 1\ \text{kg}$
$K_s = 5\ \text{N/m}$
$f_v = 1\ \text{N-s/m}$
$f(t) = u(t)\,\text{N}$

K_s M $x(t)$ $f(t)$ f_v

7. The system shown in Figure P4.5 has a unit step input. Find the output response as a function of time. Assume the system is underdamped. Notice that the result will be Equation 4.27.

Figure P4.5

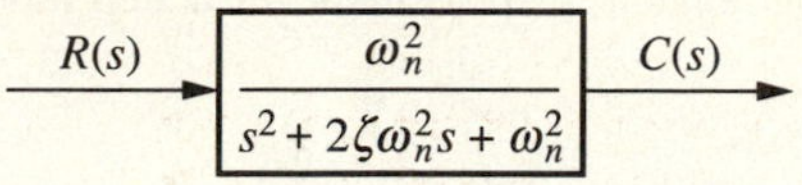

$$R(s) \longrightarrow \boxed{\dfrac{\omega_n^2}{s^2 + 2\zeta\omega_n^2 s + \omega_n^2}} \longrightarrow C(s)$$

8. Derive the relationship for damping ratio as a function of percent overshoot, Equation 4.38.

9. Calculate the exact response of each system of Problem 4, using Laplace transform techniques, and compare the results to those obtained in that problem.

10. Find the damping ratio and natural frequency for each second-order system of Problem 4 and show that the value of the damping ratio conforms to the type of response (i.e., underdamped, overdamped, etc.) predicted in that problem.

11. A system has a damping ratio of 0.5, a natural frequency of 100 rad/s, and a dc gain of 1. Find the response of the system to a unit step input.

12. For each of the second-order systems below, find ζ, ω_n, T_s, T_p, T_r, and $\%OS$.

   a. $T(s) = \dfrac{120}{s^2 + 12s + 120}$

   b. $T(s) = \dfrac{0.01}{s^2 + 0.002s + 0.01}$

   c. $T(s) = \dfrac{10^9}{s^2 + 6280s + 10^9}$

13. For each pair of second-order system specifications below, find the location of the second-order pair of poles.
   a. $\%OS = 10\%$; $T_s = 0.5$ second
   b. $\%OS = 15\%$; $T_p = 0.25$ second
   c. $T_s = 5$ seconds; $T_p = 2$ seconds

14. Given the system shown in Figure P4.5, find the location of the poles if the percent overshoot is 30% and the settling time is 0.05 seconds.

15. Find J and K in the rotational system shown in Figure P4.6 to yield a 30% overshoot and a settling time of 4 seconds for a step input in torque.

Figure P4.6

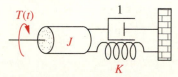

16. Given the system shown in Figure P4.7, find the damping, D, to yield a 30% overshoot in output angular displacement for a step input in torque.

Figure P4.7

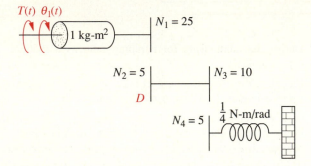

17. For the system shown in Figure P4.8, find N_1/N_2 so that the settling time for a step torque input is 16 seconds.

Figure P4.8

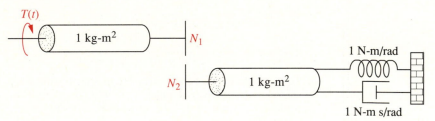

18. Consider the system shown in Figure P4.9. A 1-pound force, $f(t)$, is applied at $t = 0$. If $M = 1$, find K and D such that the response is characterized by a 20% overshoot and a damped frequency of oscillation of 10 rad/s.

Figure P4.9

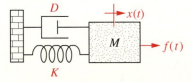

19. If $v(t)$ is a step voltage in the network shown in Figure P4.10, find the value of the resistor such that a 20% overshoot in voltage will be seen across the capacitor.

20. Derive the unit step response for each transfer function in Example 4.8.

Figure P4.10

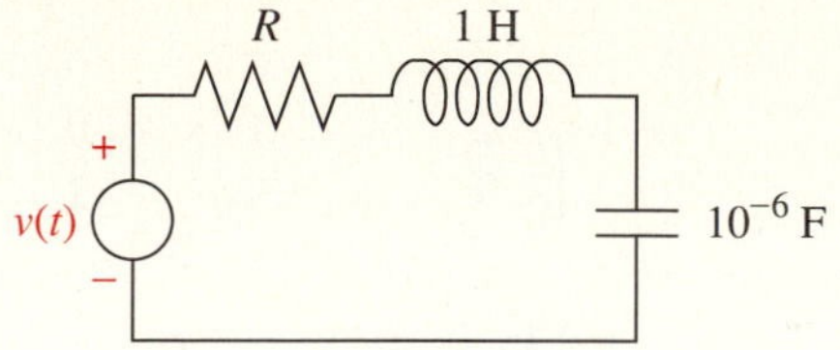

21. Find the percent overshoot, settling time, rise time, and peak time for

$$T(s) = \frac{14.145}{(s^2 + 1.204s + 2.829)(s + 5)}$$

22. For each of the following response functions, determine if pole-zero cancellation can be approximated. If there is pole-zero cancellation, find percent overshoot, settling time, rise time, and peak time.

   a. $C(s) = \dfrac{(s + 3)}{s(s + 2)(s^2 + 3s + 10)}$

   b. $C(s) = \dfrac{(s + 2.5)}{s(s + 2)(s^2 + 4s + 20)}$

   c. $C(s) = \dfrac{(s + 2.1)}{s(s + 2)(s^2 + s + 5)}$

   d. $C(s) = \dfrac{(s + 2.01)}{s(s + 2)(s^2 + 5s + 20)}$

*23. A system is represented by the state and output equations given below. Without solving the state equation, find the poles of the system.

$$\dot{\mathbf{x}} = \begin{bmatrix} -3 & -2 \\ -1 & -2 \end{bmatrix} \mathbf{x} + \begin{bmatrix} 1 \\ 1 \end{bmatrix} u(t)$$

$$y = [1 \quad 2]\mathbf{x}$$

*24. A system is represented by the state and output equations given below. Without solving the state equation, find
   a. The characteristic equation
   b. The poles of the system

$$\dot{\mathbf{x}} = \begin{bmatrix} 0 & 1 & 2 \\ 0 & 3 & 4 \\ 1 & 3 & 2 \end{bmatrix} \mathbf{x} + \begin{bmatrix} 0 \\ 0 \\ 1 \end{bmatrix} u(t)$$

$$y = [1 \quad 1 \quad 0]\mathbf{x}$$

*25. Given the following state-space representation of a system, find $Y(s)$.

$$\dot{\mathbf{x}} = \begin{bmatrix} 1 & 2 \\ -3 & -1 \end{bmatrix} \mathbf{x} + \begin{bmatrix} 1 \\ 1 \end{bmatrix} \sin 3t$$

$$y = [1 \quad 2]\mathbf{x}; \qquad \mathbf{x}(0) = \begin{bmatrix} 2 \\ 1 \end{bmatrix}$$

*26. Given the following system represented in state space, solve for $Y(s)$, using the Laplace transform method for solution of the state equation.

$$\dot{\mathbf{x}} = \begin{bmatrix} 0 & 1 & 0 \\ -2 & -4 & 1 \\ 0 & 0 & -6 \end{bmatrix} \mathbf{x} + \begin{bmatrix} 0 \\ 0 \\ 1 \end{bmatrix} e^{-t}$$

$$y = \begin{bmatrix} 1 & 0 & 0 \end{bmatrix} \mathbf{x}; \qquad \mathbf{x}(0) = \begin{bmatrix} 0 \\ 0 \\ 0 \end{bmatrix}$$

*27. Solve the following state equation and output equation for $y(t)$, where $u(t)$ is the unit step. Use the Laplace transform method.

$$\dot{\mathbf{x}} = \begin{bmatrix} -2 & 0 \\ -1 & -1 \end{bmatrix} \mathbf{x} + \begin{bmatrix} 1 \\ 1 \end{bmatrix} u(t)$$

$$y = \begin{bmatrix} 0 & 1 \end{bmatrix} \mathbf{x}; \qquad \mathbf{x}(0) = \begin{bmatrix} 1 \\ 0 \end{bmatrix}$$

*28. Solve for $y(t)$ for the following system represented in state space, where $u(t)$ is the unit step. Use the Laplace transform approach to solve the state equation.

$$\dot{\mathbf{x}} = \begin{bmatrix} -4 & 1 & 0 \\ 0 & -5 & 1 \\ 0 & 0 & -2 \end{bmatrix} \mathbf{x} + \begin{bmatrix} 0 \\ 0 \\ 1 \end{bmatrix} u(t)$$

$$y = \begin{bmatrix} 1 & 1 & 0 \end{bmatrix} \mathbf{x}; \qquad \mathbf{x}(0) = \begin{bmatrix} 0 \\ 0 \\ 0 \end{bmatrix}$$

*29. Using classical (not Laplace) methods only, solve for the state-transition matrix, the state vector, and the output for the system represented below.

$$\dot{\mathbf{x}} = \begin{bmatrix} 0 & 1 \\ -1 & -5 \end{bmatrix} \mathbf{x}; \qquad y = \begin{bmatrix} 1 & 2 \end{bmatrix} \mathbf{x}; \qquad \mathbf{x}(0) = \begin{bmatrix} 1 \\ 0 \end{bmatrix}$$

*30. Using classical (not Laplace) methods only, solve for the state-transition matrix, the state vector, and the output for the system represented below, where $u(t)$ is the unit step.

$$\dot{\mathbf{x}} = \begin{bmatrix} 0 & 1 \\ -1 & 0 \end{bmatrix} \mathbf{x} + \begin{bmatrix} 0 \\ 1 \end{bmatrix} u(t)$$

$$y = \begin{bmatrix} 3 & 2 \end{bmatrix} \mathbf{x}; \qquad \mathbf{x}(0) = \begin{bmatrix} 0 \\ 0 \end{bmatrix}$$

*31. Solve for $y(t)$ for the following system represented in state space, where $u(t)$ is the unit step. Use the classical approach to solve the state equation.

$$\dot{\mathbf{x}} = \begin{bmatrix} -2 & 1 & 0 \\ 0 & 0 & 1 \\ 0 & -6 & -1 \end{bmatrix} \mathbf{x} + \begin{bmatrix} 1 \\ 0 \\ 0 \end{bmatrix} u(t)$$

$$y = \begin{bmatrix} 1 & 0 & 0 \end{bmatrix} \mathbf{x}; \qquad \mathbf{x}(0) = \begin{bmatrix} 0 \\ 0 \\ 0 \end{bmatrix}$$

*32. Simulate the following system and plot the step response. Verify the expected values of percent overshoot, peak time, and settling time.

$$T(s) = \frac{1}{s^2 + 0.8s + 1}$$

*33. Simulate the following system and plot the output, $y(t)$, for a step input.

$$\dot{\mathbf{x}} = \begin{bmatrix} 0 & 1 & 0 \\ -10 & -7 & 1 \\ 0 & 0 & -2 \end{bmatrix} \mathbf{x} + \begin{bmatrix} 0 \\ 0 \\ 1 \end{bmatrix} u(t)$$

$$y(t) = \begin{bmatrix} 1 & 1 & 0 \end{bmatrix} \mathbf{x}; \qquad \mathbf{x}(0) = \begin{bmatrix} -1 \\ 0 \\ 0 \end{bmatrix}$$

34. A human responds to a visual cue with a physical response, as shown in Figure P4.11. The transfer function that relates the output physical response, $P(s)$, to the input visual command, $V(s)$, is $G(s) = \dfrac{P(s)}{V(s)} = \dfrac{(s + 0.5)}{(s + 2)(s + 5)}$ (Stefani, 1973). Do the following:

  a. Evaluate the output response for a unit step input using the Laplace transform.
  *b. Represent the transfer function in state-space.
  *c. Using the state-space representation, simulate the step response and obtain a plot of the output response.

Figure P4.11 Steps in Determining the Transfer Function that Relates the Output Physical Response to the Input Visual Command

Step 1: Light source on   Step 2: Recognize light source   Step 3: Respond to light source

Figure P4.12 An Unmanned Free-Swimming Submersible (UFSS) Vehicle (Courtesy of Naval Research Laboratory)

35. An Unmanned Free-Swimming Submersible (UFSS) vehicle is shown in Figure P4.12. The depth of the vehicle is controlled as follows. During forward motion, an elevator surface on the vehicle is deflected by a selected amount. This deflection causes the vehicle to rotate about the pitch axis. The pitch of the vehicle creates a vertical force that causes the vehicle to submerge or rise. The transfer function,

$$\frac{\theta(s)}{\delta_e(s)} = \frac{-.0125(s + 0.435)}{(s + 1.23)(s^2 + 0.226s + 0.0169)}$$

relates pitch angle, $\theta(s)$, to elevator surface angle, $\delta_e(s)$ (Johnson, 1980).

a. Using Laplace transforms, find the analytical expression for the response of the pitch angle to a step input in elevator-surface deflection.

b. Plot your results.

c. Using only the second-order poles shown in the transfer function, predict percent overshoot, rise time, peak time, and settling time.

d. If your results match the plot, why shouldn't you rely on part (c) in general? If your results do not match the plot, explain why not?

36. *Chapter-objective problem:* Given the schematic for the azimuth position control system shown in Figure P2.36, assume an open-loop system (feedback path disconnected).

a. Predict the open-loop angular velocity response of the power amplifier, motor, and load to a step voltage at the input to the power amplifier.

b. Find the damping ratio and natural frequency of the open-loop system.

c. Derive the open-loop, angular-velocity response of the power amplifier, motor, and load to a step-voltage input using transfer functions.

*d. Obtain the open-loop state equations and simulate the open-loop system on a digital computer.

BIBLIOGRAPHY

Johnson, H., et al. *Unmanned Free-Swimming Submersible (UFSS) System Description*. NRL Memorandum Report 4393. Naval Research Laboratory, Washington, D.C., 1980.

Kuo, B. C. *Automatic Control Systems*, 5th ed. Prentice Hall, Englewood Cliffs, N.J., 1987.

Philips, C. L., and Nagle, H. T. *Digital Control Systems Analysis and Design*. Prentice Hall, Englewood Cliffs, N.J., 1984.

Sawusch, M. R., and Summers, T. A. *1001 Things to Do with Your Macintosh™*. TAB Books, Blue Ridge Summit, Pa., 1984.

Stefani, R. T. Modeling Human Response Characteristics. COED Application Note No. 33, Computers in Education Division of ASEE, 1973.

Timothy, L. K., and Bona, B. E. *State Space Analysis: An Introduction*. McGraw-Hill, New York, 1968.

5

REPRESENTATION AND REDUCTION OF MULTIPLE SUBSYSTEMS

5.1 Introduction

Up to this point we have been working with individual subsystems represented by a block with its input and output. More complicated systems, however, are represented by the interconnection of many subsystems. Since the response of a single transfer function can be calculated, we want to represent multiple subsystems as a single transfer function. We can then apply the analytical techniques of the previous chapters and obtain transient response information about the entire system.

In this chapter, multiple subsystems are represented in two ways: (1) as block diagrams and (2) as signal-flow graphs. Although neither of these representations is limited to a particular analysis and design technique, block diagrams are usually used for frequency domain analysis and design, whereas signal-flow graphs are used for state-space analysis.

We will develop techniques to reduce each representation to a single transfer function. Block diagram algebra will be used for the reduction of block diagrams, and Mason's rule will be used for the reduction of signal-flow graphs. Again, it must be emphasized that these methods are typically used as described. As we shall see, however, either method can be used for frequency domain or state-space analysis and design.

Chapter Objective

Given the azimuth position control system of Figure 2.2, the student will be able to (1) find the single, closed-loop transfer function that models the system from the input angular rotation of the potentiometer to the output angular rotation of the load, using both block diagram reduction and Mason's rule; (2) find a state-space representation for the closed-loop system; (3) for a simplified system model, predict the percent overshoot, settling time, and peak time of the closed-loop system for a step input; (4) calculate the step response for the closed-loop system; and (5) for the simplified model, design the system gain to meet a transient response requirement.

5.2 Block Diagram Representation and Reduction of Multiple Subsystems

As we already know, subsystems are represented as a block with an input, output, and transfer function. Many systems are composed of multiple subsystems as in Figure 5.1. When multiple subsystems are interconnected, a few more schematic elements must be added to the block diagram. These new elements are *summing junctions* and *pickoff points*. All component parts of a block diagram for a linear, time-invariant system are shown in Figure 5.2. The characteristic of the summing junction shown in Figure 5.2(*c*) is that the output signal, $C(s)$, is the algebraic sum of the input signals, $R_1(s)$, $R_2(s)$, and $R_3(s)$. The figure shows three inputs, but any number can be present. A pickoff point is shown in Figure 5.2(*d*). A pickoff point distributes the input signal, $R(s)$, undiminished, to several output points.

We will now examine some common topologies for interconnecting subsystems and derive the single transfer function representation for each of them. These common topologies will form the basis for reducing more complicated systems to a single block.

Figure 5.1 The Space Shuttle Consists of Multiple Subsystems. Can You Identify Those That Are, or Are Part of, a Control System? (Courtesy of NASA)

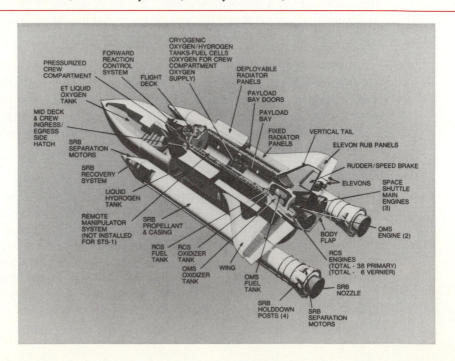

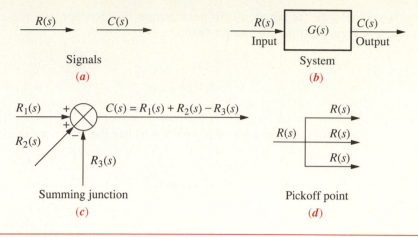

Signals

(*a*)

System

(*b*)

$C(s) = R_1(s) + R_2(s) - R_3(s)$

Summing junction

(*c*)

Pickoff point

(*d*)

Figure 5.2 Component Parts of a Block Diagram for a Linear, Time-Invariant System

Cascade Form

Figure 5.3(*a*) shows an example of cascape subsystems. Intermediate signal values are shown at the output of each subsystem. Each signal is derived from the product of the input times the transfer function. The equivalent transfer function, $G_e(s)$, shown in Figure 5.3(*b*), is the output transform divided by the input transform from Figure 5.3(*a*), or

$$G_e(s) = G_3(s)G_2(s)G_1(s) \tag{5.1}$$

which is the product of the subsystems' transfer functions.

Equation 5.1 was derived under the assumption that interconnected subsystems do not load adjacent subsystems. That is to say, a particular subsystem's output remains the same whether or not the subsequent subsystem is connected. If there is a change in the output, the subsequent subsystem loads the previous subsystem, and the equivalent transfer function is not the product of the individ-

Figure 5.3 (*a*) Cascaded Subsystems; (*b*) Equivalent Transfer Function

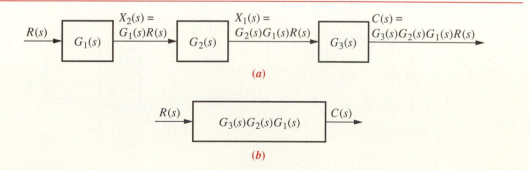

(*a*)

(*b*)

ual transfer functions. To demonstrate this concept, we present the network of Figure 5.4(a). Its transfer function is

$$G_1(s) = \frac{V_1(s)}{V_i(s)} = \frac{\dfrac{1}{R_1 C_1}}{s + \dfrac{1}{R_1 C_1}} \tag{5.2}$$

Similarly, the network of Figure 5.4(b) has the following transfer function:

$$G_2(s) = \frac{V_2(s)}{V_1(s)} = \frac{\dfrac{1}{R_2 C_2}}{s + \dfrac{1}{R_2 C_2}} \tag{5.3}$$

If the networks are placed in cascade, as in Figure 5.4(c), the reader can verify that the transfer function found using loop or node equations is

$$G(s) = \frac{V_2(s)}{V_i(s)} = \frac{\dfrac{1}{R_1 C_1 R_2 C_2}}{s^2 + \left(\dfrac{1}{R_1 C_1} + \dfrac{1}{R_2 C_2} + \dfrac{1}{R_2 C_1}\right)s + \dfrac{1}{R_1 C_1 R_2 C_2}} \tag{5.4}$$

Figure 5.4 Loading in Cascaded Systems

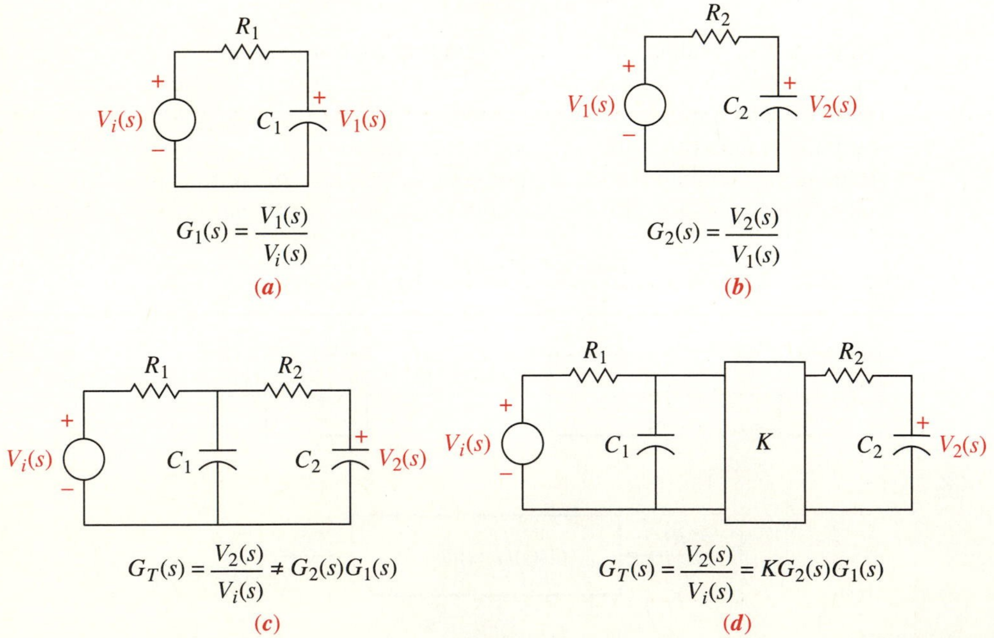

$$G_1(s) = \frac{V_1(s)}{V_i(s)}$$

(a)

$$G_2(s) = \frac{V_2(s)}{V_1(s)}$$

(b)

$$G_T(s) = \frac{V_2(s)}{V_i(s)} \neq G_2(s)G_1(s)$$

(c)

$$G_T(s) = \frac{V_2(s)}{V_i(s)} = KG_2(s)G_1(s)$$

(d)

But, using Equation 5.1,

$$G(s) = G_2(s)G_1(s) = \frac{\dfrac{1}{R_1C_1R_2C_2}}{s^2 + \left(\dfrac{1}{R_1C_1} + \dfrac{1}{R_2C_2}\right)s + \dfrac{1}{R_1C_1R_2C_2}} \tag{5.5}$$

Equations 5.4 and 5.5 are not the same: Equation 5.4 has one more term for the coefficient of s in the denominator and is correct.

One way to prevent loading is to use an amplifier between the two networks, as shown in Figure 5.4(d). The amplifier has a high-impedance input, so that it does not load the previous network. At the same time it has a low-impedance output, so that it looks like a pure voltage source to the subsequent network. With the amplifier included, the equivalent transfer function is the product of the transfer functions and the gain, K, of the amplifier.

Parallel Form

Figure 5.5 shows an example of parallel subsystems. Again, by writing the output of each subsystem, we can find the equivalent transfer function. Parallel subsystems are characterized by a common input and an output formed by the algebraic sum of the outputs from all of the subsystems. The equivalent transfer function, $G_e(s)$, is the output transform divided by the input transform from Figure 5.5(a), or

$$G_e(s) = \pm G_1(s) \pm G_2(s) \pm G_3(s) \tag{5.6}$$

Figure 5.5 (a) Parallel Subsystems; (b) Equivalent Transfer Function

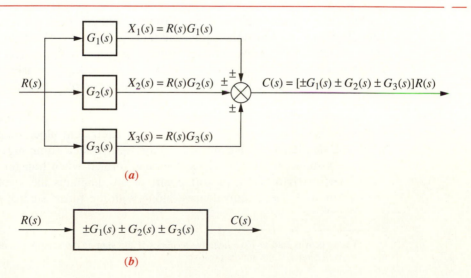

(a)

(b)

which is the algebraic sum of the subsystems' transfer functions; it appears in Figure 5.5(*b*).

Feedback Form

The third topology is the feedback form, which will be seen repeatedly as we progress through subsequent chapters. The feedback system forms the basis for our study of control systems engineering. In Chapter 1 we defined the differences between open-loop and closed-loop systems. We also pointed out the distinct advantage of closed-loop, or feedback control, systems over open-loop systems. As we move ahead, we will focus on the analysis and design of feedback systems.

Let us derive the transfer function that represents the system from its input to its output. The typical feedback system, described in detail in Chapter 1, is shown in Figure 5.6(*a*); a simplified model is shown in Figure 5.6(*b*).[1] Directing our attention to the simplified model,

$$E(s) = R(s) \mp C(s)H(s) \tag{5.7}$$

But since $C(s) = E(s)G(s)$,

$$E(s) = \frac{C(s)}{G(s)} \tag{5.8}$$

Substituting Equation 5.8 into Equation 5.7 and solving for the transfer function, $C(s)/R(s) = G_e(s)$, we obtain the equivalent, or *closed-loop,* transfer function shown in Figure 5.6 (*c*),

$$G_e(s) = \frac{G(s)}{1 \pm G(s)H(s)} \tag{5.9}$$

The product, $G(s)H(s)$, in Equation 5.9 is called the *open-loop transfer function,* or *loop gain*.

So far, we have explored three different configurations for multiple subsystems. For each, we found the equivalent transfer function. Since these three forms are combined into complex arrangements in physical systems, recognizing these topologies is one prerequisite to obtaining the equivalent transfer function of a complex system. In this section we will reduce complex systems composed of multiple subsystems to single transfer functions.

Before we begin, we must tie together a few loose ends. Let us look at Figure 5.7 and Figure 5.8. Figure 5.7 shows equivalent block diagrams formed when transfer functions are moved left or right past a summing junction, and Figure 5.8 shows equivalent block diagrams formed when transfer functions are moved left or right past a pickoff point. In the diagrams the symbol ≡ means "equivalent to." These equivalences, along with the forms studied earlier in this

---

[1]The system is said to have *negative feedback* if the sign at the summing junction is negative and *positive feedback* if the sign is positive.

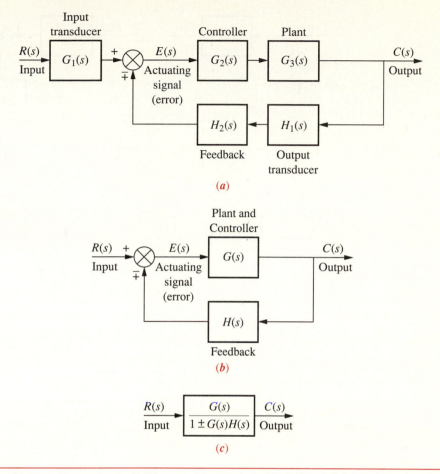

Figure 5.6 (**a**) Feedback Control System; (**b**) Feedback Control System, Simplified Model; (**c**) Equivalent Transfer Function

section, can be used to reduce a block diagram to a single transfer function. In each case of Figures 5.7 and 5.8, the equivalency can be verified by tracing the signals at the input through to the output and recognizing that the output signal is identical in both cases. For example, in Figure 5.7(a) both signals, $R(s)$ and $X(s)$, are multiplied by $G(s)$ before reaching the output. Hence, both block diagrams are equivalent. In Figure 5.7(b), $R(s)$ is multiplied by $G(s)$ before reaching the output, but $X(s)$ is not. Hence, both block diagrams in Figure 5.7(b) are equivalent since the same statement can be made about each. For pickoff points, similar reasoning yields similar results for the block diagrams of Figure 5.8(a) and (b).

Let us now put the whole story together with some examples of block diagram reduction.

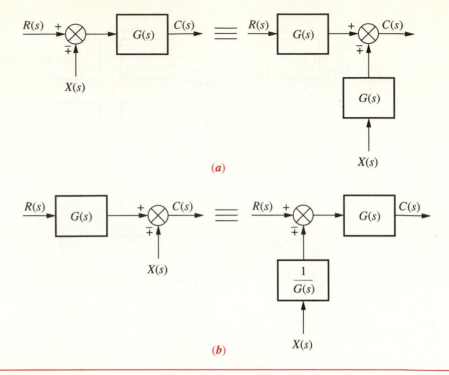

(a)

(b)

Figure 5.7 Block Diagram Algebra for Summing Junctions:
(**a**) Equivalent Forms for Moving a Block to the Left Past
a Summing Junction; (**b**) Equivalent Forms for Moving a
Block to the Right Past a Summing Junction

Example 5.1   Reduce block diagrams by recognizing familiar forms.

Problem   Reduce the block diagram shown in Figure 5.9 to a single transfer function.

Solution   The solution to the problem consists of following the steps outlined in
Figure 5.10. First, the three summing junctions can be collapsed into a single summing
junction, as shown in Figure 5.10(*a*).

   Second, recognize that the three feedback functions, $H_1(s)$, $H_2(s)$, and $H_3(s)$,
are connected in parallel. They are fed from a common signal source, and their
outputs are summed together. The equivalent function is $H_1(s) - H_2(s) + H_3(s)$. Also
recognize that $G_2(S)$ and $G_3(s)$ are connected in cascade. Thus, the equivalent transfer
function is the product, $G_3(s)G_2(s)$. The results of these steps are shown in Figure
5.10(*b*).

   Finally, the feedback system is reduced and multiplied by $G_1(s)$ to yield the equivalent transfer function shown in Figure 5.10(*c*).

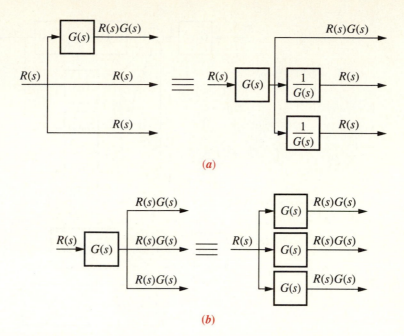

Figure 5.8 Block Diagram Algebra for Pickoff Points: (*a*) Equivalent Forms for Moving a Block to the Left Past a Pickoff Point; (*b*) Equivalent Forms for Moving a Block to the Right Past a Pickoff Point

Figure 5.9 Block Diagram for Example 5.1

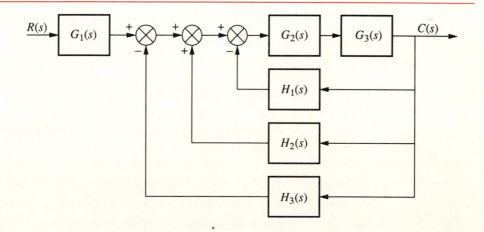

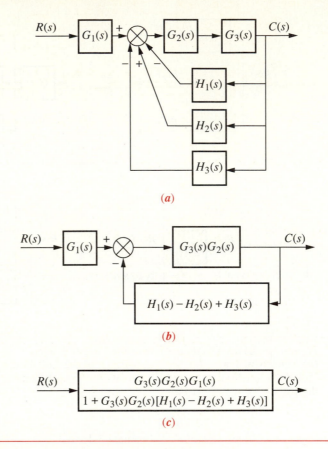

Figure 5.10 Steps in the Solution of Example 5.1:
(**a**) Collapse Summing Junctions; (**b**) Form Equivalent
Cascaded System in the Forward Path and Form Equivalent
Parallel System in the Feedback Path; (**c**) Form Equivalent
Feedback System and Multiply by Cascaded $G_1(s)$

In the previous problem we used cascade, parallel, and feedback equivalents
to reduce the block diagram. In the next example we reduce the block diagram
by moving transfer functions through summing junctions and pickoff points.

Example 5.2   Reduce block diagrams by moving blocks past summing junctions and pickoff points.

Problem   Reduce the system shown in Figure 5.11 to a single transfer function.

Solution   In this example we make use of the equivalent forms shown in Figures
5.7 and 5.8. First, move $G_2(s)$ to the left past the pickoff point to create parallel
subsystems, and reduce the feedback system consisting of $G_3(s)$ and $H_3(s)$. This
result is shown in Figure 5.12(*a*).

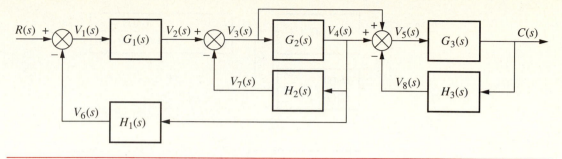

Figure 5.11 Block Diagram for Example 5.2

Second, reduce the parallel pair consisting of $1/G_2(s)$ and unity, and push $G_1(s)$ to the right past the summing junction, creating parallel subsystems in the feedback. These results are shown in Figure 5.12(b).

Third, collapse the summing junctions, add the two feedback elements together, and combine the last two cascaded blocks. Figure 5.12(c) on page 222 shows these results.

Figure 5.12 Steps in the Block Diagram Reduction for Example 5.2

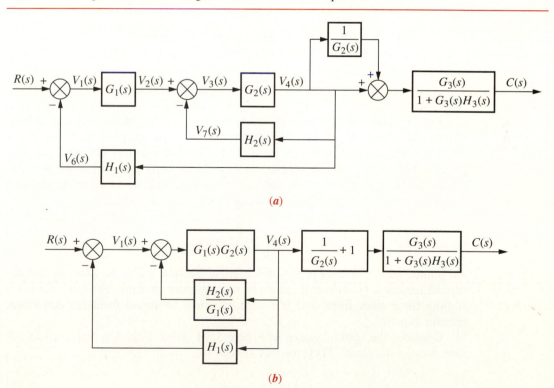

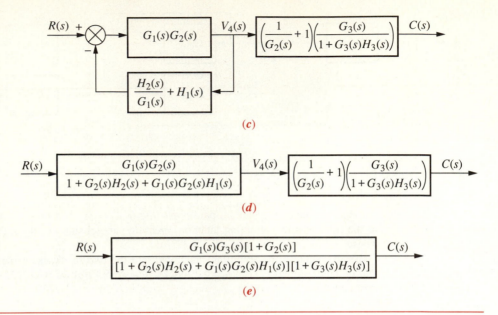

$$(c)$$

$$(d)$$

$$(e)$$

Figure 5.12 (continued) Steps in the Block Diagram
Reduction for Example 5.2

Fourth, use the feedback formula to obtain Figure 5.12(*d*).
Finally, multiply the two cascaded blocks and obtain the final result shown in
Figure 5.12(*e*).

In this section we examined the equivalency of several block diagram configurations containing signals, systems, summing junctions, and pickoff points. These
configurations were the cascade, parallel, and feedback forms. During block diagram reduction, we first attempt to produce these easily recognized forms, after
which the block diagram can be reduced to a single transfer function. In the next
section we examine some applications of block diagram reduction.

5.3 Analysis and Design of Feedback Systems

An immediate application of the principles of Section 5.2 is the analysis and design of feedback systems that reduce to second-order systems. Percent overshoot,
settling time, peak time, and rise time can then be found from the equivalent
transfer function.

Consider the system shown in Figure 5.13. From Equation 5.9, the closed-
loop transfer function, $T(s)$, for this system is

$$T(s) = \frac{K}{s^2 + as + K} \tag{5.10}$$

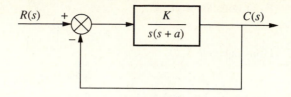

Figure 5.13 Second-Order Feedback Control System

As K varies, the poles move through the three ranges of operation of a second-order system: overdamped, critically damped, and underdamped. For example, for K between 0 and $a^2/4$, the poles of the system are real and are located at

$$s_{1,2} = -\frac{a}{2} \pm \frac{\sqrt{a^2 - 4K}}{2} \qquad (5.11)$$

As K increases, the poles move along the real axis, and the system remains overdamped until $K = a^2/4$. At that gain, both poles are real and equal, and the system is critically damped.

For gains above $a^2/4$, the system is underdamped, with complex poles located at

$$s_{1,2} = -\frac{a}{2} \pm j \frac{\sqrt{4K - a^2}}{2} \qquad (5.12)$$

Now as K increases, the real part remains constant, and the imaginary part increases. Thus, the peak time decreases, while the settling time remains constant. Let us look at two examples.

Example 5.3   Find the peak time, percent overshoot, and settling time for a feedback control system.

Problem   For the system shown in Figure 5.14, find the peak time, percent overshoot, and settling time.

Solution   The closed-loop transfer function found from Equation 5.9 is

$$T(s) = \frac{25}{s^2 + 5s + 25} \qquad (5.13)$$

Figure 5.14 Feedback System for Example 5.3

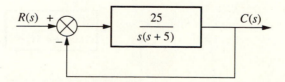

From Equation 4.17,

$$\omega_n = \sqrt{25} = 5 \qquad (5.14)$$

From Equation 4.20,

$$2\zeta\omega_n = 5 \qquad (5.15)$$

Substituting Equation 5.14 into 5.15 and solving for ζ yields

$$\zeta = 0.5 \qquad (5.16)$$

Using the values for ζ and ω_n along with Equations 4.33, 4.37, and 4.41, we find, respectively,

$$T_p = \frac{\pi}{\omega_n \sqrt{1 - \zeta^2}} = 0.726 \text{ seconds} \qquad (5.17)$$

$$\%OS = e^{-\zeta\pi/\sqrt{1-\zeta^2}} \times 100 = 16.303 \qquad (5.18)$$

$$T_s = \frac{4}{\zeta\omega_n} = 1.6 \text{ seconds} \qquad (5.19)$$

In the next example we design the gain of a feedback control system to meet transient response requirements.

Example 5.4 Design the gain of a feedback control system to meet a percent overshoot specification.

Problem Design the value of gain, K, for the feedback control system of Figure 5.15 so that the system will respond with a 10% overshoot.

Solution The closed-loop transfer function of the system is

$$T(s) = \frac{K}{s^2 + 5s + K} \qquad (5.20)$$

From Equation 5.20,

$$2\zeta\omega_n = 5 \qquad (5.21)$$

and

$$\omega_n = \sqrt{K} \qquad (5.22)$$

Figure 5.15 Feedback System for Example 5.4

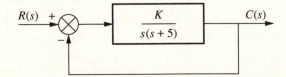

Thus

$$\zeta = \frac{5}{2\sqrt{K}} \qquad (5.23)$$

Since percent overshoot is a function only of ζ, Equation 5.23 shows that the percent overshoot is a function of K.

A 10% overshoot implies that $\zeta = 0.591$. Substituting this value for the damping ratio into Equation 5.23 and solving for K yields

$$K = 17.892 \qquad (5.24)$$

Although we are able to design for percent overshoot for this system, notice that we cannot change the settling time because the real parts of the poles of Equation 5.20 do not change with K. More sophisticated methods, such as those covered in Chapters 9 and 11, would have to be used to design both percent overshoot and settling time.

In the next section we will look into an alternative to the block diagram representation that is typically used for state-space analysis. We will define signal-flow graphs and apply them to the state-space representation.

*5.4 Signal-Flow Graph Representation and Conversion from Block Diagrams

Signal-flow graphs are an alternate system representation. Unlike block diagrams, which consist of blocks, signals, summing junctions, and pickoff points, a signal-flow graph consists only of *branches,* which represent systems, and *nodes,* which represent signals. These elements are shown in Figure 5.16(*a*) and (*b*), respectively. A system is represented by a line with an arrow showing the direction of signal flow through the system. Adjacent to the line we write the transfer function. A signal is a node with the signal's name written adjacent to the node.

Figure 5.16(*c*) shows the interconnection of the systems and the signals. Each signal is the sum of signals flowing into it. For example, the signal $X(s) =$

Figure 5.16 Signal-Flow Graph Component Parts: (*a*) System; (*b*) Signal; (*c*) Interconnection of Systems and Signals

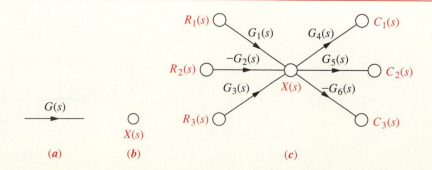

(*a*)   (*b*)   (*c*)

$R_1(s)G_1(s) - R_2(s)G_2(s) + R_3(s)G_3(s)$. The signal $C_2(s) = X(s)G_5(s) = R_1(s)G_1(s)G_5(s) - R_2(s)G_2(s)G_5(s) + R_3(s)G_3(s)G_5(s)$. The signal, $C_3(s) = -X(s)G_6(s) = -R_1(s)G_1(s)G_6(s) + R_2(s)G_2(s)G_6(s) - R_3(s)G_3(s)G_6(s)$. Notice that the summing of negative signals is handled by associating the negative sign with the system and not with a summing junction, as in the case of block diagrams.

To show the parallel between block diagrams and signal-flow graphs, we will take some of the block diagram forms from Section 5.2 and convert them to signal-flow graphs. In each case we will first convert the signals to nodes and then interconnect the nodes with systems.

Example 5.5   Convert block diagrams to signal-flow graphs.

Problem   Convert the cascaded, parallel, and feedback forms of the block diagrams shown in Figures 5.3, 5.5, and 5.6, respectively, into signal-flow graphs.

Solution   In each case we start by drawing the signal nodes for that system. Next, we interconnect the signal nodes with system branches. The signal nodes for the cas-

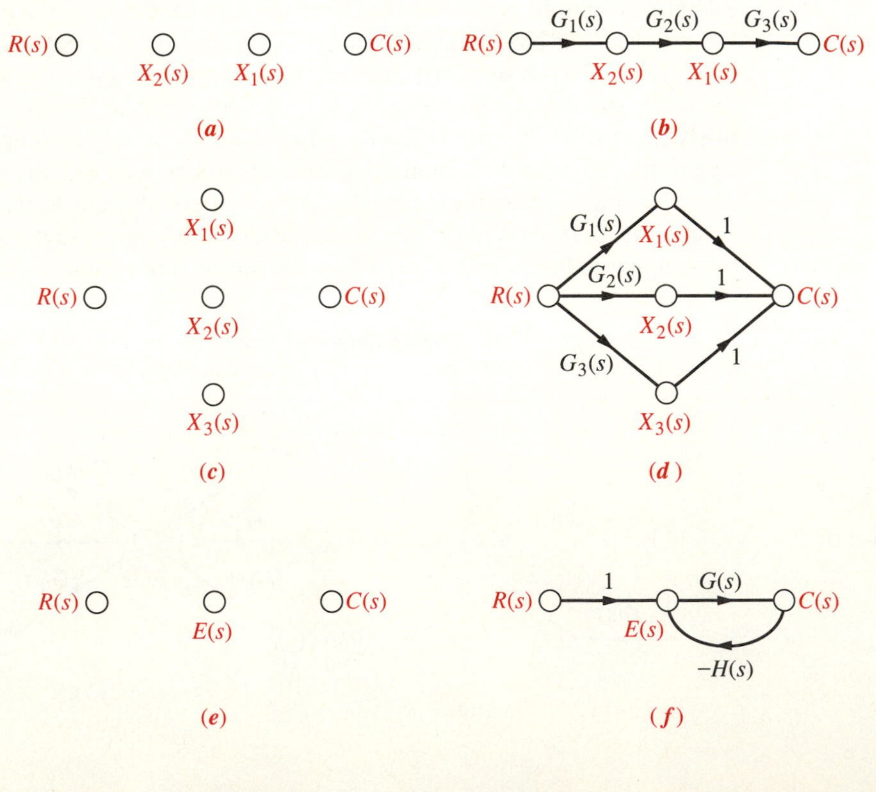

Figure 5.17 Building Signal-Flow Graphs: (**a**) Cascaded System Nodes (from Figure 5.3); (**b**) Cascaded System Signal-Flow Graph; (**c**) Parallel System Nodes (from Figure 5.5); (**d**) Parallel System Signal-Flow Graph; (**e**) Feedback System Nodes (from Figure 5.6(**b**)); (**f**) Feedback System Signal-Flow Graph

caded, parallel, and feedback forms are shown in Figure 5.17(*a*), (*c*) and (*e*), respectively. The interconnection of the nodes with branches that represent the subsystems is shown in Figure 5.17(*b*), (*d*), and (*f*) for the cascaded, parallel, and feedback forms, respectively.

In the next example we start with a more complicated block diagram and end with the equivalent signal-flow graph.

Example 5.6

Convert a block diagram to a signal-flow graph.

Problem Convert the block diagram of Figure 5.11 to a signal-flow graph.

Solution Begin by drawing the signal nodes, as shown in Figure 5.18(*a*). Next, interconnect the nodes, showing the direction of signal flow and identifying each transfer function. The result is shown in Figure 5.18(*b*). Notice that the negative signs at the summing junctions of the block diagram are represented by the negative transfer functions of the signal-flow graph. Finally, if desired, simplify the signal-flow graph to the one shown in Figure 5.18(*c*) on page 228 by eliminating signals that have a single flow in and a single flow out, such as V_2, V_6, V_7, and V_8.

Figure 5.18 Signal-Flow Graph Development for Example 5.6: (**a**) Signal Nodes; (**b**) Signal-Flow Graph

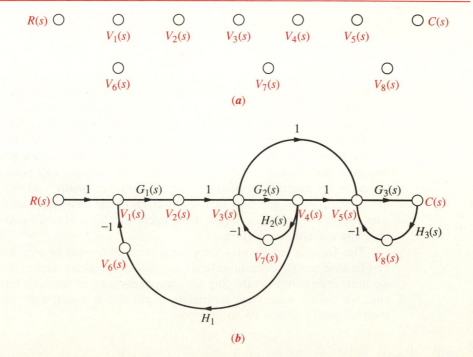

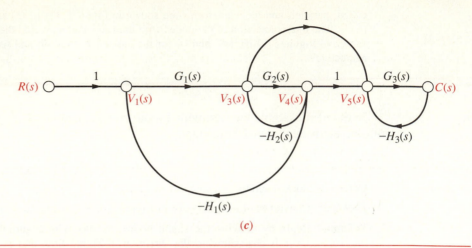

Figure 5.18 (continued) Signal-Flow Graph Development
for Example 5.6: (*c*) Simplified Signal-Flow Graph

Earlier in this chapter, we showed how to reduce block diagrams to single transfer functions. Now that we have discussed signal-flow graphs, we are ready to show a technique for reducing these graphs to single transfer functions that relate the output of a system to its input.

*5.5 Mason's Rule for Reduction of Signal-Flow Graphs

The block diagram reduction technique we studied in Section 5.2 requires successive application of fundamental relationships in order to arrive at the system transfer function. On the other hand, Mason's rule for reducing a signal-flow graph to a single transfer function requires the application of one formula. The formula was derived by S. J. Mason when he related the signal-flow graph to the simultaneous equations that can be written from the graph (Mason, 1953).

In general, it can be complicated to implement the formula without making mistakes. Specifically, the existence of what we will later call nontouching loops increases the complexity of the formula. However, many systems do not have nontouching loops and thus lend themselves to the easy application of Mason's gain formula. For these systems, the student may find Mason's rule easier to use than block diagram reduction.

The formula has several component parts that must be evaluated. We must first be sure that the definitions of the component parts are well understood. Then, we must exert care in evaluating the component parts of Mason's formula. To that end, we discuss some basic definitions applicable to signal-flow graphs; then we state Mason's rule and do an example.

Definitions

Loop gain: The product of branch gains found by traversing a path that starts at a node and ends at the same node without passing through any other node more than once and following the direction of the signal flow. For examples of loop gains, see Figure 5.19. There are four loop gains:

1. $G_2(s)H_1(s)$ (5.25a)
2. $G_4(s)H_2(s)$ (5.25b)
3. $G_4(s)G_5(s)H_3(s)$ (5.25c)
4. $G_4(s)G_6(s)H_3(s)$ (5.25d)

Forward-path gain: The product of gains found by traversing a path from the input node to the output node of the signal-flow graph. Examples of forward-path gains are also shown in Figure 5.19. There are two forward-path gains:

1. $G_1(s)G_2(s)G_3(s)G_4(s)G_5(s)G_7(s)$ (5.26a)
2. $G_1(s)G_2(s)G_3(s)G_4(s)G_6(s)G_7(s)$ (5.26b)

Nontouching loops: Loops that do not have any nodes in common. In Figure 5.19, loop $G_2(s)H_1(s)$ does not touch loops $G_4(s)H_2(s)$, $G_4(s)G_5(s)H_3(s)$, or $G_4(s)G_6(s)H_3(s)$.

Nontouching-loop gain: The product of loop gains from nontouching loops taken two, three, four, etc. at a time. In Figure 5.19, as an example, the product of the loop gain $G_2(s)H_1(s)$ times the loop gain $G_4(s)H_2(s)$ is a nontouching-loop gain taken two at a time. In summary, all three of the nontouching-loop gains taken two at a time are

1. $[G_2(s)H_1(s)][G_4(s)H_2(s)]$ (5.27a)
2. $[G_2(s)H_1(s)][G_4(s)G_5(s)H_3(s)]$ (5.27b)
3. $[G_2(s)H_1(s)][G_4(s)G_6(s)H_3(s)]$ (5.27c)

The product of loop gains $[G_4(s)G_5(s)H_3(s)][G_4(s)G_6(s)H_3(s)]$ is not a nontouching-loop gain since these two loops have nodes in common. In our exam-

Figure 5.19 Sample Signal-Flow Graph for Demonstrating Mason's Rule

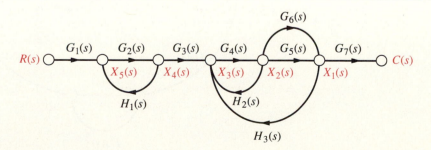

ple there are no nontouching-loop gains taken three at a time since three non-touching loops do not exist in the example.

We are now ready to state Mason's rule.

Mason's Rule

The transfer function, $C(s)/R(s)$, of a system represented by a signal-flow graph is:

$$G(s) = \frac{C(s)}{R(s)} = \frac{\sum_k T_k \Delta_k}{\Delta} \tag{5.28}$$

where

k = number of forward paths

T_k = the kth forward-path gain

Δ = $1 - \sum$ loop gains + $\sum$ nontouching-loop gains taken two at a time $- \sum$ nontouching-loop gains taken three at a time + $\sum$ nontouching-loop gains taken four at a time $- \cdots$

Δ_k = $\Delta - \sum$ loop gains touching the kth forward path. In other words, Δ_k is formed by eliminating from Δ those loop gains that touch the kth forward path.

Notice the alternating signs for each component part of Δ. The following example will help clarify Mason's Rule.

Example 5.7   Find the transfer function of a signal-flow graph using Mason's rule.

Problem Find the transfer function, $C(s)/R(s)$, for the signal-flow graph in Figure 5.20.

Figure 5.20 Signal-Flow Graph for Example 5.7

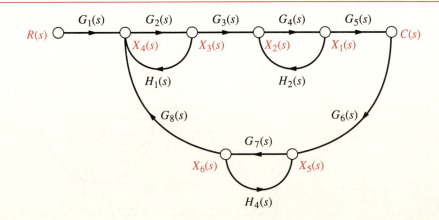

Solution First, identify the forward-path gains. In this example there is only one, as follows:

forward-path gain:

$$G_1(s)G_2(s)G_3(s)G_4(s)G_5(s) \tag{5.29}$$

Second, identify the closed-loop gains. There are four, as follows:

closed-loop gains:

1. $G_2(s)H_1(s)$ (5.30a)
2. $G_4(s)H_2(s)$ (5.30b)
3. $G_7(s)H_4(s)$ (5.30c)
4. $G_2(s)G_3(s)G_4(s)G_5(s)G_6(s)G_7(s)G_8(s)$ (5.30d)

Third, identify the nontouching loops taken two at a time. From Equation 5.30 and Figure 5.20, we can see that loop 1 does not touch loop 2, loop 1 does not touch loop 3, and loop 2 does not touch loop 3. Notice that loops 1, 2, and 3 all touch loop 4. Thus the combinations of nontouching loops taken two at a time are as follows:

nontouching loops taken two at a time:

Loop 1 and loop 2: $G_2(s)H_1(s)G_4(s)H_2(s)$ (5.31a)
Loop 1 and loop 3: $G_2(s)H_1(s)G_7(s)H_4(s)$ (5.31b)
Loop 2 and loop 3: $G_4(s)H_2(s)G_7(s)H_4(s)$ (5.31c)

Finally, the nontouching loops taken three at a time are as follows:

nontouching loops taken three at a time:

Loops 1, 2, and 3: $G_2(s)H_1(s)G_4(s)H_2(s)G_7(s)H_4(s)$ (5.32)

Now, from Equation 5.28 and its definitions, we form Δ and Δ_k. Hence

$$\begin{aligned}\Delta = 1 &- [G_2(s)H_1(s) + G_4(s)H_2(s) + G_7(s)H_4(s) \\ &\qquad + G_2(s)G_3(s)G_4(s)G_5(s)G_6(s)G_7(s)G_8(s)] \\ &+ [G_2(s)H_1(s)G_4(s)H_2(s) + G_2(s)H_1(s)G_7(s)H_4(s) \\ &\qquad + G_4(s)H_2(s)G_7(s)H_4(s)] \\ &- [G_2(s)H_1(s)G_4(s)H_2(s)G_7(s)H_4(s)]\end{aligned} \tag{5.33}$$

We form Δ_k by eliminating from Δ those loop gains that touch the kth forward path as follows:

portion of Δ not touching the forward path:

$$\Delta_1 = 1 - G_7(s)H_4(s) \tag{5.34}$$

Expressions 5.29, 5.33, and 5.34, are now substituted into Equation 5.28, yielding the transfer function:

$$G(s) = \frac{T_1\Delta_1}{\Delta} = \frac{[G_1(s)G_2(s)G_3(s)G_4(s)G_5(s)][1 - G_7(s)H_4(s)]}{\Delta} \tag{5.35}$$

Since there is only one forward path, $G(s)$ consists of only one term rather than a sum of terms, each coming from a forward path.

*5.6 Signal-Flow Graphs of State Equations

In this section we will draw signal-flow graphs from state equations. At first, this process will help us to visualize state variables. Later, we will draw signal-flow graphs first and then write alternate representations of a system in state space.

Let us return to the system of Figure 3.7 and the phase-variable form of the state equations, rewritten here in Equations 5.36, and draw a signal-flow graph of the system.

$$\dot{x}_1 = \qquad\qquad x_2 \qquad\qquad\qquad (5.36a)$$

$$\dot{x}_2 = \qquad\qquad\qquad x_3 \qquad\qquad (5.36b)$$

$$\dot{x}_3 = -24x_1 - 26x_2 - 9x_3 + 24r \qquad (5.36c)$$

$$y = \quad x_1 \qquad\qquad\qquad\qquad (5.36d)$$

First, identify three nodes, as in Figure 5.21(a), to be the three state variables, x_1, x_2, and x_3. Also identify a node as the input, r, and another node as the output, y.

The first of the three state equations, $\dot{x}_1 = x_2$, is modeled from Figure 5.21(b) by seeing that the derivative of the state variable x_1, which is x_2, would appear to the left at the input to an integrator. Remember that division by s in the frequency domain is equivalent to integration in the time domain. Similarly, the second equation, $\dot{x}_2 = x_3$, is added in Figure 5.21(c). The last of the state equations, $\dot{x}_3 = -24x_1 - 26x_2 - 9x_3 + 24r$, is added in Figure 5.21(d), where $\dot{x}_3$ is added and appears at the input of an integrator whose output is x_3. Finally, model the output, $y = x_1$.

Figure 5.21 Stages for the Development of a Signal-Flow Graph for the System of Equation 5.36

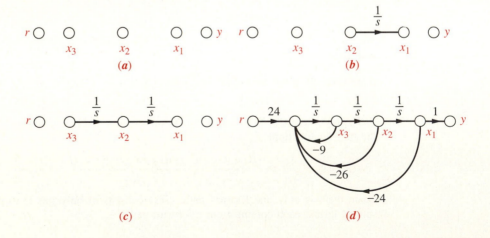

Figure 5.21(d) shows the phase-variable representation of the system of Figure 3.7. The state variables are outputs of integrators. As we proceed through the next section and talk about other representations of the same system in state space, the signal-flow model will help us visualize the process when we compute alternate representations of the same system. We will see that even though a system can be the same with respect to its input and output terminals, the state-space representations can be many and varied.

*5.7 Alternate Representations in State Space

In the last section, the system of Figure 3.7 was represented in state space in the phase-variable form along with the associated signal-flow graph shown in Figure 5.21(d). System modeling in state space can take on many different forms other than the phase-variable form. Although each of these forms yield the same output for a given input, an engineer may prefer a particular form on the basis of several considerations. For example, one set of state variables, along with its unique representation, can model actual physical variables of a system, whereas another set of state variables may not represent any system variables.

Another motive driving the choice of a particular set of state variables and the state-space model is the ease of solution. As we will see, a particular choice of state variables can decouple the system of simultaneous differential equations. Here, each equation is written in terms of only one state variable, and the solution is effected by solving n first-order differential equations individually.

Another reason for a particular choice of state variables is the ease of modeling. Certain choices may make it easy to immediately convert the subsystem to the state-variable representation by using recognizable features of the model. The engineer learns quickly how to write the state and output equations and draw the signal-flow graph, both by inspection. These converted subsystems, then, generate the definition of the state variables. We will now look at several representative forms and show how to generate the state-space representation for each.

Cascade Representation

Until now, we have shown that systems can be represented in state space where the state variables were chosen to be the phase variables, that is, those variables that are successive derivatives of each other. This choice is by no means the only choice. Returning to the system of Figure 3.7, the transfer function can be represented alternately as

$$\frac{C(s)}{R(s)} = \frac{24}{(s+2)(s+3)(s+4)} \tag{5.37}$$

Figure 5.22 shows a block diagram representation of this system formed by cascading each term of Equation 5.37. The output of each first-order system block has been labeled as a state variable. These state variables are not the phase variables.

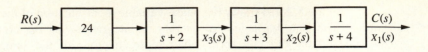

Figure 5.22 Representation of the System of Figure 3.7 as Cascaded First-Order Systems

We will now show how the signal-flow graph can be used to obtain a state-space representation of this system. In order to write the state equations with our new set of state variables, it is helpful to first draw a signal-flow graph, using Figure 5.22 as a guide. The signal flow for each first-order system of Figure 5.22 can be found by transforming each block into an equivalent differential equation. Each first-order block is of the form

$$\frac{C_i(s)}{R_i(s)} = \frac{1}{(s + a_i)} \tag{5.38}$$

Cross-multiplying, we get

$$(s + a_i)C_i(s) = R_i(s) \tag{5.39}$$

After taking the inverse Laplace transform, we have

$$\frac{dc_i}{dt} + a_i c_i = r_i(t) \tag{5.40}$$

Solving for dc_i/dt yields

$$\frac{dc_i}{dt} = -a_i c_i + r_i(t) \tag{5.41}$$

Figure 5.23(a) shows the implementation of Equation 5.41 as a signal-flow graph. Here again, a node was assumed for c_i at the output of an integrator, and its derivative was formed at the input.

Cascading the transfer functions shown in Figure 5.23(a), we arrive at the system representation shown in Figure 5.23(b).[2] Now write the state equations for the new representation of the system. Remember again that the derivative of a state variable will be at the input to each integrator.

$$\dot{x}_1 = -4x_1 + x_2 \tag{5.42a}$$

$$\dot{x}_2 = \qquad -3x_2 + x_3 \tag{5.42b}$$

$$\dot{x}_3 = \qquad \qquad -2x_3 + 24r \tag{5.42c}$$

---

[2]Note that nodes $X_3(s)$ and the following node cannot be merged, or else the input to the first integrator would be changed by the feedback from $X_2(s)$, and the signal $X_3(s)$ would be lost. A similar argument can be made for $X_2(s)$ and the following node.

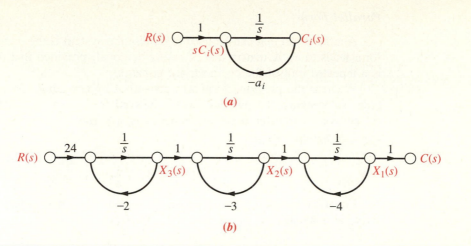

Figure 5.23 (*a*) First-Order Subsystem; (*b*) Signal-Flow Graph for the System of Figure 5.22

The output equation is written by inspection from Figure 5.23(*b*).

$$y = c(t) = x_1 \tag{5.43}$$

The state-space representation is now completed by rewriting Equations 5.42 and 5.43 in vector-matrix form:

$$\dot{\mathbf{x}} = \begin{bmatrix} -4 & 1 & 0 \\ 0 & -3 & 1 \\ 0 & 0 & -2 \end{bmatrix} \mathbf{x} + \begin{bmatrix} 0 \\ 0 \\ 24 \end{bmatrix} r \tag{5.44a}$$

$$y = \begin{bmatrix} 1 & 0 & 0 \end{bmatrix} \mathbf{x} \tag{5.44b}$$

Comparing Equations 5.44 with Figure 5.23(*b*), you can form a vivid picture of the meaning of some of the component parts of the state equation. For the following discussion, please refer back to the general form of the state and output equations, 3.18 and 3.19.

For example, the **B** matrix is the input coupling matrix since it contains the terms that couple the input, $r(t)$ to the system. In particular, the constant 24 appears in both the signal-flow diagram at the input, as shown in Figure 5.23(*b*), and within the input coupling matrix in Equations 5.44. Also, the **C** matrix is the output coupling matrix since it contains the constant that couples the state variable, x_1, to the output, $c(t)$. Finally, the **A** matrix is the system matrix since it contains the terms relative to the internal system itself. In the form of Equations 5.44, the system matrix actually contains the system poles along the diagonal.

Compare Equations 5.44 to the phase-variable representation in Equations 3.59. In that representation, the coefficients of the system's characteristic polynomial appeared along the last row, whereas in our current representation the roots of the characteristic equation, the system poles, appear along the diagonal.

Parallel Form

Another form that can be used to represent a system is the parallel form. This form leads to an **A** matrix that is purely diagonal, provided that no system pole is a repeated root of the characteristic equation.

Whereas the previous form was arrived at by cascading the individual first-order subsystems, the parallel form is derived from a partial fraction expansion of the system transfer function. Performing a partial fraction expansion on our example system, we get

$$\frac{C(s)}{R(s)} = \frac{24}{(s+2)(s+3)(s+4)} = \frac{12}{(s+2)} - \frac{24}{(s+3)} + \frac{12}{(s+4)} \qquad (5.45)$$

Equation 5.45 represents the sum of the individual first-order subsystems. To arrive at a signal-flow graph, first solve for C(s).

$$C(s) = R(s)\frac{12}{(s+2)} - R(s)\frac{24}{(s+3)} + R(s)\frac{12}{(s+4)} \qquad (5.46)$$

and recognize that $C(s)$ is the sum of three terms. Each term is a first-order subsystem with $R(s)$ as the input. Formulating this idea as a signal-flow graph renders the representation shown in Figure 5.24.

Once again, we use the signal-flow graph as an aid to obtaining the state equations. By inspection, the state variables are the outputs of each integrator, where the derivatives of the state variables exist at the integrator inputs. We write the state equations by summing the signals at the integrator inputs:

$$\dot{x}_1 = -2x_1 \qquad\qquad\quad + 12r(t) \qquad (5.47a)$$

$$\dot{x}_2 = \qquad -3x_2 \qquad -24r(t) \qquad (5.47b)$$

$$\dot{x}_3 = \qquad\qquad -4x_3 + 12r(t) \qquad (5.47c)$$

Figure 5.24 Signal-Flow Representation of Equation 5.45

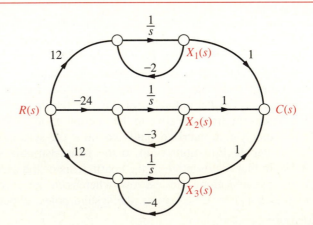

The output equation is found by summing the signals that give $c(t)$:

$$y = c(t) = x_1 + x_2 + x_3 \tag{5.48}$$

In vector-matrix form, Equations 5.47 and 5.48 become

$$\dot{\mathbf{x}} = \begin{bmatrix} -2 & 0 & 0 \\ 0 & -3 & 0 \\ 0 & 0 & -4 \end{bmatrix} \dot{\mathbf{x}} + \begin{bmatrix} 12 \\ -24 \\ 12 \end{bmatrix} r(t) \tag{5.49}$$

and

$$y = \begin{bmatrix} 1 & 1 & 1 \end{bmatrix} \mathbf{x} \tag{5.50}$$

Thus, our third representation of the system of Figure 3.7 yields a diagonal system matrix. What is the advantage of this representation? Each equation is a first-order differential equation in only one variable. Thus, we would solve these equations independently. The equations are said to be *decoupled*.

If the denominator of the transfer function has repeated real roots, the parallel form can still be derived from a partial fraction expansion. However, the system matrix will not be diagonal. Let us look at an example. Assume the system shown in Equation 5.51,

$$\frac{C(s)}{R(s)} = \frac{(s + 3)}{(s + 1)^2 (s + 2)} \tag{5.51}$$

which can be expanded as partial fractions:

$$\frac{C(s)}{R(s)} = \frac{2}{(s + 1)^2} - \frac{1}{(s + 1)} + \frac{1}{(s + 2)} \tag{5.52}$$

Proceeding as before, the signal-flow graph for Equation 5.52 is shown in Figure 5.25. The term $-1/(s + 1)$ was formed by creating the signal flow from

Figure 5.25 Signal-Flow Representation of Equation 5.52

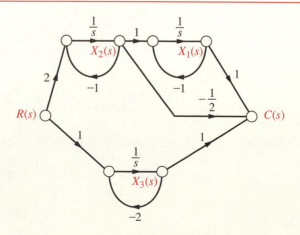

$X_2(s)$ to $C(s)$. Now the state and output equations can be written by inspection from Figure 5.25 as follows:

$$\dot{x}_1 = -2\,x_1 \quad + x_2 \tag{5.53a}$$

$$\dot{x}_2 = \qquad\quad -x_2 \qquad +2\,r(t) \tag{5.53b}$$

$$\dot{x}_3 = \qquad\qquad\quad -2\,x_3 \;+ r(t) \tag{5.53c}$$

$$y = c(t) = \quad x_1 - \frac{1}{2}\,x_2 + x_3 \tag{5.53d}$$

or, in vector-matrix form,

$$\dot{\mathbf{x}} = \begin{bmatrix} -1 & 1 & 0 \\ 0 & -1 & 0 \\ 0 & 0 & -2 \end{bmatrix} \mathbf{x} + \begin{bmatrix} 0 \\ 2 \\ 1 \end{bmatrix} r(t) \tag{5.54a}$$

$$y = \begin{bmatrix} 1 & -\dfrac{1}{2} & 1 \end{bmatrix} \mathbf{x} \tag{5.54b}$$

This system matrix, although not diagonal, has the system poles along the diagonal. Notice the 1 off the diagonal for the case of the repeated root. The form of the system matrix is known as the *Jordan canonical form*.

Dual Phase-Variable Form

Another form especially useful for systems with finite zeros is the dual phase-variable form. As an example, the system shown in Equation 5.55 will be represented in this form. Starting with the transfer function,

$$\frac{C(s)}{R(s)} = \frac{s^2 + 7s + 2}{s^3 + 9s^2 + 26s + 24} \tag{5.55}$$

divide all terms in the numerator and denominator by the highest power of s, s^3, and obtain

$$\frac{C(s)}{R(s)} = \frac{\dfrac{1}{s} + \dfrac{7}{s^2} + \dfrac{2}{s^3}}{1 + \dfrac{9}{s} + \dfrac{26}{s^2} + \dfrac{24}{s^3}} \tag{5.56}$$

Cross-multiplying yields

$$\left[\frac{1}{s} + \frac{7}{s^2} + \frac{2}{s^3}\right] R(s) = \left[1 + \frac{9}{s} + \frac{26}{s^2} + \frac{24}{s^3}\right] C(s) \tag{5.57}$$

Combining terms of like powers of integration gives

$$C(s) = \frac{1}{s}\,[R(s) - 9C(s)] + \frac{1}{s^2}\,[7R(s) - 26C(s)]$$

$$+ \frac{1}{s^3}\,[2R(s) - 24C(s)] \tag{5.58}$$

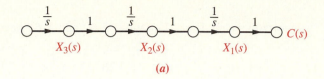

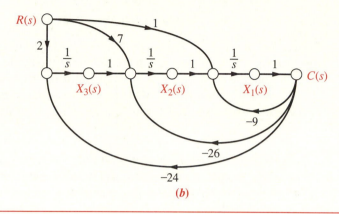

Figure 5.26 Signal-Flow Graph for Dual Phase Variables:
(*a*) Planning; (*b*) Implementation

Or

$$c(s) = \frac{1}{s}\left[[R(s) - 9C(s)] + \frac{1}{s}\left([7R(s) - 26C(s)] + \frac{1}{s}[2R(s) - 24C(s)] \right) \right]$$

(5.59)

Equation 5.58 or 5.59 can be used to draw the signal-flow graph. Start with three integrations, as shown in Figure 5.26(*a*).

Using Equation 5.58, the first term tells us that output $C(s)$ is formed, in part, by integrating $[R(s) - 9C(s)]$. We thus form $[R(s) - 9C(s)]$ at the input to the integrator closest to the output, $C(s)$. The second term of Equation 5.58 tells us that the term $[7R(s) - 26C(s)]$ must be integrated twice. Now form $[7R(s) - 26C(s)]$ at the input to the second integrator. Finally, the last term of Equation 5.58 must be integrated three times. Form $[2R(s) - 24C(s)]$ at the input to the first integrator.

Identifying the state variables as the outputs of the integrators, we write the following state equations:

$$\dot{x}_1 = -9x_1 + x_2 \quad\quad\quad + r(t)$$ (5.60*a*)

$$\dot{x}_2 = -26x_1 \quad\quad + x_3 + 7r(t)$$ (5.60*b*)

$$\dot{x}_3 = -24x_1 \quad\quad\quad\quad + 2r(t)$$ (5.60*c*)

The output equation from Figure 5.26(*b*) is

$$y = c(t) = x_1 \tag{5.61}$$

In vector-matrix form, Equations 5.60 and 5.61 become

$$\dot{\mathbf{x}} = \begin{bmatrix} -9 & 1 & 0 \\ -26 & 0 & 1 \\ -24 & 0 & 0 \end{bmatrix} \mathbf{x} + \begin{bmatrix} 1 \\ 7 \\ 2 \end{bmatrix} r(t) \tag{5.62a}$$

$$y = [1 \quad 0 \quad 0] \mathbf{x} \tag{5.62b}$$

Notice that the form of Equations 5.62 is similar to the phase-variable form, except that the coefficients of the denominator of the transfer function are in the first column, and the coefficients of the numerator form the input coupling matrix, **B**.

We conclude this section with an example that demonstrates the application of the previously discussed forms to a feedback control system.

Example 5.8 Represent feedback control systems in state space.

Problem Represent the feedback control system shown in Figure 5.27 in state space. Model the forward transfer function in cascade form.

Solution First, we model the forward transfer function in the cascade form. The gain of 100, the pole at -2, and the pole at -3 are shown cascaded in Figure 5.28(*a*). The zero at -5 was obtained using the method for implementing zeros for a system represented in phase-variable form discussed in Section 3.5.

Next, add the feedback and input paths, as shown in Figure 5.28(*b*). Now, by inspection, write the state equations:

$$\dot{x}_1 = -3x_1 \quad +x_2 \tag{5.63a}$$

$$\dot{x}_2 = \quad\quad -2x_2 + 100(r - c) \tag{5.63b}$$

But, from Figure 5.28(*b*),

$$c = 5x_1 + (x_2 - 3x_1) = 2x_1 + x_2 \tag{5.64}$$

Substituting Equation 5.64 into 5.63*b*, we find that the state equations for the system are

$$\dot{x}_1 = \quad -3x_1 \quad + x_2 \tag{5.65a}$$

$$\dot{x}_2 = -200x_1 - 102x_2 + 100r \tag{5.65b}$$

Figure 5.27 Feedback Control System for Example 5.8

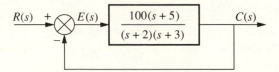

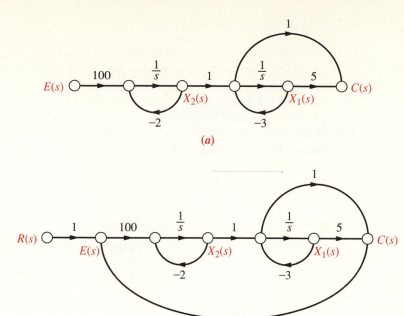

Figure 5.28 Steps in Drawing the Signal-Flow Graph for the Feedback System of Figure 5.27: (*a*) Forward Transfer Function; (*b*) Complete System

The output equation is the same as Equation 5.64, or

$$y = c(t) = 2x_1 + x_2 \tag{5.66}$$

In vector-matrix form,

$$\dot{\mathbf{x}} = \begin{bmatrix} -3 & 1 \\ -200 & -102 \end{bmatrix} \mathbf{x} + \begin{bmatrix} 0 \\ 100 \end{bmatrix} r \tag{5.67a}$$

$$y = \begin{bmatrix} 2 & 1 \end{bmatrix} \mathbf{x} \tag{5.67b}$$

In this section we saw that besides the phase-variable form, state equations can also be represented in dual phase-variable, cascade, and parallel forms. Before we summarize the concepts of this chapter, let us look at one more topic, which will show us how to make direct transformations between state variables by using matrix transformations.

*5.8 Similarity Transformations

In Section 5.7 we saw that systems can be represented with different state variables even though the transfer function relating the output to the input remains the same. The various forms of the state equations were found by manipulating the transfer function, drawing a signal-flow graph, and then writing the state equations from the signal-flow graph. These systems are called *similar* systems. Although their state-space representations are different, similar systems have the same transfer function and hence the same poles and eigenvalues.

The question now arises whether we can make transformations among similar systems from one set of state equations to another without using the transfer function and signal-flow graphs.

Expressing Any Vector in Terms of Basis Vectors

Let us begin the discussion by reviewing the representation of vector quantities in space. In Chapter 3, we learned that the state variables form the axes of the state space. Using a second-order system as an example, Figure 5.29 shows two sets of axes, x_1x_2 and z_1z_2.[3] Thus a state vector, $\mathbf{x}$, in state space can be written either in terms of the state variables or axes, x_1 and x_2, or if we call it $\mathbf{z}$, the state variables or axes, z_1 and z_2. In other words, the same vector is expressed in terms of different state variables. From this discussion we begin to see that the transformation from one set of state equations to another may be simply the transformation from one set of axes to another set of axes. Let us look further into this possibility by first clarifying the ways in which vectors can be represented in space.

Unit vectors $\mathbf{U_{x_1}}$, and $\mathbf{U_{x_2}}$, which are collinear with the axes x_1 and x_2, form linearly independent vectors called *basis* vectors for the space x_1x_2. Any vector in the space can be written in two ways. First, any vector in the space can be written as a linear combination of the basis vectors. This linear combination implies vector summation of the basis vectors to form that vector. Second, any vector can be written in terms of its components along the axes. Summarizing these two ways of writing a vector, we have

$$\mathbf{x} = x_1\mathbf{U_{x_1}} + x_2\mathbf{U_{x_2}} = \begin{bmatrix} x_1 \\ x_2 \end{bmatrix} \tag{5.68}$$

Similarly, the same vector, which will now be called $\mathbf{z}$, can be written in terms of the basis vectors in the z_1z_2 space,

$$\mathbf{z} = z_1\mathbf{U_{z_1}} + z_2\mathbf{U_{z_2}} = \begin{bmatrix} z_1 \\ z_2 \end{bmatrix} \tag{5.69}$$

---

[3]These axes are shown to be *orthogonal* (90° to each other) for clarity. In general, the axes need be only linearly independent and are not necessarily at 90°. Linear independence precludes collinear axes.

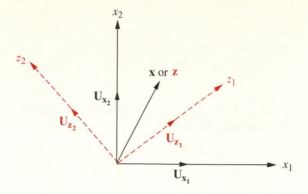

Figure 5.29 State-Space Transformations

Vector Transformations

What is the relationship between the components of **x** and **z** in Equations 5.68 and 5.69? In other words, how do we transform vector **x** into vector **z** and vice versa? To begin, we realize that unit vectors $\mathbf{U_{z_1}}$, and $\mathbf{U_{z_2}}$, which are collinear with z_1 and z_2 and are basis vectors for the space $z_1 z_2$, can also be written in terms of the basis vectors of the $x_1 x_2$ space. Hence,

$$\mathbf{U_{z_1}} = p_{11}\mathbf{U_{x_1}} + p_{21}\mathbf{U_{x_2}} \qquad (5.70a)$$

$$\mathbf{U_{z_2}} = p_{12}\mathbf{U_{x_1}} + p_{22}\mathbf{U_{x_2}} \qquad (5.70b)$$

Substituting Equations 5.70 into Equation 5.69 and realizing that the vectors **z** and **x** are the same yields **x** in terms of the components of **z** , or

$$\mathbf{x} = (z_1 p_{11} + z_2 p_{12})\mathbf{U_{x_1}} + (z_1 p_{21} + z_2 p_{22})\mathbf{U_{x_2}} = \begin{bmatrix} z_1 p_{11} + z_2 p_{12} \\ z_1 p_{21} + z_2 p_{22} \end{bmatrix} \qquad (5.71)$$

which is equivalent to

$$\mathbf{x} = \begin{bmatrix} p_{11} & p_{12} \\ p_{21} & p_{22} \end{bmatrix} \begin{bmatrix} z_1 \\ z_2 \end{bmatrix} = \mathbf{Pz} \qquad (5.72)$$

and

$$\mathbf{z} = \mathbf{P}^{-1}\mathbf{x} \qquad (5.73)$$

We can think of Equation 5.72 as a transformation that takes **z** in the $z_1 z_2$ plane and transforms it to **x** in the $x_1 x_2$ plane. Hence, if we can find **P**, we can make the transformation between the two state-space representations.

Finding the Transformation Matrix, **P**

We can find the transformation matrix, **P**, from Equations 5.70. Since we know all vector quantities in the equation, we can then solve for the p_{ij}'s. Notice that the columns of **P** are the coordinates of the basis vectors of the $z_1 z_2$ space expressed as linear combinations of the basis vectors of the $x_1 x_2$ space, as shown in Equations 5.70. Thus the first column of **P** is $\mathbf{U}_{z_1}$ and the second column is $\mathbf{U}_{z_2}$. Partitioning **P**, we get

$$\mathbf{P} = [\mathbf{U}_{z_1} \mathbf{U}_{z_2}] \tag{5.74}$$

Let us look at an example that demonstrates the transformation of a vector from one space to another.

Example 5.9   Transform a vector to a new set of basis vectors.

Problem   Transform the vector

$$\mathbf{x} = \begin{bmatrix} 1 \\ 2 \\ 2 \end{bmatrix} \tag{5.75}$$

expressed with its basis vectors,

$$\mathbf{U}_{x_1} = \begin{bmatrix} 1 \\ 0 \\ 0 \end{bmatrix}; \qquad \mathbf{U}_{x_2} = \begin{bmatrix} 0 \\ 1 \\ 0 \end{bmatrix}; \qquad \mathbf{U}_{x_3} = \begin{bmatrix} 0 \\ 0 \\ 1 \end{bmatrix} \tag{5.76}$$

to a vector expressed in the following system:

$$\mathbf{U}_{z_1} = \begin{bmatrix} 0 \\ 1/\sqrt{2} \\ 1/\sqrt{2} \end{bmatrix}; \qquad \mathbf{U}_{z_2} = \begin{bmatrix} 0 \\ -1/\sqrt{2} \\ 1/\sqrt{2} \end{bmatrix}; \qquad \mathbf{U}_{z_3} = \begin{bmatrix} 1 \\ 0 \\ 0 \end{bmatrix} \tag{5.77}$$

Solution   Using Equation 5.69 as a guide, the vector **z** can be written in terms of the basis vectors, $\mathbf{U}_{z_i}$.

$$\mathbf{z} = z_1 \mathbf{U}_{z_1} + z_2 \mathbf{U}_{z_2} + z_3 \mathbf{U}_{z_3} \tag{5.78}$$

Substituting the values of each $\mathbf{U}_{z_i}$ expressed in Equation 5.77 as components of the basis vectors $\mathbf{U}_{x_i}$, Equation 5.78 is transformed to the components of **x**,

$$\mathbf{x} = z_1 \begin{bmatrix} 0 \\ 1/\sqrt{2} \\ 1/\sqrt{2} \end{bmatrix} + z_2 \begin{bmatrix} 0 \\ -1/\sqrt{2} \\ 1/\sqrt{2} \end{bmatrix} + z_3 \begin{bmatrix} 1 \\ 0 \\ 0 \end{bmatrix}$$

$$= \begin{bmatrix} 0z_1 + 0z_2 + z_3 \\ (1/\sqrt{2})z_1 - (1/\sqrt{2})z_2 + 0z_3 \\ (1/\sqrt{2})z_1 + (1/\sqrt{2})z_2 + 0z_3 \end{bmatrix} \tag{5.79}$$

which can be written as

$$\mathbf{x} = \begin{bmatrix} 0 & 0 & 1 \\ 1/\sqrt{2} & -1/\sqrt{2} & 0 \\ 1/\sqrt{2} & 1/\sqrt{2} & 0 \end{bmatrix} \begin{bmatrix} z_1 \\ z_2 \\ z_3 \end{bmatrix} = \mathbf{Pz} \tag{5.80}$$

As we predicted, the columns of **P** are the basis vectors of the $z_1 z_2$ space (Equations 5.77).

Also,

$$\mathbf{z} = \mathbf{P}^{-1}\mathbf{x} = \begin{bmatrix} 0 & 0.707 & 0.707 \\ 0 & -0.707 & 0.707 \\ 1 & 0 & 0 \end{bmatrix} \begin{bmatrix} 1 \\ 2 \\ 2 \end{bmatrix} = \begin{bmatrix} 2.83 \\ 0 \\ 1 \end{bmatrix} \tag{5.81}$$

In summary, the vector $\mathbf{x} = [\,1 \quad 2 \quad 2\,]^T$ in the $x_1 x_2$ space transforms into $\mathbf{z} = [\,2.83 \quad 0 \quad 1\,]^T$ in the $z_1 z_2$ space; thus, $\mathbf{x}$ and $\mathbf{z}$ are the same vector expressed in different coordinate systems.

Now that we are able to transform a state vector into different basis systems, let us see how to transform the state-space representation between basis systems.

Transforming the State Equations

We have seen that the same state vector can be expressed in terms of different basis vectors. This conversion amounts to selecting a different set of state variables to represent the same system transfer function.

Let us now convert a state-space representation with state vector $\mathbf{x}$ into a state-space representation with a state vector $\mathbf{z}$. Assume the state-space representation shown in Equation 5.82.

$$\dot{\mathbf{x}} = \mathbf{Ax} + \mathbf{Bu} \tag{5.82a}$$

$$\mathbf{y} = \mathbf{Cx} + \mathbf{Du} \tag{5.82b}$$

Let $\mathbf{x} = \mathbf{Pz}$ from Equation 5.72. Hence

$$\mathbf{P\dot{z}} = \mathbf{APz} + \mathbf{Bu} \tag{5.83a}$$

$$\mathbf{y} = \mathbf{CPz} + \mathbf{Du} \tag{5.83b}$$

Premultiplying the state equation by $\mathbf{P}^{-1}$ yields

$$\dot{\mathbf{z}} = \mathbf{P}^{-1}\mathbf{APz} + \mathbf{P}^{-1}\mathbf{Bu} \tag{5.84a}$$

$$\mathbf{y} = \mathbf{CPz} + \mathbf{Du} \tag{5.84b}$$

The equations in (5.84) are an alternate representation of a system in state space. The transformed system matrix is $\mathbf{P}^{-1}\mathbf{AP}$, the input coupling matrix is $\mathbf{P}^{-1}\mathbf{B}$, the output matrix is $\mathbf{CP}$, and the feedforward matrix remains $\mathbf{D}$.

We will show now that the transfer function, $T(s) = Y(s)/U(s)$, which relates the output of the system to its input for the system represented by Equation 5.84, is the same as the system of Equation 5.82 if $\mathbf{y}$ and $\mathbf{u}$ are scalars, $y(t)$ and $u(t)$.

From Equation 3.73, the transfer function for the system of Equation 5.82 is

$$T(s) = \frac{Y(s)}{U(s)} = \mathbf{C}(s\mathbf{I} - \mathbf{A})^{-1}\mathbf{B} + \mathbf{D} \tag{5.85}$$

The transfer function of the system of Equations 5.84 can be found by substituting its equivalent output, system, input coupling, and feedforward matrix into Equation 5.85. Hence, the transfer function for the system of Equation 5.84 is

$$T(s) = \frac{Y(s)}{U(s)} = \mathbf{CP}(s\mathbf{I} - \mathbf{P}^{-1}\mathbf{AP})^{-1}\mathbf{P}^{-1}\mathbf{B} + \mathbf{D} \qquad (5.86)$$

Making successive use of the matrix inverse theorem, $(\mathbf{MN})^{-1} = \mathbf{N}^{-1}\mathbf{M}^{-1}$, we find

$$T(s) = \mathbf{CP}\big[\mathbf{P}(s\mathbf{I} - \mathbf{P}^{-1}\mathbf{AP})\big]^{-1}\mathbf{B} + \mathbf{D}$$
$$= \mathbf{C}\big[\mathbf{P}(s\mathbf{I} - \mathbf{P}^{-1}\mathbf{AP})\mathbf{P}^{-1}\big]^{-1}\mathbf{B} + \mathbf{D} \qquad (5.87)$$

Since $(s\mathbf{I} - \mathbf{P}^{-1}\mathbf{AP})\mathbf{P}^{-1} = (s\mathbf{P}^{-1} - \mathbf{P}^{-1}\mathbf{A})$,

$$T(s) = \mathbf{C}\big[\mathbf{P}(s\mathbf{P}^{-1} - \mathbf{P}^{-1}\mathbf{A})\big]^{-1}\mathbf{B} + \mathbf{D}$$
$$= \mathbf{C}\big[(s\mathbf{I} - \mathbf{A})\big]^{-1}\mathbf{B} + \mathbf{D} \qquad (5.88)$$

which is identical to Equation 5.85. Since the transfer function is the same, the system's poles and zeros remain the same through the transformation.

We can show more formally that the eigenvalues do not change under a similarity transformation. The characteristic equation for the system prior to the transformation is $\det(s\mathbf{I} - \mathbf{A}) = 0$. After the transformation, the characteristic equation is $\det(s\mathbf{I} - \mathbf{P}^{-1}\mathbf{AP}) = 0$. But, $\mathbf{I} = \mathbf{P}^{-1}\mathbf{P}$. Therefore the characteristic equation after the transformation can be written as

$$\det(s\mathbf{P}^{-1}\mathbf{P} - \mathbf{P}^{-1}\mathbf{AP}) = \det\big[\mathbf{P}^{-1}(s\mathbf{I} - \mathbf{A})\mathbf{P}\big] = 0 \qquad (5.89)$$

Since the determinant of the product of matrices is the product of the determinants,

$$\det\big[\mathbf{P}^{-1}(s\mathbf{I} - \mathbf{A})\mathbf{P}\big] = \det(\mathbf{P}^{-1})\det(s\mathbf{I} - \mathbf{A})\det(\mathbf{P}) = 0 \qquad (5.90)$$

But,

$$\det(\mathbf{P}^{-1})\det(\mathbf{P}) = \det(\mathbf{I}) = 1 \qquad (5.91)$$

Hence,

$$\det(s\mathbf{I} - \mathbf{P}^{-1}\mathbf{AP}) = \det(s\mathbf{I} - \mathbf{A}) = 0 \qquad (5.92)$$

Let us look at an example that demonstrates the transformation of a state vector and the representation of a system in state space.

Example 5.10   Perform similarity transformations on a state vector and state equations.

Problem   Given the system represented in state space by Equations 5.93,

$$\dot{\mathbf{x}} = \begin{bmatrix} 0 & 1 & 0 \\ 0 & 0 & 1 \\ -2 & -5 & -7 \end{bmatrix}\mathbf{x} + \begin{bmatrix} 0 \\ 0 \\ 1 \end{bmatrix}u \qquad (5.93a)$$

$$y = \begin{bmatrix} 1 & 0 & 0 \end{bmatrix}\mathbf{x} \qquad (5.93b)$$

transform the system to a new set of state variables, **z**, where the new state variables are related to the original state variables, **x**, as follows:

$$z_1 = 2x_1 \tag{5.94a}$$

$$z_2 = 3x_1 + 2x_2 \tag{5.94b}$$

$$z_3 = x_1 + 4x_2 + 5x_3 \tag{5.94c}$$

Solution From Equations 5.73 and 5.94,

$$\mathbf{z} = \begin{bmatrix} 2 & 0 & 0 \\ 3 & 2 & 0 \\ 1 & 4 & 5 \end{bmatrix} \mathbf{x} = \mathbf{P}^{-1}\mathbf{x} \tag{5.95}$$

Using Equation 5.84 as a guide,

$$\mathbf{P}^{-1}\mathbf{AP} = \begin{bmatrix} 2 & 0 & 0 \\ 3 & 2 & 0 \\ 1 & 4 & 5 \end{bmatrix} \begin{bmatrix} 0 & 1 & 0 \\ 0 & 0 & 1 \\ -2 & -5 & -7 \end{bmatrix} \begin{bmatrix} 0.5 & 0 & 0 \\ -0.75 & 0.5 & 0 \\ 0.5 & -0.4 & 0.2 \end{bmatrix}$$

$$= \begin{bmatrix} -1.5 & 1 & 0 \\ -1.25 & 0.7 & 0.4 \\ -2.5 & 0.4 & -6.2 \end{bmatrix} \tag{5.96}$$

$$\mathbf{P}^{-1}\mathbf{B} = \begin{bmatrix} 2 & 0 & 0 \\ 3 & 2 & 0 \\ 1 & 4 & 5 \end{bmatrix} \begin{bmatrix} 0 \\ 0 \\ 1 \end{bmatrix} = \begin{bmatrix} 0 \\ 0 \\ 5 \end{bmatrix} \tag{5.97}$$

$$\mathbf{CP} = [1 \quad 0 \quad 0] \begin{bmatrix} 0.5 & 0 & 0 \\ -0.75 & 0.5 & 0 \\ 0.5 & -0.4 & 0.2 \end{bmatrix} = [0.5 \quad 0 \quad 0] \tag{5.98}$$

Therefore, the transformed system is

$$\dot{\mathbf{z}} = \begin{bmatrix} -1.5 & 1 & 0 \\ -1.25 & 0.7 & 0.4 \\ -2.5 & 0.4 & -6.2 \end{bmatrix} \mathbf{z} + \begin{bmatrix} 0 \\ 0 \\ 5 \end{bmatrix} u \tag{5.99a}$$

$$y = [0.5 \quad 0 \quad 0] \mathbf{z} \tag{5.99b}$$

In this section we have thus far talked about transforming systems between basis vectors in a different state space. One major advantage of finding these similar systems is apparent in the transformation to a system that has a diagonal matrix. We explore this possibility next.

Diagonalization of a System Matrix

In Section 5.7 we saw that the parallel form of a signal-flow graph can yield a diagonal system matrix. A diagonal system matrix has the advantage that each state equation is a function of only one state variable. Hence, each differential equation can be solved independently of the other equations. We say that the equations, are *decoupled*.

Rather than using partial fraction expansion and signal-flow graphs, we can decouple a system using matrix transformations. If we find the correct matrix, **P**, then the transformed system matrix, $\mathbf{P}^{-1}\mathbf{AP}$, will be a diagonal matrix. Thus, we are looking for a transformation to another state space that yields a diagonal matrix in that space. This new state space also has basis vectors that lie along its state variables. We give a special name to any vectors that are collinear with the basis vectors of the new system that yields a diagonal system matrix: they are called *eigenvectors*. Thus, the coordinates of the eigenvectors form the columns of the transformation matrix, **P**, as we demonstrated in Equation 5.74.

First, let us formally define eigenvectors from another perspective and then show that they posses the property just described. Then we will define eigenvalues. Finally, we will show how to diagonalize a matrix.

Definitions

Eigenvector: The eigenvectors of the matrix **A** are all vectors, $\mathbf{x_i} \neq \mathbf{0}$, which under the transformation **A** become multiples of themselves; that is,

$$\mathbf{Ax_i} = \lambda_i \mathbf{x_i} \tag{5.100}$$

where λ_i's are constants.

Figure 5.30 shows this definition of eigenvectors. If **Ax** is not collinear with **x** after the transformation, as in Figure 5.30(*a*), then **x** is not an eigenvector. If **Ax** is collinear with **x** after the transformation, as in Figure 5.30(*b*), then **x** is an eigenvector.

Eigenvalue: The eigenvalues of the matrix **A** are the values of λ_i that satisfy Equation 5.100 for $\mathbf{x_i} \neq \mathbf{0}$.

To find the eigenvectors, we rearrange Equation 5.100. The eigenvectors $\mathbf{x_i}$ satisfy

$$\mathbf{0} = (\lambda_i \mathbf{I} - \mathbf{A})\mathbf{x_i} \tag{5.101}$$

Figure 5.30 Definition of an Eigenvector: (*a*) **x** Is Not an Eigenvector Since the Transformation **Ax** Is Not Collinear with **x**; (*b*) **x** Is an Eigenvector Since the Transformation **Ax** Is Collinear with **x**

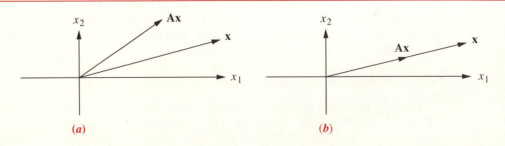

Solving for $\mathbf{x_i}$ by premultiplying both sides by $(\lambda_i \mathbf{I} - \mathbf{A})^{-1}$ yields

$$\mathbf{x_i} = (\lambda_i \mathbf{I} - \mathbf{A})^{-1}\mathbf{0} = \frac{\text{adj}\,(\lambda_i \mathbf{I} - \mathbf{A})}{\det\,(\lambda_i \mathbf{I} - \mathbf{A})}\mathbf{0} \tag{5.102}$$

Since $\mathbf{x_i} \neq \mathbf{0}$, a nonzero solution exists if

$$\det\,(\lambda_i \mathbf{I} - \mathbf{A}) = 0 \tag{5.103}$$

from which λ_i, the eigenvalues, can be found.

We are now ready to show how to find the eigenvectors $\mathbf{x_i}$. First, we find the eigenvalues λ_i, using $\det\,(\lambda_i \mathbf{I} - \mathbf{A}) = 0$, and then we use Equation 5.100 to find the eigenvectors.

Example 5.11   Find the eigenvectors of a matrix.

Problem   Find the eigenvectors of the matrix

$$\mathbf{A} = \begin{bmatrix} -3 & 1 \\ 1 & -3 \end{bmatrix} \tag{5.104}$$

Solution   The eigenvectors $\mathbf{x_i}$ satisfy Equation 5.101. First, use $\det\,(\lambda_i \mathbf{I} - \mathbf{A}) = 0$ to find the eigenvalues λ_i for Equation 5.101.

$$\begin{aligned} \det\,(\lambda \mathbf{I} - \mathbf{A}) &= \left\| \begin{bmatrix} \lambda & 0 \\ 0 & \lambda \end{bmatrix} - \begin{bmatrix} -3 & 1 \\ 1 & -3 \end{bmatrix} \right\| \\ &= \begin{vmatrix} \lambda + 3 & -1 \\ -1 & \lambda + 3 \end{vmatrix} \\ &= \lambda^2 + 6\lambda + 8 \end{aligned} \tag{5.105}$$

from which the eigenvalues are $\lambda = -2$, and -4.

Using Equation 5.100 successively with each eigenvalue, we have

$$\mathbf{Ax_i} = \lambda\mathbf{x_i}$$

$$\begin{bmatrix} -3 & 1 \\ 1 & -3 \end{bmatrix}\begin{bmatrix} x_1 \\ x_2 \end{bmatrix} = -2\begin{bmatrix} x_1 \\ x_2 \end{bmatrix} \tag{5.106}$$

or

$$-3x_1 + x_2 = -2x_1 \tag{5.107a}$$

$$x_1 - 3x_2 = -2x_2 \tag{5.107b}$$

from which $x_1 = x_2$. Thus,

$$\mathbf{x} = \begin{bmatrix} c \\ c \end{bmatrix} \tag{5.108}$$

Using the other eigenvalue, -4, we have

$$\mathbf{x} = \begin{bmatrix} c \\ -c \end{bmatrix} \tag{5.109}$$

Using Equations 5.108 and 5.109, one choice of eigenvectors is

$$\mathbf{x_1} = \begin{bmatrix} 1 \\ 1 \end{bmatrix} \quad \text{and} \quad \mathbf{x_2} = \begin{bmatrix} 1 \\ -1 \end{bmatrix} \tag{5.110}$$

We now show that if the eigenvectors of the matrix **A** are chosen as the basis vectors of a transformation, **P**, the resulting system matrix will be diagonal. Let the transformation matrix **P** consist of the eigenvectors of **A**, $\mathbf{x_i}$.

$$\mathbf{P} = [\mathbf{x_1}, \mathbf{x_2}, \mathbf{x_3}, \ldots, \mathbf{x_n}] \tag{5.111}$$

Since $\mathbf{x_i}$ are eigenvectors, $\mathbf{Ax_i} = \lambda_i \mathbf{x_i}$, which can be written equivalently as a set of equations expressed by

$$\mathbf{AP} = \mathbf{PD} \tag{5.112}$$

where **D** is a diagonal matrix consisting of λ_i's, the eigenvalues, along the diagonal, and **P** is as defined in Equation 5.111. Solving Equation 5.112 for **D** by premultiplying by $\mathbf{P}^{-1}$, we get

$$\mathbf{D} = \mathbf{P}^{-1}\mathbf{AP} \tag{5.113}$$

which is the system matrix of Equation 5.84.

In summary, under the transformation **P**, where **P** consists of the eigenvectors of the system matrix, the transformed system will be diagonal, with the eigenvalues of the system along the diagonal. The transformed system is identical to that previously obtained using partial fraction expansion of the transfer function with distinct real roots.

In Example 5.11, we found eigenvectors of a second-order system. Let us demonstrate diagonalization by continuing with this problem and diagonalizing the system matrix.

Example 5.12   Decouple or diagonalize a system represented in state space.

Problem   Given the system of Equations 5.114, find the diagonal system that is similar.

$$\dot{\mathbf{x}} = \begin{bmatrix} -3 & 1 \\ 1 & -3 \end{bmatrix} \mathbf{x} + \begin{bmatrix} 1 \\ 2 \end{bmatrix} u(t) \tag{5.114a}$$

$$y = [2 \quad 3]\mathbf{x} \tag{5.114b}$$

Solution   First find the eigenvalues and the eigenvectors. This step was performed in Example 5.11. Next, form the transformation matrix **P**, whose columns consist of the eigenvectors.

$$\mathbf{P} = \begin{bmatrix} 1 & 1 \\ 1 & -1 \end{bmatrix} \tag{5.115}$$

Finally, form the similar system's system matrix, input coupling matrix, and output matrix, respectively.

$$\mathbf{P}^{-1}\mathbf{AP} = \begin{bmatrix} 1/2 & 1/2 \\ 1/2 & -1/2 \end{bmatrix}\begin{bmatrix} -3 & 1 \\ 1 & -3 \end{bmatrix}\begin{bmatrix} 1 & 1 \\ 1 & -1 \end{bmatrix} = \begin{bmatrix} -2 & 0 \\ 0 & -4 \end{bmatrix} \tag{5.116a}$$

$$\mathbf{P}^{-1}\mathbf{B} = \begin{bmatrix} 1/2 & 1/2 \\ 1/2 & -1/2 \end{bmatrix}\begin{bmatrix} 1 \\ 2 \end{bmatrix} = \begin{bmatrix} 3/2 \\ -1/2 \end{bmatrix} \tag{5.116b}$$

$$\mathbf{CP} = \begin{bmatrix} 2 & 3 \end{bmatrix}\begin{bmatrix} 1 & 1 \\ 1 & -1 \end{bmatrix} = \begin{bmatrix} 5 & -1 \end{bmatrix} \tag{5.116c}$$

Substituting Equations 5.116 into 5.84, we get

$$\dot{\mathbf{z}} = \begin{bmatrix} -2 & 0 \\ 0 & -4 \end{bmatrix}\mathbf{z} + \begin{bmatrix} 3/2 \\ -1/2 \end{bmatrix}r(t) \tag{5.117a}$$

$$y = \begin{bmatrix} 5 & -1 \end{bmatrix}\mathbf{z} \tag{5.117b}$$

Notice that the system matrix is diagonal with the eigenvalues along the diagonal.

In this section we learned how to move between different state-space representations of the same system via matrix transformations rather than transfer function manipulation and signal-flow graphs. These different representations are called *similar*. The characteristics of similar systems are that the transfer functions relating the output to the input are the same, as are as the eigenvalues and poles. A particularly useful transformation was converting a system with distinct, real eigenvalues to a diagonal system matrix.

We now summarize the concepts of block diagram and signal-flow representations of systems, first through chapter-objective demonstration problems and then in a written summary.

5.9 Chapter-Objective Demonstration Problems

This chapter shows us that physical subsystems can be modeled mathematically with transfer functions and then interconnected to form a feedback system. The interconnected mathematical models can then be reduced to a single transfer function that represents the system from input to output. This transfer function, the closed-loop transfer function, is then used to determine the system response.

The chapter-objective demonstration problem will use our ongoing antenna azimuth control system to show how to reduce the azimuth control's subsystems to a single, closed-loop transfer function in order to analyze and design the transient response characteristics.

Example 5.13   Reduce a feedback control system to a single transfer function that relates the output to the input in order to analyze and design the closed-loop response.

Problem   Given the schematic for the antenna azimuth position control system shown in Figure 2.36,

a. Find the closed-loop transfer function using block diagram reduction.

∗b. Represent each subsystem with a signal-flow graph and find the state-space representation of the closed-loop system from the signal-flow graph.

∗c. Use the signal-flow graph found in part (**b**) along with Mason's rule to find the closed-loop transfer function.

d. Replace the power amplifier with a transfer function of unity and evaluate the closed-loop peak time, percent overshoot, and settling time for $K = 1000$.

e. For the system of part (**d**), derive the expression for the closed-loop step response of the system.

f. For the simplified model of part (**d**), find the value of K that yields a 10% overshoot.

Solution   Each subsystem's transfer function was evaluated in the chapter-objective demonstration problem in Chapter 2. We first assemble them into the closed-loop, feedback control system block diagram shown in Figure 5.31(*a*).

a. The steps taken to reduce the block diagram to a single, closed-loop transfer function relating the output angular displacement to the input angular displacement are shown in Figure 5.31(*a*) through (*d*). In Figure 5.31(*b*) the input potentiometer was pushed to the right past the summing junction, creating a unity feedback system. In Figure 5.31(*c*) all the blocks of the forward transfer function are multiplied together, forming the equivalent forward transfer function. Finally, the feedback formula is applied, yielding the closed-loop transfer function in Figure 5.31(*d*).

∗**b.** In order to obtain the signal-flow graph of each subsystem, we use the state equations derived in the chapter-objective demonstration problem of Chapter 3 to obtain the signal-flow graph. The signal-flow graph for the power amplifier is drawn from the state equations of Equations 3.86 and 3.87, and the signal-flow graph of the motor is drawn from the state equation of Equation 3.97. Other subsystems are pure gains. The signal-flow graph for Figure 5.31(*a*) is shown in Figure 5.32.

The state equations are written from Figure 5.32. First, define the state variables as the outputs of the integrators. Hence the state vector is

$$\mathbf{x} = \begin{bmatrix} x_1 \\ x_2 \\ e_a \end{bmatrix} \tag{5.118}$$

Using Figure 5.32, we write the state equations by inspection:

$$\dot{x}_1 = \qquad\qquad +x_2 \tag{5.119a}$$

$$\dot{x}_2 = \qquad -1.71x_2 +2.083e_a \tag{5.119b}$$

$$\dot{e}_a = -3.18Kx_1 \qquad -100e_a + 31.8K\theta_i \tag{5.119c}$$

along with the output equation,

$$y = \theta_o = 0.1x_1 \tag{5.120}$$

where $1/\pi = 0.318$.

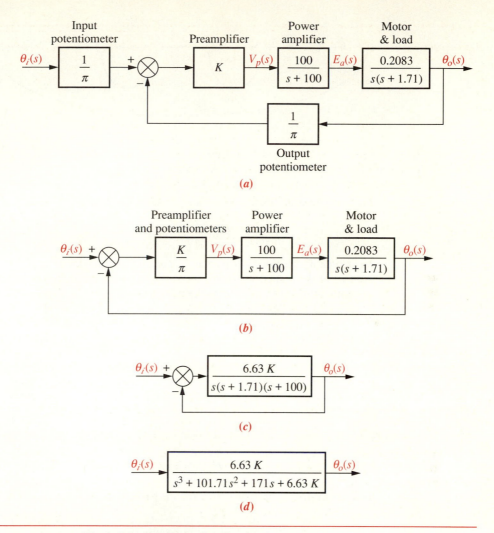

Figure 5.31 Block Diagram Reduction for the Antenna Azimuth Position Control System: (*a*) Original; (*b*) Pushing Input Potentiometer to the Right Past the Summing Junction; (*c*) Showing Equivalent Forward Transfer Function; (*d*) Final Closed-Loop Transfer Function

In vector-matrix form,

$$\dot{\mathbf{x}} = \begin{bmatrix} 0 & 1 & 0 \\ 0 & -1.71 & 2.083 \\ -3.18K & 0 & -100 \end{bmatrix} \mathbf{x} + \begin{bmatrix} 0 \\ 0 \\ 31.8K \end{bmatrix} \theta_i \qquad (5.121a)$$

$$y = [0.1 \quad 0 \quad 0] \mathbf{x} \qquad (5.121b)$$

*c. We now apply Mason's rule to Figure 5.32 to derive the closed-loop transfer function of the antenna azimuth control system. First find the forward-path gains.

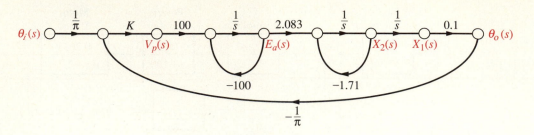

Figure 5.32 Signal-Flow Graph for the Antenna Azimuth Control System

From Figure 5.32, there is only one forward-path gain:

$$T_1 = \left(\frac{1}{\pi}\right)(K)(100)\left(\frac{1}{s}\right)(2.083)\left(\frac{1}{s}\right)\left(\frac{1}{s}\right)(0.1) = \frac{6.63K}{s^3} \qquad (5.122)$$

Next, identify the closed-loop gains. There are three: the power amplifier loop, $G_{L1}(s)$, with e_a at the output; the motor loop, $G_{L2}(s)$, with x_2 at the output; and the entire system loop, $G_{L3}(s)$, with θ_o at the output.

$$G_{L1}(s) = \frac{-100}{s} \qquad (5.123a)$$

$$G_{L2}(s) = \frac{-1.71}{s} \qquad (5.123b)$$

$$G_{L3}(s) = (K)(100)\left(\frac{1}{s}\right)(2.083)\left(\frac{1}{s}\right)\left(\frac{1}{s}\right)(0.1)\left(\frac{-1}{\pi}\right) = \frac{-6.63K}{s^3} \qquad (5.123c)$$

Only $G_{L1}(s)$ and $G_{L2}(s)$ are nontouching loops. Thus, the nontouching loop gain is

$$G_{L1}(s)G_{L2}(s) = \frac{171}{s^2} \qquad (5.124)$$

Forming Δ and Δ_k in Equation 5.28, we have

$$\Delta = 1 - [G_{L1}(s) + G_{L2}(s) + G_{L3}(s)] + [G_{L1}(s)G_{L2}(s)]$$

$$= 1 + \frac{100}{s} + \frac{1.71}{s} + \frac{6.63K}{s^3} + \frac{171}{s^2} \qquad (5.125)$$

and

$$\Delta_1 = 1 \qquad (5.126)$$

Substituting Equations 5.122, 5.125, and 5.126 into Equation 5.28, we obtain the closed-loop transfer function as

$$T(s) = \frac{C(s)}{R(s)} = \frac{T_1 \Delta_1}{\Delta} = \frac{6.63K}{s^3 + 101.71s^2 + 171s + 6.63K} \qquad (5.127)$$

d. Replacing the power amplifier with unity gain and letting the preamplifier gain, K, in Figure 5.31(b) equal 1000 yields a forward transfer function, $G(s)$, of

$$G(s) = \frac{66.3}{s(s + 1.71)} \qquad (5.128)$$

Using the feedback formula to evaluate the closed-loop transfer function, we obtain

$$T(s) = \frac{66.3}{s^2 + 1.71s + 66.3} \tag{5.129}$$

From the denominator, $\omega_n = 8.14$, $\zeta = 0.105$. Using Equations 4.33, 4.37, and 4.41, the peak time $= 0.388$ seconds, the percent overshoot $= 71.77\%$, and the settling time $= 4.68$ seconds.

e. The Laplace transform of the step response is found by first multiplying Equation 5.129 by $1/s$, a unit-step input, and expanding into partial fractions:

$$C(s) = \frac{66.3}{s(s^2 + 1.71s + 66.3)} = \frac{1}{s} - \frac{s + 1.71}{s^2 + 1.71s + 66.3}$$

$$= \frac{1}{s} - \frac{(s + 0.856) + 0.106(8.097)}{(s + 0.855)^2 + (8.097)^2} \tag{5.130}$$

Taking the inverse Laplace transform, we find

$$c(t) = 1 - e^{-0.855t}(\cos 8.097t + 0.106 \sin 8.097t) \tag{5.131}$$

f. For the simplified model, we have

$$G(s) = \frac{0.0663K}{s(s + 1.71)} \tag{5.132}$$

from which the closed-loop transfer function is calculated to be

$$T(s) = \frac{0.0663K}{s^2 + 1.71s + 0.0663K} \tag{5.133}$$

From Equation 4.38, a 10% overshoot yields $\zeta = 0.591$. Using the denominator of Equation 5.133, $\omega_n = \sqrt{0.0663K}$ and $2\zeta\omega_n = 1.71$. Thus,

$$\zeta = \frac{1.71}{2\sqrt{0.0663K}} = 0.591 \tag{5.134}$$

from which $K = 31.57$.

In our next example we return to the Unmanned Free-Swimming Submersible vehicle (Johnson, 1980) briefly described in Problem 35 of Chapter 4 (see Figure 5.33). We will model the pitch-angle control that is used as part of the depth control system and represent the pitch control loop in state space.

Example 5.14   Represent a closed-loop system in state space.

Problem   Consider the block diagram of the pitch control loop of the Unmanned Free-Swimming Submersible vehicle in Figure 5.34. The pitch angle, θ, is controlled by a commanded pitch angle, θ_c, which along with pitch-angle and pitch-rate

Figure 5.33 Electronics Carriage and Free-Swimming Submersible Vehicle Dome (Courtesy of Naval Research Laboratory)

Figure 5.34 Pitch Control Loop for the Unmanned Free-Swimming Submersible Vehicle (Adapted from Johnson, H., et al. *Unmanned Free-Swimming Submersible (UFSS) System Description. NRL Memorandum* Report 4393. Naval Research Laboratory, Washington, D.C., 1980)

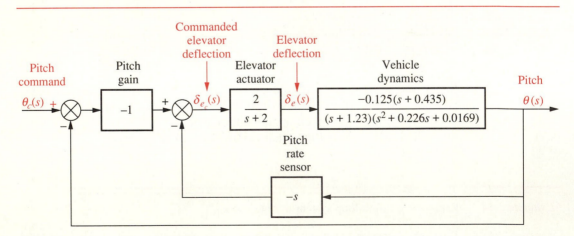

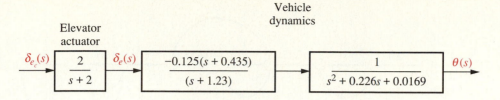

Figure 5.35 Block Diagram Representation of the Unmanned Free-Swimming Submersible Vehicle's Elevator and Vehicle Dynamics from Which Signal-Flow Representation Can Be Drawn (Adapted from Johnson, H., et al. *Unmanned Free-Swimming Submersible (UFSS) System Description.* NRL Memorandum Report 4393. Naval Research Laboratory, Washington, D.C., 1980)

feedback determines the elevator deflection, δ_e, that acts through the vehicle dynamics to determine the pitch angle.

    a. Draw the signal-flow graph for each subsystem, making sure that pitch angle, pitch rate, and elevator deflection are represented as state variables.

    b. Use the signal-flow graph obtained in **(a)** to represent the pitch control loop in state space.

Solution

    a. The vehicle dynamics are split into two transfer functions, from which the signal-flow graph is drawn. Figure 5.35 shows the division along with the elevator actuator. Each block is drawn in phase-variable form to meet the requirement that particular system variables be state variables. This result is shown in Figure 5.36(*a*). The feedback paths are then added to complete the signal-flow graph. The completed signal-flow graph is shown in Figure 5.36(*b*).

    b. By inspection, the derivatives of the state variables x_1 through x_4 are written as

$$\dot{x}_1 = x_2 \tag{5.135a}$$

$$\dot{x}_2 = -0.0169x_1 - 0.226x_2 + 0.435x_3 - 1.23x_3 - 0.125x_4 \tag{5.135b}$$

$$\dot{x}_3 = -1.23x_3 \qquad -0.125x_4 \tag{5.135c}$$

$$\dot{x}_4 = +2x_1 \quad +2x_2 \qquad -2x_4 + 2\theta_c \tag{5.135d}$$

Finally, the output $y = x_1$.

    In vector-matrix form, the state and output equations are

$$\dot{\mathbf{x}} = \begin{bmatrix} 0 & 1 & 0 & 0 \\ -0.0169 & -0.226 & -0.795 & -0.125 \\ 0 & 0 & -1.23 & -0.125 \\ 2 & 2 & 0 & -2 \end{bmatrix} \mathbf{x} + \begin{bmatrix} 0 \\ 0 \\ 0 \\ 2 \end{bmatrix} \theta_c \tag{5.136a}$$

$$y = \begin{bmatrix} 1 & 0 & 0 & 0 \end{bmatrix} \mathbf{x} \tag{5.136b}$$

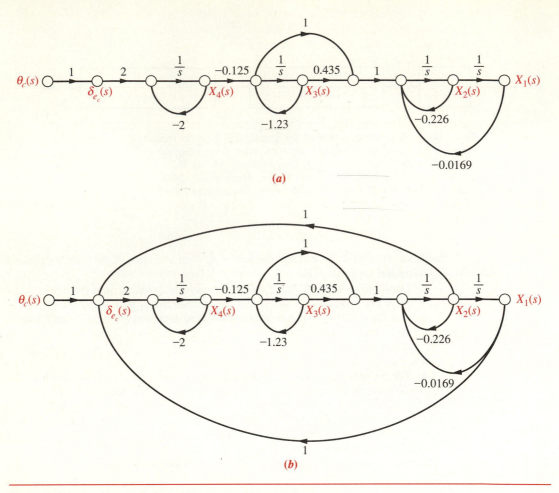

Figure 5.36 Signal-Flow Graph Representation of the Unmanned
Free-Swimming Submersible Vehicle's Pitch-Control System: (*a*) without
Position and Rate Feedback; (*b*) with Position and Rate Feedback
(Note: Explicitly required variables are: $x_1 = \theta$, $x_2 = d\theta/dt$, and
$x_4 = \delta_e$)

5.10 Summary

One objective of this chapter is to learn how to represent multiple subsystems via
block diagrams or signal-flow graphs. Another objective is to be able to reduce
either the block diagram representation or the signal-flow graph representation to
a single transfer function.

We saw that the block diagram of a linear, time-invariant system consisted
of signals, systems, summing junctions, and pickoff points. These block diagram

elements were assembled into some basic forms: cascade, parallel, and feedback. Some basic operations were then derived: moving systems across summing junctions and moving systems across pickoff points. Once we recognized the basic forms and the basic operations, we were able to reduce a complicated block diagram to a single transfer function relating input to output. Then we applied the methods of Chapter 4 for analyzing and designing a second-order system for transient behavior. We saw that the adjustment of the gain of a feedback control system gave us partial control of the transient response.

The signal-flow representation of linear, time-invariant systems only consists of nodes, which represent signals, and lines with arrows, which represent subsystems. Summing junctions and pickoff points are implicit in the signal-flow graph. The signal-flow graphs are helpful in visualizing the meaning of the state variables. Also, signal-flow graphs can be drawn first as an aid to obtaining the state equations for a system.

Mason's rule was used to derive the system's transfer function from the signal-flow graph. Mason's rule, a formula, replaced block diagram reduction techniques. Mason's rule may seem complicated; however, the technique becomes simplified if there are no nontouching loops. For these cases the transfer function can be written by inspection, and labor is saved over the block diagram reduction technique.

Finally, we saw that there are alternate forms of representing systems in state space using different sets of state variables. In the last two chapters we have covered phase-variable, dual phase-variable, cascade, and parallel forms. One reason for choosing a particular representation is that one set of state variables can have a different physical meaning than another set. Another reason is the ease with which particular state equations can be solved.

In the next chapter, we discuss system stability. Without stability we cannot begin to design a system for the desired transient response. We will find out how to tell whether a system is stable and what effect parameter values have on a system's stability.

REVIEW QUESTIONS

1. Name the four component parts of a block diagram for a linear, time-invariant system.
2. Name three basic forms for interconnecting subsystems.
3. For each of the forms in Question 2 above, state (respectively) how the equivalent transfer function is found.
4. Besides knowing the basic forms as discussed in Questions 2 and 3 above, what other equivalents must you know in order to perform block diagram reduction?
5. For a simple second-order feedback control system of the type shown in Figure 5.13, describe the effect that variations of forward-path gain, K, have on the transient response.

6. For a simple, second-order feedback control system of the type shown in Figure 5.13, describe the changes in damping ratio as the gain, K, is increased over the underdamped region.

*7. Name the two component parts of a signal-flow graph.

*8. How are summing junctions shown on a signal-flow graph?

*9. If a forward path touched all closed loops, what would be the value of Δ_k?

*10. Name five representations of systems in state space.

*11. Which two forms of the state-space representation are found using the same method?

*12. Which form of the state-space representation leads to a diagonal matrix?

*13. When the system matrix is diagonal, what quantities lie along the diagonal?

*14. What terms lie along the diagonal for a system represented in Jordan canonical form?

*15. What is the advantage of having a system represented in a form that has a diagonal system matrix?

*16. Give two reasons for wanting to represent a system by alternate forms.

*17. For what kind of system would you use dual phase-variable form?

*18. Describe state-vector transformations from the perspective of different basis.

*19. What is the definition of an eigenvector?

*20. Based upon your definition of an eigenvector, what is an eigenvalue?

*21. What is the significance of using eigenvectors as basis vectors for a system transformation?

PROBLEMS

1. Reduce the block diagram shown in Figure P5.1 to a single transfer function, $T(s) = C(s)/R(s)$.

Figure P5.1

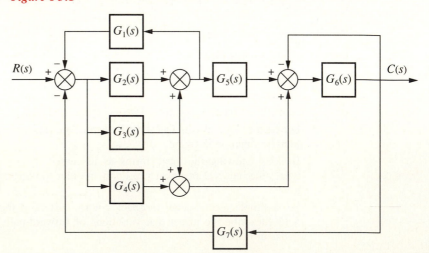

2. Find the closed-loop transfer function, $T(s) = C(s)/R(s)$, for the system shown in Figure P5.2, using block diagram reduction.

Figure P5.2

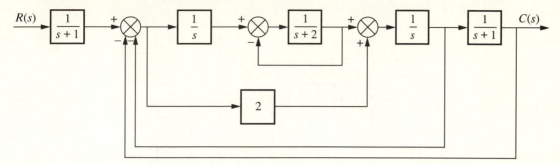

3. Find the equivalent transfer function, $T(s) = C(s)/R(s)$, for the system shown in Figure P5.3.

Figure P5.3

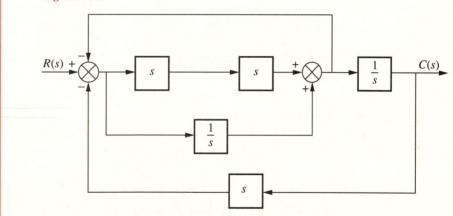

4. Reduce the system shown in Figure P5.4 to a single transfer function, $T(s) = C(s)/R(s)$.

Figure P5.4

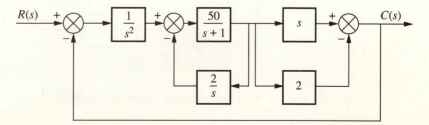

5. Find the transfer function, $T(s) = C(s)/R(s)$, for the system shown in Figure P5.5.

Figure P5.5

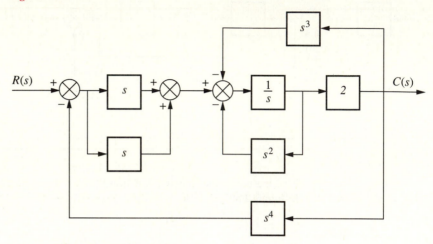

6. Find the transfer function, $T(s) = C(s)/R(s)$, for the system shown in Figure P5.6.

Figure P5.6

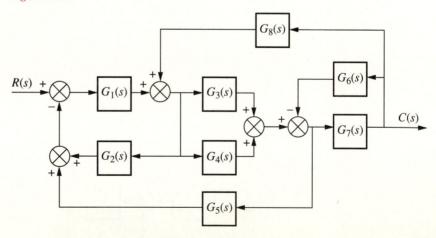

7. Reduce the block diagram shown in Figure P5.7 to a single block, $T(s) = C(s)/R(s)$.

Figure P5.7

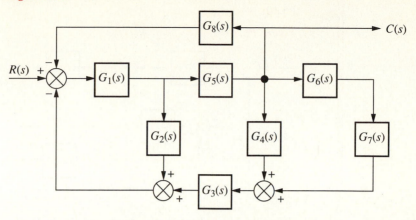

8. For the block diagram shown in Figure P5.8, find the closed-loop transfer function, $T(s) = C(s)/R(s)$.

Figure P5.8

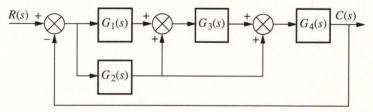

9. Find the unity feedback system that is equivalent to the system shown in Figure P5.9.

Figure P5.9

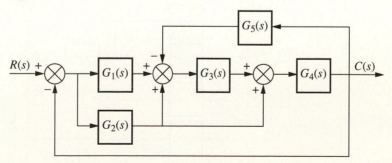

10. Find the unity feedback system that is equivalent to the system shown in Figure P5.10.

Figure P5.10

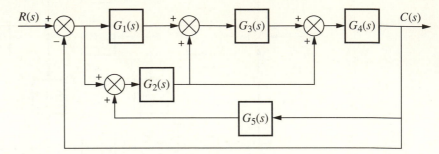

11. Given the block diagram of a system shown in Figure P5.11, find the transfer function, $G(s) = \theta_{22}(s)/\theta_{11}(s)$.

Figure P5.11

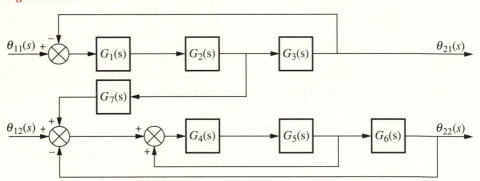

12. Reduce the block diagram shown in Figure P5.12 to a single transfer function, $T(s) = C(s)/R(s)$.

Figure P5.12

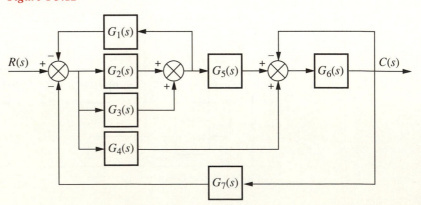

13. Reduce the system shown in Figure P5.13 to a single equivalent transfer function, $T(s) = E_o(s)/E_i(s)$.

Figure P5.13

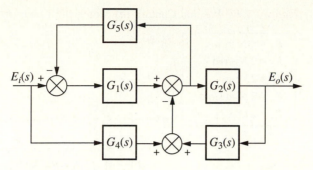

14. For the system shown in Figure P5.14, find the percent overshoot, settling time, and peak time for a step input if the system's response is underdamped. (Is it? Why?)

Figure P5.14

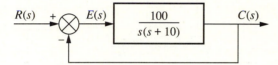

15. For the system shown in Figure P5.15, find the output, $c(t)$, if the input, $r(t)$, is a unit step.

Figure P5.15

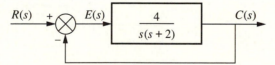

16. For the system shown in Figure P5.16, find K and α to yield a settling time of 0.5 second and a 40% overshoot.

Figure P5.16

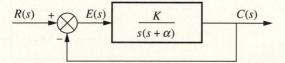

17. The motor and load shown in Figure P.5.17(a) are used as part of a unity feedback system shown in Figure P5.17(b). Find the value of the coefficient of viscous damping, D_L, that must be used in order to yield a closed-loop transient response having a 20% overshoot.

Figure P5.17 Position Control: (**a**) Motor and Load; (**b**) Block Diagram

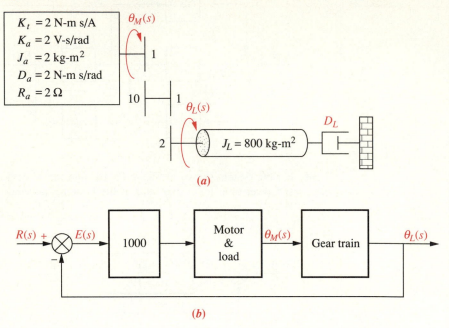

(**a**)

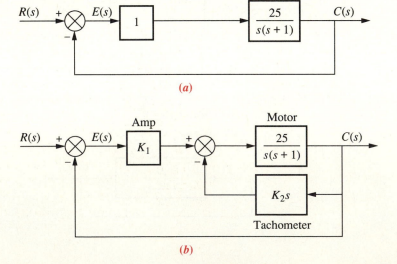

(**b**)

18. Assume that the motor whose transfer function is shown in Figure P5.18(a) is used as the forward path of a closed-loop, unity-feedback system.

Figure P5.18 (**a**) Position Control; (**b**) Position Control with Tachometer

a. Calculate the percent overshoot and settling time that could be expected.

b. You want to improve the response found in part (**a**). Since the motor constants cannot be changed, and you cannot use a different motor, an amplifier and a tachometer (voltage generator) are inserted into the loop, as shown in Figure P5.18(*b*). Find the values of K_1 and K_2 to yield a 25% overshoot and a settling time of 0.2 second.

19. For the system shown in Figure P5.19, find ζ, ω_n, percent overshoot, peak time, and settling time.

Figure P5.19

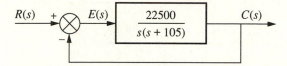

20. The system shown in Figure P5.20 will have its transient response altered by the addition of a tachometer. Design K and K_2 in the system shown in the figure to yield a damping ratio of 0.5. The natural frequency of the system before the addition of the tachometer is 10 rad/s.

Figure P5.20

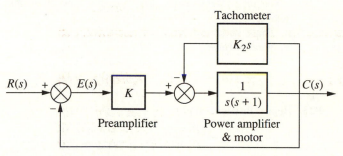

21. A motor and generator are set up to drive a load, as shown in Figure P5.21. If the generator output voltage is $E_g = K_f i_f$, where i_f is the generator's field current, find the transfer function, $G(s) = \theta_o(s)/E_i(s)$. For the generator, $K_f = 2\ \Omega$. For the motor, $K_t = 1$ N-m/A, and $K_a = 1$ V-s/rad.

Figure P5.21

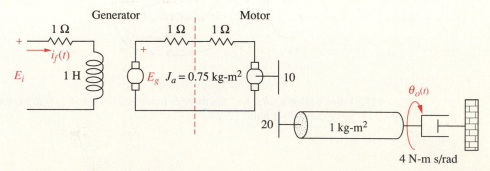

*22. Label signals and draw a signal-flow graph for the block diagrams shown in the following problems:

a. 1

b. 2

c. 6

*23. Draw a signal-flow graph for each of the state equations below:

a. $\dot{\mathbf{x}} = \begin{bmatrix} 0 & 1 & 0 \\ 0 & 0 & 1 \\ -2 & -4 & -6 \end{bmatrix} \mathbf{x} + \begin{bmatrix} 0 \\ 0 \\ 1 \end{bmatrix} r(t)$

$y = \begin{bmatrix} 1 & 1 & 0 \end{bmatrix} \mathbf{x}$

b. $\dot{\mathbf{x}} = \begin{bmatrix} -2 & 1 & 0 \\ 0 & -3 & 1 \\ -3 & -4 & -5 \end{bmatrix} \mathbf{x} + \begin{bmatrix} 0 \\ 0 \\ 1 \end{bmatrix} r(t)$

$y = \begin{bmatrix} 0 & 1 & 0 \end{bmatrix} \mathbf{x}$

c. $\dot{\mathbf{x}} = \begin{bmatrix} 1 & -1 & 1 \\ 0 & -2 & 1 \\ -2 & -4 & -6 \end{bmatrix} \mathbf{x} + \begin{bmatrix} 1 \\ 0 \\ 1 \end{bmatrix} r(t)$

$y = \begin{bmatrix} 0 & 1 & 1 \end{bmatrix} \mathbf{x}$

*24. Given the system below, draw a signal-flow graph and represent the system in state space in the following forms:

a. Phase-variable form

b. Parallel form

$$G(s) = \frac{5}{s(s+6)(s+8)}$$

*25. Repeat Problem 24 for

$$G(s) = \frac{10}{(s+1)(s+3)(s+4)(s+6)}$$

*26. Using Mason's rule, find the transfer function, $T(s) = C(s)/R(s)$, for the system represented in Figure P5.22.

Figure P5.22

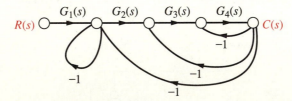

*27. Using Mason's rule, find the transfer function, $T(s) = C(s)/R(s)$, for the system represented by Figure P5.23.

Figure P5.23

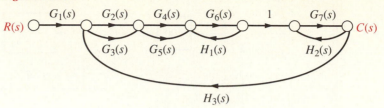

*28. Use Mason's rule to find the transfer function of Problem 3.

29. Use block diagram reduction to find the transfer function of Figure 5.20 and compare your answer with that obtained by Mason's rule.

*30. Represent the following systems in state space in Jordan canonical form. Draw the signal-flow graphs.

 a. $G(s) = \dfrac{(s + 2)}{(s + 5)^2(s + 7)^2}$

 b. $G(s) = \dfrac{(s + 3)}{(s + 2)^2(s + 4)(s + 5)}$

*31. Represent the systems below in state space in dual phase-variable form. Draw the signal-flow graphs.

 a. $G(s) = \dfrac{s + 3}{s^2 + 2s + 7}$

 b. $G(s) = \dfrac{s^3 + 2s^2 + 7s + 1}{s^4 + 3s^3 + 5s^2 + 6s + 4}$

*32. Repeat Problem 31 and represent each system in phase-variable form.

*33. Represent the feedback control systems shown in Figure P5.24 on pages 269 and 270 in state space. When possible, represent the open-loop transfer functions separately in cascade and complete the feedback loop with the signal path from output to input. Draw your signal-flow graph (as closely as possible) to be in one-to-one correspondence to the block diagrams.

Figure P5.24

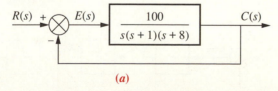

(a)

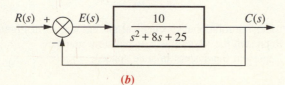

(b)

Figure P5.24 (continued)

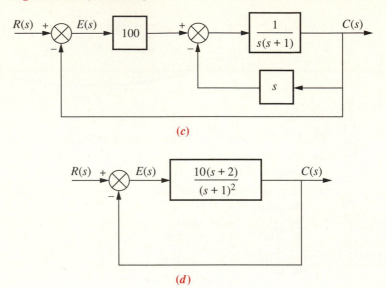

(c)

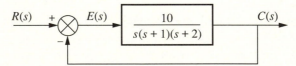

(d)

*34. You are given the system shown in Figure P5.25.
   a. Represent the system in state space in phase-variable form.
   b. Represent the system in state space in any other form besides phase-variable.

Figure P5.25

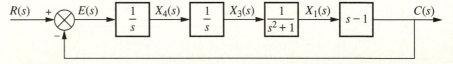

*35. Represent the system shown in Figure P5.26 in state space where $x_1(t)$, $x_3(t)$, and $x_4(t)$, as shown, are among the state variables, $c(t)$ is the output, and $x_2(t)$ is internal to $X_1(s)/X_3(s)$.

Figure P5.26

*36. Consider the rotational mechanical system shown in Figure P5.27.
   a. Represent the system as a signal-flow graph.
   b. Represent the system in state space.

Figure P5.27

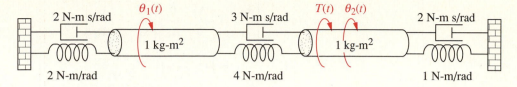

*37. Consider the cascaded subsystems shown in Figure P5.28. If $G_1(s)$ is represented in state space as

$$\dot{\mathbf{x}}_1 = \mathbf{A}_1\mathbf{x}_1 + \mathbf{B}_1 r$$

$$y_1 = \mathbf{C}_1\mathbf{x}_1$$

and $G_2(s)$ is represented in state space as

$$\dot{\mathbf{x}}_2 = \mathbf{A}_2\mathbf{x}_2 + \mathbf{B}_2 y_1$$

$$y_2 = \mathbf{C}_2\mathbf{x}_2$$

show that the entire system can be represented in state space as

$$\begin{bmatrix} \dot{\mathbf{x}}_1 \\ \dot{\mathbf{x}}_2 \end{bmatrix} = \begin{bmatrix} \mathbf{A}_1 & \mathbf{0} \\ \mathbf{B}_2\mathbf{C}_1 & \mathbf{A}_2 \end{bmatrix} \begin{bmatrix} \mathbf{x}_1 \\ \mathbf{x}_2 \end{bmatrix} + \begin{bmatrix} \mathbf{B}_1 \\ \mathbf{0} \end{bmatrix} r$$

$$y_2 = \begin{bmatrix} \mathbf{0} & \mathbf{C}_2 \end{bmatrix} \begin{bmatrix} \mathbf{x}_1 \\ \mathbf{x}_2 \end{bmatrix}$$

Figure P5.28

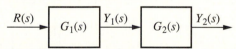

*38. Consider the parallel subsystems shown in Figure P5.29. If $G_1(s)$ is represented in state space as

$$\dot{\mathbf{x}} = \mathbf{A}_1\mathbf{x}_1 + \mathbf{B}_1 r$$

$$y_1 = \mathbf{C}_1\mathbf{x}_1$$

and $G_2(s)$ is represented in state space as

$$\dot{\mathbf{x}}_2 = \mathbf{A}_2\mathbf{x}_2 + \mathbf{B}_2 r$$

$$y_2 = \mathbf{C}_2\mathbf{x}_2$$

show that the entire system can be represented in state space as

$$\begin{bmatrix} \dot{\mathbf{x}}_1 \\ \hline \dot{\mathbf{x}}_2 \end{bmatrix} = \begin{bmatrix} \mathbf{A}_1 & 0 \\ \hline 0 & \mathbf{A}_2 \end{bmatrix} \begin{bmatrix} \mathbf{x}_1 \\ \hline \mathbf{x}_2 \end{bmatrix} + \begin{bmatrix} \mathbf{B}_1 \\ \hline \mathbf{B}_2 \end{bmatrix} r$$

$$y = \begin{bmatrix} \mathbf{C}_1 & \mathbf{C}_2 \end{bmatrix} \begin{bmatrix} \mathbf{x}_1 \\ \hline \mathbf{x}_2 \end{bmatrix}$$

Figure P5.29

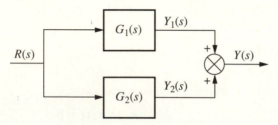

*39. Consider the subsystems shown in Figure P5.30 and connected to form a feedback system. If $G(s)$ is represented in state space as

$$\dot{\mathbf{x}}_1 = \mathbf{A}_1\mathbf{x}_1 + \mathbf{B}_1\mathbf{e}$$

$$y = \mathbf{C}_1\mathbf{x}_1$$

and $H(s)$ is represented in state space as

$$\dot{\mathbf{x}}_2 = \mathbf{A}_2\mathbf{x}_2 + \mathbf{B}_2 y$$

$$p = \mathbf{C}_2\mathbf{x}_2$$

show that the closed-loop system can be represented in state space as

$$\begin{bmatrix} \dot{\mathbf{x}}_1 \\ \hline \dot{\mathbf{x}}_2 \end{bmatrix} = \begin{bmatrix} \mathbf{A}_1 & -\mathbf{B}_1\mathbf{C}_2 \\ \hline \mathbf{B}_2\mathbf{C}_1 & \mathbf{A}_2 \end{bmatrix} \begin{bmatrix} \mathbf{x}_1 \\ \hline \mathbf{x}_2 \end{bmatrix} + \begin{bmatrix} \mathbf{B}_1 \\ \hline 0 \end{bmatrix} r$$

$$y = \begin{bmatrix} \mathbf{C}_1 & 0 \end{bmatrix} \begin{bmatrix} \mathbf{x}_1 \\ \hline \mathbf{x}_2 \end{bmatrix}$$

Figure P5.30

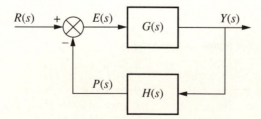

*40. Given the system represented in state space as follows:

$$\dot{\mathbf{x}} = \begin{bmatrix} 1 & 0 & -2 \\ 0 & 3 & 1 \\ -5 & -2 & -3 \end{bmatrix} \mathbf{x} + \begin{bmatrix} 1 \\ 0 \\ 2 \end{bmatrix} r(t)$$

$$y = [1 \quad 3 \quad 2] \mathbf{x}$$

convert the system to one where the new state vector, z, is

$$\mathbf{z} = \begin{bmatrix} 1 & 3 & -2 \\ 4 & -1 & 0 \\ 2 & 5 & 1 \end{bmatrix} \mathbf{x}$$

*41. Repeat Problem 40 for the following system:

$$\dot{\mathbf{x}} = \begin{bmatrix} 1 & -1 & 1 \\ 2 & 1 & 3 \\ -2 & -1 & -3 \end{bmatrix} \mathbf{x} + \begin{bmatrix} 7 \\ 1 \\ -2 \end{bmatrix} r(t)$$

$$y = [1 \quad -3 \quad 4] \mathbf{x}$$

and the following state-vector transformation:

$$\mathbf{z} = \begin{bmatrix} 4 & -1 & 0 \\ 2 & 3 & -2 \\ 8 & 5 & 1 \end{bmatrix} \mathbf{x}$$

*42. Diagonalize the following system.

$$\dot{\mathbf{x}} = \begin{bmatrix} -5 & -5 & 4 \\ 2 & 0 & -2 \\ 0 & -2 & -1 \end{bmatrix} \mathbf{x} + \begin{bmatrix} -1 \\ 2 \\ -2 \end{bmatrix} r(t)$$

$$y = [-1 \quad 1 \quad 2] \mathbf{x}$$

*43. Repeat Problem 42 for the following system.

$$\dot{\mathbf{x}} = \begin{bmatrix} -10 & -3 & 7 \\ 18.25 & 6.25 & -11.75 \\ -7.25 & -2.25 & 5.75 \end{bmatrix} \mathbf{x} + \begin{bmatrix} 1 \\ 3 \\ 2 \end{bmatrix} r(t)$$

$$y = [1 \quad -2 \quad 4] \mathbf{x}$$

44. During ascent, the space shuttle is steered by commands generated by the computer's guidance calculations. These commands are in the form of vehicle attitude, attitude rates, and attitude accelerations obtained through measurements made by the vehicle's inertial measuring unit, rate gyro assembly, and accelerometer assembly, respectively. The ascent digital autopilot uses the errors between the actual and commanded attitude, rates, and accelerations to position the space shuttle main engines and the solid rocket boosters to effect the desired vehicle attitude. The space shuttle's attitude control system employs the same method in

Figure P5.31 Simplified Pitch Control System for the Space Shuttle (Source: Rockwell International, *Space Shuttle GN&C Operations Manual,* 1988)

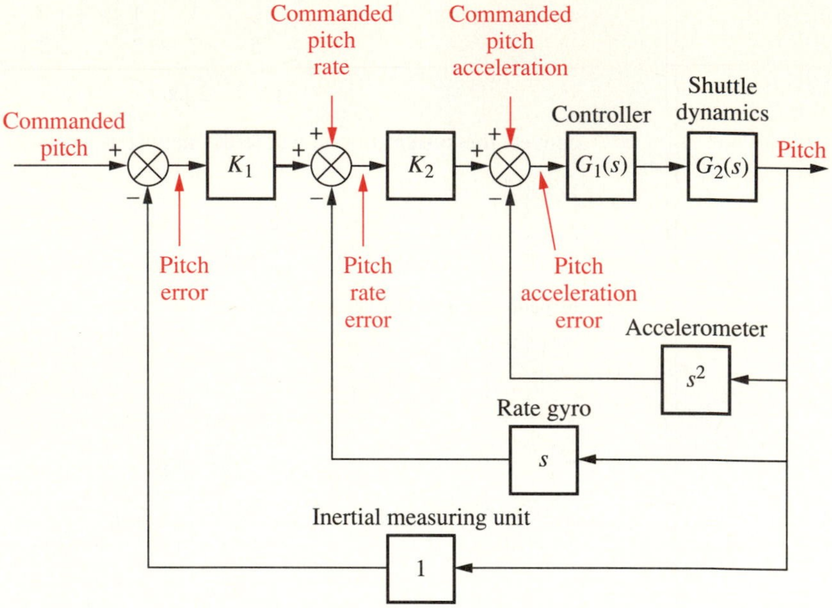

the pitch, roll, and yaw control systems. A simplified model of the pitch control system is shown in Figure P5.31.[4]

a. Find the closed-loop transfer function relating actual pitch to commanded pitch. Assume all other inputs are zero.

b. Find the closed-loop transfer function relating actual pitch rate to commanded pitch rate. Assume all other inputs are zero.

c. Find the closed-loop transfer function relating actual pitch acceleration to commanded pitch acceleration. Assume all other inputs are zero.

*45. An AM radio modulator generates the product of a carrier waveform and a message waveform, as shown in Figure P5.32 (Kurland, 1971). Represent the system in state space if the carrier is a sinusoid of frequency $\omega = a^2$, and the message is a sinusoid of frequency b^2. Note that this system is nonlinear because of the multiplier.

46. A model for human eye movement consists of the closed-loop system shown in Figure P5.33, where an object's position is the input, and the eye position is the output. The brain sends signals to the muscles that move the eye. These signals consist of the difference between the object's position and position and rate information from the eye sent by the muscle spindles. The eye motion is modeled as an inertia and viscous damping and assumes no elasticity (spring) (Milhorn, 1966). Assuming that the delays in the brain and nervous system are negligible, find the closed-loop transfer function for the eye position control.

---

[4]Source of background information for this problem: Rockwell International.

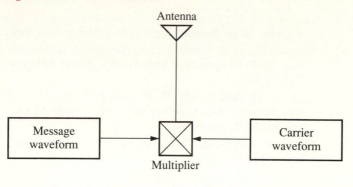

Figure P5.32 AM Modulator

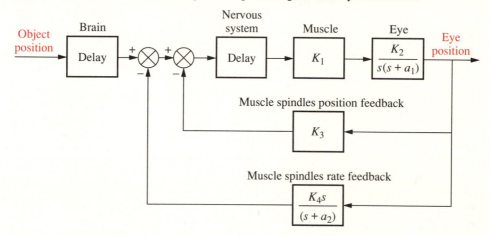

Figure P5.33 Feedback Control System Representing Human Eye Movement

47. *Chapter-objective problem:* Consider the schematic for the antenna azimuth position control system shown in Figure P2.36.

    a. Find the closed-loop transfer function, using block diagram reduction.

    *****b.** Represent each subsystem with a signal-flow graph and find the state-space representation of the closed-loop system from the signal-flow graph.

    *****c.** Use the signal-flow graph found in part (**b**) along with Mason's rule to find the closed-loop transfer function.

    d. Replace the power amplifier with a transfer function of unity and evaluate the closed-loop percent overshoot, settling time, and peak time for $K = 5$.

    e. For the system of part (**d**), derive the expression for the closed-loop step response of the system.

    f. For the simplified model in part (**d**), find the value of preamplifier gain, K, to yield 15% overshoot.

BIBLIOGRAPHY

Hostetter, G. H.; Savant, C. J., Jr.; and Stefani, R. T. *Design of Feedback Control Systems*. 2d ed. Saunders College Publishing, New York, 1989.

Johnson, H., et al. *Unmanned Free-Swimming Submersible (UFSS) System Description*. NRL Memorandum Report 4393. Naval Research Laboratory, Washington, D.C., 1980.

Kurland, M., and Papson, T. P. Analog Computer Simulation of Linear Modulation Systems, *Transactions of the Analog/Hybrid Computer Educational Society,* January 1971, pp. 9–18.

Mason, S. J. Feedback Theory—Some Properties of Signal-Flow Graphs. *Proc. IRE,* September 1953, pp. 1144–1156.

Milhorn, H. T., Jr. *The Application of Control Theory to Physiological Systems.* W. B. Saunders, Philadelphia, 1966.

Timothy, L. K., and Bona, B. E. *State Space Analysis: An Introduction*. McGraw-Hill, New York, 1968.

6

STABILITY

6.1 Introduction

In Chapter 1 we said that three requirements enter into the design of a control system; transient response, stability, and steady-state errors. Thus far we have covered transient response, which we will revisit in Chapter 8. We are now ready to discuss the next requirement, stability.

Stability is the most important system specification. If a system is unstable, transient response and steady-state errors are moot points. An unstable system cannot be designed for a specific transient response or steady-state error requirement. What then is *stability*? There are many definitions for stability, depending upon the kind of system or the point of view. In this section we limit ourselves to linear, time-invariant systems.

A linear, time-invariant system is *stable* if the natural response approaches zero as time approaches infinity. Since the total response of a system is the sum of the forced and natural responses, as shown in Equation 6.1, the definition of stability implies that only the forced response remains as the natural response approaches zero.

$$c(t) = c_{\text{forced}}(t) + c_{\text{natural}}(t) \tag{6.1}$$

A linear, time-invariant system is *unstable* if the natural response grows without bound as time approaches infinity. Finally, a linear, time-invariant system is *marginally stable* if the natural response neither decays nor grows, but remains constant or oscillates as time approaches infinity.

Physically, an unstable system whose natural response grows without bound can cause damage to the system, to adjacent property, or to human life. Many times, systems are designed with limit stops to prevent total runaway. From the perspective of the time response plot, instability is displayed by transients that grow without bound and, consequently, a total response that does not approach a steady-state value or other forced response.[1] How do we determine if a system is stable?

---

[1]Care must be taken here to distinguish between transients growing without bound and a forced response, such as a ramp or exponential increase, that also grows without bound. Systems whose forced response approaches infinity are perfectly stable as long as the natural response approaches zero.

Recall from our study of system poles that poles in the left half-plane yield either pure exponential decay or damped sinusoidal natural responses. These natural responses decay to zero as time approaches infinity. Thus, if the closed-loop system poles are in the left half of the s-plane and hence have a negative real part, the system is stable. That is, *stable systems have closed-loop transfer functions with poles only in the left half-plane.*

Figure 6.1 (**a**) A Stable System, Closed-Loop Poles, and Response; (**b**) an Unstable System, Closed-Loop Poles, and Response

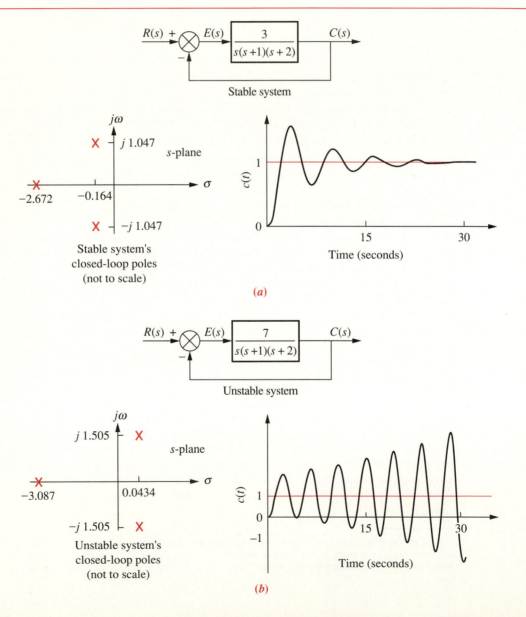

On the other hand, poles in the right half-plane yield either pure exponentially increasing or exponentially increasing sinusoidal natural responses. These natural responses approach infinity as time approaches infinity. Thus, if the closed-loop system poles are in the right half of the s-plane and hence have a positive real part, the system is unstable. Also, poles of multiplicity greater than one on the imaginary axis lead to the sum of responses of the form, $At^n \cos(\omega t + \phi)$, where $n = 1, 2, \ldots,$ which also approaches infinity as time approaches infinity. Thus, *unstable systems have closed-loop transfer functions with at least one pole in the right half-plane and/or poles of multiplicity greater than one on the imaginary axis*.

Finally, a system that has imaginary axis poles of multiplicity 1 yields pure sinusoidal oscillations as a natural response. These responses neither increase nor decrease in amplitude. We call this type of system *marginally stable*. Thus, *marginally stable systems have closed-loop transfer functions with only imaginary axis poles of multiplicity 1 and poles in the left half-plane*.

As an example, the unit step response of the stable system of Figure 6.1(a) is compared to the unit step response of the unstable system of Figure 6.1(b). The responses, also shown in Figure 6.1, show that while the oscillations for the stable system diminish, those for the unstable system increase without bound. Also notice that the stable system's response in this case approaches a steady-state value of unity.

Since a linear, time-invariant system is stable if the closed-loop system poles exist only in the left half of the s-plane, it would appear to be a simple matter to factor the denominator of the closed-loop system, find the poles, and determine that all poles are in the left half of the s-plane. Unfortunately, a typical problem that arises is shown in Figure 6.2. Although we know the poles of the forward transfer function in Figure 6.2(a), we do not know the location of the poles of the equivalent closed-loop system of Figure 6.2(b) without factoring or otherwise solving for the roots.

Figure 6.2 Common Cause of the Difficulty in Determining Closed-Loop Poles: (**a**) Original System; (**b**) the Equivalent System

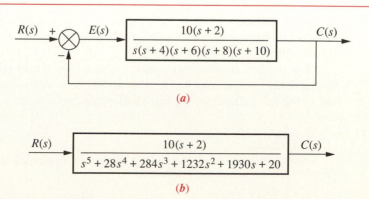

(a)

(b)

However, under certain conditions, we can draw some conclusions about the stability of the system. First, if the closed-loop transfer function has only left-half-plane poles, then the factors of the denominator of the closed-loop system transfer function consist of the product of terms such as $(s + a_i)$, where a_i is real and positive, or complex with a positive real part. Thus, the product of such terms is a polynomial with all positive coefficients.[2] No term of the polynomial can be missing since that would imply cancellation between positive and negative coefficients or imaginary axis roots in the factors, which is not the case. Thus, a sufficient condition for a system to be unstable is that all signs of the coefficients of the denominator of the closed-loop transfer function are not the same. If powers of s are missing, then the system is either unstable or, at best, marginally stable. Unfortunately, if all coefficients of the denominator are positive and not missing, we do not have definitive information about the system's pole locations.

If the method described in the previous paragraph is not sufficient, then a computer can be used to determine the stability by calculating the root locations of the denominator of the closed-loop transfer function. Today, even some hand-held calculators can evaluate the roots of a polynomial. There is, however, another method to test for stability without having to solve for the roots of the polynomial. This method will be discussed in detail after we state the chapter objective.

Chapter Objective

Given the azimuth position control system of Figure 2.2, the student will be able to determine the range of preamplifier gain to keep the system stable.

6.2 Routh-Hurwitz Criterion

In this section we learn a method that yields stability information without the need to solve for the closed-loop system poles. Using this method, we can tell how many closed-loop system poles are in the left half-plane, in the right half-plane, and on the $j\omega$-axis. (Notice that we say *how many,* not *where.*) We will be able to find the number of poles in each section of the s-plane, but we will not be able to find out their coordinates. The method is called the *Routh-Hurwitz criterion* for stability (Routh, 1905).

The method requires two steps: (1) generate a data table called a *Routh table,* and (2) interpret the Routh table to tell how many closed-loop system poles are in the left half-plane, the right half-plane, and on the $j\omega$-axis. In this section

---

[2]The coefficients can also be made all negative by multiplying the polynomial by -1. This operation does not change the root location.

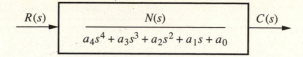

Figure 6.3 An Equivalent Closed-Loop Transfer Function

we make and interpret a basic Routh table. In the next section we consider two special cases that can arise when generating this data table.

Generating a Basic Routh Table

Look at the equivalent closed-loop transfer function shown in Figure 6.3. Since we are interested in the system poles, we focus our attention on the denominator. We first create the Routh table shown in Table 6.1. Begin by labeling the rows with powers of s from the highest power of the denominator of the closed-loop transfer function to s^0. Next, start with the coefficient of the highest power of s in the denominator and list, horizontally in the first row, every other coefficient. In the second row list horizontally, starting with the next highest power of s, every coefficient that was skipped in the first row.

The remaining entries are filled in as follows. Each entry is a negative determinant of entries in the previous two rows divided by the entry in the first column directly above the calculated row. The left-hand column of the determinant is always the first column of the previous two rows, and the right-hand column is the elements of the column above and to the right. The table is complete when all of the rows are completed down to s^0. Table 6.2 is the completed Routh table.

Table 6.1 Beginning Layout for the Routh Table

s^4	a_4	a_2	a_0
s^3	a_3	a_1	0
s^2			
s^1			
s^0			

Table 6.2 Completed Routh Table

s^4	a_4	a_2	a_0
s^3	a_3	a_1	0
s^2	$\dfrac{-\begin{vmatrix} a_4 & a_2 \\ a_3 & a_1 \end{vmatrix}}{a_3} = b_1$	$\dfrac{-\begin{vmatrix} a_4 & a_0 \\ a_3 & 0 \end{vmatrix}}{a_3} = b_2$	$\dfrac{-\begin{vmatrix} a_4 & 0 \\ a_3 & 0 \end{vmatrix}}{a_3} = 0$
s^1	$\dfrac{-\begin{vmatrix} a_3 & a_1 \\ b_1 & b_2 \end{vmatrix}}{b_1} = c_1$	$\dfrac{-\begin{vmatrix} a_3 & 0 \\ b_1 & 0 \end{vmatrix}}{b_1} = 0$	$\dfrac{-\begin{vmatrix} a_3 & 0 \\ b_1 & 0 \end{vmatrix}}{b_1} = 0$
s^0	$\dfrac{-\begin{vmatrix} b_1 & b_2 \\ c_1 & 0 \end{vmatrix}}{c_1} = d_1$	$\dfrac{-\begin{vmatrix} b_1 & 0 \\ c_1 & 0 \end{vmatrix}}{c_1} = 0$	$\dfrac{-\begin{vmatrix} b_1 & 0 \\ c_1 & 0 \end{vmatrix}}{c_1} = 0$

Example 6.1   Practice making a Routh table.

Problem   Make the Routh table for the system shown in Figure 6.4(a).

Solution   The first step is to find the equivalent closed-loop system because we want to test the denominator of this function, not the given forward transfer function, for pole location. Using the feedback formula, we obtain the equivalent system of Figure 6.4(b). The Routh-Hurwitz criterion will be applied to this denominator. First, label the rows with powers of s from s^3 down to s^0 in a vertical column, as shown in Table 6.3. Next, form the first row of the table, using the coefficients of the denominator of

Figure 6.4 (**a**) Feedback System for Example 6.1;
(**b**) Equivalent Closed-Loop System

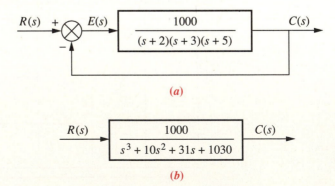

(**a**)

(**b**)

Table 6.3 Completed Routh Table for Example 6.1

s^3	1	31	0
s^2	~~10~~ 1	~~1030~~ 103	0
s^1	$-\dfrac{\begin{vmatrix} 1 & 31 \\ 1 & 103 \end{vmatrix}}{1} = -72$	$-\dfrac{\begin{vmatrix} 1 & 0 \\ 1 & 0 \end{vmatrix}}{1} = 0$	$-\dfrac{\begin{vmatrix} 1 & 0 \\ 1 & 0 \end{vmatrix}}{1} = 0$
s^0	$-\dfrac{\begin{vmatrix} 1 & 103 \\ -72 & 0 \end{vmatrix}}{-72} = 103$	$-\dfrac{\begin{vmatrix} 1 & 0 \\ -72 & 0 \end{vmatrix}}{-72} = 0$	$-\dfrac{\begin{vmatrix} 1 & 0 \\ -72 & 0 \end{vmatrix}}{-72} = 0$

the closed-loop transfer function. Start with the coefficient of the highest power and skip every other power of s. Now form the second row with the coefficients of the denominator skipped in the previous step. Subsequent rows are formed with determinants, as shown in Table 6.2.

For convenience, any row of the Routh table can be multiplied by a positive constant without changing the values of the rows below. This fact can be proven by examining the expressions for the entries and verifying that any multiplicative constant from a previous row cancels out. In the second row of Table 6.3, for example, the row was multiplied by $1/10$. We see later that care must be taken not to multiply the row by a negative constant.

Now that we know how to generate the Routh table, let us see how to interpret it.

Interpreting the Basic Routh Table

The basic Routh table applies to systems with poles in the left and right half-planes. Systems with imaginary poles and the kind of Routh table that results will be discussed in the next section. Simply stated, the Routh-Hurwitz criterion declares that *the number of roots of the polynomial that are in the right half-plane is equal to the number of sign changes in the first column.*

If the closed-loop transfer function has all poles in the left half of the s-plane, then the system is stable. Thus, a system is stable if there are no sign changes in the first column of the Routh table. For example, Table 6.3 has two sign changes in the first column. The first sign change occurs from 1 in the s^2 row to -72 in the s^1 row. The second sign change occurs from -72 in the s^1 row to 103 in the s^0 row. Thus, the system of Figure 6.4 is unstable since two poles exist in the right half-plane.

Now that we have described how to generate and interpret a basic Routh table, let us look at two special cases that can arise.

6.3 Routh-Hurwitz Criterion: Special Cases

Two special cases can occur: (1) the Routh table sometimes will have a zero *only in the first column* of a row, or (2) the Routh table sometimes will have an *entire row* that consists of zeros. Let us examine the first case.

A Zero Only in the First Column

If the first element of a row is zero, division by zero would be required to form the next row. To avoid this phenomenon, an epsilon, ϵ, is assigned to replace the zero in the first column. The value ϵ is then allowed to approach zero from either the positive or negative side, after which the signs of the entries in the first column can be determined. Let us look at an example.

Example 6.2

Find and interpret the Routh table when there is a zero only in the first column of a row.

Problem   Determine the stability of the closed-loop transfer function of Equation 6.2.

$$T(s) = \frac{10}{s^5 + 2s^4 + 3s^3 + 6s^2 + 5s + 3} \tag{6.2}$$

Solution   The solution is shown in Table 6.4. We form the Routh table by using the denominator of Equation 6.2. Begin by assembling the Routh table down to the row where a zero appears *only* in the first column (the s^3 row). Next, replace the zero by a small number, ϵ, and complete the table. To begin the interpretation, we must first assume a sign, positive or negative, for the quantity ϵ. Table 6.5 shows the first column of Table 6.4, along with the resulting signs for a choice of ϵ positive and ϵ negative.

Table 6.4 Completed Routh Table for Example 6.2

s^5	1	3	5
s^4	2	6	3
s^3	$\cancel{0}\ \epsilon$	$\dfrac{7}{2}$	0
s^2	$\dfrac{6\epsilon - 7}{\epsilon}$	3	0
s^1	$\dfrac{42\epsilon - 49 - 6\epsilon^2}{12\epsilon - 14}$	0	0
s^0	3	0	0

Table 6.5 Determining the Signs in the First Column of a Routh Table That Has a Row Where the First Element is Zero

Label	First Column	$\epsilon = +$	$\epsilon = -$
s^5	1	+	+
s^4	2	+	+
s^3	$0 \; \epsilon$	+	−
s^2	$\dfrac{6\epsilon - 7}{\epsilon}$	−	+
s^1	$\dfrac{42\epsilon - 49 - 6\epsilon^2}{12\epsilon - 14}$	+	+
s^0	3	+	+

If ϵ is chosen to be positive, then Table 6.5 will show a sign change from the s^3 row to the s^2 row, and there will be another sign change from the s^2 row to the s^1 row. Hence, the system is unstable and has two poles in the right half-plane.

Alternately, we could choose ϵ to be negative. Table 6.5 then will show a sign change from the s^4 row to the s^3 row. Another sign change will occur from the s^3 row to the s^2 row. Our result is exactly the same as that for a positive choice for ϵ. Thus the system is unstable, with two poles in the right half-plane.

We now look at the second special case.

An Entire Row Is Zero

Sometimes while making a Routh table, we find that an entire row consists of zeros due to the existence of an even polynomial that is a factor of the original polynomial. This case must be handled differently from the case of a zero in only the first column of a row. First, an example will demonstrate how to construct the Routh table. We then discuss how to interpret the Routh table. Finally, we offer a complete example that shows how to construct and interpret a Routh table where an entire row of zeros is present.

Example 6.3

Find the Routh table when there are zeros in an entire row.

Problem Determine the number of right-half-plane poles in the closed-loop transfer function of Equation 6.3.

$$T(s) = \frac{10}{s^5 + 7s^4 + 6s^3 + 42s^2 + 8s + 56} \tag{6.3}$$

Table 6.6 Routh Table for Example 6.3

s^5	1	6	8
s^4	~~7~~ 1	~~42~~ 6	~~56~~ 8
s^3	~~0~~ ~~4~~ 1	~~0~~ ~~12~~ 3	~~0~~ ~~0~~ 0
s^2	3	8	0
s^1	$\dfrac{1}{3}$	0	0
s^0	8	0	0

Solution Start by forming the Routh table for the denominator of Equation 6.3 (see Table 6.6). At the second row, we multiply through by 1/7 for convenience. We stop at the third row, since the entire row consists of zeros, and use the following procedure. First, we return to the row immediately above the row of zeros and form an auxiliary polynomial, using the entries in that row as coefficients. The polynomial will start with the power of s in the label column and continue by skipping every other one and diminishing in power. Thus the polynomial formed for this example is

$$P(s) = s^4 + 6s^2 + 8 \tag{6.4}$$

Next, we differentiate the polynomial with respect to s and obtain

$$\frac{dP(s)}{ds} = 4s^3 + 12s + 0 \tag{6.5}$$

Finally, we use the coefficients of Equation 6.5 to replace the row of zeros. Again, for convenience, the third row is multiplied by 1/4 after replacing the zeros.

The remainder of the table is formed in a straightforward manner by following the standard form shown in Table 6.2. Table 6.6 shows that all entries in the first column are positive. Hence, there are no right-half-plane poles.

Let us look further into the case that yields an entire row of zeros. An entire row of zeros will appear in the Routh table when a purely even or purely odd polynomial is a factor of the original polynomial. For example, $s^4 + 5s^2 + 7$ is an even polynomial; it has only even powers of s. Even polynomials only have roots that are symmetrical about the origin. This symmetry can occur under several conditions of root position: (1) the roots are symmetrical and real, (2) the roots are symmetrical and imaginary, (3) the roots are quadrantal. Figure 6.5 shows each of these cases. Each case or combination of these cases will generate an even polynomial.

It is this even polynomial that causes the row of zeros to appear. Thus, the row of zeros tells us of the existence of an even polynomial whose roots are

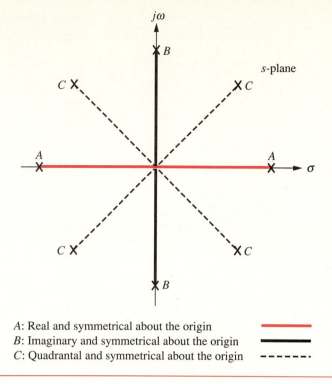

A: Real and symmetrical about the origin

B: Imaginary and symmetrical about the origin

C: Quadrantal and symmetrical about the origin

Figure 6.5 Root Positions to Generate Even Polynomials
(Pattern A, B, C, or Any Combination)

symmetric about the origin. Some of these roots could be on the $j\omega$-axis. On the other hand, since $j\omega$ roots are symmetric about the origin, if we do not have a row of zeros, we cannot possibly have $j\omega$ roots.

Another characteristic of the Routh table for the case in question is that the row previous to the row of zeros contains the even polynomial that is a factor of the original polynomial. Finally, everything from the row containing the even polynomial down to the end of the Routh table is a test of only the even polynomial. Let us put these facts together in an example.

Example 6.4

Generate and interpret a Routh table that has a complete row of zeros.

Problem For the transfer function of Equation 6.6, tell how many poles are in the right half-plane, the left half-plane, and on the $j\omega$-axis.

$$T(s) = \frac{20}{s^8 + s^7 + 12s^6 + 22s^5 + 39s^4 + 59s^3 + 48s^2 + 38s + 20} \quad (6.6)$$

Solution Use the denominator and form the Routh table in Table 6.7. For convenience, the s^6 row is multiplied by $1/10$, and the s^5 row is multiplied by $1/20$. At

Table 6.7 Routh Table for Example 6.4

s^8	1	12	39	48	20
s^7	1	22	59	38	0
s^6	~~−10~~ −1	~~−20~~ −2	~~10~~ 1	~~20~~ 2	0
s^5	~~20~~ 1	~~60~~ 3	~~40~~ 2	0	0
s^4	1	3	2	0	0
s^3	~~0~~ ~~4~~ 2	~~0~~ ~~6~~ 3	~~0~~ ~~0~~ 0	0	0
s^2	$\dfrac{3}{2}$ 3	~~2~~ 4	0	0	0
s^1	$\dfrac{1}{3}$	0	0	0	0
s^0	4	0	0	0	0

the s^3 row, we obtain a row of zeros. Moving back one row to s^4, we extract the even polynomial, $P(s)$, as

$$P(s) = s^4 + 3s^2 + 2 \tag{6.7}$$

This polynomial will divide evenly into the denominator of Equation 6.6 and thus is a factor. Taking the derivative with respect to s to obtain the coefficients that replace the row of zeros in the s^3 row, we find

$$\frac{dP(s)}{ds} = 4s^3 + 6s + 0 \tag{6.8}$$

Replace the row of zeros with 4, 6, and 0 and multiply the row by $1/2$ for convenience. Finally, continue the table to the s^0 row, using the standard procedure. How do we now interpret this Routh table?

Since all entries from the even polynomial at the s^4 row down to the s^0 row are a test of the even polynomial, we begin to draw some conclusions about the roots of the even polynomial. No sign changes exist from the s^4 row down to the s^0 row. Thus, the even polynomial does not have right-half-plane poles. Since there are no right-half-plane poles, no left-half-plane poles are present because of the requirement for symmetry. Hence, the even polynomial, Equation 6.7, must have all four of its poles on the $j\omega$-axis.[3] These results are summarized in the first column of Table 6.8.

---

[3] A necessary condition for stability is that the $j\omega$ roots must have unit multiplicity. The even polynomial must be checked for multiple $j\omega$ roots. For this case, the existence of multiple $j\omega$ roots would lead to a perfect, fourth-order square polynomial. Since Equation 6.7 is not a perfect square, the four $j\omega$ roots are distinct.

Table 6.8 Summary of Pole Locations
for Example 6.4

Even *(4th-order)*	*Rest* *(4th-order)*	*Total* *(8th-order)*
0 rhp	2 rhp	2 rhp
0 lhp	2 lhp	2 lhp
4 $j\omega$	0 $j\omega$	4 $j\omega$

Note: rhp = right half-plane, lhp = left
half-plane.

The remaining roots of the total polynomial are evaluated from the s^8 row down to the s^4 row. We notice two sign changes: one from the s^7 row to the s^6 row and the other from the s^6 row to the s^5 row. Thus, the rest of the polynomial must have two roots in the right half-plane. These results are included in Table 6.8 under *Rest*. The final tally is the sum of roots from each component part, the even polynomial and the rest of the total polynomial as shown under *Total* in Table 6.8. Thus, the system has two poles in the right half-plane, two poles in the left half-plane, and four poles on the $j\omega$-axis; it is unstable because of the existence of right-half-plane poles.

We now summarize what we have learned about polynomials that generate entire rows of zeros in the Routh table. These polynomials have a purely even factor with roots that are symmetrical about the origin. The even polynomial appears in the Routh table in the row directly above the row of zeros. Every entry in the table from the even polynomial's row to the end of the chart applies only to the even polynomial. Therefore, the number of sign changes from the even polynomial to the end of the table equals the number of right-half-plane roots of the even polynomial. Because of the symmetry of roots about the origin, the even polynomial must have the same number of left-half-plane roots as it does right-half-plane roots. Having accounted for the roots in the right and left half-planes, we know the remaining roots must be on the $j\omega$-axis.

Every entry in the Routh table from the beginning of the chart to the even polynomial applies only to the remaining factor of the original polynomial after the even polynomial has been extracted. For this factor, the number of sign changes, from the beginning of the table down to the even polynomial, equals the number of right-half-plane roots. The remaining roots are left-half-plane roots. There can be no $j\omega$ roots contained in the rest of the polynomial.

Let us demonstrate the usefulness of the Routh-Hurwitz criterion with a few additional examples.

6.4 Additional Examples Using the Routh-Hurwitz Criterion

The previous two sections have introduced the Routh-Hurwitz criterion. Now we need to demonstrate the method's application to a number of analysis and design problems.

Example 6.5

Find the number of closed-loop poles in each of the three regions of the s-plane.

Problem   Find the number of poles in the left half-plane, the right half-plane, and on the $j\omega$-axis for the system of Figure 6.6.

Solution   First, find the closed-loop transfer function shown in Equation 6.9.

$$T(s) = \frac{200}{s^4 + 6s^3 + 11s^2 + 6s + 200} \tag{6.9}$$

The Routh table for the denominator of Equation 6.9 is shown in Table 6.9. For clarity, we leave most zero cells blank. At the s^1 row there is a negative coefficient; thus, there are two sign changes. The system is unstable since it has two right-half-plane poles and two left-half-plane poles. The system cannot have $j\omega$ poles since a row of zeros did not appear in the Routh table.

Figure 6.6 Feedback Control System for Example 6.5

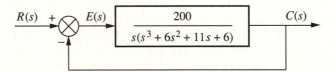

Table 6.9 Routh Table for Example 6.5

s^4	1	11	200
s^3	~~6~~ 1	~~6~~ 1	
s^2	~~10~~ 1	~~200~~ 20	
s^1	−19		
s^0	20		

The next example demonstrates the occurrence of a zero in only the first column of a row.

Example 6.6 Find the number of closed-loop poles in each of the three regions of the *s*-plane.

Problem Find the number of poles in the left half-plane, the right half-plane, and on the *jω*-axis for the system of Figure 6.7.

Solution The closed-loop transfer function is shown in Equation 6.10.

$$T(s) = \frac{1}{2s^5 + 3s^4 + 2s^3 + 3s^2 + 2s + 1} \tag{6.10}$$

Form the Routh table shown in Table 6.10, using the denominator of Equation 6.10. A zero appears in the first column of the s^3 row. Since the entire row is not zero, simply replace the zero with a small quantity, ϵ, and continue the table. Permitting ϵ to be a small, positive quantity, we find that the first term of the s^2 row is negative. Thus, there are two sign changes, and the system is unstable, with two poles in the right half-plane. The remaining poles are in the left half-plane.

Figure 6.7 Feedback Control System for Example 6.6

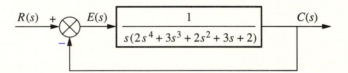

Table 6.10 Routh Table for Example 6.6

s^5	2	2	2
s^4	3	3	1
s^3	$\cancel{0}\ \epsilon$	$\dfrac{4}{3}$	
s^2	$\dfrac{3\epsilon - 4}{\epsilon}$	1	
s^1	$\dfrac{12\epsilon - 16 - 3\epsilon^2}{9\epsilon - 12}$		
s^0	1		

In the next example we see an entire row of zeros appear along with the possibility of imaginary roots.

Example 6.7

Find the number of closed-loop poles in each of the three regions of the s-plane.

Problem Find the number of poles in the left half-plane, the right half-plane, and at the $j\omega$-axis for the system of Figure 6.8. Draw conclusions about the stability of the closed-loop system.

Solution The closed-loop transfer function for the system of Figure 6.8 is

$$T(s) = \frac{128}{s^8 + 3s^7 + 10s^6 + 24s^5 + 48s^4 + 96s^3 + 128s^2 + 192s + 128} \tag{6.11}$$

Using the denominator, form the Routh table, as shown in Table 6.11. A row of zeros appears in the s^5 row. Thus, the closed-loop transfer function denominator must have an even polynomial as a factor. Return to the s^6 row and form the even polynomial:

$$P(s) = s^6 + 8s^4 + 32s^2 + 64 \tag{6.12}$$

Figure 6.8 Feedback Control System for Example 6.7

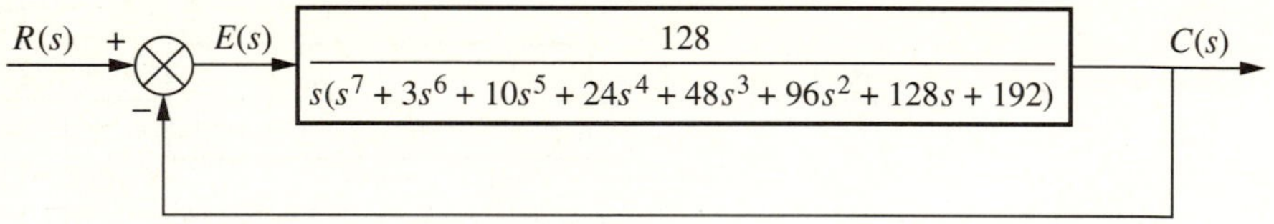

Table 6.11 Routh Table for Example 6.7

s^8	1	10	48	128	128
s^7	~~3~~ 1	~~24~~ 8	~~96~~ 32	~~192~~ 64	
s^6	~~2~~ 1	~~16~~ 8	~~64~~ 32	~~128~~ 64	
s^5	~~0~~ ~~6~~ 3	~~0~~ ~~32~~ 16	~~0~~ ~~64~~ 32	~~0~~ ~~0~~ 0	
s^4	$\frac{8}{3}$ 1	$\frac{64}{3}$ 8	~~64~~ 24		
s^3	~~-8~~ -1	~~-40~~ -5			
s^2	~~3~~ 1	~~24~~ 8			
s^1	3				
s^0	8				

Table 6.12 Summary of Pole
Locations for Example 6.7

Even (6th-order)	Rest (2nd-order)	Total (8th-order)
2 rhp	0 rhp	2 rhp
2 lhp	2 lhp	4 lhp
2 $j\omega$	0 $j\omega$	2 $j\omega$

Note: rhp = right half-plane, lhp = left half-plane.

Differentiate this polynomial with respect to s to form the coefficients that will replace the row of zeros:

$$\frac{dP(s)}{ds} = 6s^5 + 32s^3 + 64s + 0 \qquad (6.13)$$

Replace the row of zeros at the s^5 row by the coefficients of Equation 6.13 and multiply through by $1/2$ for convenience. Then complete the table.

We note that there are two sign changes from the even polynomial at the s^6 row down to the end of the table. Hence, the even polynomial has two right-half-plane poles. Because of the symmetry about the origin, the even polynomial must have an equal number of left-half-plane poles. Therefore, the even polynomial has two left-half-plane poles. Since the even polynomial is of sixth order, the two remaining poles must be on the $j\omega$-axis.

There are no sign changes from the beginning of the table down to the even polynomial at the s^6 row. Therefore, the rest of the polynomial has no right-half-plane poles. The results are summarized in Table 6.12. The system has two poles in the right half-plane, four poles in the left half-plane, and two poles on the $j\omega$-axis, which are of unit multiplicity. Thus the closed-loop system is unstable because of the right-half-plane poles.

The Routh-Hurwitz criterion gives vivid proof that changes in the gain of a feedback control system result in differences in transient response because of changes in closed-loop pole locations. The next example demonstrates this concept. We will see that for systems such as that shown in Figure 6.9, gain variations can move poles from stable regions of the s-plane on to the $j\omega$-axis and then into the right half-plane.

Example 6.8   Show the effect of gain upon the stability of a closed-loop feedback control system.

Problem   Find the range of gain, K, for the system of Figure 6.10 that will cause the system to be stable, unstable, and marginally stable. Assume $K > 0$.

Solution   First, find the closed-loop transfer function shown in Equation 6.14.

$$T(s) = \frac{K}{s^3 + 18s^2 + 77s + K} \qquad (6.14)$$

Next, form the Routh table, as shown in Table 6.13.

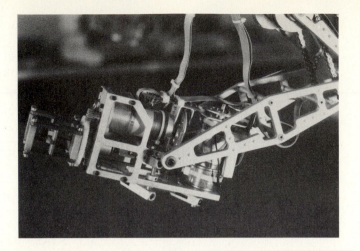

Figure 6.9 Robotic Arm Showing Mechanisms for Wrist Rotation and Grip (Photo by Mark E. Van Dusen)

Figure 6.10 Feedback Control System for Example 6.8

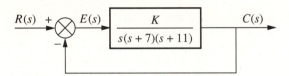

Table 6.13 Routh Table for Example 6.8

s^3	1	77
s^2	18	K
s^1	$\dfrac{1386 - K}{18}$	
s^0	K	

Table 6.14 Routh Table for Example 6.8 with $K = 1386$

s^3	1	77
s^2	18	1386
s^1	$\not{0}$ 36	
s^0	1386	

Since K is assumed positive, we see that all elements in the first column are always positive except the s^1 row. This entry can be positive, zero, or negative, depending upon the value of K. If $K < 1386$, all terms in the first column will be positive, and since there are no sign changes, the system will have three poles in the left half-plane and be *stable*.

If $K > 1386$, the s^1 term in the first column is negative. There are two sign changes, indicating that the system has two right-half-plane poles and one left-half-plane pole, which makes the system *unstable*.

If $K = 1386$, we have an entire row of zeros, which could signify $j\omega$ poles. Returning to the s^2 row and replacing K with 1386, we form the even polynomial shown in Equation 6.15:

$$P(s) = 18s^2 + 1386 \tag{6.15}$$

Differentiating with respect to s, we have

$$\frac{dP(s)}{ds} = 36s + 0 \tag{6.16}$$

Replacing the row of zeros with the coefficients of Equation 6.16, we obtain the Routh-Hurwitz table shown in Table 6.14 for the case of $K = 1386$.

Since there are no sign changes from the even polynomial (s^2 row) down to the bottom of the table, the even polynomial has its two roots on the $j\omega$-axis of unit multiplicity. Since there are no sign changes above the even polynomial, the remaining root is in the left half-plane. Therefore the system is *marginally stable*.

The Routh-Hurwitz criterion is often used in limited applications to factor polynomials containing even factors. Let us look at an example.

Example 6.9   Use the Routh-Hurwitz criterion to factor polynomials that contain even factors.

Problem   Factor the polynomial in Equation 6.17.

$$s^4 + 3s^3 + 30s^2 + 30s + 200 \tag{6.17}$$

Solution   Form the Routh table of Table 6.15. We find that the s^1 row is a row of zeros. Now form the even polynomial at the s^2 row:

$$P(s) = s^2 + 10 \tag{6.18}$$

Table 6.15 Routh Table for Example 6.9

s^4	1	30	200
s^3	$\cancel{3}$ 1	$\cancel{30}$ 10	
s^2	$\cancel{20}$ 1	$\cancel{200}$ 10	
s^1	$\cancel{0}$ 2	$\cancel{0}$ 0	
s^0	10		

This polynomial is differentiated with respect to s in order to complete the Routh table. However, since this polynomial is a factor of the original polynomial in Equation 6.17, dividing Equation 6.17 by 6.18 yields $(s^2 + 3s + 20)$ as the other factor. Hence,

$$
\begin{aligned}
s^4 + 3s^3 + 30s^2 + 30s + 200 &= (s^2 + 10)(s^2 + 3s + 20) \\
&= (s + j3.1623)(s - j3.1623) \\
&\quad \times (s + 1.5 + j4.213)(s + 1.5 - j4.213) \quad (6.19)
\end{aligned}
$$

Up to this point, we have examined stability from the s-plane viewpoint. Now we look at stability from the perspective of state space.

*6.5 Stability via Eigenvalues for the State-Space Representation

In Section 4.9, we mentioned that the values of the system's poles were equal to the eigenvalues of the system matrix, $\mathbf{A}$. We stated that the eigenvalues of the matrix $\mathbf{A}$ were solutions of the equation $\det(s\mathbf{I} - \mathbf{A}) = 0$, which also yielded the poles of the transfer function. Eigenvalues appeared again in Section 5.8, where they were formally defined and used to diagonalize a matrix. Let us now formally show that the eigenvalues and the system poles have the same values.

Reviewing Section 5.8, the eigenvalues of a matrix, $\mathbf{A}$, are values of λ that permit a nontrivial solution (other than $\mathbf{0}$) for $\mathbf{x}$ in the equation

$$\mathbf{Ax} = \lambda\mathbf{x} \tag{6.20}$$

In order to solve for those values of λ that do indeed permit a solution for $\mathbf{x}$, we rearrange Equation 6.20 as follows:

$$\lambda\mathbf{x} - \mathbf{Ax} = \mathbf{0} \tag{6.21}$$

or

$$(\lambda\mathbf{I} - \mathbf{A})\mathbf{x} = \mathbf{0} \tag{6.22}$$

Solving for **x** yields

$$\mathbf{x} = (\lambda \mathbf{I} - \mathbf{A})^{-1}\mathbf{0} \tag{6.23}$$

or

$$\mathbf{x} = \frac{\text{adj}\,(\lambda \mathbf{I} - \mathbf{A})}{\det\,(\lambda \mathbf{I} - \mathbf{A})}\mathbf{0} \tag{6.24}$$

We see that all solutions will be the null vector except for the occurrence of zero in the denominator. Since this is the only condition where elements of **x** will be $0/0$, or indeterminant, it is the only case where a nonzero solution is possible.

The values of λ are calculated by forcing the denominator to zero:

$$\det\,(\lambda \mathbf{I} - \mathbf{A}) = 0 \tag{6.25}$$

This equation determines those values of λ for which a nonzero solution for **x** in Equation 6.20 exists. In Section 5.8 we defined **x** to be *eigenvectors,* and the values of λ were defined to be the *eigenvalues* of the matrix **A**.

Let us now relate the eigenvalues of the system matrix, **A**, to the system's poles. In Chapter 3 we derived the equation of the system transfer function from the state equations (Equation 3.73). The system transfer function has $\det\,(s\mathbf{I} - \mathbf{A})$ in the denominator because of the presence of $(s\mathbf{I} - \mathbf{A})^{-1}$. Thus,

$$\det\,(s\mathbf{I} - \mathbf{A}) = 0 \tag{6.26}$$

is the characteristic equation for the system from which the system poles can be found.

Since Equations 6.25 and 6.26 are identical other than a change in variable name, we conclude that the eigenvalues of the matrix **A** are identical to the system's poles. Thus, we can determine the stability of a system represented in state space by finding the eigenvalues of the system matrix, **A,** and determining their locations on the s-plane.

Example 6.10   Determine the poles of a system represented in state space.

Problem   Given the system defined by Equations 6.27, find out how many poles are in the left half-plane, in the right half-plane, and on the $j\omega$-axis.

$$\dot{\mathbf{x}} = \begin{bmatrix} 0 & 1 & 0 \\ 0 & 0 & 1 \\ -10 & -5 & -2 \end{bmatrix}\mathbf{x} + \begin{bmatrix} 10 \\ 0 \\ 0 \end{bmatrix}u(t) \tag{6.27a}$$

$$y = \begin{bmatrix} 1 & 0 & 0 \end{bmatrix}\mathbf{x} \tag{6.27b}$$

Solution   First form $(s\mathbf{I} - \mathbf{A})$:

$$(s\mathbf{I} - \mathbf{A}) = \begin{bmatrix} s & 0 & 0 \\ 0 & s & 0 \\ 0 & 0 & s \end{bmatrix} - \begin{bmatrix} 0 & 1 & 0 \\ 0 & 0 & 1 \\ -10 & -5 & -2 \end{bmatrix} = \begin{bmatrix} s & -1 & 0 \\ 0 & s & -1 \\ 10 & 5 & s+2 \end{bmatrix} \tag{6.28}$$

Table 6.16 Routh Table for
Example 6.10

s^3	1	5
s^2	~~2~~ 1	~~10~~ 5
s^1	~~0~~ 2	~~0~~ 0
s^0	5	

Now find the det $(s\mathbf{I} - \mathbf{A})$:

$$\det(s\mathbf{I} - \mathbf{A}) = s^3 + 2s^2 + 5s + 10 \qquad (6.29)$$

Using this polynomial, form the Routh table, as shown in Table 6.16.

A row of zeros appeared at the s^1 row; thus, a second-order, even polynomial can be extracted at the s^2 row. A test on the even polynomial shows that there are no right-half-plane poles since there are no sign changes below the even polynomial. Thus, the second-order even polynomial has two $j\omega$ roots. The remaining third pole for the system is in the left half-plane since there are no sign changes above the even polynomial. In summary, the system has one left-half-plane pole and two $j\omega$ poles. It is therefore marginally stable.

6.6 Chapter-Objective Demonstration Problems

This chapter covered the elements of stability. We saw that stable systems have their closed-loop poles in the left half of the s-plane. As the loop gain is changed, the locations of the poles are also changed, creating the possibility that the poles can move into the right half of the s-plane, which yields instability. Proper gain settings are essential for the stability of closed-loop systems. The following chapter-objective problems demonstrate the proper setting of the loop gain to ensure stability. The antenna azimuth control system and the Unmanned Free-Swimming Submersible vehicle serve as examples.

Example 6.11   Determine the range of loop gain required to keep a closed-loop system stable.

Problem   Given the schematic for the antenna azimuth position control system shown in Figure 2.36, find the range of preamplifier gain required to keep the closed-loop system stable.

Solution   The closed-loop transfer function was derived in Chapter 5 (see Figure 5.31):

$$T(s) = \frac{6.63K}{s^3 + 101.71s^2 + 171s + 6.63K} \qquad (6.30)$$

Table 6.17 Routh Table for Example 6.11

s^3	1	171
s^2	101.71	6.63K
s^1	17392.41 − 6.63K	0
s^0	6.63K	

Using the denominator, create the Routh table shown in Table 6.17. The third row of the table shows that a row of zeros occurs if $K = 2623.29$. This value of K makes the system marginally stable. Therefore, there will be no sign changes in the first column if $0 < K < 2623.29$. We conclude that, for stability, $0 < K < 2623.29$.

The pitch control system for the Unmanned Free-Swimming Submersible (UFSS) vehicle is part of the depth control system. In the next example, we find the range of pitch gain that will keep the pitch-control loop stable.

Example 6.12 Find the range of gain required to keep a closed-loop system stable.

Problem The pitch-control loop for the UFSS vehicle (Johnson, 1980) is shown in Figure 5.34. If the pitch gain is made variable with gain $-K$, find the range of K that ensures that the closed-loop pitch control system is stable.

Solution The first step is to reduce Figure 5.34 to a single, closed-loop transfer function. The equivalent forward transfer function, $G_e(s)$, is

$$G_e(s) = \frac{0.25K(s + 0.435)}{s^4 + 3.456s^3 + 3.457s^2 + 0.719s + 0.0416} \tag{6.31}$$

With unity feedback, the closed-loop transfer function, $T(s)$, is

$$T(s) = \frac{0.25K(s + 0.435)}{s^4 + 3.456s^3 + 3.457s^2 + (0.719 + 0.25K)s + (0.0416 + 0.109K)} \tag{6.32}$$

The denominator of Equation 6.32 is now used to form the Routh table shown in Table 6.18.

Looking at the first column, the s^4 and s^3 rows are positive. Thus, all elements of the first column must be positive for stability. For the first column of the s^2 row to be positive, $-\infty < K < 44.912$. For the first column of the s^1 row to be positive, the numerator must be positive since the denominator is positive from the previous step. The solution to the quadratic term in the numerator yields roots of $K = -4.685$ and 25.869. Thus, for a positive numerator, $-4.685 < K < 25.869$. Finally, for the first column of the s^0 row to be positive, $-0.382 < K < \infty$. Using all three conditions, stability will be ensured if $-0.382 < K < 25.869$.

Table 6.18 Routh Table for Example 6.12

s^4	1	3.457	$0.0416 + 0.109K$
s^3	3.456	$0.719 + 0.25K$	
s^2	$11.228 - 0.25K$	$0.144 + 0.377K$	
s^1	$\dfrac{-0.0625K^2 + 1.324K + 7.575}{11.228 - 0.25K}$		
s^0	$0.144 + 0.377K$		

Note: Some rows have been multiplied by a positive constant for convenience.

In our chapter-objective demonstration problems, we calculated the ranges of gain to ensure stability. The reader should be aware that, although these ranges yield stability, setting gain within these limits may not yield the desired transient response or steady-state error characteristics. In Chapters 9 and 11 we will explore design techniques that yield more flexibility in obtaining desired characteristics than a simple gain adjustment. We conclude this chapter with a summary of the concepts of stability. In the next chapter we look at steady-state errors, the last of the three control system requirements.

6.7 Summary

In this chapter we explored the concepts of system stability from both the classical and state-space viewpoints. We found that for linear systems, stability is based upon a natural response that decays to zero as time approaches infinity. On the other hand if the natural response increases without bound, the forced response is overpowered by the natural response, and we lose control. This condition is known as *instability*. A third possibility exists: the natural response may neither decay nor grow without bound, but oscillate. In this case the system is said to be *marginally stable*.

Mathematically, stability for linear, time-invariant systems can be determined from the location of the closed-loop poles. If the poles are only in the left half-plane, the system is stable. If any poles are in the right half-plane, the system is unstable. Finally, if the poles are on the $j\omega$-axis and in the left half-plane, the system is marginally stable as long as the poles on the $j\omega$-axis are of unit multiplicity; it is unstable if there are any multiple $j\omega$ poles. Unfortunately, although the open-loop poles may be known, we found that in higher-order systems it is difficult to find the closed-loop poles without a computer program.

The Routh-Hurwitz criterion lets us find how many poles are in each of the sections of the s-plane without giving us the coordinates of the poles. Just knowing that there are poles in the right half-plane is enough to determine that a system is unstable. Under certain limited conditions, when an even polynomial is present, the Routh table can be used to factor the system's characteristic equation.

Obtaining stability from the state-space representation of a system is based upon the same concept—the location of the roots of the characteristic equation. These roots are equivalent to the eigenvalues of the system matrix and can be found by solving $\det(s\mathbf{I} - \mathbf{A}) = 0$. Again, the Routh-Hurwitz criterion can be applied to this polynomial. The point here is that the state-space representation of a system need not be converted back to a transfer function in order to investigate stability.

REVIEW QUESTIONS

1. What part of the output response is responsible for determining the stability of a linear system?

2. What happens to the response named in Question 1 that creates instability?

3. What would happen to a physical system that becomes unstable?

4. Where do system poles have to be to ensure that a system is not unstable?

5. What does the Routh-Hurwitz criterion tell us?

6. Under what conditions would the Routh-Hurwitz criterion easily tell us the actual location of the system's closed-loop poles?

7. What causes a zero to show up only in the first column of the Routh table?

8. What causes an entire row of zeros to show up in the Routh table?

9. Why do we sometimes multiply a row of a Routh table by a positive constant?

10. Why do we not multiply a row of a Routh table by a negative constant?

11. If a Routh table has two sign changes above the even polynomial and five sign changes below the even polynomial, how many right-half-plane poles does the system have?

12. Does the presence of an entire row of zeros always mean that the system has $j\omega$ poles?

13. If a seventh-order system has a row of zeros at the s^3 row and two sign changes below the s^4 row, how many $j\omega$ poles does the system have?

*14. Is it true that the eigenvalues of the system matrix are the same as the closed-loop poles?

*15. How do we find the eigenvalues?

PROBLEMS

1. Tell how many roots of the following polynomial are in the right half-plane, the left half-plane, and on the $j\omega$-axis.

$$P(s) = s^5 + 2s^4 + 2s^3 + 4s^2 + s + 2$$

2. Using the Routh table, tell how many poles of the following function are in the right half-plane, the left half-plane, and on the $j\omega$-axis.

$$T(s) = \frac{s^2 + 7s + 10}{s^6 + 2s^4 - s^2 - 2}$$

3. The closed-loop transfer function of a system is

$$T(s) = \frac{s + 5}{s^5 - s^4 + 3s^3 - 3s^2 + 2s - 1}$$

Determine how many closed-loop poles lie in the right half-plane, the left half-plane, and on the $j\omega$-axis.

4. How many poles are in the right half-plane, the left half-plane, and on the $j\omega$-axis for the open-loop system of Figure P6.1?

Figure P6.1

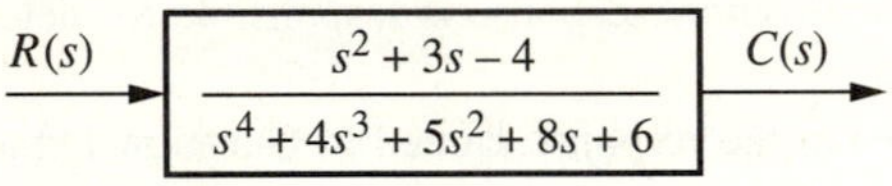

$$R(s) \longrightarrow \boxed{\dfrac{s^2 + 3s - 4}{s^4 + 4s^3 + 5s^2 + 8s + 6}} \longrightarrow C(s)$$

5. How many poles are in the right half-plane, the left half-plane, and on the $j\omega$-axis for the open-loop system of Figure P6.2.?

Figure P6.2

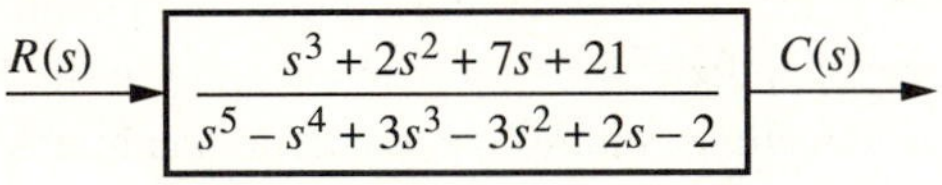

$$R(s) \longrightarrow \boxed{\dfrac{s^3 + 2s^2 + 7s + 21}{s^5 - s^4 + 3s^3 - 3s^2 + 2s - 2}} \longrightarrow C(s)$$

6. Determine if the unity feedback system of Figure P6.3 is stable if

$$G(s) = \frac{240}{(s + 1)(s + 2)(s + 3)(s + 4)}$$

Figure P6.3

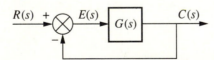

7. Consider the unity feedback system of Figure P6.3 with

$$G(s) = \frac{1}{4s^2(s^2 + 1)}$$

Using the Routh-Hurwitz criterion, find the region of the s-plane where the poles of the closed-loop system are located.

8. Given the unity feedback system of Figure P6.3 with

$$G(s) = \frac{4}{s(s^6 - 2s^5 + 2s^4 - 4s^3 - s^2 + 2s - 2)}$$

find out how many poles of the closed-loop transfer function lie in the right half-plane, the left half-plane, and on the $j\omega$-axis.

9. Using the Routh-Hurwitz criterion and the unity feedback system of Figure P6.3 with

$$G(s) = \frac{1}{2s^3 + s^2 + 2s}$$

tell whether or not the closed-loop system is stable.

10. Given the unity feedback system of Figure P6.3 with

$$G(s) = \frac{20}{s^5 + s^4 + 5s^3 + 5s^2 + 20s}$$

tell where the closed-loop poles are located (i.e., right half-plane, left half-plane, $j\omega$-axis).

11. Given the unity feedback system of Figure P6.3 with

$$G(s) = \frac{-3}{s(s^4 - s^3 + 4s^2 - 4s + 3)}$$

find out how many closed-loop poles are in the right half-plane, the left half-plane, and on the $j\omega$-axis.

12. Consider the following Routh table. Notice that the s^5 row was originally all zeros. Tell how many roots of the original polynomial were in the right half-plane, the left half-plane, and on the $j\omega$-axis.

s^7	1	2	−1	−2
s^6	1	2	−1	−2
s^5	3	4	−1	0
s^4	1	−1	−3	0
s^3	7	8	0	0
s^2	−15	−21	0	0
s^1	−9	0	0	0
s^0	−21	0	0	0

13. For the system of Figure P6.4, tell where the closed-loop poles are located (i.e., right half-plane, left half-plane, $j\omega$-axis). Notice that there is positive feedback.

Figure P6.4

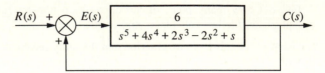

14. Determine if the unity feedback system of Figure P6.3 with

$$G(s) = \frac{K(s^2 + 1)}{(s + 1)(s + 2)}$$

can be unstable.

15. For the unity feedback system of Figure P6.3 with

$$G(s) = \frac{K(s + 6)}{s(s + 1)(s + 3)}$$

determine the range of K to ensure stability.

16. For the unity feedback system of Figure P6.3 with

$$G(s) = \frac{K(s - 1)(s + 2)(s + 3)}{(s^2 + 1)}$$

find the range of K to keep the system stable.

17. For the unity feedback system of Figure P6.3 with

$$G(s) = \frac{K(s + 1)}{s(s + 2)(s + 3)(s + 4)}$$

determine the range of K for stability.

18. Find the range of K for stability for the unity feedback system of Figure P6.3 with

$$G(s) = \frac{K(s + 1)(s - 1)}{(s^2 + 1)}$$

19. For the unity feedback system of Figure P6.3 with

$$G(s) = \frac{K(s + 1)}{s^4(s + 2)}$$

find the range of K for stability.

20. Given the unity feedback system of Figure P6.3 with

$$G(s) = \frac{K(s + 1)(s + 2)}{s(s + 3)(s + 4)}$$

find the range of K for stability.

21. Find the range of gain, K, to ensure stability in the unity feedback system of Figure P6.3 with

$$G(s) = \frac{K(s^2 + 1)}{(s - 1)(s + 2)(s + 3)}$$

22. Find the range of gain, K, to ensure stability in the unity feedback system of Figure P6.3 with

$$G(s) = \frac{K(s - 1)(s + 2)(s + 3)}{(s^2 + 1)}$$

23. Find the range of gain, K, to ensure stability in the unity feedback system of Figure P6.3 with

$$G(s) = \frac{K(s + 2)}{(s^2 + 1)(s + 4)(s - 1)}$$

24. Using the Routh-Hurwitz criterion, find the value of K that will yield oscillations for the unity feedback system of Figure P6.3 with

$$G(s) = \frac{K}{(s + 1)(s + 2)(s + 3)}$$

25. Find the range of K for which the system of Figure P6.5 is stable.

Figure P6.5

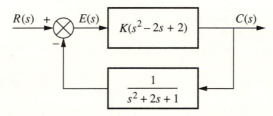

26. Using the Routh-Hurwitz criterion, find the following:
   a. The range of K that will ensure stability in the unity feedback system of Figure P6.3 with

$$G(s) = \frac{K(s + 1)^2}{s(s - 1)^2}$$

   b. The value of K that will cause oscillation
   c. The frequency of oscillation for part (**b**)

27. Given the unity feedback system of Figure P6.3 with

$$G(s) = \frac{K s(s + 2)}{(s^2 - 4s + 8)(s + 3)}$$

   a. Find the range of K for stability.
   b. Find the frequency of oscillation when the system is marginally stable.

28. For the unity feedback system of Figure P6.3 with

$$G(s) = \frac{K(s + 2)}{(s^2 + 1)(s + 4)(s - 1)}$$

find the range of K for which there will be only two closed-loop, right-half-plane poles.

29. For the unity feedback system of Figure P6.3 with

$$G(s) = \frac{K}{(s + 1)^3(s + 4)}$$

a. Find the range of K for stability.
b. Find the frequency of oscillation when the system is marginally stable.

30. Given the unity feedback system of Figure P6.3 with

$$G(s) = \frac{K}{(s + 10)(s^2 + 4s + 5)}$$

a. Find the range of K for stability.
b. Find the frequency of oscillation when the system is marginally stable.

31. Using the Routh-Hurwitz criterion and the unity feedback system of Figure P6.3 with

$$G(s) = \frac{K}{s(s + 1)(s + 2)(s + 5)}$$

a. Find the range of K for stability.
b. Find the value of K for marginal stability.
c. Find the actual location of the closed-loop poles when the system is marginally stable.

32. Find the range of K to keep the system shown in Figure P6.6 stable.

Figure P6.6

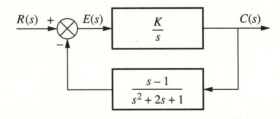

33. Find the value of K in the system of Figure P6.7 that will place the closed-loop poles as shown.

34. The closed-loop transfer function of a system is

$$T(s) = \frac{s^2 + K_1 s + K_2}{s^4 + K_1 s^3 + K_2 s^2 + 5s + 1}$$

Determine the range of K_1 in order for the system to be stable. What is the relationship between K_1 and K_2 for stability?

Figure P6.7 Closed-Loop System with Pole Plot

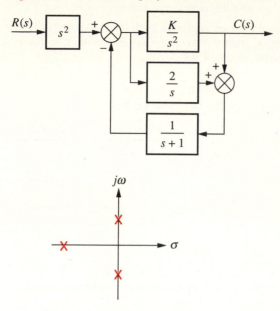

35. For the transfer function below, find the constraints on K_1 and K_2 such that the function will have only two $j\omega$ poles.

$$T(s) = \frac{K_1 s + K_2}{s^4 + K_1 s^3 + s^2 + K_2 s + 1}$$

36. The Unmanned Free-Swimming Submersible vehicle (Johnson, 1980) steers via the control system shown in Figure P6.8. A heading command is the input. The input along with feedback from the ship heading and heading rate are used to

Figure P6.8 Unmanned Free-Swimming Submersible Vehicle Heading Control (Adapted from Johnson, H., et al. *Unmanned Free-Swimming Submersible (UFSS) System Description*. NRL Memorandum Report 4393. Naval Research Laboratory, Washington, D.C., 1980)

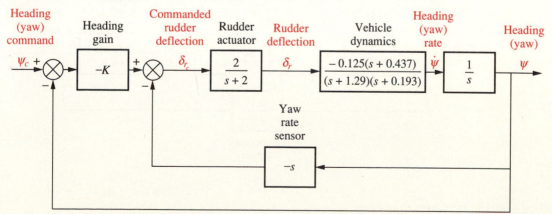

generate a rudder command that steers the ship. Find the range of heading gain that ensures the vehicle's stability.

37. A model for an airplane's pitch loop is shown in Figure P6.9. Find the range of gain, K, that will keep the system stable. Can the system ever be unstable for positive values of K?

Figure P6.9 Aircraft Pitch Loop Model

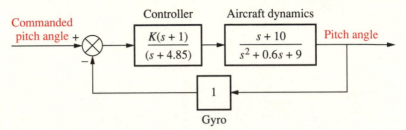

38. A common application in control systems is the regulation of the temperature of a chemical process (see Figure P6.10). The flow of chemical reactant to a process is controlled by an actuator and valve. The reactant causes the temperature in the vat to change. This temperature is then sensed and compared to a desired set-point temperature in a closed-loop, where in turn the flow of reactant is adjusted to yield the desired temperature. In Chapter 9 we will learn how a PID controller is used to improve the performance of such process-control systems. Figure P6.10 shows the control system prior to the addition of the PID controller. The PID controller is replaced by the shaded box with a gain of unity. For this system, prior to the design of the PID controller, find the range of amplifier gain, K, to keep the system stable.

Figure P6.10 Block Diagram of a Chemical Process-Control System

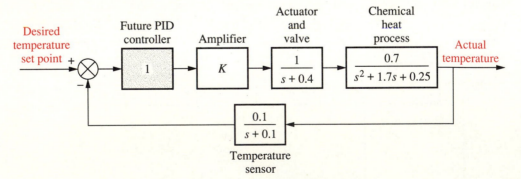

*39. A system is represented in state space as

$$\dot{\mathbf{x}} = \begin{bmatrix} 0 & 1 & 2 \\ 3 & 1 & -4 \\ 1 & 1 & 3 \end{bmatrix} \mathbf{x} + \begin{bmatrix} 0 \\ 1 \\ 0 \end{bmatrix} u$$

$$y = \begin{bmatrix} 1 & 1 & 0 \end{bmatrix} \mathbf{x}$$

Determine how many eigenvalues are in the right half-plane, the left half-plane, and on the $j\omega$-axis.

*40. Repeat Problem 39 for

$$\dot{\mathbf{x}} = \begin{bmatrix} 0 & 1 & 0 \\ 0 & 1 & -4 \\ -1 & 1 & 3 \end{bmatrix} \mathbf{x} = \begin{bmatrix} 0 \\ 0 \\ 1 \end{bmatrix} u$$

$$y = \begin{bmatrix} 0 & 0 & 1 \end{bmatrix} \mathbf{x}$$

41. *Chapter-objective problem:* Given the schematic for the antenna azimuth position control system shown in Figure P2.36, find the range of preamplifier gain required to keep the closed-loop system stable.

BIBLIOGRAPHY

D'Azzo, J., and Houpis, C. H. *Linear Control System Analysis and Design*. 3d ed. McGraw-Hill, New York, 1988.

Dorf, R. C. *Modern Control Systems*. 5th ed. Addison-Wesley, Reading, Mass., 1989.

Hostetter, G. H.; Savant, C. J., Jr.; and Stefani, R. T. *Design of Feedback Control Systems*. 2d ed. Saunders College Publishing, New York, 1989.

Johnson, H., et al. *Unmanned Free-Swimming Submersible (UFSS) System Description*. NRL Memorandum Report 4393. Naval Research Laboratory, Washington, D.C., 1980.

Routh, E. J. *Dynamics of a System of Rigid Bodies*. 6th ed. Macmillan, London, 1905.

Timothy, L. K., and Bona, B. E. *State Space Analysis: An Introduction*. McGraw-Hill, New York, 1968.

7

STEADY-STATE
ERRORS

7.1 Introduction

In Chapter 1 we saw that control systems analysis and design focuses on three specifications: (1) transient response, (2) stability, and (3) steady-state errors. Elements of transient analysis were derived in Chapter 4 for first- and second-order systems. These concepts are revisited in Chapter 8, where they are extended to higher-order systems. Stability was covered in Chapter 6, where we saw that forced responses were overpowered by natural responses that increase without bound if the system is unstable. Now, we are ready to examine steady-state errors. We will define the errors and derive methods of controlling them.

Steady-state error is the difference between the input and the output for a prescribed test input as $t \rightarrow \infty$. Table 1.1 shows test inputs used for control systems engineering. Let us focus on the last three entries in the table—steps, ramps, and parabolic waveforms—that are used for steady-state error analysis and design.

Step inputs represent constant position and are thus useful in determining the ability of a control system to position itself with respect to a stationary target such as a satellite in geostationary orbit (see Figure 7.1). An antenna position control is an example of a system that can be tested for accuracy using step inputs.

Ramp inputs represent constant velocity by their linearly increasing amplitude. These waveforms can be used to test a system's ability to follow a linearly increasing input, which is tantamount to tracking a constant-velocity target. For example, a control system that tracks a satellite that moves across the sky at a constant angular velocity, as shown in Figure 7.1, would be tested with a ramp input to evaluate the steady-state error between the satellite's angular position and that of the control system.

Finally, parabolas, whose second derivatives are constant, represent constant acceleration and can be used to represent accelerating targets such as the missile in Figure 7.1 to determine the steady-state error performance.

Since we are concerned with the difference between the input and output of a feedback control system after the steady state has been reached, our discussion will be limited to stable systems, where the natural response approaches zero as $t \rightarrow \infty$. Unstable systems represent loss of control in the steady state and are not acceptable for use at all. We will see that the expressions we derive to calculate

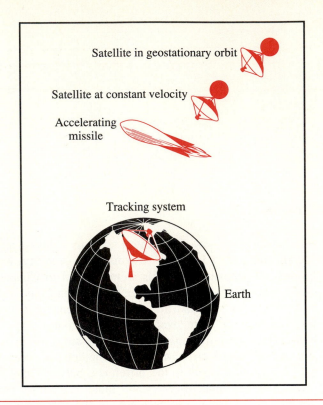

Figure 7.1 Different Target Types Require Different Test Inputs for Steady-State Error Analysis and Design

the steady-state error can be applied erroneously to an unstable system. Thus, the engineer must check the system for stability while performing steady-state error analysis and design. However, in order to focus on the topic, we assume that all examples and problems associated with this chapter are stable systems. For practice, you may want to test some of the systems for stability. Let us examine the concept of steady-state errors.

In Figure 7.2(*a*) a step, or constant-position, input and two possible outputs are shown. Output 1 has zero steady-state error, and output 2 has a finite steady-state error, $e_2(\infty)$. A similar example is shown in Figure 7.2 (*b*), where a ramp, or constant-velocity, input is compared with output 1, which has zero steady-state error, and output 2, which has a finite steady-state error, $e_2(\infty)$, as measured vertically between the input and output 2 after the transients have died down. For the ramp input, another possibility exists. If the output's slope is different than that of the input, output 3, shown in Figure 7.2(*b*), results. Here the steady-state error is infinite as measured vertically between the input and output 3 after the transients have died down.

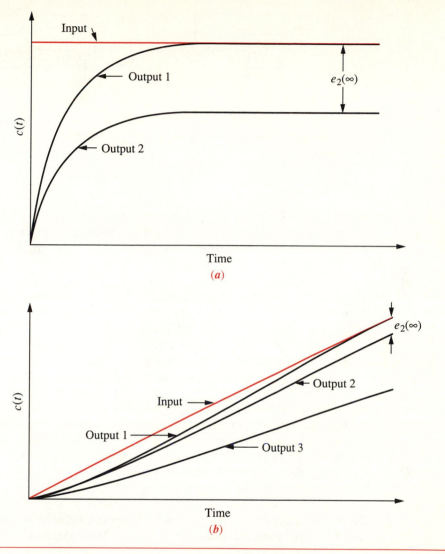

Figure 7.2 Demonstration of Steady-State Error: (*a*) Step Input; (*b*) Ramp Input

Let us now look at the error from the perspective of the most general block diagram. Since the error is the difference between the input and output of a system, we assume a closed-loop transfer function, $T(s)$, and form the error, $E(s)$, by taking the difference between the input and the output, as shown in Figure 7.3(*a*). Here we are interested in the steady-state, or final, value of $e(t)$. For unity feedback systems, $E(s)$ appears as shown in Figure 7.3(*b*). In this

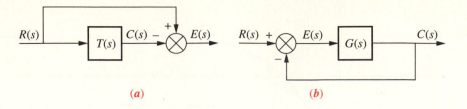

Figure 7.3 Definition of Error for Closed-Loop Control
Systems: (**a**) General Representation; (**b**) Alternate
Representation for Unity Feedback Systems

chapter we will first study and derive expressions for the steady-state error for
unity feedback systems and then expand to non-unity feedback systems. Before we
state our objectives and begin our study of steady-state errors for unity feedback
systems, let us look at the sources of the errors with which we will deal.

Many steady-state errors in control systems arise from nonlinear sources, such
as backlash in gears or a motor that will not move unless the input voltage exceeds
a threshold. Nonlinear behavior as a source of steady-state errors, although a
viable topic for study, is beyond the scope of a text on linear control systems.
The steady-state errors we study here are errors that arise from the configuration
of the system itself and the type of applied input.

For example, look at the system of Figure 7.4(*a*), where $R(s)$ is the input,
$C(s)$ is the output, and $E(s) = R(s) - C(s)$ is the error. Consider a step input.
In the steady state, if $c(t)$ equals $r(t)$, $e(t)$ will be zero. But with a pure gain,
K, the error, $e(t)$, cannot be zero if $c(t)$ is to be finite and nonzero. Thus, by
virtue of the configuration of the system (a pure gain of K in the forward path),
an error must exist. If we call $c_{\text{steady-state}}$ the steady-state value of the output and
$e_{\text{steady-state}}$ the steady-state value of the error, then

$$c_{\text{steady-state}} = K e_{\text{steady-state}} \tag{7.1}$$

Thus, the larger the value of K, the smaller the value of $e_{\text{steady-state}}$ would have
to be to yield a similar value of $c_{\text{steady-state}}$. The conclusion that we can draw is

Figure 7.4 (**a**) System with a Finite Steady-State Error for a Step Input;
(**b**) System with Zero Steady-State Error for a Step Input

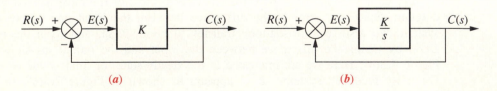

that with a pure gain in the forward path, there will always be a steady-state error for a step input. This error will diminish as the value of K increases.

If the forward-path gain is replaced by an integrator, as shown in Figure 7.4(b), there will be zero error in the steady state for a step input. The reasoning is as follows: as $c(t)$ increases, $e(t)$ will decrease, since $e(t) = r(t) - c(t)$. This decrease will continue until there is zero error, but there will still be a value for $c(t)$ since an integrator can have a constant output without any input. For example, a motor can be represented simply as an integrator. A voltage applied to the motor will cause rotation. When the applied voltage is removed, the motor will stop and remain at its present output position. Since it does not return to its initial position, we have an angular displacement output without an input to the motor. Therefore, a system similar to Figure 7.4(b), which uses a motor in the forward path, can have zero steady-state error for a step input.

We have qualitatively examined two cases to show how a system can be expected to exhibit various steady-state error characteristics, depending upon the system configuration. We now formalize the concepts and derive the relationships that exist between the steady-state errors and the system configuration generating these errors.

Chapter Objective

Given the azimuth position control system of Figure 2.2, the student will be able to derive the steady-state errors for step, ramp, and parabolic inputs. We now derive the steady-state errors for unity feedback systems.

7.2 Derivation of the Steady-State Errors for a Unity Feedback Control System

Consider the feedback control system shown in Figure 7.3(b). Since the feedback, $H(s)$, equals 1, the system has unity feedback. The implication here is that $E(s)$ is actually the error between the input, $R(s)$, and the output, $C(s)$. Thus, if we solve for $E(s)$, we will indeed have an expression for the error. We will then apply the final value theorem, Item 11 in Table 2.2, to evaluate the steady-state error.

Writing $E(s)$ from Figure 7.3(b), we obtain

$$E(s) = R(s) - C(s) \tag{7.2}$$

But

$$C(s) = E(s)G(s) \tag{7.3}$$

Finally, substituting Equation 7.3 into Equation 7.2 and solving for $E(s)$ yields

$$E(s) = \frac{R(s)}{1 + G(s)} \tag{7.4}$$

Now, applying the final value theorem,[1] which states that

$$e(\infty) = \lim_{s \to 0} sE(s) \qquad (7.5)$$

we obtain

$$e(\infty) = \lim_{s \to 0} \frac{sR(s)}{1 + G(s)} \qquad (7.6)$$

Equation 7.6 allows us to calculate the steady-state error, $e(\infty)$, given the input, $R(s)$, and the system, $G(s)$.[2] We will now substitute several inputs for $R(s)$ and then draw conclusions about the relationships that exist between the open-loop system, $G(s)$, and the nature of the steady-state error, $e(\infty)$.

The three test signals we use to establish specifications for a control system's steady-state error characteristics are shown in Table 7.1. Let us take each input and evaluate its effect on the steady-state error by using Equation 7.6.

Step Input

Using Equation 7.6 with $R(s) = 1/s$, we find

$$e(\infty) = \lim_{s \to 0} \frac{s(1/s)}{1 + G(s)} = \frac{1}{1 + \lim_{s \to 0} G(s)} \qquad (7.7)$$

The term $\lim_{s \to 0} G(s)$ is the dc gain of the forward transfer function, since s, the frequency variable, is approaching zero. In order to have zero steady-state error,

$$\lim_{s \to 0} G(s) = \infty \qquad (7.8)$$

---

[1] The final value theorem is derived from the Laplace transform of the derivative. Thus

$$\mathscr{L}[\dot{f}(t)] = \int_{0-}^{\infty} \dot{f}(t)e^{-st}\,dt = sF(s) - f(0-)$$

As $s \to 0$,

$$\int_{0-}^{\infty} \dot{f}(t)\,dt = f(\infty) - f(0-) = \lim_{s \to 0} sF(s) - f(0-)$$

or

$$f(\infty) = \lim_{s \to 0} sF(s)$$

For finite steady-state errors, the final value theorem is only valid if $F(s)$ has poles only in the left half-plane and, at most, one pole at the origin. However, correct results that yield steady-state errors that are infinite can be obtained if $F(s)$ has more than one pole at the origin (see D'Azzo and Houpis, 1988). If $F(s)$ has poles in the right half-plane or poles on the imaginary axis other than at the origin, the final value theorem is invalid.

[2] Valid only if (1) $E(s)$ has poles only in the left half-plane and at the origin, and (2) the closed-loop transfer function, $T(s)$, is stable. Notice that by using Equation 7.6, numerical results can be obtained for unstable systems. These results, however, are meaningless.

Table 7.1 Test Waveforms for Evaluating Steady-State Errors for Control Systems

Waveform	Name	Time function	Laplace transform
$r(t)$	Step, or constant position	1	$\dfrac{1}{s}$
$r(t)$	Ramp, or constant velocity	t	$\dfrac{1}{s^2}$
$r(t)$	Parabolic, or constant acceleration	$\dfrac{1}{2}t^2$	$\dfrac{1}{s^3}$

Hence, to satisfy Equation 7.8, $G(s)$ must take on the following form:

$$G(s) \equiv \frac{(s + z_1)(s + z_2) \cdots}{s^n(s + p_1)(s + p_2) \cdots} \tag{7.9}$$

and for the limit to be infinite, the denominator must be equal to zero as s goes to zero. Thus, $n \geq 1$; that is to say, at least one power of s must exist in the denominator. Since division by s in the frequency domain is integration in the time domain (see Table 2.2, Item 10), we are also saying that at least one pure integration must be present in the forward path. The steady-state response for this case of zero steady-state error is similar to that shown in Figure 7.2(a), output 1.

If there are no integrations, then $n = 0$. Using Equation 7.9, we have

$$\lim_{s \to 0} G(s) = \frac{z_1 z_2 \cdots}{p_1 p_2 \cdots} \tag{7.10}$$

which is finite and yields a finite error, from Equation 7.7. Figure 7.2(a), output 2, is an example of this case of finite steady-state error.

In summary, for a step input to a unity feedback system, the steady-state error will be zero if there is at least one pure integration in the forward path. If there are no integrations, then there will be a non-zero finite error. This result is comparable to our qualitative discussion in Section 7.1, where we found that a pure gain yields a constant steady-state error for a step input, but an integrator yields zero error for the same type of input. We now repeat the development for a ramp input.

Ramp Input

Using Equation 7.6, with $R(s) = 1/s^2$, we obtain

$$e(\infty) = \lim_{s \to 0} \frac{s(1/s^2)}{1 + G(s)} = \lim_{s \to 0} \frac{1}{s + sG(s)} = \frac{1}{\lim\limits_{s \to 0} sG(s)} \tag{7.11}$$

To have zero steady-state error for a ramp input, we must have

$$\lim_{s \to 0} sG(s) = \infty \tag{7.12}$$

To satisfy Equation 7.12, $G(s)$ must take the same form as Equation 7.9, except that $n \geq 2$. In other words, there must be at least two integrations in the forward path. An example of zero steady-state error for a ramp input is shown in Figure 7.2(b), output 1.

If only one integration exists in the forward path, then, assuming Equation 7.9,

$$\lim_{s \to 0} sG(s) = \frac{z_1 z_2 \cdots}{p_1 p_2 \cdots} \tag{7.13}$$

which is finite rather than infinite. Using Equation 7.11, we find that this configuration leads to a constant error, as shown in Figure 7.2(b), output 2.

If there are no integrations in the forward path, then

$$\lim_{s \to 0} sG(s) = 0 \tag{7.14}$$

and the steady-state error would be infinite and lead to diverging ramps, as shown in Figure 7.2(b), output 3. Finally, we repeat the development for a parabolic input.

Parabolic Input

Using Equation 7.6, with $R(s) = 1/s^3$, we obtain

$$e(\infty) = \lim_{s \to 0} \frac{s(1/s^3)}{1 + G(s)} = \lim_{s \to 0} \frac{1}{s^2 + s^2 G(s)} = \frac{1}{\lim\limits_{s \to 0} s^2 G(s)} \tag{7.15}$$

In order to have zero steady-state error for a parabolic input, we must have

$$\lim_{s \to 0} s^2 G(s) = \infty \tag{7.16}$$

To satisfy Equation 7.16, $G(s)$ must take on the same form as Equation 7.9, except that $n \geq 3$. In other words, there must be at least three integrations in the forward path.

If there are only two integrations in the forward path, then

$$\lim_{s \to 0} s^2 G(s) = \frac{z_1 z_2 \cdots}{p_1 p_2 \cdots} \tag{7.17}$$

which is finite rather than infinite. Using Equation 7.15, we find that this configuration leads to a constant error.

If there is only one or less integration in the forward path, then

$$\lim_{s \to 0} s^2 G(s) = 0 \tag{7.18}$$

and the steady-state error is infinite. Two examples demonstrate these concepts.

Example 7.1

Find the steady-state errors for step, ramp, and parabolic inputs to a system with no integrations.

Problem   Find the steady-state errors for inputs of $5u(t)$, $5tu(t)$, and $5t^2 u(t)$ to the system shown in Figure 7.5. The function $u(t)$ is the unit step.

Solution   First, we verify that the system is indeed stable. Next, for the input $5u(t)$, whose Laplace transform is $5/s$, the steady-state error will be 5 times larger than that given by Equation 7.7, or

$$e(\infty) = \frac{5}{1 + \lim_{s \to 0} G(s)} = \frac{5}{1 + 20} = \frac{5}{21} \tag{7.19}$$

which implies a response similar to output 2 of Figure 7.2(a).

For the input $5tu(t)$, whose Laplace transform is $5/s^2$, the steady-state error will be 5 times larger than that given by Equation 7.11, or

$$e(\infty) = \frac{5}{\lim_{s \to 0} s G(s)} = \frac{5}{0} = \infty \tag{7.20}$$

which implies a response similar to output 3 of Figure 7.2(b).

For the input $5t^2 u(t)$, whose Laplace transform is $10/s^3$, the steady-state error will be 10 times larger than that given by Equation 7.15, or

$$e(\infty) = \frac{10}{\lim_{s \to 0} s^2 G(s)} = \frac{10}{0} = \infty \tag{7.21}$$

Figure 7.5 Feedback Control System for Example 7.1

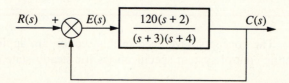

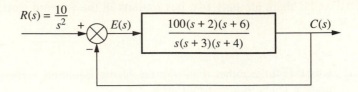

$$R(s) = \frac{10}{s^2}$$

$$E(s)$$

$$\frac{100(s+2)(s+6)}{s(s+3)(s+4)}$$

$$C(s)$$

Figure 7.6 Feedback Control System for Example 7.2

Example 7.2

Find the steady-state errors for step, ramp, and parabolic inputs to a system with one integration.

Problem   Find the steady-state errors for inputs of $5u(t)$, $5tu(t)$, and $5t^2u(t)$ to the system shown in Figure 7.6. The function $u(t)$ is the unit step.

Solution   First, verify that the system is indeed stable. Next, note that since there is an integration in the forward path, the steady-state errors for some of the input waveforms will be less than that found in Example 7.1. For the input $5u(t)$, whose Laplace transform is $5/s$, the steady-state error will be 5 times larger than that given by Equation 7.7, or

$$e(\infty) = \frac{5}{1 + \lim_{s \to 0} G(s)} = \frac{5}{\infty} = 0 \tag{7.22}$$

which implies a response similar to output 1 of Figure 7.2(a). Notice that the integration in the forward path yields zero error for a step input rather than the finite error found in Example 7.1.

For the input $5tu(t)$, whose Laplace transform is $5/s^2$, the steady-state error will be 5 times larger than that given by Equation 7.11, or

$$e(\infty) = \frac{5}{\lim_{s \to 0} s G(s)} = \frac{5}{100} = \frac{1}{20} \tag{7.23}$$

which implies a response similar to output 2 of Figure 7.2(b). Notice that the integration in the forward path yields a finite error for a ramp input rather than the infinite error found in Example 7.1.

For the input, $5t^2u(t)$, whose Laplace transform is $10/s^3$, the steady-state error will be 10 times larger than that given by Equation 7.15, or

$$e(\infty) = \frac{10}{\lim_{s \to 0} s^2 G(s)} = \frac{10}{0} = \infty \tag{7.24}$$

Notice that the integration in the forward path does not yield any improvement in steady-state error over that found in Example 7.1 for a parabolic input.

The concepts developed in this section are applied in the next section as we define control system specifications for steady-state errors.

7.3 Definition of Static Error Constants and System Type

Static Error Coefficients

In the previous section we derived the following relationships for steady-state error. For a step input, $u(t)$,

$$e(\infty) = \frac{1}{1 + \lim_{s \to 0} G(s)} \tag{7.25}$$

For a ramp input, $tu(t)$,

$$e(\infty) = \frac{1}{\lim_{s \to 0} sG(s)} \tag{7.26}$$

For a parabolic input, $\frac{1}{2}t^2 u(t)$,

$$e(\infty) = \frac{1}{\lim_{s \to 0} s^2 G(s)} \tag{7.27}$$

The three terms in the denominator that are taken to the limit determine the steady-state error. We call these limits *static error constants*. Individually, their names are

Position constant, K_p, where

$$K_p = \lim_{s \to 0} G(s) \tag{7.28}$$

Velocity constant, K_v, where

$$K_v = \lim_{s \to 0} sG(s) \tag{7.29}$$

Acceleration constant, K_a, where

$$K_a = \lim_{s \to 0} s^2 G(s) \tag{7.30}$$

As we have seen, these quantities, depending upon the form of $G(s)$, can assume values of zero, finite constant, or infinity. Since the static error constant appears in the denominator of the steady-state error (Equations 7.25 through 7.27), the value of the steady-state error decreases as the static error constant increases.

In Section 7.2 we evaluated the steady-state error by using the final value theorem. An alternate method makes use of the static error constants. A few examples follow.

Example 7.3

Evaluate static error constants and find the steady-state error.

Problem   For each system of Figure 7.7, evaluate the static error constants and find the expected error for the standard position, velocity, and acceleration inputs.

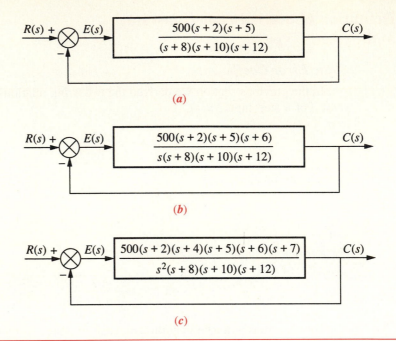

Figure 7.7 Feedback Control Systems for Example 7.3

Solution First, verify that all systems shown are indeed stable. Next, for Figure 7.7(*a*),

$$K_p = \lim_{s \to 0} G(s) = \frac{500 \times 2 \times 5}{8 \times 10 \times 12} = 5.208 \tag{7.31}$$

$$K_v = \lim_{s \to 0} sG(s) = 0 \tag{7.32}$$

$$K_a = \lim_{s \to 0} s^2 G(s) = 0 \tag{7.33}$$

Thus, for a position input,

$$e(\infty) = \frac{1}{1 + K_p} = 0.161 \tag{7.34}$$

For a velocity input,

$$e(\infty) = \frac{1}{K_v} = \infty \tag{7.35}$$

For an acceleration input,

$$e(\infty) = \frac{1}{K_a} = \infty \tag{7.36}$$

Now, for Figure 7.7(*b*),

$$K_p = \lim_{s \to 0} G(s) = \infty \tag{7.37}$$

$$K_v = \lim_{s \to 0} sG(s) = \frac{500 \times 2 \times 5 \times 6}{8 \times 10 \times 12} = 31.25 \tag{7.38}$$

and

$$K_a = \lim_{s \to 0} s^2 G(s) = 0 \tag{7.39}$$

Thus, for a position input,

$$e(\infty) = \frac{1}{1 + K_p} = 0 \tag{7.40}$$

For a velocity input,

$$e(\infty) = \frac{1}{K_v} = \frac{1}{31.25} = 0.032 \tag{7.41}$$

For an acceleration input,

$$e(\infty) = \frac{1}{K_a} = \infty \tag{7.42}$$

Finally, for Figure 7.7(*c*),

$$K_p = \lim_{s \to 0} G(s) = \infty \tag{7.43}$$

$$K_v = \lim_{s \to 0} sG(s) = \infty \tag{7.44}$$

and

$$K_a = \lim_{s \to 0} s^2 G(s) = \frac{500 \times 2 \times 4 \times 5 \times 6 \times 7}{8 \times 10 \times 12} = 875 \tag{7.45}$$

Thus, for a position input,

$$e(\infty) = \frac{1}{1 + K_p} = 0 \tag{7.46}$$

For a velocity input,

$$e(\infty) = \frac{1}{K_v} = 0 \tag{7.47}$$

For an acceleration input,

$$e(\infty) = \frac{1}{K_a} = \frac{1}{875} = 1.14 \times 10^{-3} \tag{7.48}$$

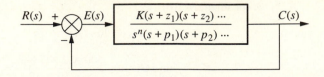

Figure 7.8 Feedback Control System for Defining System Type

Definition of System Type

The values of the static error constants, again, depend upon the form of $G(s)$, especially the number of pure integrations in the forward path. Since steady-state errors are dependent upon the number of integrations in the forward path, we give a name to this system's attribute. Given the system in Figure 7.8, we define *system type* to be the value of n in the denominator, or equivalently, the number of pure integrations in the forward path. Therefore, a system with $n = 0$ is a Type 0 system. If $n = 1$ or $n = 2$, the corresponding system is a Type 1 or Type 2 system, respectively.

Table 7.2 ties together the concepts of steady-state error, static error constants, and system type. The table shows the static error constants and the steady-state errors as functions of input waveform and system type.

In this section we defined steady-state errors, static error constants, and system type. Now the specifications for a control system's steady-state errors will be formulated, followed by some examples.

Table 7.2 Summary of Relationships between Input, System Type, Static Error Coefficients, and Steady-State Errors

Input	Steady-state error formula	Type 0		Type 1		Type 2	
		Static error constant	Error	Static error constant	Error	Static error constant	Error
Step, $u(t)$	$\dfrac{1}{1 + K_p}$	$K_p =$ Constant	$\dfrac{1}{1 + K_p}$	$K_p = \infty$	0	$K_p = \infty$	0
Ramp, $tu(t)$	$\dfrac{1}{K_v}$	$K_v = 0$	∞	$K_v =$ Constant	$\dfrac{1}{K_v}$	$K_v = \infty$	0
Parabola, $\dfrac{1}{2}t^2u(t)$	$\dfrac{1}{K_a}$	$K_a = 0$	∞	$K_a = 0$	∞	$K_v =$ Constant	$\dfrac{1}{K_a}$

7.4 Steady-State Error Specifications and Examples

Static error constants can be used to specify the steady-state error characteristics of control systems such as that shown in Figure 7.9. Just as damping ratio, ζ, settling time, T_s, peak time, T_p, and percent overshoot, $\%OS$, are used as specifications for a control system's transient response, so the position constant, K_p, velocity constant, K_v, and acceleration constant, K_a, can be used as specifications for a control system's steady-state errors. We will soon see that a wealth of information is contained within the specification of a static error constant.

For example, if a control system has the specification $K_v = 1000$, we can draw several conclusions:

1. The system is stable.
2. The system is of Type 1, since only Type 1 systems have K_v's that are finite constants. Recall that $K_v = 0$ for Type 0 systems, whereas $K_v = \infty$ for Type 2 systems.
3. A ramp input is the test signal. Since K_v is specified as a finite constant, and the steady-state error for a ramp input is inversely proportional to K_v, we know the input is a ramp.
4. The steady-state error between the input ramp and the output ramp is $1/K_v$ per unit of input slope.

Figure 7.9 Steady-State Error Is an Important Design Consideration for the Hellfire Missile. Shown Here Is the Missile's Laser Seeker Guidance System Undergoing Fine Tuning (Courtesy of Rockwell International)

Example 7.4 Interpret the steady-state error specification.

Problem What information is contained in the specification $K_p = 1000$?

Solution The system is stable. The system is Type 0, since only a Type 0 system has a finite K_p. Type 1 and Type 2 systems have $K_p = \infty$. The input test signal is a step, since K_p is specified. Finally, the error per unit step is

$$e(\infty) = \frac{1}{1 + K_p} = \frac{1}{1 + 1000} = \frac{1}{1001} \tag{7.49}$$

Our examples up to this point have been analysis problems. Let us now look at a design problem where we specify the steady-state error and then find the gain to accomplish the task.

Example 7.5 Design the gain of a simple control system to meet a steady-state error specification.

Problem Given the control system in Figure 7.10, find the value of K so that there is 10% error in the steady state.

Solution Since the system is Type 1, the error stated in the problem must apply to a ramp input; only a ramp yields a finite error in a Type 1 system. Thus,

$$e(\infty) = \frac{1}{K_v} = 0.1 \tag{7.50}$$

Therefore,

$$K_v = 10 = \lim_{s \to 0} s\,G(s) = \frac{K \times 5}{6 \times 7 \times 8} \tag{7.51}$$

which yields

$$K = 672 \tag{7.52}$$

Applying the Routh-Hurwitz criterion, we see that the system is stable at this gain.

This example completes our discussion of unity feedback systems. In the remaining sections, we deal with the steady-state errors for disturbances and the steady-state errors for feedback control systems in which the feedback is not unity.

Figure 7.10 Feedback Control System for Example 7.5

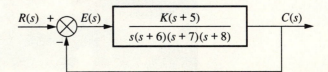

7.5 The Steady-State Error for Disturbances

Feedback control systems are used to compensate for disturbances or unwanted inputs that enter a system. The advantage of using feedback is that, regardless of these disturbances, the system can be designed to follow the input with small or zero error, as we now demonstrate. Figure 7.11 shows a feedback control system with a disturbance, $D(s)$, injected between the controller and the plant. We now rederive the expression for steady-state error with the disturbance included.

The transform of the output is given by

$$C(s) = E(s)G_1(s)G_2(s) + D(s)G_2(s) \tag{7.53}$$

But

$$C(s) = R(s) - E(s) \tag{7.54}$$

Substituting Equation 7.54 into Equation 7.53 and solving for $E(s)$, we obtain

$$E(s) = \frac{1}{1 + G_1(s)G_2(s)}R(s) - \frac{G_2(s)}{1 + G_1(s)G_2(s)}D(s) \tag{7.55}$$

where we can think of $\dfrac{1}{1 + G_1(s)G_2(s)}$ as a transfer function relating $E(s)$ to $R(s)$ and $-\dfrac{G_2(s)}{1 + G_1(s)G_2(s)}$ as a transfer function relating $E(s)$ to $D(s)$.

To find the steady-state value of the error, we apply the final value theorem[3] to Equation 7.55 and obtain

$$e(\infty) = \lim_{s \to 0} sE(s) = \lim_{s \to 0} \frac{s}{1 + G_1(s)G_2(s)}R(s) - \lim_{s \to 0} \frac{sG_2(s)}{1 + G_1(s)G_2(s)}D(s)$$

$$= e_R(\infty) + e_D(\infty) \tag{7.56}$$

where

$$e_R(\infty) = \lim_{s \to 0} \frac{s}{1 + G_1(s)G_2(s)}R(s)$$

---

[3] Remember that the final value theorem can only be applied if the system is stable, i.e., the roots of $[1 + G_1(s)G_2(s)]$ are in the left half-plane.

Figure 7.11 Feedback Control System Showing Disturbance

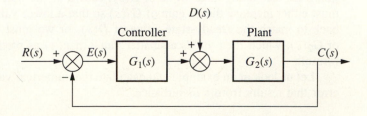

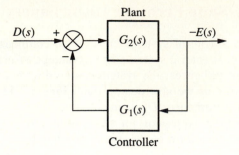

Figure 7.12 The System of Figure 7.11 Rearranged to Show the Disturbance as the Input and the Error as the Output, with $R(s) = 0$

and

$$e_D(\infty) = -\lim_{s \to 0} \frac{sG_2(s)}{1 + G_1(s)G_2(s)} D(s)$$

The first term, $e_R(\infty)$, is the steady-state error due to $R(s)$, which we have already obtained. The second term, $e_D(\infty)$, is the steady-state error due to the disturbance. Let us explore the conditions on $e_D(\infty)$ that must exist to reduce the error due to the disturbance.

At this point we must make some assumptions about $D(s)$, the controller, and the plant. First, we assume a step disturbance, $D(s) = 1/s$. Substituting this value into the second term, $e_D(\infty)$, of Equation 7.56, the steady-state error component due to a step disturbance is found to be

$$e_D(\infty) = -\frac{1}{\displaystyle\lim_{s \to 0} \frac{1}{G_2(s)} + \lim_{s \to 0} G_1(s)} \tag{7.57}$$

This equation shows that the steady-state error produced by a step disturbance can be reduced by increasing the dc gain of $G_1(s)$ or decreasing the dc gain of $G_2(s)$.

This concept is shown in Figure 7.12, where the system of Figure 7.11 has been rearranged so that the disturbance, $D(s)$, is depicted as the input, and the error, $E(s)$, is depicted as the output, with $R(s)$ set equal to zero. If we want to minimize the steady-state value of $E(s)$, shown as the output in Figure 7.12, we must either increase the dc gain of $G_1(s)$ so that a lower value of $E(s)$ will be fed back to match the steady-state value of $D(s)$, or we must decrease the dc value of $G_2(s)$, which then yields a smaller value of $e(\infty)$ as predicted by the feedback formula.

Let us look at an example and calculate the numerical value of the steady-state error that results from a disturbance.

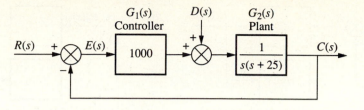

Figure 7.13 Feedback Control System for Example 7.6

Example 7.6   Find the steady-state error component due to a step disturbance.

Problem   Find the steady-state error component due to a step disturbance for the system of Figure 7.13.

Solution   The system is stable. Using Figure 7.12 and Equation 7.57, we find

$$e_D(\infty) = -\frac{1}{\displaystyle\lim_{s\to 0}\frac{1}{G_2(s)} + \lim_{s\to 0}G_1(s)} = -\frac{1}{0 + 1000} = -\frac{1}{1000} \tag{7.58}$$

The result shows that the steady-state error produced by the step disturbance is inversely proportional to the dc gain of $G_1(s)$. The dc gain of $G_2(s)$ is infinite in this example.

Now let us examine the steady-state errors for non-unity feedback systems.

7.6 Steady-State Errors for Non-Unity Feedback Systems

Control systems often do not have unity feedback because of the compensation used to improve performance or because of the physical model for the system. The feedback path can be a pure gain other than unity or have some dynamic representation.

In order to derive a method for handling non-unity feedback systems, take the non-unity feedback control system shown in Figure 7.14(*a*) and form a unity feedback system by adding and subtracting unity feedback paths, as shown in Figure 7.14(*b*). In this figure, we call the plant's actuating signal $E_a(s)$ and realize that the actual error is $E(s) = R(s) - C(s)$. Next, combine $H(s)$ with the negative unity feedback, as shown in Figure 7.14(*c*). Finally, combine the feedback system consisting of $G(s)$ and $[H(s) - 1]$, leaving an equivalent forward path and a unity feedback, as shown in Figure 7.14(*d*).

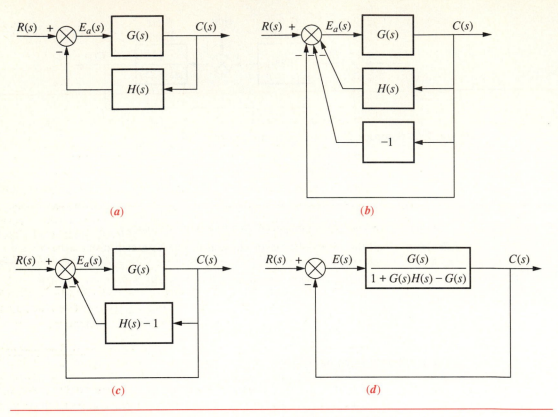

Figure 7.14 Steps in Forming an Equivalent Unity Feedback System from a General Non-Unity Feedback System

The following example summarizes the concepts of steady-state error, system type, and static error constants for non-unity feedback systems.

Example 7.7   Find the system type, static error constants, and steady-state error for a non-unity feedback system.

Problem   For the system shown in Figure 7.15, find the system type, the appropriate error constant associated with the system type, and the steady-state error for a unit step input.

Solution   After determining that the system is indeed stable, one may impulsively declare the system to be Type 1. This may not be the case, since there is a non-unity feedback element, and the plant's actuating signal is not the difference between the

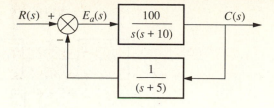

Figure 7.15 Non-Unity Feedback Control System for
Example 7.7

input and the output. The first step in solving the problem is to convert the system of
Figure 7.15 into an equivalent unity feedback system. Using the equivalent forward
transfer function of Figure 7.14(d) along with

$$G(s) = \frac{100}{s(s + 10)} \tag{7.59}$$

and

$$H(s) = \frac{1}{(s + 5)} \tag{7.60}$$

we find

$$G_e(s) = \frac{G(s)}{1 + G(s)H(s) - G(s)} = \frac{100(s + 5)}{s^3 + 15s^2 - 50s - 400} \tag{7.61}$$

Thus, the system is Type 0, since there are no pure integrations in Equation 7.61.
The appropriate static error constant is then K_p, whose value is

$$K_p = \lim_{s \to 0} G_e(s) = \frac{100 \times 5}{-400} = -\frac{5}{4} \tag{7.62}$$

The steady-state error, $e(\infty)$, is

$$e(\infty) = \frac{1}{1 + K_p} = \frac{1}{1 - \frac{5}{4}} = -4 \tag{7.63}$$

The negative value for steady-state error implies that the output step is larger than the
input step.

To conclude our discussion of steady-state error for systems with non-unity
feedback, let us look at the general system of Figure 7.16, which has both a
disturbance and non-unity feedback. We will derive a general equation for the

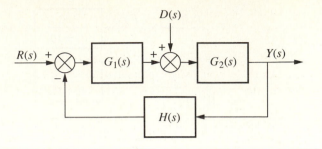

Figure 7.16 Non-Unity Feedback Control with Disturbance

steady-state error and then determine the parameters of the system in order to drive the error to zero for step inputs and step disturbances.[4]

The steady-state error for this system, $e(\infty) = r(\infty) - y(\infty)$, is

$$e(\infty) = \lim_{s \to 0} sE(s) = \lim_{s \to 0} s \left\{ \left[1 - \frac{G_1(s)G_2(s)}{1 + G_1(s)G_2(s)H(s)} \right] R(s) - \left[\frac{G_2(s)}{1 + G_1(s)G_2(s)H(s)} \right] D(s) \right\} \quad (7.64)$$

Now, limiting the discussion to step inputs and step disturbances, where $R(s) = D(s) = 1/s$, Equation 7.64 becomes,

$$e(\infty) = \lim_{s \to 0} sE(s) = \left\{ \left[1 - \frac{\lim\limits_{s \to 0}[G_1(s)G_2(s)]}{\lim\limits_{s \to 0}[1 + G_1(s)G_2(s)H(s)]} \right] - \left[\frac{\lim\limits_{s \to 0} G_2(s)}{\lim\limits_{s \to 0}[1 + G_1(s)G_2(s)H(s)]} \right] \right\} \quad (7.65)$$

For zero error,

$$\frac{\lim\limits_{s \to 0}[G_1(s)G_2(s)]}{\lim\limits_{s \to 0}[1 + G_1(s)G_2(s)H(s)]} = 1 \quad \text{and} \quad \frac{\lim\limits_{s \to 0} G_2(s)}{\lim\limits_{s \to 0}[1 + G_1(s)G_2(s)H(s)]} = 0 \quad (7.66)$$

Equations 7.66 can always be satisfied if (1) the system is stable, (2) $G_1(s)$ is a Type 1 system, (3) $G_2(s)$ is a Type 0 system, and (4) $H(s)$ is a Type 0 system with a dc gain of unity.

In this section we applied steady-state error analysis to non-unity feedback systems. One method for determining the steady-state error is to force a unity feedback path around the system and then to absorb the original $G(s)$ and $H(s)$

---

[4]The details of the derivation are included as a problem at the end of this chapter.

into an equivalent forward-path transfer function. When non-unity feedback is present, the plant's actuating signal is not the actual error or difference between the input and the output.

We concluded our discussion by deriving a general expression for the steady-state error of a non-unity feedback system with a disturbance. We used this equation to determine the attributes of the subsystems so that there was zero error for step inputs and step disturbances.

Before concluding this chapter, we will discuss a topic that is not only significant for steady-state errors, but is generally useful throughout the control systems design process.

7.7 Sensitivity

During the design process, the engineer may want to consider the extent to which changes in system parameters affect the behavior of a system. Ideally, parameter changes due to heat or other causes should not appreciably affect a system's performance. The degree to which changes in system parameters affect system transfer functions, and hence performance, is called *sensitivity*. Zero sensitivity (i.e., changes in a system parameter have no effect upon the transfer function) is ideal. The greater the sensitivity, the less desirable is the effect of a parameter change.

For example, assume the function, $F = \dfrac{K}{K + a}$. If $K = 10$ and $a = 100$, then $F = 0.091$. If parameter a triples to 300, then $F = 0.032$. Thus a fractional change in parameter a of $\dfrac{300 - 100}{100} = 2$, (i.e., a 200% change) yields a change in the function F of $\dfrac{0.032 - 0.091}{0.091} = -0.65$ (i.e., -65% change). Thus, the function F has a reduced sensitivity to changes in parameter a. As we proceed, we will see that another advantage of feedback is that in general it affords reduced sensitivity to parameter changes.

Based upon the previous discussion, let us formalize a definition of sensitivity: *Sensitivity* is the ratio of the fractional change in the function to the fractional change in the parameter as the fractional change of the parameter approaches zero. That is,

$$S_{F:P} = \lim_{\Delta P \to 0} \frac{\text{Fractional change in the function, } F}{\text{Fractional change in the parameter, } P}$$

$$= \lim_{\Delta P \to 0} \frac{\Delta F / F}{\Delta P / P}$$

$$= \lim_{\Delta P \to 0} \frac{P}{F} \frac{\Delta F}{\Delta P}$$

which reduces to

$$S_{F:P} = \frac{P}{F} \frac{\delta F}{\delta P} \tag{7.67}$$

Let us now apply the definition to a feedback control system.

Example 7.8 Find the sensitivity of a closed-loop transfer function to a change in a parameter.

Problem Given the system of Figure 7.17, calculate the sensitivity of the closed-loop transfer function to changes in the parameter a. How would you reduce the sensitivity?

Solution The closed-loop transfer function is

$$T(s) = \frac{K}{s^2 + as + K} \tag{7.68}$$

The sensitivity is given by

$$S_{T:a} = \frac{a}{T} \frac{\delta T}{\delta a} = \frac{a}{\left(\dfrac{K}{s^2 + as + K}\right)} \left(\frac{-Ks}{(s^2 + as + K)^2}\right) = \frac{-as}{s^2 + as + K} \tag{7.69}$$

which is a function of the value of s. For any value of s, however, an increase in K reduces the sensitivity of the closed-loop transfer function to changes in the parameter a.

In the next example we apply the concept of sensitivity to steady-state error.

Example 7.9 Find the sensitivity of the steady-state error to variations of a parameter.

Problem For the system of Example 7.8, find the sensitivity of the steady-state error to changes in parameter K and parameter a with ramp inputs.

Solution The steady-state error for the system of Example 7.8 is

$$e(\infty) = \frac{1}{K_v} = \frac{a}{K} \tag{7.70}$$

Figure 7.17 Feedback Control System for Example 7.8

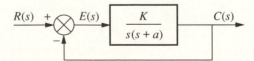

The sensitivity of $e(\infty)$ to changes in parameter a is

$$S_{e:a} = \frac{a}{e}\frac{\delta e}{\delta a} = \frac{a}{a/K}\left[\frac{1}{K}\right] = 1 \tag{7.71}$$

The sensitivity of $e(\infty)$ to changes in parameter K is

$$S_{e:K} = \frac{K}{e}\frac{\delta e}{\delta K} = \frac{K}{a/K}\left[\frac{-a}{K^2}\right] = -1 \tag{7.72}$$

Thus, changes in either parameter a or parameter K are directly reflected in $e(\infty)$, and there is no reduction or increase in sensitivity. The negative sign indicates a decrease in $e(\infty)$ for an increase in K. Both of these results could have been obtained directly from Equation 7.70 since $e(\infty)$ is directly proportional to parameter a and inversely proportional to parameter K.

Let us now look at an example where the sensitivity is not as obvious.

Example 7.10 Find the sensitivity of the steady-state error to variations of a parameter.

Problem Find the sensitivity of the steady-state error to changes in parameter K and parameter a for the system shown in Figure 7.18 with a step input.

Solution The steady-state error for this Type 0 system is

$$e(\infty) = \frac{1}{1 + K_p} = \frac{1}{1 + \dfrac{K}{ab}} = \frac{ab}{ab + K} \tag{7.73}$$

The sensitivity of $e(\infty)$ to changes in parameter a is

$$S_{e:a} = \frac{a}{e}\frac{\delta e}{\delta a} = \frac{a}{\left(\dfrac{ab}{ab + K}\right)}\frac{(ab + K)b - ab^2}{(ab + K)^2} = \frac{K}{ab + K} \tag{7.74}$$

The sensitivity of $e(\infty)$ to changes in parameter K is

$$S_{e:K} = \frac{K}{e}\frac{\delta e}{\delta K} = \frac{K}{\left(\dfrac{ab}{ab + K}\right)}\frac{-ab}{(ab + K)^2} = \frac{-K}{ab + K} \tag{7.75}$$

Figure 7.18 Feedback Control System for Example 7.10

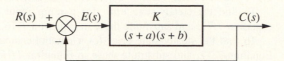

Equations 7.74 and 7.75 show that the sensitivity to changes in parameter K and parameter a is less than unity for positive a and b. Thus, feedback in this case yields reduced sensitivity to variations in both parameters.

In this section we defined sensitivity and showed that in some cases feedback reduces the sensitivity of a system's steady-state error to changes in system parameters. The concept of sensitivity can be applied to other measures of control system performance as well; it is not limited to the sensitivity of the steady-state error performance. We summarize this chapter in the next two sections.

7.8 Chapter-Objective Demonstration Problems

This chapter showed us how to find steady-state errors for step, ramp, and parabolic inputs to a closed-loop feedback control system. We also learned how to evaluate the gain to meet a steady-state error requirement. Two chapter-objective demonstration problems will be presented. The first uses our ongoing antenna azimuth control system to summarize the concepts. The second example deals with the focusing system for a video laser disc.

Example 7.11   Find the steady-state error for an antenna azimuth control system and evaluate the gain to meet a steady-state error requirement.

Problem   For the antenna azimuth control system shown in Figure 2.36
   a. Find the steady-state error in terms of gain, K, for step, ramp, and parabolic inputs.
   b. Find the value of gain, K, to yield a 10% error in the steady state.

Solution
   a. The simplified block diagram for the system is shown in Figure 5.31(c). The steady-state error is given by the expression

$$e(\infty) = \lim_{s \to 0} sE(s) = \lim_{s \to 0} \frac{sR(s)}{1 + G(s)} \tag{7.76}$$

From Figure 5.31(c),

$$G(s) = \frac{6.63K}{s(s + 1.71)(s + 100)} \tag{7.77}$$

To find the steady-state error for a step input, use $R(s) = 1/s$ along with Equation 7.77, and substitute these in Equation 7.76. The result is $e(\infty) = 0$.

To find the steady-state error for a ramp input, use $R(s) = 1/s^2$ along with Equation 7.77, and substitute these in Equation 7.76. The result is $e(\infty) = 25.79/K$.

To find the steady-state error for a parabolic input, use $R(s) = 1/s^3$ along with Equation 7.77, and substitute these in Equation 7.76. The result is $e(\infty) = \infty$.

b. Since the system is Type 1, a 10% error in the steady-state must refer to a ramp input. This is the only input that yields a finite, nonzero error. Hence, for a unit ramp input,

$$e(\infty) = 0.1 = \frac{1}{K_v} = \frac{(1.71)(100)}{6.63K} = \frac{25.79}{K} \tag{7.78}$$

from which $K = 257.9$. The student should verify that the value of K is within the range of gains that ensures system stability. From the chapter-objective demonstration problem in the last chapter, the range of gain for stability was found to be $0 < K < 2623.29$. Hence, the system is stable for a gain of 257.9.

As a final example, let us look at a video laser disc focusing system for recording.

Example 7.12   Design the gain of a system to meet a steady-state error requirement.

Problem   In order to record on a video laser disc, a 0.5 μm laser spot must be focused on the recording medium in order to burn pits that represent the program material. The small laser spot requires that the focusing lens be positioned to an accuracy of $\pm 0.1\,\mu$m. A model of the feedback control system for the focusing lens is shown in Figure 7.19(a). The detector detects the distance between the focusing lens and the video disc by measuring the degree of focus as follows. Reflected laser light from the disc is split into two beams, each focused on one of two elements of a split photodiode. When the beam is in focus on the laser disc, the output from each element of the photodetector is equal. When the beam is out of focus, one detector element will output a larger voltage. A simplified model for the detector is a straight line relating the differential voltage output from the two elements to the nominal distance between the lens and the laser disc. A linearized plot of the detector input-output relationship is shown in Figure 7.19(b) (Isailović, 1985). Assume that a warp on the disc yields a worst-case disturbance in the focus of $10t^2\,\mu$m. Find the value of $K_1K_2K_3$ in order to meet the focusing accuracy required by the system.

Figure 7.19 Video Laser Disc Recording: (a) Control System for Focusing Write Beam

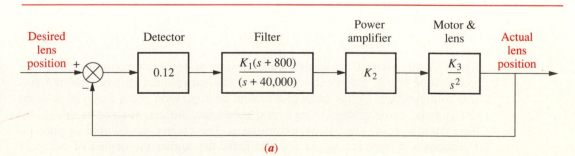

(a)

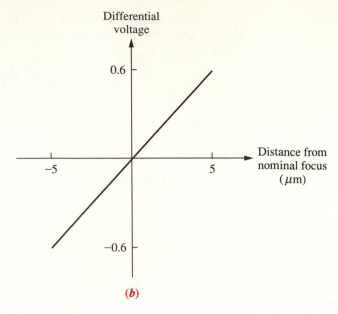

Note: Position is measured in micrometers, μm

Figure 7.19 (continued) Video Laser Disc Recording:
(*b*) Linearized Transfer Function for Focus Detector

Solution Since the system is Type 2, it can respond to acceleration inputs with finite error. We can assume that the disturbance has the same effect as an input of $10t^2 \mu$m. The Laplace transform of $10t^2$ is $20/s^3$, or 20 units greater than the unit acceleration that was used to derive the general equation of the error for an acceleration input. Thus, $e(\infty) = 20/K_a$. But $K_a = \lim_{s \to 0} s^2 G(s)$.

From Figure 7.19(*a*), $K_a = 0.0024 K_1 K_2 K_3$. Also, from the problem statement, the error must be no greater than 0.1μm. Hence, $e(\infty) = 416.67/K_1 K_2 K_3 = 0.1$. Thus, $K_1 K_2 K_3 \geq 4166.7$ and the system is stable.

7.9 Summary

This chapter covered the analysis and design of feedback control systems for steady-state errors. The steady-state errors studied resulted strictly from the system configuration. On the basis of a system configuration and a group of selected test signals, namely, steps, ramps, and parabolas, we can analyze or design for the system's steady-state error performance. The greater the number of pure integrations a system has in the forward path, the higher the degree of accuracy assuming the system is stable.

The steady-state errors depend upon the type of test input. Applying the final value theorem to stable systems, the steady-state error for unit step inputs is

$$e(\infty) = \frac{1}{1 + \lim_{s \to 0} G(s)} \qquad (7.79)$$

The steady-state error for ramp inputs of unit velocity is

$$e(\infty) = \frac{1}{\lim_{s \to 0} s G(s)} \qquad (7.80)$$

and for parabolic inputs of unit acceleration it is

$$e(\infty) = \frac{1}{\lim_{s \to 0} s^2 G(s)} \qquad (7.81)$$

The terms taken to the limit in Equations 7.79 through 7.81 are called *static error constants*. Beginning with Equation 7.79, the terms in the denominator taken to the limit are called the *position constant, velocity constant,* and *acceleration constant,* respectively. The static error constants are the steady-state error specifications for control systems. By specifying a static error constant, one is stating the number of pure integrations in the forward path, the test signal used, and the expected steady-state error.

Another definition covered in this chapter was that of *system type*. The system type is the number of pure integrations in the forward path, assuming a unity feedback system. Increasing the system type decreases the steady-state error as long as the system remains stable.

Since the steady-state error is, for the most part, inversely proportional to the static error constant, the larger the static error constant, the smaller the steady-state error. Increasing system gain increases the static error constant. Thus, in general, increasing system gain decreases the steady-state error as long as the system remains stable.

Non-unity feedback systems were handled by deriving an equivalent unity feedback system whose steady-state error characteristics followed all previous development.

Finally, we saw how feedback decreases a system's steady-state error caused by disturbances. With feedback, the effect of a disturbance can be reduced by system gain adjustments.

REVIEW QUESTIONS

1. Name two sources of steady-state errors.
2. A position control tracking with a constant difference in velocity would yield how much position error in the steady-state?

3. Name the test inputs used to evaluate steady-state error.

4. How many integrations in the forward path are required in order for there to be zero steady-state error for each of the test inputs listed in Question 3?

5. Increasing system gain has what effect upon the steady-state error?

6. For a step input, the steady-state error is approximately the reciprocal of the static error constant if what condition holds true?

7. What is the exact relationship between the static error constants and the steady-state errors for ramp and parabolic inputs?

8. What information is contained in the specification $K_p = 10,000$?

9. Define *system type*.

10. The forward transfer function of a control system has three poles at -1, -2, and -3. What is the system type?

11. What effect does feedback have upon disturbances?

12. For a step input disturbance at the input to the plant, describe the effect of controller and plant gain upon minimizing the effect of the disturbance.

13. Is the forward-path actuating signal the system error if the system has a non-unity feedback element?

14. How are non-unity feedback systems analyzed and designed for steady-state errors?

15. Define, in words, *sensitivity*, and describe the goal of feedback-control-system engineering as it applies to sensitivity.

PROBLEMS

1. For the unity feedback system shown in Figure P7.1, where

$$G(s) = \frac{500(s + 7)(s + 2)(s + 9)}{s^2(s + 27)(s^2 + 2s + 10)}$$

find the steady-state errors for the following test inputs: $25u(t)$, $37tu(t)$, $50t^2u(t)$.

Figure P7.1

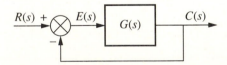

2. For the unity feedback system shown in Figure P7.1, where

$$G(s) = \frac{10(s + 2)(s + 4)(s + 6)}{s^2(s + 5)(s + 7)}$$

find the steady-state error if the input is $20t^2$.

3. For the system shown in Figure P7.2, what steady-state error can be expected for an input of $15u(t)$?

Figure P7.2

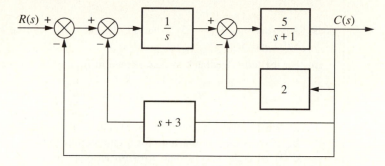

4. For the unity feedback system shown in Figure P7.1, where

$$G(s) = \frac{200}{(s + 10)(s^2 + 4s + 5)}$$

find the steady-state error for inputs of $35u(t)$, $70tu(t)$, and $80t^2u(t)$.

5. Find the steady-state error in acceleration for the unity feedback system shown in Figure P7.1, where

$$G(s) = \frac{10(s + 1)}{s(s + 2)}$$

if the input is $(1/6)t^3$. The acceleration error, $e_a(\infty)$, is defined as

$$e_a(\infty) = \left(\frac{d^2r}{dt^2} - \frac{d^2c}{dt^2} \right)\Big|_{t\to\infty}$$

where r is the system input, and c is the system output.

6. An input of $t^3u(t)$ is applied to the input of a Type 3, unity feedback system, as shown in Figure P7.1, where

$$G(s) = \frac{30(s + 1)(s + 2)(s + 3)}{s^3(s + 5)(s + 10)}$$

Find the steady-state error in position.

7. The steady-state error in velocity of a system is defined to be

$$\left(\frac{dr}{dt} - \frac{dc}{dt} \right)\Big|_{t\to\infty}$$

where r is the system input, and c is the system output. Find the steady-state error in velocity for an input of $t^3u(t)$ to a unity feedback system with a forward transfer function of

$$G(s) = \frac{100(s + 1)(s + 2)}{s^2(s + 3)(s + 10)}$$

8. What is the steady-state error for a step input of 10 units applied to the unity feedback system of Figure P7.1, where

$$G(s) = \frac{(s + 1)(s + 2)(s + 3)}{(s + 6)(s + 7)(s + 8)}$$

9. A system has $K_p = 3$. What steady-state error can be expected for inputs of $8u(t)$ and $8tu(t)$?

10. For the unity feedback system shown in Figure P7.1, where

$$G(s) = \frac{1250}{s(s + 50)}$$

a. What is the expected percent overshoot for a unit step input?
b. What is the settling time for a unit step input?
c. What is the steady-state error for an input of $5u(t)$?
d. What is the steady-state error for an input of $5tu(t)$?
e. What is the steady-state error for an input of $5t^2u(t)$?

11. Given the unity feedback system shown in Figure P7.1, where

$$G(s) = \frac{100(s + 2)(s + 9)}{s(s + 18)(s + \alpha)(s + 10)}$$

find the value of α to yield a $K_v = 1000$.

12. For the unity feedback system of Figure P7.1, where

$$G(s) = \frac{K(s + 2)(s + 4)(s + 6)}{s^2(s + 5)(s + 7)}$$

find the value of K to yield a static error constant of 10,000.

13. For the system shown in Figure P7.3,
   a. Find K_p, K_v, and K_a.
   b. Find the steady-state error for an input of $50u(t)$, $50tu(t)$, and $50t^2u(t)$.
   c. State the system type.

Figure P7.3

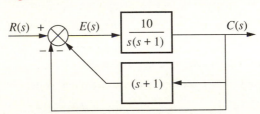

14. A Type 3, unity feedback system has $r(t) = t^3$ applied to its input. Find the steady-state position error for this input if the forward transfer function is

$$G(s) = \frac{1000(s^2 + 4s + 20)(s^2 + 20s + 15)}{s^3(s + 2)(s + 10)}$$

15. Find the system type for the system of Figure P7.4.

Figure P7.4

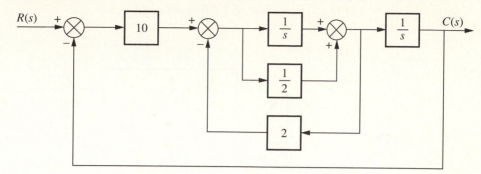

16. The steady-state error is defined to be the difference in position between input and output as time approaches infinity. Let us define a steady-state velocity error, which is the difference in velocity between input and output. Derive an expression for the error in velocity, $\dot{e}(\infty) = \dot{r}(\infty) - \dot{c}(\infty)$, and complete the table below for the error in velocity.

		Type		
		0	1	2
Input	Step			
	Ramp			
	Parabola			

17. For the system shown in Figure P7.5,
    a. What value of K will yield a steady-state error in position of 0.01 for an input of $(1/10)t$?
    b. What is the K_v for the value of K found in (**a**)?

Figure P7.5

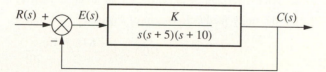

18. Given the unity feedback system of Figure P7.1, where

$$G(s) = \frac{K(s + a)}{s(s + 1)(s + 10)}$$

find the value of K_a so that a ramp input of slope 15 will yield an error of 0.003 in the steady state when compared to the output.

19. Given the system of Figure P7.6, design the value of K so that for an input of $100tu(t)$, there will be a 0.01 error in the steady state.

Figure P7.6

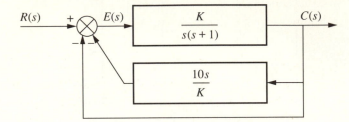

20. Find the value of K for the unity feedback system shown in Figure P7.1, where

$$G(s) = \frac{K(s + 2)}{s^2(s + 4)}$$

if the input is $10t^2u(t)$, and the desired steady-state error is 0.01 for this input.

21. The unity feedback system of Figure P7.1, where

$$G(s) = \frac{K(s^2 + 3s + 30)}{s^n(s + 5)}$$

is to have $1/6000$ error between an input of $10tu(t)$ and the output in the steady state.
   a. Find K and n to meet the specification.
   b. What are K_p, K_v, and K_a?

22. For the unity feedback system of Figure P7.1, where

$$G(s) = \frac{K(s^2 + 2s + 5)}{(s + 2)^2(s + 3)}$$

   a. Find the system type.
   b. What error can be expected for an input of $10u(t)$?
   c. What error can be expected for an input of $10tu(t)$?

23. For the unity feedback system of Figure P7.1, where

$$G(s) = \frac{K(s + 10)}{s(s + 2)(s + 20)}$$

find the value of K to yield a steady-state error of 0.1 for a ramp input of $50tu(t)$.

24. Given the unity feedback system of Figure P7.1, where

$$G(s) = \frac{K(s + 12)}{(s + 5)(s + 10)}$$

find the value of K to yield a steady-state error of 0.5 for an input of $50u(t)$.

25. For the unity feedback system of Figure P7.1, where

$$G(s) = \frac{K}{s(s + 4)(s + 8)}$$

find the minimum possible position error if a unit ramp is applied. What places the constraint upon the error?

26. The unity feedback system of Figure P7.1, where

$$G(s) = \frac{K(s + \alpha)}{(s + \beta)^2}$$

is to be designed to meet the following specifications: steady-state error for a unit step input $= 0.1$; damping ratio $= 0.5$; natural frequency $= \sqrt{10}$. Find K, α, and β.

27. A second-order, unity feedback system is to follow a velocity input with the following specifications: the steady-state output position shall differ from the input position by 0.01 of the input velocity; the natural frequency of the closed-loop system shall be 10 rad/s. Find
   a. The system type
   b. The exact expression for the forward-path transfer function
   c. The closed-loop system's damping ratio

28. The unity feedback system of Figure P7.1, where

$$G(s) = \frac{K(s + \alpha)}{s(s + \beta)}$$

is to be designed to meet the following requirements: the position error for a unit ramp input equals $1/10$; the closed-loop poles will be located at $-1 \pm j1$. Find K, α, and β in order to meet the specifications.

29. Given the unity feedback control system of Figure P7.1, where

$$G(s) = \frac{K}{s^n(s + a)}$$

find the values of n, K, and α in order to meet specifications of 10% overshoot and $K_v = 100$.

30. The following specification applies to a position control: $K_v = 10$. On hand is an amplifier with a variable gain, K_1, with which to drive a motor. Two one-turn pots to convert shaft position into voltage are also available, where $\pm 3\pi$ volts are placed across the pots. A motor is available whose transfer function is

$$\frac{\theta_m(s)}{E_a(s)} = \frac{k}{s(s + \alpha)}$$

where $\theta_m(s)$ is the motor armature position, and $E_a(s)$ is the armature voltage. The components are interconnected as shown in Figure P7.7.

The transfer function of the motor is found experimentally as follows. The motor and load are driven separately by applying a large, short square wave (i.e., unit impulse) to the armature. An oscillograph of the response shows that the mo-

Figure P7.7 Position Control System

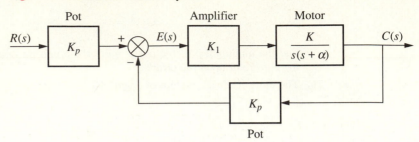

tor reached 63% of its final value output at 0.5 seconds after application of the impulse. Furthermore, with 10 volts dc applied to the armature, the constant output speed was 100 rad/s. Draw the completed block diagram of the system, specifying the transfer function of each component part of the block diagram.

31. Given the unity feedback control system of Figure P7.1, where

$$G(s) = \frac{K}{s(s + a)}$$

find K and a to yield $K_v = 1000$ and a 20% overshoot.

32. The system of Figure P7.8 is to have the following specifications: $K_v = 10$; $\zeta = 0.5$. Find the values of K_1 and K_f required for the specifications of the system to be met.

Figure P7.8

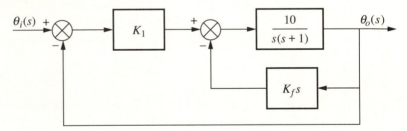

33. Find the total steady-state error due to a unit step input and a unit step disturbance in the system of Figure P7.9.

Figure P7.9

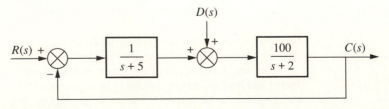

34. Design the values of K_1 and K_2 in the system of Figure P7.10 to meet the following specifications:

    a. The steady-state error component due to a unit step disturbance is -0.000012.

    b. The steady-state error component due to a unit ramp input is 0.003.

Figure P7.10

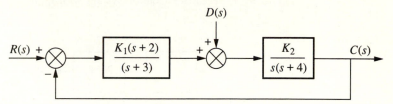

35. For each of the systems shown in Figure P7.11, find the following:

    a. The system type

    b. The appropriate static error constant

    c. The input waveform to yield a constant error

    d. The error for a unit input of the waveform found in **(c)**

Figure P7.11 Closed-Loop System with Non-Unity Feedback

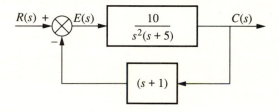

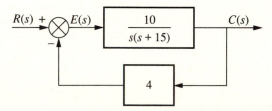

36. Derive Equation 7.64.

37. Given the system shown in Figure P7.12, do the following:
   a. Derive the expression for the error, $E(s)$, in terms of $R(s)$ and $D(s)$.
   b. Derive the steady-state error, $e(\infty)$, if $R(s)$ and $D(s)$ are unit step functions.
   c. Determine the attributes of $G_1(s)$, $G_2(s)$, and $H(s)$ necessary for the steady-state error to become zero.

Figure P7.12 Feedback Control System with Input and Disturbance

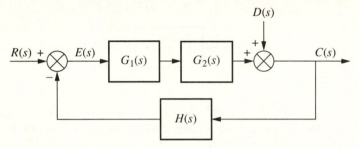

38. Given the system shown in Figure P7.13, find the sensitivity of the steady-state error to parameter a. Assume a step input. Plot the sensitivity as a function of parameter a.

Figure P7.13

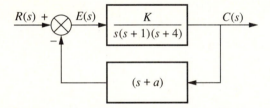

39. For the system shown in Figure P7.14, find the sensitivity of the steady-state error for changes in K_1 and in K_2, when $K_1 = 100$ and $K_2 = 0.1$. Assume step inputs for both the input and the disturbance.

Figure P7.14 Feedback Control System with Disturbance

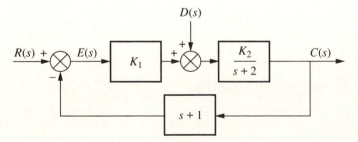

40. A small boat is circling a large ship that is using a tracking radar. The speed of the small boat is 20 knots, and it is circling the large ship at a distance of 1 nautical mile, as shown in Figure P7.15(*a*). A simplified model of the tracking system is shown in Figure P7.15(*b*). Find the value of K so that the small boat is kept in the center of the radar beam with no more than 0.1 degree error.

Figure P7.15 Small Boat Tracked by Large Ship's Radar: (*a*) Physical Arrangement; (*b*) Block Diagram of Tracking System

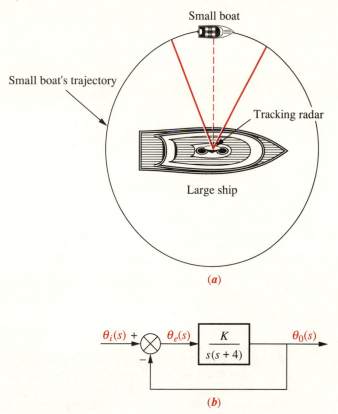

Small boat

Small boat's trajectory

Tracking radar

Large ship

(*a*)

$$\theta_i(s) \xrightarrow{+} \bigotimes \xrightarrow{\theta_e(s)} \boxed{\dfrac{K}{s(s+4)}} \xrightarrow{\theta_0(s)}$$

(*b*)

41. An automobile guidance system yields an actual output distance, $X(s)$, for a desired input distance, $X_c(s)$, as shown in Figure P7.16(*a*). Any difference, $X_e(s)$, between the commanded distance and the actual distance is converted into a velocity command, $V_c(s)$, by the controller and applied to the vehicle accelerator. The vehicle responds to the velocity command with a velocity, $V(s)$, and a displacement, $X(s)$, is realized. The velocity control, $G_2(s)$, is itself a closed-loop system, as shown in Figure P7.16(*b*). Here the difference, $V_e(s)$, between the commanded velocity, $V_c(s)$, and the actual vehicle velocity, $V(s)$, drives a motor that displaces the automobile's accelerator by $Y_c(s)$ (Stefani, 1978). Find

Figure P7.16 Automobile Guidance System: (*a*) Displacement Control System; (*b*) Velocity Control Loop

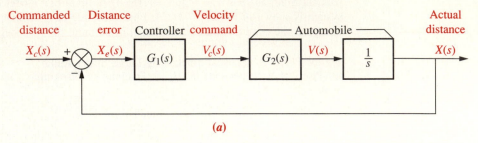

(*a*)

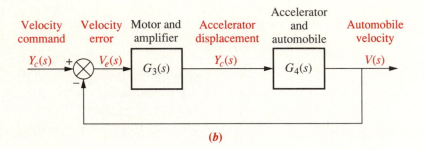

(*b*)

the steady-state error for the velocity control loop if the motor and amplifier,

$G_3(s) = \dfrac{K}{s(s+1)}$. Assume $G_4(s)$ to be a first-order system, where a maximum possible 1-foot displacement of the accelerator linkage yields a steady-state velocity of 100 miles/hour, with the automobile reaching 60 miles/hour in 10 seconds.

42. A simplified block diagram of a meter used to measure oxygen concentration is shown in Figure P7.17. The meter uses the paramagnetic properties of a stream of oxygen. A small body is placed in a stream of oxygen whose concentration is $R(s)$, and it is subjected to a magnetic field. The torque on the body, $K_1 R(s)$, due to the magnetic field is a function of the concentration of the oxygen. The displacement of the body, $\theta(s)$, is detected, and a voltage, $C(s)$, is developed

Figure P7.17 Block Diagram of a Paramagnetic Oxygen Analyzer

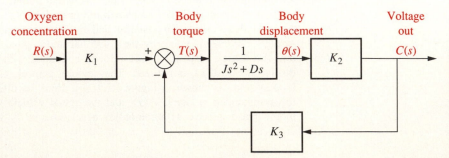

proportional to the displacement. This voltage is used to develop an electrostatic field that places a torque, $K_3 C(s)$, on the body opposite to that developed by the magnetic field. When the body comes to rest, the output voltage represents the strength of the magnetic torque, which in turn is related to the concentration of the oxygen (Chesmond, 1982).

Find the steady-state error between the output voltage and the input oxygen concentration. How would you reduce the error to zero?

43. *Chapter-objective problem:* For the antenna azimuth control system shown in Figure P2.36,
   a. Find the steady-state errors in terms of gain, K, for step, ramp, and parabolic inputs.
   b. Find the value of gain, K, to yield a 20% error in the steady state.

BIBLIOGRAPHY

Chesmond, C. J. *Control System Technology.* E. Arnold, London, 1982.

D'Azzo, J. J., and Houpis, C. H. *Feedback Control System Analysis and Design Conventional and Modern.* 3d ed. McGraw-Hill, New York, 1988.

Hostetter, G. H.; Savant, C. J., Jr.; and Stefani, R. T. *Design of Feedback Control Systems.* 2d ed. Saunders College Publishing, New York, 1989.

Isailović, J. *Videodisc and Optical Memory Systems.* Prentice Hall, Englewood Cliffs, N.J., 1985.

Stefani, R. T. Design and Simulation of an Automobile Guidance and Control System, *Transactions, Computers in Education Division of ASEE.* January 1978, pp. 1–9.

8

ROOT LOCUS TECHNIQUES FOR STABILITY AND TRANSIENT RESPONSE

8.1 Introduction

Root locus is a powerful method of analysis and design for stability and transient response (Evans, 1948; 1950). Feedback control systems are difficult to comprehend from a qualitative point of view and hence they rely heavily upon mathematics. The root locus covered in this chapter is a graphic technique that gives us this qualitative description of a control system's performance that we are looking for and also serves as a powerful quantitative tool that yields more information than the methods already discussed.

Up to this point, gains and other system parameters were designed to yield a desired transient response for only first- and second-order systems. Even though the root locus can be used to solve the same kind of problem, its real power lies in its ability to provide solutions for systems of order higher than two. For example, under the right conditions, a fourth-order system's parameters can be designed to yield a given percent overshoot and settling time using the concepts learned in Chapter 4.

The root locus can be used to qualitatively describe the performance of a system as various parameters are changed. For example, the effect of varying gain upon percent overshoot, settling time, and peak time can be vividly displayed. Furthermore, the qualitative description can then be verified with quantitative analysis.

Besides transient response, the root locus also gives a graphic representation of a system's stability. We can clearly see ranges of stability, ranges of instability, and the conditions that cause a system to break into oscillation.

Before discussing root locus let us review two concepts that we will need for the ensuing discussion: (1) the control system problem, and (2) complex numbers and their representation as vectors.

The Control System Problem

We have previously encountered the control system problem in Chapter 6: Whereas the poles of the open-loop transfer function are easily found (typically they are known by inspection and do not change with changes in system gain), the poles of the closed-loop transfer function are more difficult to find (typically they cannot be found without factoring the closed-loop system's characteristic polynomial, the denominator of the closed-loop transfer function) and, further, the closed-loop poles change with changes in system gain.

A typical closed-loop feedback control system is shown in Figure 8.1(a). The open-loop transfer function was defined in Chapter 5 to be $KG(s)H(s)$. Ordinarily, we can determine the poles of $KG(s)H(s)$ since these poles arise from simple cascaded first- or second-order subsystems. Further, variations in K do not affect the pole location of this function. On the other hand, we cannot determine the poles of $T(s) = \dfrac{KG(s)}{1 + KG(s)H(s)}$ unless we factor the denominator. Also, the poles of $T(s)$ change with K. Let us demonstrate.

Letting

$$G(s) = \frac{N_G(s)}{D_G(s)} \tag{8.1}$$

and

$$H(s) = \frac{N_H(s)}{D_H(s)} \tag{8.2}$$

then

$$T(s) = \frac{KN_G(s)D_H(s)}{D_G(s)D_H(s) + KN_G(s)N_H(s)} \tag{8.3}$$

Figure 8.1 (**a**) Closed-Loop System; (**b**) Equivalent Transfer Function

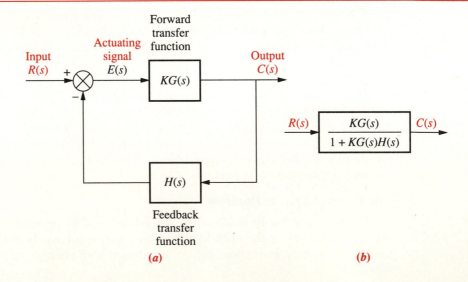

where N and D are factored polynomials and signify numerator and denominator terms, respectively. We observe the following: Typically we know the factors of the numerators and denominators of $G(s)$ and $H(s)$. Also, the zeros of $T(s)$ consist of the zeros of $G(s)$ and the poles of $H(s)$. The poles of $T(s)$ are not immediately known and, in fact, can change with K. For example, if $G(s) = \dfrac{(s + 1)}{s(s + 2)}$, and $H(s) = \dfrac{(s + 3)}{(s + 4)}$, the poles of $KG(s)H(s)$ are 0, -2, and -4. The zeros of $KG(s)H(s)$ are -1 and -3. Now, $T(s) = \dfrac{K(s + 1)(s + 4)}{s^3 + (6 + K)s^2 + (8 + 4K)s + 3K}$.

Thus, the zeros of $T(s)$ consist of the zeros of $G(s)$ and the poles of $H(s)$. The poles of $T(s)$ are not immediately known without factoring the denominator, and they are a function of K. Since the system's transient response and stability are dependent upon the poles of $T(s)$, we have no knowledge of the system's performance unless we factor the denominator for specific values of K. The root locus will be used to give us a vivid picture of the poles of $T(s)$ as K varies.

Complex Numbers and Their Vector Representation

Any *complex number*, $\sigma + j\omega$, described in cartesian coordinates can be graphically represented by a vector, as shown in Figure 8.2(a). The complex

Figure 8.2 Vector Representation of Complex Numbers:
(**a**) $s = \sigma + j\omega$; (**b**) $(s + a)$; (**c**) Alternate Representation
of $(s + a)$; (**d**) $(s + a)|_{s \to 5 + j2}$

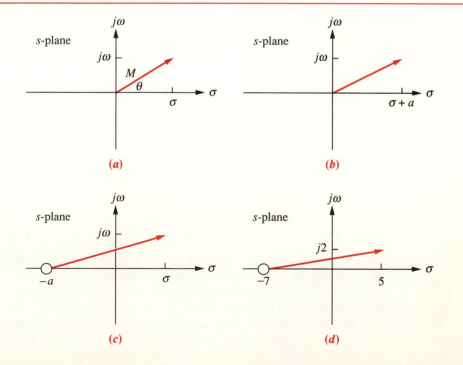

number also can be described in polar form with magnitude M and angle θ, as $M \angle \theta$. If the complex number is substituted into a complex function, $F(s)$, another complex number will result. For example if $F(s) = (s + a)$, then substituting the complex number $s = \sigma + j\omega$ yields $F(s) = (\sigma + a) + j\omega$, another complex number. This number is shown in Figure 8.2(b). Notice that $F(s)$ has a zero at $-a$. If we translate the vector a units to the left, as in Figure 8.2(c), we have an alternate representation of the complex number that originates at the zero of $F(s)$ and terminates on the point $s = \sigma + j\omega$.

We conclude that $(s + a)$ *is a complex number and can be represented by a vector drawn from the zero of the function to the point s*. For example, $(s + 7)|_{s \rightarrow 5 + j2}$ is a complex number drawn from the zero of the function, -7, to the point s, which is $5 + j2$, as shown in Figure 8.2(d).

Now, let us apply the concepts to a complicated function. Assume a function,

$$F(s) = \frac{\displaystyle\prod_{i=1}^{m} (s + z_i)}{\displaystyle\prod_{j=1}^{n} (s + p_j)} = \frac{\prod \text{Numerator's complex factors}}{\prod \text{Denominator's complex factors}} \qquad (8.4)$$

where the symbol $\prod$ means "product", m = number of zeros, and n = number of poles. Each factor in the numerator and each factor in the denominator is a complex number that can be represented as a vector. The function defines the complex arithmetic to be performed in order to evaluate $F(s)$ at any point, s. Since each complex factor can be thought of as a vector, the magnitude, M, of $F(s)$ at any point, s, is

$$M = \frac{\prod \text{Zero lengths}}{\prod \text{Pole lengths}} = \frac{\displaystyle\prod_{i=1}^{m} |(s + z_i)|}{\displaystyle\prod_{j=1}^{n} |(s + p_j)|} \qquad (8.5)$$

where a zero length, $|(s + z_i)|$, is the magnitude of the vector drawn from the zero of $F(s)$ at $-z_i$ to the point s, and a pole length, $|(s + p_j)|$, is the magnitude of the vector drawn from the pole of $F(s)$ at $-p_j$ to the point s. The angle, θ, of $F(s)$ at any point, s, is

$$\theta = \sum \text{Zero angles} - \sum \text{Pole angles}$$

$$= \sum_{i=1}^{m} \angle (s + z_i) - \sum_{j=1}^{n} \angle (s + p_j) \qquad (8.6)$$

where a zero angle is the angle, measured from the positive extension of the real axis, of a vector drawn from the zero of $F(s)$ at $-z_i$ to the point s, and a pole angle is the angle, measured from the positive extension of the real axis, of the vector drawn from the pole of $F(s)$ at $-p_j$ to the point s.

As a demonstration of the above concept, consider the following example.

Example 8.1   Evaluate a complex function using vector representations.

Problem   Given $F(s)$ in Equation 8.7, find $F(s)$ at the point $s = -3 + j4$.

$$F(s) = \frac{(s + 1)}{s(s + 2)} \tag{8.7}$$

Solution   The problem is graphically depicted in Figure 8.3, where each vector, $(s + \alpha)$, of the function is shown terminating on the selected point $s = -3 + j4$. The vector originating at the zero at -1 is

$$\sqrt{20}\angle116.57° \tag{8.8}$$

The vector originating at the pole at the origin is

$$5\angle126.87° \tag{8.9}$$

The vector originating at the pole at -2 is

$$\sqrt{17}\angle104.04° \tag{8.10}$$

Substituting Equations 8.8 through 8.10 into Equations 8.5 and 8.6 yields

$$M\angle\theta = \frac{\sqrt{20}}{5\sqrt{17}}\angle116.57° - 126.87° - 104.04° = 0.217\angle - 114.34° \tag{8.11}$$

as the result for evaluating $F(s)$ at the point $-3 + j4$.

We are now ready to state the chapter objective and begin our discussion of the root locus.

Figure 8.3 Vector Representation of Equation 8.7

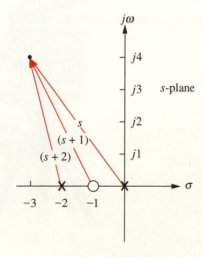

Chapter Objective

Given the third-order azimuth position control system shown in Figure 2.2, where the power amplifier is represented as having a pole rather than as a pure gain, the student will be able to determine the differential preamplifier gain required to meet a percent overshoot requirement for the closed-loop system.

8.2 Defining the Root Locus

Consider the feedback control system shown in Figure 8.4(*a*). A camera system is designed to follow a transmitter worn by the subject. An imbalance between two sensors mounted on the tripod causes the system to balance out the error and seek the source of energy emitted by the transmitter. The root locus technique can be used to analyze and design the effect of loop gain upon the system's transient response and stability.

Assume the block diagram representation of the system as shown in Figure 8.4(*b*) where the closed-loop poles of the system change location as the gain, K, is varied. Table 8.1, which was formed using the quadratic formula on the

Figure 8.4 (*a*) The Automatic Cameraman for Camcorders™ Mounted between the Camera and Tripod (Courtesy of Visionary Products)

(*a*)

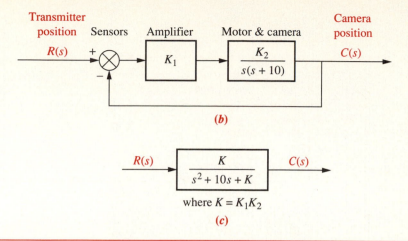

Figure 8.4 (continued) (**b**) Block Diagram; (**c**) Closed-Loop Transfer Function

denominator of the transfer function shown in Figure 8.4(c), shows the variation of pole location for different values of gain, K. The data of Table 8.1 is graphically displayed in Figure 8.5(a), which shows each pole and its gain.

As the gain, K, increases in Table 8.1 and Figure 8.5(a), the closed-loop pole, which is at -10 for $K = 0$, moves toward the right, and the closed-loop pole, which is at 0 for $K = 0$, moves toward the left. They meet at -5, break away from the real axis and move into the complex plane. One closed-loop pole moves upward while the other moves downward. We cannot tell which pole moves up or which moves down. In Figure 8.5(b) we have removed the individual

Table 8.1 Pole Location as a Function of Gain for the System of Figure 8.4

K	Pole 1	Pole 2
0	-10	0
5	-9.47	-0.53
10	-8.87	-1.13
15	-8.16	-1.84
20	-7.24	-2.76
25	-5	-5
30	$-5 + j2.24$	$-5 - j2.24$
35	$-5 + j3.16$	$-5 - j3.16$
40	$-5 + j3.87$	$-5 - j3.87$
45	$-5 + j4.47$	$-5 - j4.47$
50	$-5 + j5$	$-5 - j5$

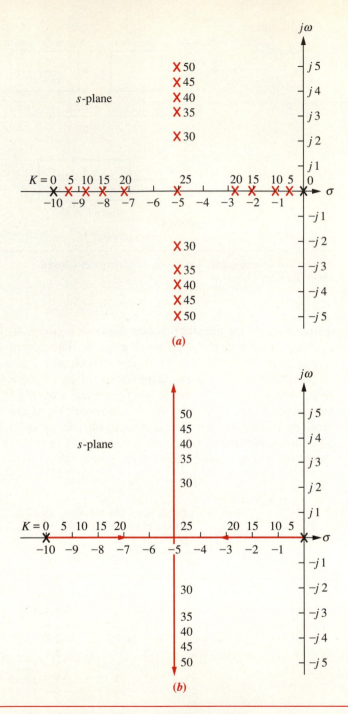

Figure 8.5 (*a*) Pole Plot from Table 8.1; (*b*) Root Locus

closed-loop pole locations and represented their path with solid lines. It is this *representation of the path of the closed-loop poles as the gain is varied* that we call a *root locus*. For most of our work, the discussion will be limited to positive gain, or $K \geq 0$.

The root locus shows the changes in the transient response as the gain, K, varies. First of all, the poles are real for gains less than 25. Thus, the system is overdamped. At a gain of 25, the poles are real and multiple and hence critically damped. For gains above 25, the system is underdamped. Even though the above conclusions were available through the analytical techniques covered in Chapter 4, the following conclusions are graphically demonstrated by the root locus.

Directing our attention to the underdamped portion of the root locus, we see that regardless of the value of gain, the real parts of the complex poles are always the same. Since the settling time is inversely proportional to the real part of the complex poles for this second-order system, the conclusion is that, regardless of the value of gain, the settling time for the system remains the same under all conditions of underdamped responses.

Also, as we increase the gain, the damping ratio diminishes, and the percent overshoot increases. The damped frequency of oscillation, which is equal to the imaginary part of the pole, also increases with an increase in gain, resulting in a reduction of the peak time. Finally, since the root locus never crosses over into the right half-plane, the system is always stable, regardless of the value of gain, and can never break into a sinusoidal oscillation.

The above conclusions for such a simple system as the one we were discussing may appear to be trivial. What we are about to see is that the analysis is applicable to systems of order higher than two. For these systems, it is difficult to tie transient response characteristics to the pole location. We will see that the root locus will allow us to make that association and will become an important technique in the analysis and design of higher-order systems.

8.3 Properties of the Root Locus

In Section 8.2 we arrived at the root locus by factoring the second-order polynomial in the denominator of the transfer function. Consider what would happen if that polynomial were of fifth or tenth order. Without a computer, factoring the polynomial would be quite a problem for numerous values of gain.

We are about to examine the properties of the root locus. From these properties we will be able to make a rapid *sketch* of the root locus for higher-order systems without having to factor the denominator of the closed-loop transfer function.

The properties of the root locus can be derived from the general control system of Figure 8.1(a). The closed-loop transfer function for the system is

$$T(s) = \frac{KG(s)}{1 + KG(s)H(s)} \tag{8.12}$$

From Equation 8.12, a pole, s, exists when the characteristic polynomial in the denominator becomes zero, or

$$KG(s)H(s) = -1 = 1\angle(2k + 1)180° \qquad k = 0, \pm1, \pm2, \pm3, \ldots \qquad (8.13)$$

where -1 is represented in polar form as, $1\angle(2k + 1)\,180°$. Alternately, a value of s is a closed-loop pole if

$$|KG(s)H(s)| = 1 \qquad (8.14)$$

and

$$\angle KG(s)H(s) = (2k + 1)\,180° \qquad (8.15)$$

Equation 8.13 implies that if a value of s is substituted into the function $KG(s)H(s)$, a complex number results. If the angle of the complex number is an odd multiple of $180°$, that value of s is a system pole for some particular value of K. What value of K? Since the angle criterion of Equation 8.15 is satisfied, all that remains is to satisfy the magnitude criterion, Equation 8.14. Thus,

$$K = \frac{1}{|G(s)||H(s)|} \qquad (8.16)$$

We have just found that a pole of the closed-loop system causes the angle of $KG(s)H(s)$, or simply $G(s)H(s)$ since K is a scalar, to be an odd multiple of $180°$. Furthermore, the magnitude of $KG(s)H(s)$ must be unity, implying that the value of K is the reciprocal of the magnitude of $G(s)H(s)$ when the pole value is substituted for s.

Let us demonstrate this relationship for the second-order system of Figure 8.4. The fact that closed-loop poles exist at -9.47 and -0.53 when the gain is 5 has already been established in Table 8.1. For this system,

$$KG(s)H(s) = \frac{K}{s(s + 10)} \qquad (8.17)$$

Substituting the pole at -9.47 for s and 5 for K yields $KG(s)H(s) = -1$. The reader can repeat the exercise for other points in Table 8.1 and show that each case yields $KG(s)H(s) = -1$.

It is helpful to visualize graphically the meaning of Equation 8.15. Let us apply the complex number concepts reviewed in Section 8.1 to the root locus of the system shown in Figure 8.6. For this system, the open-loop transfer function is

$$KG(s)H(s) = \frac{K(s + 3)(s + 4)}{(s + 1)(s + 2)} \qquad (8.18)$$

The closed-loop transfer function, $T(s)$, is

$$T(s) = \frac{K(s + 3)(s + 4)}{(1 + K)s^2 + (3 + 7K)s + (2 + 12K)} \qquad (8.19)$$

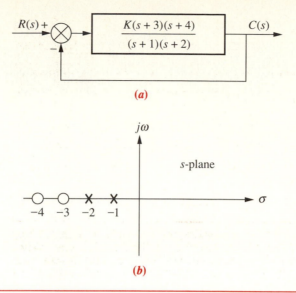

Figure 8.6 (**a**) Example System; (**b**) Pole-Zero Plot of $G(s)$

If point s is a closed-loop system pole for some value of gain, K, then s must satisfy Equations 8.14 and 8.15.

Consider the point $-2 + j3$. If this point is a closed-loop pole for some value of gain, then the angles of the zeros minus the angles of the poles must equal an odd multiple of 180°. From Figure 8.7,

$$\theta_1 + \theta_2 - \theta_3 - \theta_4 = 56.31° + 71.57° - 90° - 108.43°$$
$$= -70.55° \qquad (8.20)$$

Therefore, $-2 + j3$ is not a point on the root locus or, alternatively, $-2 + j3$ is not a closed-loop pole for any gain.

If the above calculations are repeated for the point $-2 + j\dfrac{\sqrt{2}}{2}$, the angles do add up to 180°. That is to say, $-2 + j\dfrac{\sqrt{2}}{2}$ is a point on the root locus for some value of gain. We now proceed to evaluate that value of gain.

From Equations 8.5 and 8.16,

$$K = \frac{1}{|G(s)H(s)|} = \frac{1}{M} = \frac{\prod \text{ Pole lengths}}{\prod \text{ Zero lengths}} \qquad (8.21)$$

Looking at Figure 8.7 with the point $-2 + j3$ replaced by $-2 + j\dfrac{\sqrt{2}}{2}$, the gain, K, is calculated as

$$K = \frac{L_3 L_4}{L_1 L_2} = \frac{\dfrac{\sqrt{2}}{2}(1.22)}{(2.12)(1.22)} = 0.33 \qquad (8.22)$$

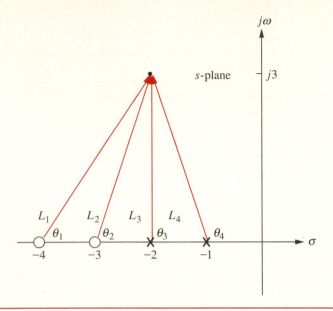

Figure 8.7 Vector Representation of $G(s)$ from Figure 8.6(a) at $-2 + j3$

Thus, the point $-2 + j\dfrac{\sqrt{2}}{2}$ is a point on the root locus for a gain of 0.33.

We summarize what we have found as follows: Given the poles and zeros of the open-loop transfer function, $KG(s)H(s)$, a point in the s-plane is on the root locus for a particular value of gain, K, if the angles of the zeros minus the angles of the poles, all drawn to the selected point on the s-plane, add up to $(2k + 1)180°$. Furthermore, gain K at that point for which the angles add up to $(2k + 1)180°$ is found by dividing the product of the pole lengths by the product of the zero lengths.

8.4 Rules for Sketching the Root Locus

It appears from our previous discussion that the root locus can be obtained by sweeping through every point in the s-plane to locate those points for which the angles, as previously described, add up to an odd multiple of 180°. Although this task is tedious without the aid of a computer, the concept can be used to develop rules that can be used to *sketch* the root locus without the effort required to *plot* the locus. Once a sketch is obtained, it is possible to accurately plot just those points that are of interest to us for a particular problem. We now examine the rules for sketching the root locus.

1. Number of branches. Each closed-loop pole moves as the gain is varied. If we define the path that one pole traverses as a *branch*, then there will be one branch for each closed-loop pole. Our first rule, then, defines the number of branches of the root locus: *The number of branches of the root locus equals the number of closed-loop poles*. As an example, look at Figure 8.5(*b*), where the two branches are shown. One originates at the origin, the other at -10.

2. Symmetry. If complex closed-loop poles do not exist in conjugate pairs, the resulting polynomial, formed by multiplying the factors containing the closed-loop poles, would have complex coefficients. Physically realizable systems cannot have complex coefficients in their transfer functions. Thus we conclude that *the root locus is symmetrical about the real axis*. An example of symmetry about the real axis is shown in Figure 8.5(*b*).

3. Real-axis segments. Let us make use of the angle property, Equation 8.15, of the points on the root locus to determine where the real-axis segments of the root locus exist. Figure 8.8 shows the poles and zeros of a general open-loop system. If an attempt is made to calculate the angular contribution of the poles and zeros at each point, P_1, P_2, P_3, and P_4, along the real axis, we observe the following: (1) at each point, the angular contribution of a pair of open-loop complex poles or zeros is zero, and (2) the contribution of the open-loop poles and open-loop zeros to the left of the respective point is zero. The conclusion is that the only contribution to the angle at any of the points comes from the open-loop, real-axis poles and zeros that exist to the right of the respective point. If we calculate the angle at each point using only the open-loop, real-axis poles and zeros to the right of each point, we note the following: (1) the angles on the real axis alternate between $0°$ and $180°$, and (2) the angle is $180°$ for regions of the real axis that exist to the left of an odd number of poles and/or zeros. The following rule summarizes the findings: *On the real axis, for $K > 0$, the root locus exists to the left of an odd number of real-axis, open-loop poles and/or open-loop zeros*.

Examine Figure 8.6(*b*). According to the rule just developed, the real-axis

Figure 8.8 Poles and Zeros of a General Open-Loop System with Test Points, P_i, on the Real Axis

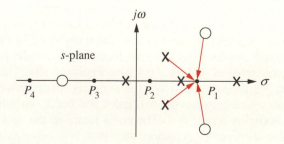

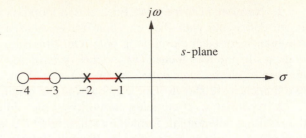

Figure 8.9 Real-Axis Segments of the Root Locus for the
System of Figure 8.6

segments of the root locus are between -1 and -2, and between -3 and -4, as
shown in Figure 8.9.

4. Starting and ending points. Where does the root locus begin (zero gain)
and end (infinite gain)? The answer to this question will enable us to expand the
sketch of the root locus beyond the real-axis segments. Consider the closed-loop
transfer function, $T(s)$, described by Equation 8.3. $T(s)$ can now be evaluated
for both large and small gains, K. As K approaches zero (small gain),

$$T(s) \approx \frac{K N_G(s) D_H(s)}{D_G(s) D_H(s) + \epsilon} \tag{8.23}$$

From Equation 8.23, we see that the closed-loop system poles at small gains
approach the combined poles of $G(s)$ and $H(s)$. We conclude that the root locus
begins at the poles of $G(s)H(s)$, the open-loop transfer function.

At high gains, where K is approaching infinity,

$$T(s) \approx \frac{K N_G(s) D_H(s)}{\epsilon + K N_G(s) N_H(s)} \tag{8.24}$$

From Equation 8.24 we see that the closed-loop system poles at large gains
approach the combined zeros of $G(s)$ and $H(s)$. Now we conclude that the root
locus ends at the zeros of $G(s)H(s)$, the open-loop transfer function.

Summarizing what we have found, *the root locus begins at the poles of
$G(s)H(s)$ and ends at the zeros of $G(s)H(s)$.* Remember that these poles and
zeros are the system's open-loop poles and zeros.

In order to demonstrate this rule, look at the system in Figure 8.6(a), whose
real axis segments have already been sketched in Figure 8.9. Using the rule just
derived, we find that the root locus begins at the poles at -1 and -2 and ends
at the zeros at -3 and -4 (see Figure 8.10). Thus, the poles start out at -1
and -2 and move through the real-axis space between the two poles. They meet
somewhere between the two poles and break out into the complex plane, moving
as complex conjugates. The poles return to the real axis somewhere between the
zeros at -3 and -4, where their path is completed as they move away from each
other and end up respectively at the two zeros of the open-loop system at -3 and
-4.

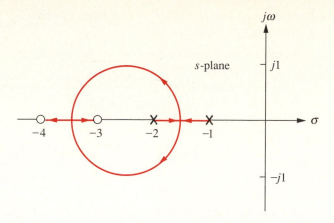

Figure 8.10 Complete Root Locus for the System of Figure 8.6

5. Behavior at infinity. Consider applying Rule 4 to the following open-loop transfer function:

$$KG(s)H(s) = \frac{K}{s(s+1)(s+2)} \tag{8.25}$$

There are three finite poles at s = 0, −1, and −2, and no finite zeros.

A function can also have *infinite* poles and zeros. If the function approaches infinity as s approaches infinity, then the function has a pole at infinity. If the function approaches zero as s approaches infinity, then the function has a zero at infinity. For example, the function $G(s) = s$ has a pole at infinity since $G(s)$ approaches infinity as s approaches infinity. On the other hand, $G(s) = 1/s$ has a zero at infinity since $G(s)$ approaches zero as s approaches infinity.

Every function of s has an equal number of poles and zeros if we include the infinite poles and zeros as well as the finite poles and zeros. In this example, Equation 8.25 contains three finite poles and three infinite zeros. To illustrate, let s approach infinity. The open-loop transfer function becomes

$$KG(s)H(s) \approx \frac{K}{s^3} = \frac{K}{s \cdot s \cdot s} \tag{8.26}$$

Each s in the denominator causes the open-loop function, $KG(s)H(s)$, to become zero as that s approaches infinity. Hence, Equation 8.26 has three zeros at infinity.

Thus, for Equation 8.25, the root locus begins at the finite poles of $KG(s)H(s)$ and ends at the infinite zeros. The question remains: Where are the infinite zeros? We must know where these zeros are in order to show the locus moving from the three finite poles to the three infinite zeros. Rule 5 helps us locate these zeros at infinity. Rule 5 also helps us locate poles at infinity for systems containing more finite zeros than finite poles.

We now derive Rule 5, which will tell us what the root locus looks like as it approaches zeros at infinity or as it moves from the poles at infinity.

*Derivation of the Behavior of the Root Locus
at Infinity (Kuo, 1987)*

Let the open-loop transfer function be represented as follows:

$$KG(s)H(s) = \frac{K(s^m + a_1 s^{m-1} + \cdots + a_m)}{(s^{m+n} + b_1 s^{m+n-1} + \cdots + b_{m+n})} \tag{8.27}$$

or

$$KG(s)H(s) = \frac{K}{\left(\dfrac{s^{m+n} + b_1 s^{m+n-1} + \cdots + b_{m+n}}{s^m + a_1 s^{m-1} + \cdots + a_m}\right)} \tag{8.28}$$

Performing the indicated division in the denominator, we obtain

$$KG(s)H(s) = \frac{K}{s^n + (b_1 - a_1)s^{n-1} + \cdots} \tag{8.29}$$

In order for a pole of the closed-loop transfer function to exist,

$$KG(s)H(s) = -1 \tag{8.30}$$

Assuming large values of s that would exist as the locus moves toward infinity, Equation 8.29 becomes

$$s^n + (b_1 - a_1)s^{n-1} = -K \tag{8.31}$$

Factoring out s^n, Equation 8.31 becomes

$$s^n\left(1 + \frac{b_1 - a_1}{s}\right) = -K \tag{8.32}$$

Taking the nth root of both sides, we have

$$s\left(1 + \frac{b_1 - a_1}{s}\right)^{1/n} = -K^{1/n} \tag{8.33}$$

If the term

$$\left(1 + \frac{b_1 - a_1}{s}\right)^{1/n} \tag{8.34}$$

is expanded into an infinite series where only the first two terms are significant,[1] we obtain

$$s\left(1 + \frac{b_1 - a_1}{ns}\right) = (-K)^{1/n} \tag{8.35}$$

Distributing the factor s on the left-hand side yields

$$s + \frac{b_1 - a_1}{n} = (-K)^{1/n} \tag{8.36}$$

---

[1]This is a good approximation since s is approaching infinity for the region applicable to the derivation.

Now, letting $s = \sigma + j\omega$ and $(-K)^{1/n} = |K^{1/n}|e^{j(2k+1)\pi/n}$, where $(-1)^{1/n} = e^{j(2k+1)\pi/n} = \cos\left(\dfrac{(2k+1)\pi}{n}\right) + j\sin\left(\dfrac{(2k+1)\pi}{n}\right)$ Equation 8.36 becomes

$$\sigma + j\omega + \frac{b_1 - a_1}{n} = |K^{1/n}|\left[\cos\frac{(2k+1)\pi}{n} + j\sin\frac{(2k+1)\pi}{n}\right] \tag{8.37}$$

where $k = 0, \pm 1, \pm 2, \pm 3, \ldots$ Setting the real and imaginary parts of both sides equal to each other, we obtain

$$\sigma + \frac{b_1 - a_1}{n} = |K^{1/n}|\cos\frac{(2k+1)\pi}{n} \tag{8.38a}$$

$$\omega = |K^{1/n}|\sin\frac{(2k+1)\pi}{n} \tag{8.38b}$$

Dividing the two equations to eliminate $|K^{1/n}|$, we obtain

$$\frac{\sigma + \dfrac{b_1 - a_1}{n}}{\omega} = \frac{\cos\dfrac{(2k+1)\pi}{n}}{\sin\dfrac{(2k+1)\pi}{n}} \tag{8.39}$$

Finally, solving for ω, we find

$$\omega = \left[\tan\frac{(2k+1)\pi}{n}\right]\left[\sigma + \frac{b_1 - a_1}{n}\right] \tag{8.40}$$

The form of this equation is that of a straight line,

$$\omega = M(\sigma - \sigma_0) \tag{8.41}$$

where the slope of the line, M, is

$$M = \tan\frac{(2k+1)\pi}{n} \tag{8.42}$$

and the σ intercept is

$$\sigma_0 = -\left[\frac{b_1 - a_1}{n}\right] \tag{8.43}$$

From the theory of equations,[2]

$$b_1 = -\sum \text{Poles} \tag{8.44a}$$

$$a_1 = -\sum \text{Zeros} \tag{8.44b}$$

Also, from Equation 8.27,

$$n = \text{number of finite poles} - \text{number of finite zeros}$$

$$= \#\text{poles} - \#\text{zeros} \tag{8.45}$$

---

[2]Given an nth-order polynomial of the form $s^n + a_{n-1}s^{n-1} + \cdots$, the coefficient, a_{n-1}, is the negative sum of the roots of the polynomial.

By examining Equation 8.41, we conclude that the root locus approaches a straight line as the locus approaches infinity. Further, this straight line intersects the σ axis at

$$\sigma_0 = \frac{\sum \text{Poles} - \sum \text{Zeros}}{\#\text{poles} - \#\text{zeros}} \tag{8.46}$$

which is obtained by substituting Equations 8.44 and 8.45 into Equation 8.43.

Let us summarize the results: *The root locus approaches straight lines as asymptotes as the locus approaches infinity. Further, the equation of the asymptotes is given by the real-axis intercept and slope as follows:*

$$\sigma_0 = \frac{\sum \text{Poles} - \sum \text{Zeros}}{\#\text{poles} - \#\text{zeros}} \tag{8.47}$$

$$M = \tan \frac{(2k + 1)\pi}{\#\text{poles} - \#\text{zeros}} \tag{8.48}$$

where $k = 0, \pm 1, \pm 2, \pm 3, \ldots$ Notice that the running index, k, in Equation 8.48 yields a multiplicity of lines that account for the many branches of a root locus that approach infinity. Let us demonstrate the concepts with an example.

Example 8.2   Sketch a root locus with asymptotes.

Problem   Sketch the root locus for the system shown in Figure 8.11.

Solution   Let us begin by calculating the asymptotes. Using Equation 8.47, the σ intercept is evaluated as

$$\sigma = \frac{(-1 - 2 - 4) - (-3)}{4 - 1} = -\frac{4}{3} \tag{8.49}$$

The slopes of the lines that intersect at $-4/3$ are given by

$$M = \tan \frac{(2k + 1)\pi}{\#\text{poles} - \#\text{zeros}} \tag{8.50a}$$

$$= \tan (\pi/3) \qquad \text{for } k = 0 \tag{8.50b}$$

$$= \tan (\pi) \qquad \text{for } k = 1 \tag{8.50c}$$

$$= \tan (5\pi/3) \qquad \text{for } k = 2 \tag{8.50d}$$

Figure 8.11 System for Example 8.2

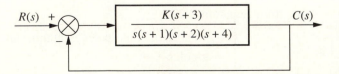

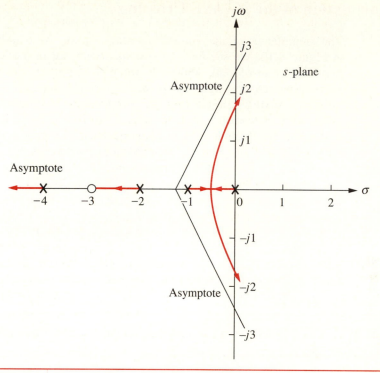

Figure 8.12 Root Locus and the Asymptotes for the System
of Figure 8.11

If the value for k continued to increase, the angles would begin to repeat. The number of lines obtained equals the difference between the number of finite poles and the number of finite zeros.

Rule 4 states that the locus begins at the open-loop poles and ends at the open-loop zeros. For the example, there are more open-loop poles than open-loop zeros. Thus, there must be zeros at infinity. The asymptotes tell us how we get to these zeros at infinity.

Figure 8.12 shows the complete root locus as well as the asymptotes that were just calculated. Notice that we have made use of all the rules learned so far. The real-axis segments lie to the left of an odd number of poles and/or zeros. The locus starts at the open-loop poles and ends at the open-loop zeros. For the example, there is only one open-loop finite zero and three infinite zeros. Rule 5, then, tells us that the three zeros at infinity are at the ends of the asymptotes.

We now discuss stability from the point of view of the root locus.

8.5 Calculation of the $j\omega$-Axis Crossing

The importance of the $j\omega$-axis crossing should be readily apparent. Looking at Figure 8.12, we see that the system's poles are in the left half-plane up to a particular value of gain. Above this value of gain, two of the closed-loop system's poles move into the right half-plane, signifying that the system is unstable. The $j\omega$-axis crossing is a point on the root locus that separates the stable operation of the system from the unstable operation. The value of ω at the axis crossing will yield the frequency of oscillation, while the gain at the $j\omega$-axis crossing will yield, for this example, the maximum positive gain for system stability. We should note here that other examples illustrate instability at small values of gain and stability at large values of gain. These systems have a root locus starting in the right half-plane (unstable at small values of gain) and ending in the left half-plane (stable for high values of gain).

To find the $j\omega$-axis crossing, we can use the Routh-Hurwitz criterion covered in Chapter 6 as follows: forcing a row of zeros in the Routh table will yield the gain; going back one row to the even polynomial equation and solving for the roots yields the frequency at the imaginary axis crossing.

Example 8.3

Find the frequency and gain at which the root locus crosses the imaginary axis.

Problem For the system of Figure 8.11, find the frequency and gain, K, for which the root locus crosses the imaginary axis. For what range of K is the system stable?

Solution The closed-loop transfer function for the system of Figure 8.11 is

$$T(s) = \frac{K(s+3)}{s^4 + 7s^3 + 14s^2 + (8+K)s + 3K} \tag{8.51}$$

Using the denominator and simplifying some of the entries by multiplying any row by a constant, we obtain the Routh array shown in Table 8.2.

A complete row of zeros yields the possibility for imaginary axis roots. For positive values of gain, those for which the root locus is plotted, only the s^1 row can yield a

Table 8.2 Routh Table for Equation 8.51

s^4	1	14	$3K$
s^3	7	$8+K$	
s^2	$90 - K$	$21K$	
s^1	$\dfrac{-K^2 - 65K + 720}{90 - K}$		
s^0	$21K$		

row of zeros. Thus,

$$-K^2 - 65K + 720 = 0 \qquad (8.52)$$

From this equation, K is evaluated as

$$K = 9.65 \qquad (8.53)$$

Forming the even polynomial by using the s^2 row with K = 9.65, we obtain

$$(90 - K)s^2 + 21K = 80.35s^2 + 202.65 = 0 \qquad (8.54)$$

and s is found to be equal to $\pm j1.59$. Thus the root locus crosses the $j\omega$-axis at $\pm j1.59$ at a gain of 9.65. We conclude that the system is stable for $0 \le K < 9.65$.

In this section we learned how to find the point where the root locus crosses the imaginary axis. In the next section we learn how to find points on the root locus accurately and calculate the gain at those points.

8.6 Plotting anc Calibrating the Root Locus

A fairly accurate sketch of the root locus can be made by using the rules from the previous sections. Many times this sketch is sufficient. At other times, more precise knowledge of the root locus is required. For example, we might want to know the exact coordinates of the root locus as it crosses the radial line representing 20% overshoot. Further, we may also want the value of gain at that point.

Consider the root locus shown in Figure 8.12. Let us assume we want to find the exact point at which the locus crosses the 0.45 damping ratio line and the gain at that point. Figure 8.13 shows the system's open-loop poles and zeros

Figure 8.13 Finding and Calibrating Exact Points on the Root Locus of Figure 8.12

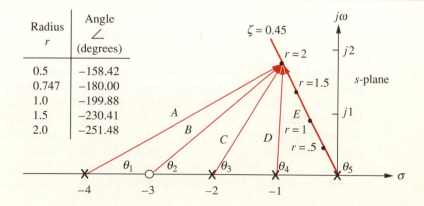

along with the $\zeta = 0.45$ line. If a few test points along the $\zeta = 0.45$ line are selected, we evaluate their angular sum and locate that point where the angles add up to an odd multiple of $180°$. It is at this point that the root locus exists. Equation 8.21 can then be used to evaluate the gain, K, at that point.

Selecting the point at radius 2 ($r = 2$) on the $\zeta = 0.45$ line, we add the angles of the zeros and subtract the angles of the poles, obtaining

$$\theta_2 - \theta_1 - \theta_3 - \theta_4 - \theta_5 = -251.48° \tag{8.55}$$

Since the sum is not equal to an odd multiple of $180°$, the point at radius $= 2$ is not on the root locus. Proceeding similarly for the points at radius $= 1.5$, 1, 0.747, and 0.5, we obtain the table shown in Figure 8.13. This table lists the points, giving their radius, r, and the sum of angles indicated by the symbol $\angle$. From the table, we see that the point at radius 0.747 is on the root locus since the angles add up to $180°$. Using Equation 8.21, the gain, K, at this point is found to be

$$K = \frac{|A||C||D||E|}{|B|} = 1.71 \tag{8.56}$$

In summary, *we search a given line for the point yielding a summation of angles (zero angles − pole angles) equal to an odd multiple of 180°.* We conclude that the point is on the root locus. The gain at that point is then found by *multiplying the pole lengths drawn to that point and dividing by the product of the zero lengths drawn to that point.*

We have yet to address the computational methods required to perform the above task. In the past, an inexpensive tool called a Spirule™ added the angles together rapidly. The tool then quickly performed the task of multiplying and dividing the lengths to obtain the gain. Today we can rely on hand-held or programmable calculators as well as personal computers to make the calculations. In Appendix C a program in Microsoft® QuickBASIC is listed that can be used to perform the calculations described in this section. The program runs on a Macintosh™, but it can be adapted for a programmable hand-held calculator or personal computer.

8.7 Real-Axis Breakaway and Break-in Points

Numerous root loci appear to break away from the real axis as the system poles move from the real axis to the complex plane. At other times, the loci appear to return to the real axis as a pair of complex poles becomes real. We illustrate this point in Figure 8.14. This locus is sketched using the first four rules: (1) number of branches, (2) symmetry, (3) real-axis segments, and (4) starting and ending points. The figure shows a root locus leaving the real axis between -1 and -2 and returning to the real axis between $+3$ and $+5$. The point where the locus leaves the real axis, $-\sigma_1$, is called the *breakaway point*, and the point where the

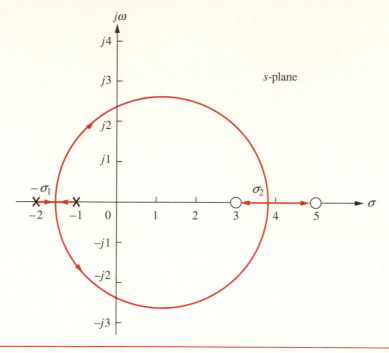

Figure 8.14 Root Locus Example Showing Real-Axis Break-away $(-\sigma_1)$ and Break-in Points (σ_2)

locus returns to the real axis, σ_2, is called the *break-in* point. Our objective is to find these two points.

As the two closed-loop poles, which are at -1 and -2 when $K = 0$, move toward each other, the gain increases from a value of zero. We conclude that the gain must be maximum along the real axis at the point where the breakaway occurs, somewhere in the region between -1 and -2. Naturally, the gain increases above this value as the poles move into the complex plane. The conclusion is that the breakaway point occurs at a point of maximum gain on the real axis between the open-loop poles.

Now let us turn our attention to the break-in point somewhere between $+3$ and $+5$ on the real axis. When the closed-loop complex pair returns to the real axis, the gain will continue to increase to infinity as the closed-loop poles move toward the open-loop zeros. It must be true, then, that the gain at the break-in point is the minimum gain found along the real axis between the two zeros.

The sketch in Figure 8.15 shows the variation of real-axis gain. The breakaway point is found at the maximum gain between -1 and -2, and the break-in point is found at the minimum gain between $+3$ and $+5$.

There are two methods for finding the points at which the root locus breaks away from and breaks into the real axis. The first method of finding the breakaway and break-in points is to maximize and minimize the gain, K, using differential

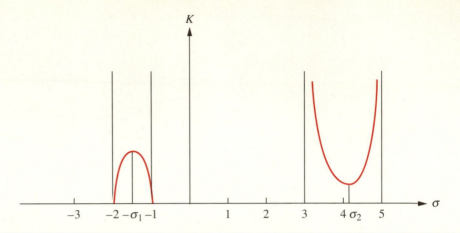

Figure 8.15 Variation of Gain along the Real Axis for the
Root Locus of Figure 8.14

calculus. For all points on the root locus, Equation 8.13 yields

$$K = -\frac{1}{G(s)H(s)} \tag{8.57}$$

For points along the real-axis segment of the root locus where breakaway and
break-in points could exist, $s = \sigma$. Hence, along the real axis, Equation 8.57
becomes

$$K = -\frac{1}{G(\sigma)H(\sigma)} \tag{8.58}$$

This equation then represents a curve of K versus σ similar to that shown in
Figure 8.15. Hence, if we differentiate Equation 8.58 with respect to σ and set
the derivative equal to zero, we can find the points of maximum and minimum
gain and hence the breakaway and break-in points. Let us demonstrate.

Example 8.4   Find the breakaway and break-in points, using differential calculus.

Problem   Find the breakaway and break-in points for the root locus of Figure 8.14,
using differential calculus.

Solution   Using the open-loop poles and zeros, we represent the open-loop system
whose root locus is shown in Figure 8.14 as follows:

$$KG(s)H(s) = \frac{K(s-3)(s-5)}{(s+1)(s+2)} = \frac{K(s^2-8s+15)}{(s^2+3s+2)} \tag{8.59}$$

But for all points along the root locus, $KG(s)H(s) = -1$, and along the real axis, $s = \sigma$. Hence,

$$\frac{K(\sigma^2 - 8\sigma + 15)}{(\sigma^2 + 3\sigma + 2)} = -1 \tag{8.60}$$

Solving for K, we find

$$K = \frac{-(\sigma^2 + 3\sigma + 2)}{(\sigma^2 - 8\sigma + 15)} \tag{8.61}$$

Differentiating K with respect to σ and setting the derivative equal to zero yields

$$\frac{dK}{d\sigma} = \frac{(11\sigma^2 - 26\sigma - 61)}{(\sigma^2 - 8\sigma + 15)^2} = 0 \tag{8.62}$$

Solving for σ, we find $\sigma = -1.45$, and 3.82, which are the breakaway and break-in points.

In the second method, the root locus program in Appendix C can be used to find the breakaway and break-in points. Simply use the program to search for the point of maximum gain between -1 and -2 and to search for the point of minimum gain between $+3$ and $+5$. Table 8.3 shows the results of the search. The locus leaves the axis at -1.45, the point of maximum gain between -1 and -2, and reenters the real axis at $+3.8$, the point of minimum gain between $+3$ and $+5$. These results are the same as those obtained using the first method.

Table 8.3 Data for Search for Breakaway and Break-in Points for the Root Locus of Figure 8.14

Real axis value	Gain	
-1.41	0.008557	
-1.42	0.008585	
-1.43	0.008605	
-1.44	0.008617	
-1.45	0.008623	← Max. gain: breakaway
-1.46	0.008622	
3.3	44.686	
3.4	37.125	
3.5	33.000	
3.6	30.667	
3.7	29.440	
3.8	29.000	← Min. gain: break-in
3.9	29.202	

Comparing the two methods, we can see that for lower order systems, the first method yields the results by a direct analytical calculation. However, if the order of the system is greater than second order, finding the point where $dK/d\sigma = 0$ will require solving a high-order polynomial with computational aids. Thus the first method reduces to a technique similar to the second method if the system is of order higher than two.

8.8 An Example

We now summarize root locus concepts with an example.

Example 8.5   Sketch a root locus and find the values of critical points.

Problem   Sketch the root locus for the system shown in Figure 8.16(*a*) and find the following:

    a. The exact point and gain where the locus crosses the 0.45 damping ratio line
    b. The exact point and gain where the locus crosses the $j\omega$-axis
    c. The breakaway point on the real axis
    d. The range of K within which the system is stable

Solution   The problem solution is shown, in part, in Figure 8.16(*b*). First sketch the root locus. Using Rule 3, the real-axis segment is found to be between -2 and -4. Rule 4 tells us that the root locus starts at the open-loop poles and ends at the open-loop zeros. These two rules alone give us the general shape of the root locus.

    a. To find the exact point where the locus crosses the $\zeta = 0.45$ line, we can use the root locus program in Appendix C to search along the line

$$\theta = 180° - \cos^{-1} 0.45 = 116.744° \tag{8.63}$$

for the point where the angles add up to an odd multiple of 180°. Searching in polar coordinates, we find that the root locus crosses the $\zeta = 0.45$ line at $3.4\angle 116.744°$ with a gain, K, of 0.417.

    b. To find the exact point where the locus crosses the $j\omega$-axis, use the root locus program to search along the line

$$\theta = 90° \tag{8.64}$$

for the point where the angles add up to an odd multiple of 180°. Searching in polar coordinates, we find that the root locus crosses the $j\omega$-axis at $\pm j3.9$ with a gain of $K = 1.5$.

    c. To find the breakaway point, use the root locus program to search the real axis between -2 and -4 for the point that yields maximum gain. Naturally, all points will have the sum of their angles equal to an odd multiple of 180°. A maximum gain of 0.024755 is found at the point -2.88. Therefore the breakaway point is between the open-loop poles on the real axis at -2.88.

    d. From the answer to **(b)**, the system is stable for K between 0 and 1.5.

We now show how to use the root locus for the design of transient response.

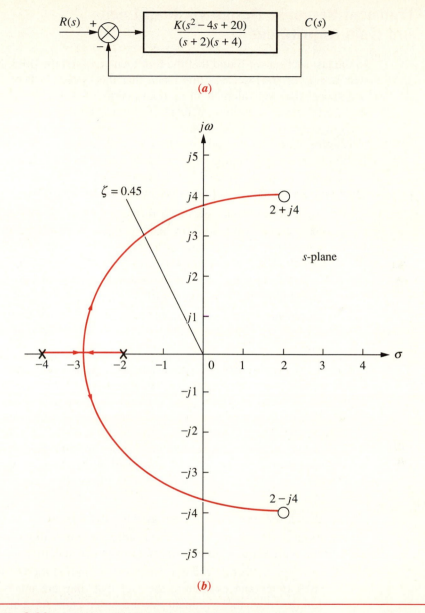

Figure 8.16 (*a*) System for Example 8.5; (*b*) Root Locus Sketch for Example 8.5

8.9 Transient Response Design via Root Locus and Gain Adjustment

In the last section we found that the root locus crossed the 0.45 damping ratio line with a gain of 0.418. Does this mean that the system will respond with 20.5% overshoot, the equivalent to a damping ratio of 0.45? It must be emphasized that the formulae describing percent overshoot, settling time, and peak time were derived only for a system with two closed-loop complex poles and no closed-loop zeros. The effect of additional poles and zeros as well as the conditions for justifying an approximation of a two-pole system were discussed in Sections 4.7 and 4.8 and apply here to closed-loop systems and their root locus. The conditions justifying a second-order approximation are restated here:

1. Higher-order poles are much further into the left half of the s-plane than the dominant second-order pair of poles. The response that results from a higher-order pole does not appreciably change the transient response expected from the dominant second-order poles.

2. Closed-loop zeros near the closed-loop second-order pole pair are nearly canceled by the close proximity of higher-order closed-loop poles.

3. Closed-loop zeros not canceled by the close proximity of higher-order closed-loop poles are far removed from the closed-loop second-order pole pair.

The first condition as it applies to the root locus is graphically shown in Figure 8.17(a) and (b). Figure 8.17(b) would yield a much better second-order approximation than Figure 8.17(a) since closed-loop pole p_3 is farther from the dominant, closed-loop second-order pair, p_1 and p_2.

The second condition is graphically shown in Figure 8.17(c) and (d). Figure 8.17(d) would yield a much better second-order approximation than Figure 8.17(c) since closed-loop pole p_3 is closer to canceling the closed-loop zero.

Summarizing the design procedure for higher-order systems, we arrive at the following:

1. Sketch the root locus for the given system.

2. Assume the system is a second-order system without any zeros and then find the gain to meet the transient response specification.

3. Justify your second-order assumption by finding the location of all higher-order poles and evaluating the fact that they are much further from the $j\omega$-axis than the dominant second-order pair. As a rule of thumb, this textbook assumes a factor of 5 times further. Also, verify that closed-loop zeros are approximately canceled by higher-order poles. If closed-loop zeros are not canceled by higher-order closed-loop poles, be sure that the zero is far removed from the dominant second-order pole pair to yield approximately the same response obtained without the finite zero.

4. If the assumptions cannot be justified, your solution will have to be simulated to be sure it meets the transient response specification. It is a good idea to simulate all solutions anyway.

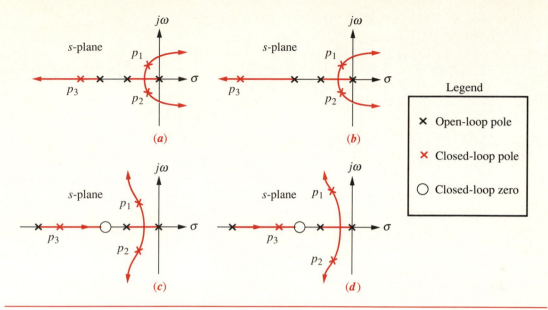

Figure 8.17 Comparison Study for Making Second-Order Approximations

We now look at a design example to show how to make a second-order approximation and then verify whether or not the approximation is valid.

Example 8.6

Design the gain of a third-order system using the root locus.

Problem Consider the system shown in Figure 8.18. Design the value of gain, K, to yield 1.52% overshoot. Also estimate the settling time, peak time, and steady-state error.

Solution The root locus is shown in Figure 8.19. Notice that this system is a third-order system with one zero. Breakaway points on the real axis can occur between 0 and -1, and between -1.5 and -10, where the gain reaches a peak. Using the root locus program and searching in these regions for the peaks in gain, breakaway points are found at -0.62 with a gain of 2.511 and at -4.4 with a gain of 28.89. A break-in point on the real axis can occur between -1.5 and -10,

Figure 8.18 System for Example 8.6

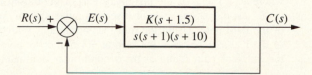

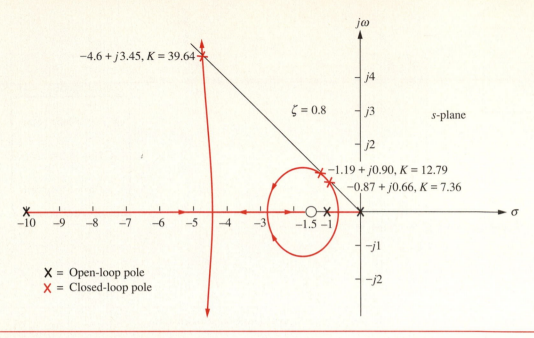

Figure 8.19 Root Locus for Example 8.6

where the gain reaches a local minimum. Using the root locus program and searching in these regions for the local minimum gain, a break-in point is found at -2.8 with a gain of 27.91.

Next, assume that the system can be approximated by a second-order, underdamped system without any zeros. A 1.52% overshoot corresponds to a damping ratio of 0.8. Sketch this damping ratio line on the root locus, as shown in Figure 8.19.

Use the root locus program to search along that 0.8 damping ratio line for the point where the angles from the open-loop poles and zeros add up to an odd multiple of 180°. This is the point where the root locus crosses the 0.8 damping ratio or 1.52 percent overshoot line. There are three points that satisfy this criterion: $-0.87 \pm j0.65$, $-1.2 \pm j0.9$, and $-4.6 \pm j3.45$ with respective gains of 7.33, 12.89, and 39.6. For each point, the settling time and peak time are evaluated using

$$T_s = \frac{4}{\zeta \omega_n} \tag{8.65}$$

where $\zeta \omega_n$ is the real part of the closed-loop pole, and also using

$$T_p = \frac{\pi}{\omega_n \sqrt{1 - \zeta^2}} \tag{8.66}$$

where $\omega_n \sqrt{1 - \zeta^2}$ is the imaginary part of the closed-loop pole.

To test our assumption of a second-order system, we must calculate the location of the third pole. Using the root locus program, search along the negative extension of the real axis between the zero at -1.5 and the pole at -10 for points that match the value of gain found at the second-order dominant poles. For each of the three

Table 8.4 Characteristics of the System of Example 8.6

Case	Closed-loop poles	Closed-loop zero	Gain	Third closed-loop pole	Settling time	Peak time	K_v
1	$-0.87 \pm j0.66$	$-1.5 + j0$	7.36	-9.25	4.60	4.76	1.1
2	$-1.19 \pm j0.90$	$-1.5 + j0$	12.79	-8.61	3.36	3.49	1.9
3	$-4.60 \pm j3.45$	$-1.5 + j0$	39.64	-1.80	0.87	0.91	5.9

crossings of the 0.8 damping ratio line, the third closed-loop pole is at -9.25, -8.6, and -1.8 respectively. The results are summarized in Table 8.4.

Finally, let us examine the steady-state error produced in each case. Note that we have little control over the steady-state error at this point. When the gain is set to meet the transient response, we have also designed the steady-state error. For the example, the steady-state error specification is given by K_v and is calculated by

$$K_v = \lim_{s \to 0} sG(s) = \frac{K(1.5)}{(1)(10)} \tag{8.67}$$

The results for each case are shown in Table 8.4.

How valid are the second-order assumptions? From Table 8.4, Cases 1 and 2 yield third closed-loop poles that are relatively far from the closed-loop zero. For these two cases there is no pole-zero cancellation, and a second-order system approximation is not valid. In Case 3, the third closed-loop pole and the closed-loop zero are relatively close to each other, and a second-order system approximation can be considered valid. In order to show this, let us make a partial fraction expansion of the closed-loop step response of Case 3 and see that the amplitude of the exponential decay is much less than the amplitude of the underdamped sinusoid. The closed-loop step response, $C_3(s)$, formed from the closed-loop poles and zeros of Case 3 is

$$
\begin{aligned}
C_3(s) &= \frac{39.64(s + 1.5)}{s(s + 1.8)(s + 4.6 + j3.45)(s + 4.6 - j3.45)} \\
&= \frac{39.64(s + 1.5)}{s(s + 1.8)(s^2 + 9.2s + 33.06)} \\
&= \frac{1}{s} + \frac{0.3}{s(s + 1.8)} - \frac{1.3(s + 4.6) + 1.6(3.45)}{(s + 4.6)^2 + 3.45^2} \tag{8.68}
\end{aligned}
$$

Thus, the amplitude of the exponential decay from the third pole is 0.3, and the amplitude of the underdamped response from the dominant poles is $\sqrt{1.3^2 + 1.6^2} = 2.06$. Hence, the dominant pole response is 6.9 times larger than the nondominant exponential response and we assume that a second-order approximation is valid.

Using a simulation program, we obtain Figure 8.20, which shows comparisons of step responses for the problem we have just solved. Cases 2 and 3 are plotted for both the third-order response and for a second-order response, assuming just the dominant pair of poles calculated in the design problem. Again, the second-order approximation was justified for Case 3, where there is a small difference in percent overshoot. The second-order approximation is not valid for Case 2. Other than the excess overshoot, Case 3 responses are similar.

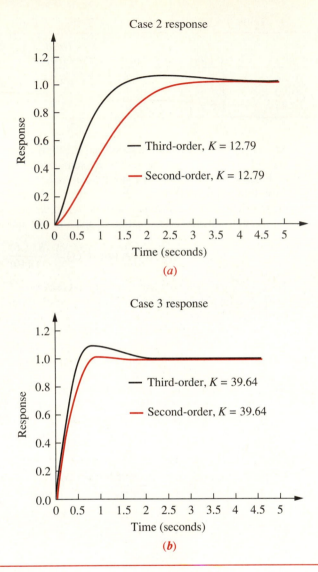

Figure 8.20 Comparison between Second- and Third-Order Responses for Example 8.6: (***a***) Case 2; (***b***) Case 3

8.10 Generalized Root Locus

Up to this point we have always drawn the root locus as a function of the forward-path gain, K. The control system designer must often know how the closed-loop poles change as a function of another parameter. For example, in Figure 8.21 the parameter of interest is the open-loop pole at $-p_1$. How can we obtain a root locus for variations of the value of p_1?

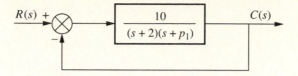

Figure 8.21 System That Requires a Root Locus Calibrated with p_1 as a Parameter

If the function $KG(s)H(s)$ is formed as

$$KG(s)H(s) = \frac{10}{(s+2)(s+p_1)} \qquad (8.69)$$

the problem is that p_1 is not a multiplying factor of the function as the gain, K, was in all of the previous problems. The solution to this dilemma is to create an equivalent system where p_1 appears as the forward-path gain. Since the closed-loop transfer function's denominator is $1 + KG(s)H(s)$, we effectively want to create an equivalent system whose denominator is $1 + p_1G(s)H(s)$.

For the system of Figure 8.21, the closed-loop transfer function is

$$T(s) = \frac{KG(s)}{1 + KG(s)H(s)} = \frac{10}{s^2 + (p_1 + 2)s + 2p_1 + 10} \qquad (8.70)$$

Isolating p_1, we have

$$T(s) = \frac{10}{s^2 + 2s + 10 + p_1(s+2)} \qquad (8.71)$$

Converting the denominator to the form $[1 + p_1G(s)H(s)]$ by dividing numerator and denominator by the term not included with p_1, $s^2 + 2s + 10$, we obtain

$$T(s) = \frac{\dfrac{10}{s^2 + 2s + 10}}{1 + \dfrac{p_1(s+2)}{s^2 + 2s + 10}} \qquad (8.72)$$

Conceptually, Equation 8.72 implies that we have a system for which

$$KG(s)H(s) = \frac{p_1(s+2)}{s^2 + 2s + 10} \qquad (8.73)$$

The root locus can now be sketched as a function of p_1 assuming the open-loop system of Equation 8.73. The final result is shown in Figure 8.22.

In this section we learned to plot the root locus as a function of any system parameter. In the next section we will learn how to plot root loci for positive feedback systems.

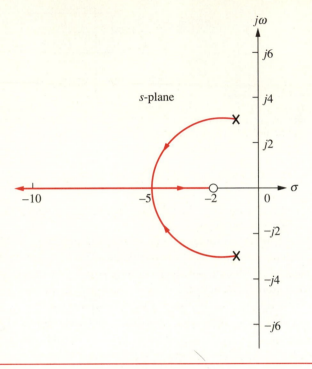

Figure 8.22 Root Locus for the System of Figure 8.21, with p_1 as a Parameter

8.11 Root Locus for Positive-Feedback Systems

The properties of the root locus were derived from the system of Figure 8.1. This is a negative-feedback system because of the negative summing of the feedback signal to the input signal. The properties of the root locus change dramatically if the feedback signal is added to the input rather than subtracted. A positive-feedback system can be thought of as a negative-feedback system with a negative value of $H(s)$. Using this concept, we find that the transfer function for the positive-feedback system shown in Figure 8.23 is

$$T(s) = \frac{KG(s)}{1 - KG(s)H(s)} \tag{8.74}$$

We now retrace the development of the root locus for the denominator of Equation 8.74. Obviously a pole, s, exists when

$$KG(s)H(s) = 1 = 1\angle k360° \qquad k = 0, \pm1, \pm2, \pm3, \ldots \tag{8.75}$$

Therefore, the root locus for positive-feedback systems consists of all points on the s-plane where the angle of $KG(s)H(S) = k360°$. How does this relationship change the rules for sketching the root locus?

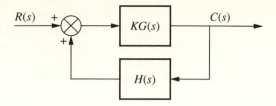

Figure 8.23 Positive-Feedback System

1. **Number of branches.** The same arguments for negative feedback apply to this rule. There is no change.

2. **Symmetry.** The same arguments for negative feedback apply to this rule. There is no change.

3. **Real-axis segments.** The development in Section 8.4 for the real-axis segments led to the fact that the sum of the angles of $G(s)H(s)$ along the real axis added up to either some odd multiple of $180°$ or multiples of $360°$. Thus, for positive-feedback systems, the root locus will exist on the real axis along those sections where the locus for negative-feedback systems does not exist. The rule follows. *Real-axis segments: On the real axis, the root locus for positive-feedback systems exists to the left of an **even** number of real-axis poles and/or zeros.* The change in the rule is the word *even*; for negative-feedback systems the locus existed to the left of an *odd* number of poles and/or zeros.

4. **Starting and ending points.** The reader will find no change in the development in Section 8.4 if Equation 8.74 is used instead of Equation 8.12. Therefore, we have the following rule. *Starting and ending points: The root locus for positive-feedback systems begins at the poles of $G(s)H(s)$ and ends at the zeros of $G(s)H(s)$.*

5. **Behavior at infinity.** The changes in the development of the asymptotes begin at Equation 8.30, since positive-feedback systems follow the relationship shown in Equation 8.75. That change yields a different slope for the asymptotes. The value of the real-axis intercept for the asymptotes remains unchanged. The student is encouraged to go through the development in detail and show that the behavior at infinity for positive-feedback systems is given by the following rule. *Behavior at infinity: The root locus approaches straight lines as asymptotes as the locus approaches infinity. Further, the equation of the asymptotes for positive feedback systems is given by the real-axis intercept and slope as follows:*

$$\sigma_0 = \frac{\sum \text{Poles} - \sum \text{Zeros}}{\#\text{poles} - \#\text{zeros}} \tag{8.76}$$

$$M = \tan \frac{k2\pi}{\#\text{poles} - \#\text{zeros}} \tag{8.77}$$

where $k = 0, \pm 1, \pm 2, \pm 3, \ldots$

The change that we see is that the numerator of Equation 8.77 is $k2\pi$ instead of $(2k + 1)\pi$. What about other calculations? The imaginary-axis crossing can be found using the root locus program. In a search of the $j\omega$-axis, you are looking for the point where the angles add up to some multiple of 360° instead of an odd multiple of 180°. The breakaway points are found by looking for the maximum value of K. The break-in points are found by looking for the minimum value of K.

When we were discussing *negative*-feedback systems, we always made the root locus plot for positive values of gain. Since *positive*-feedback systems can also be thought of as *negative*-feedback systems with negative gain, the rules developed in this section apply equally to *negative*-feedback systems with negative gain. Let us look at an example.

Example 8.7   Find the root locus for a positive-feedback system.

Problem   Sketch the root locus as a function of negative gain, K, for the system shown in Figure 8.11.

Figure 8.24 (*a*) Equivalent Positive-Feedback System for Example 8.7; (*b*) Root Locus

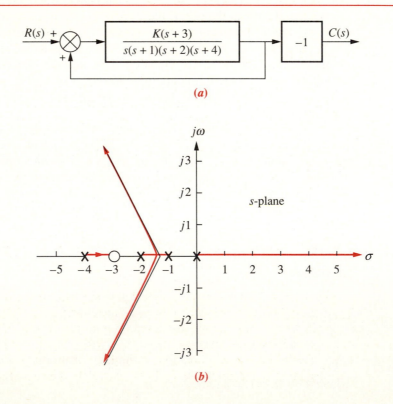

(*a*)

(*b*)

Solution The equivalent positive feedback system found by pushing -1, associated with K, to the right past the pickoff point is shown in Figure 8.24(a). Therefore, as the gain of the equivalent system goes through positive values of K, the root locus will be equivalent to that generated by the gain, K, of the original system in Figure 8.11 as it goes through negative values.

The root locus exists on the real axis to the left of an even number of real open-loop poles and/or zeros. Therefore, the locus exists on the entire positive extension of the real axis, between -1 and -2, and between -3 and -4. Using Equation 8.47, the σ intercept is found to be

$$\sigma = \frac{(-1 - 2 - 4) - (-3)}{4 - 1} = -\frac{4}{3} \tag{8.78}$$

The slopes of the lines that intersect at $-4/3$ are given by

$$M = \tan \frac{k2\pi}{\#\text{poles} - \#\text{zeros}} \tag{8.79a}$$

$$= \tan(0) \qquad \text{for } k = 0 \tag{8.79b}$$

$$= \tan(2\pi/3) \qquad \text{for } k = 1 \tag{8.79c}$$

$$= \tan(4\pi/3) \qquad \text{for } k = 2 \tag{8.79d}$$

The final root locus sketch is shown in Figure 8.24(b).

8.12 Pole Sensitivity

The root locus is a plot of the closed-loop poles as a system parameter is varied. Typically the system parameter is gain. Any change in the parameter changes the closed-loop poles and, subsequently, the performance of the system. Many times the parameter changes against our wishes due to heat or other environmental conditions. We would like to find out the extent to which changes in parameter values affect the performance of our system.

The root locus exhibits a nonlinear relationship between gain and pole location. Along some sections of the root locus (1) very small changes in gain yield very large changes in pole location and hence performance; along other sections of the root locus (2) very large changes in gain yield very small changes in pole location. In the first case, we say that the system has a high sensitivity to changes in gain. In the second case, the system has a low sensitivity to changes in gain. We prefer systems with low sensitivity to changes in gain.

In Section 7.7 we defined sensitivity as the ratio of the fractional change in a function to the fractional change in a parameter as the change in the parameter approaches zero. Applying the same definition to the closed-loop poles of a system that vary with a parameter, we define *root sensitivity* to be the ratio of the fractional change in a closed-loop pole to the fractional change in a system parameter such as gain. Using Equation 7.67, we calculate the sensitivity of a closed-loop pole, s, to gain, K:

$$S_{s:K} = \frac{K}{s} \frac{\delta s}{\delta K} \tag{8.80}$$

where s is the current pole location, and K is the current gain. Using Equation 8.80 and converting the partials to finite increments, the actual change in the closed-loop poles can be approximated as

$$\Delta s = s(S_{s:K}) \frac{\Delta K}{K} \tag{8.81}$$

where Δs is the change in pole location, and $\Delta K / K$ is the fractional change in the gain, K. Let us demonstrate with an example. We will begin with the characteristic equation from which $\delta s / \delta K$ can be found. Then, using Equation 8.80 with the current closed-loop pole, s, and its associated gain, K, we can find the sensitivity.

Example 8.8

Find the root sensitivity of a closed-loop system to changes in gain.

Problem   Find the root sensitivity of the system in Figure 8.4 at $s = -9.47$ and $-5 + j5$. Also calculate the change in the pole location for a 10% change in K.

Solution   The system's characteristic equation, found from the closed-loop transfer function denominator, is $s^2 + 10s + K = 0$. Differentiating with respect to K, we have

$$2s \frac{\delta s}{\delta K} + 10 \frac{\delta s}{\delta K} + 1 = 0 \tag{8.82}$$

from which

$$\frac{\delta s}{\delta K} = \frac{-1}{2s + 10} \tag{8.83}$$

Substituting into Equation 8.80, the sensitivity is found to be

$$S_{s:K} = \frac{K}{s} \frac{-1}{2s + 10} \tag{8.84}$$

For $s = -9.47$, Table 8.1 shows $K = 5$. Substituting these values into Equation 8.84 yields $S_{s:K} = -0.059$. The change in the pole location for a 10% change in K can be found using Equation 8.81, with $s = -9.47$, $\Delta K / K = 0.1$, and $S_{s:K} = -0.059$. Hence, $\Delta s = 0.056$, or the pole will move to the right by 0.056 units for a 10% change in K.

For $s = -5 + j5$, Table 8.1 shows $K = 50$. Substituting these values into Equation 8.84 yields $S_{s:K} = \dfrac{1}{1 + j1} = \dfrac{1}{\sqrt{2}} \angle -45°$. The change in the pole location for a 10% change in K can be found using Equation 8.81, with $s = -5 + j5$, $\Delta K / K = 0.1$, and $S_{s:K} = \dfrac{1}{\sqrt{2}} \angle -45°$. Hence, $\Delta s = j0.5$, or the pole will move vertically by 0.5 units for a 10% change in K.

In summary, then, at $K = 5$, $S_{s:K} = -0.059$. At $K = 50$, $S_{s:K} = \dfrac{1}{\sqrt{2}} \angle -45°$.

Comparing magnitudes, we conclude that the root locus is less sensitive to changes in gain at the lower value of K. Notice that root sensitivity is a complex quantity possessing both the magnitude and direction information from which the change in poles can be calculated.

In this section we discussed the effect that changes of parameters would have on the root locus. We saw that the amount of change is not constant everywhere on the root locus. Some closed-loop poles are less sensitive to parameter changes than others.

8.13 Chapter-Objective Demonstration Problems

The main thrust of this chapter is to demonstrate design through gain adjustment of higher-order systems (higher than two), such as the one shown in Figure 8.25. Specifically, we are interested in determining the value of gain required to meet

Figure 8.25 Automatic Paper Roll Handling Vehicle Uses "Infloor" Wire as Part of a Guidance System (Courtesy of Control Engineering Co.)

transient response requirements such as percent overshoot, settling time, and peak time. The following chapter-objective demonstration problem emphasizes this design procedure using the root locus.

Example 8.9

Determine the loop gain required to meet a transient response requirement for a higher-order system.

Problem Given the schematic for the antenna azimuth position control system shown in Figure 2.36, find the preamplifier gain required for 25% overshoot.

Solution The block diagram for the system was derived in Section 5.9 and is shown in Figure 5.31(c), where $G(s) = \dfrac{6.63K}{s(s + 1.71)(s + 100)}$.

First, a sketch of the root locus is made to orient the designer. The real-axis segments are between the origin and -1.71, and from -100 to infinity. The locus begins at the open-loop poles, which are all on the real axis at the origin, -1.71 and -100. The locus then moves toward the zeros at infinity by following asymptotes that, from Equations 8.47 and 8.48, intersect the real axis at -33.9 at angles of $60°$, $180°$, and $-60°$. A portion of the root locus is shown in Figure 8.26.

From Equation 4.38, 25% overshoot corresponds to a damping ratio of 0.404. Now

Figure 8.26 A Portion of the Root Locus for the Antenna Azimuth Control System

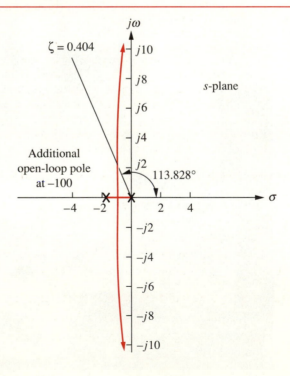

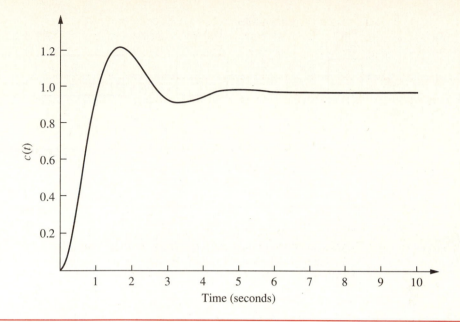

Figure 8.27 Step Response of the Gain-Adjusted Antenna
Azimuth Position Control System

draw a radial line from the origin at an angle of $\cos^{-1} \zeta = 113.828°$. The intersection of this line with the root locus locates the system's dominant, second-order closed-loop poles. Using the root locus program in Appendix C to search the radial line for $180°$ yields the closed-loop dominant poles as $2.063 \angle 113.828° = -0.833 \pm j1.887$. The gain value yields $6.63K = 425.713$, from which $K = 64.21$.

Checking our second-order assumption, the third pole must be to the left of the open-loop pole at -100 and is thus greater than five times the real part of the dominant pole pair, which is -0.833. The second-order approximation is thus valid.

The computer simulation of the closed-loop system's step response in Figure 8.27 shows that the design requirement of 25% overshoot is met.

In the next example, we apply the root locus to the Unmanned Free-Swimming Submersible (UFSS) vehicle pitch control loop. In Figure 5.34, the pitch control loop is shown with both rate and position feedback. In the example that follows, we will plot the root locus without the rate feedback and then with the rate feedback. We will see the stabilizing effect that rate feedback has upon the system.

Example 8.10   Use the root locus to determine transient response.

Problem   Consider the pitch control loop for the UFSS vehicle shown in Figure 8.28 (Johnson, 1980).

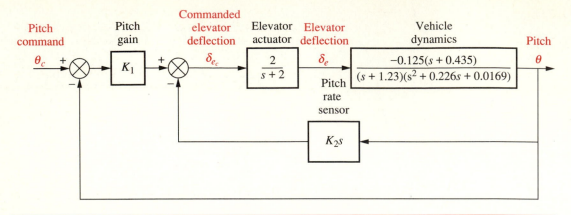

Figure 8.28 Pitch Control Loop for the UFSS Vehicle (Adapted from Johnson, H., et al. *Unmanned Free-Swimming Submersible (UFSS) System Description*. NRL Memorandum Report 4393. Naval Research Laboratory, Washington, D.C., 1980)

a. If $K_2 = 0$ (i.e., no rate feedback), plot the root locus for the system as a function of pitch gain, K_1, and estimate the settling time and peak time with the system responding in closed-loop with 20% overshoot.

b. Let $K_2 = K_1$ (i.e., add rate feedback) and repeat part **(a)**.

Solution

a. Letting $K_2 = 0$ in Figure 8.28, the open-loop transfer function is

$$G(s)H(s) = \frac{-0.25K_1(s + 0.435)}{(s + 1.23)(s + 2)(s^2 + 0.226s + 0.0169)} \tag{8.85}$$

from which the root locus is plotted in Figure 8.29. Searching along the 20% overshoot line, we find the dominant second-order poles to be $-0.202 \pm j0.394$ with a gain of $K = -0.25K_1 = 0.706$, or $K_1 = -2.824$.

From the real part of the dominant pole, the settling time is estimated to be $T_s = 4/0.202 = 19.8$ seconds. From the imaginary part of the dominant pole, the peak time is estimated to be $T_p = \pi/0.394 = 7.97$ seconds. Since our estimates are based upon a second-order assumption, we now test our assumption by finding the third closed-loop pole location between -0.435 and -1.23, and the fourth closed-loop pole location between -2 and infinity. Searching each of these regions for a gain of $K = 0.706$, we find the third and fourth poles at -0.784 and -2.27, respectively. The third pole at -0.784 may not be close enough to the zero at -0.435, and thus the system should be simulated. The fourth pole at -2.27 is 11 times further from the imaginary axis than the dominant poles and thus meets the requirement of at least 5 times the real part of the dominant poles.

A computer simulation of the step response for the system, which is shown in Figure 8.30, shows a 29% overshoot above a final value of 0.88, approximately 20-second settling time, and a peak time of approximately 7.5 seconds.

b. Adding rate feedback by letting $K_2 = K_1$ in Figure 8.28, we proceed to find the new open-loop transfer function. Pushing K_1 to the right past the summing junction,

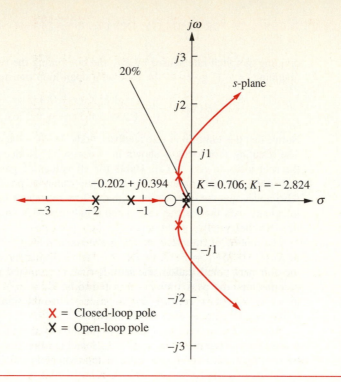

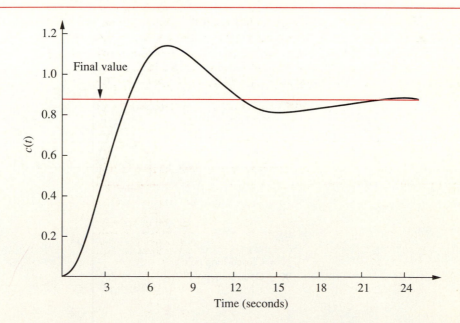

Figure 8.29 Root Locus of Pitch Control Loop without Rate Feedback, UFSS Vehicle

Figure 8.30 Computer Simulation of Step Response of Pitch Control Loop without Rate Feedback, UFSS Vehicle

dividing the pitch rate sensor by K_1, and combining the two resulting feedback paths obtaining $(s + 1)$ gives us the following open-loop transfer function:

$$G(s)H(s) = \frac{-0.25K_1(s + 0.435)(s + 1)}{(s + 1.23)(s + 2)(s^2 + 0.226s + 0.0169)} \qquad (8.86)$$

Notice that the addition of rate feedback adds a zero to the open-loop transfer function. The resulting root locus is shown in Figure 8.31. Notice that this root locus, unlike the root locus in part (a), is stable for all values of gain since the locus does not enter the right half of the s-plane for any value of positive gain, $K = -0.25K_1$. Also notice that the intersection with the 20% overshoot line is much further from the imaginary axis than is the case without rate feedback, resulting in a faster response time for the system.

The root locus intersects the 20% overshoot line at $-1.024 \pm j1.998$ with a gain of $K = -0.25K_1 = 5.17$, or $K_1 = -20.68$. Using the real and imaginary parts of the dominant pole location, the settling time is predicted to be $T_s = 4/1.024 = 3.9$ seconds, and the peak time is estimated to be $T_p = \pi/1.998 = 1.57$ seconds. The new estimates show considerable improvement in the transient response as compared to the system without the rate feedback.

Now we test our second-order approximation by finding the location of the third and fourth poles between -0.435 and -1. Searching this region for a gain of $K = 5.17$, we locate the third and fourth poles at approximately -0.5 and -0.91. Since the zero

Figure 8.31 Root Locus of Pitch Control Loop with Rate Feedback, UFSS Vehicle

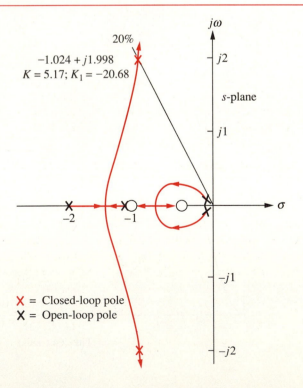

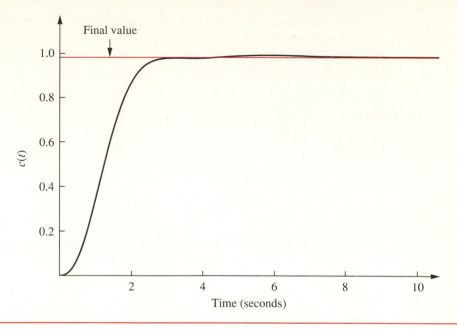

Figure 8.32 Computer Simulation of Step Response of
Pitch Control Loop with Rate Feedback, UFSS Vehicle

at -1 is a zero of $H(s)$, the reader can verify that this zero is not a zero of the closed-loop transfer function. Thus, although there may be pole-zero cancellation between the closed-loop pole at -0.5 and the closed-loop zero at -0.435, there is no *closed-loop* zero to cancel the closed-loop pole at -0.91.[3] Our second-order approximation is not valid.

A computer simulation of the system with rate feedback is shown in Figure 8.32. Although the response shows that our second-order approximation is invalid, it still represents a considerable improvement in performance over the system without rate feedback; the percent overshoot is small, and the settling time is about 6 seconds instead of about 20 seconds.

We have summarized the chapter with two examples showing the use and application of the root locus. We have seen how to plot a root locus and estimate the transient response by making a second-order approximation. In Example 8.10, we saw that the second-order approximation held when rate feedback was not used. When rate feedback was used, an open-loop zero from $H(s)$ was introduced. Since this open-loop zero was not a closed-loop zero, there was no pole-zero cancellation, and a second-order approximation could not be justified. In this case,

---

[3]The zero at -1 shown on the root locus plot of Figure 8.31 is an open-loop zero since it comes from the numerator of $H(s)$.

however, the transient response with rate feedback did represent an improvement in transient response over the system without rate feedback. In subsequent chapters we will show why rate feedback yields an improvement. We will also show other methods of improving the transient response.

8.14 Summary

In this chapter, we examined a powerful tool for the design and analysis of control systems. The root locus empowers us with both qualitative and quantitative information about the stability and transient response of feedback control systems. The root locus allows us to find the poles of the closed-loop system starting from the open-loop system's poles and zeros. It is basically a graphic root-solving technique.

We looked at ways to sketch the root locus rapidly, even for higher-order systems. The sketch gave us qualitative information about changes in the transient response as parameters were varied. From the locus, we were able to determine whether a system was unstable for any range of gain.

Next, we developed the criterion for determining whether a point in the s-plane was on the root locus: the angles from the zeros minus the angles from the poles drawn to the point in the s-plane add up to an odd multiple of $180°$.

The computer program shown in Appendix C helps us to search rapidly for points on the root locus. This program allows us to find points and gains to meet certain transient response specifications as long as we are able to justify a second-order assumption for higher-order systems.

Our method of design in this chapter is gain adjustment. We are limited to transient responses governed by the poles on the root locus. Transient responses represented by pole locations outside of the root locus cannot be obtained by a simple gain adjustment. Further, once the transient response has been established, the gain is set and so is the steady-state error performance. In other words, by a simple gain adjustment, we have to trade off between a specified transient response and a specified steady-state error. Transient response and steady-state error cannot be designed independently with a simple gain adjustment.

We also learned how to plot the root locus against other system parameters other than gain. In order to make this root locus plot, we must first convert the closed-loop transfer function into an equivalent transfer function that has the desired system parameter in the same position as the gain.

The chapter concluded with a discussion of positive-feedback systems and how to plot the root loci for these systems. The next chapter extends the concept of the root locus to the design of compensation networks. These networks have as an advantage the separate design of transient performance and steady-state error performance.

REVIEW QUESTIONS

1. What is a root locus?

2. Describe two ways of obtaining the root locus.

3. If $KG(s)H(s) = 5\angle 180°$, for what value of gain is s a point on the root locus?

4. Do the zeros of a system change with a change in gain?

5. Where are the zeros of the closed-loop transfer function?

6. What are two ways of finding out where the root locus crosses the imaginary axis?

7. How can you tell from the root locus if a system is unstable?

8. How can you tell from the root locus if the settling time does not change over a region of gain?

9. How can you tell from the root locus that the natural frequency does not change over a region of gain?

10. How would you determine whether or not a root locus plot crossed the real axis?

11. Describe the conditions that must exist for all closed-loop poles and zeros in order to make a second-order approximation.

12. What rules for plotting the root locus are the same whether the system is a positive- or negative-feedback system?

13. Briefly describe how the zeros of the open-loop system affect the root locus and the transient response.

PROBLEMS

1. For each of the root loci shown in Figure P8.1, tell whether or not the sketch can be a root locus. If the sketch cannot be a root locus, explain why. Give *all* reasons.

Figure P8.1

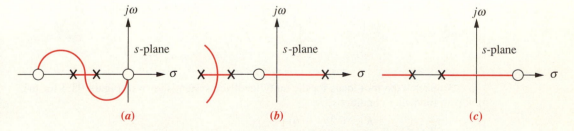

(a)   (b)   (c)

Figure P8.1 (continued)

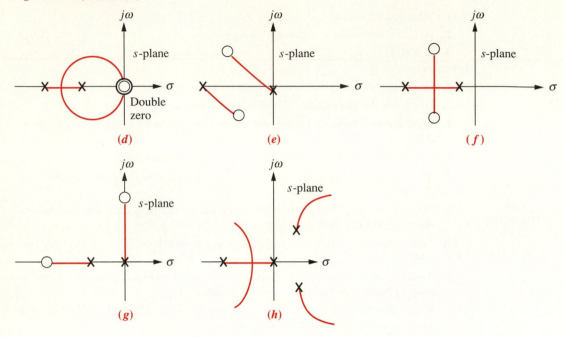

(d)   (e)   (f)

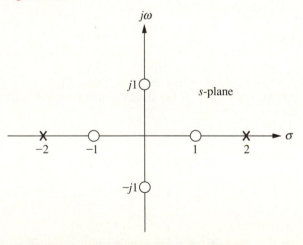

(g)   (h)

2. Sketch the root locus for the open-loop system shown in Figure P8.2.

Figure P8.2

3. Sketch the root locus for the unity feedback system shown in Figure P8.3 for the following conditions.

a. $G(s) = \dfrac{K(s+2)(s+6)}{s^2 + 8s + 25}$

Figure P8.3

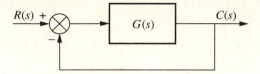

b. $G(s) = \dfrac{K(s^2 + 4)}{(s^2 + 1)}$

c. $G(s) = \dfrac{K(s^2 + 1)}{s^2}$

d. $G(s) = \dfrac{K}{(s + 1)^3(s + 4)}$

e. $G(s) = \dfrac{K(s^2 + 1)}{s^2}$

4. For the open-loop pole-zero plot shown in Figure P8.4, sketch the root locus and find the breakaway point.

Figure P8.4

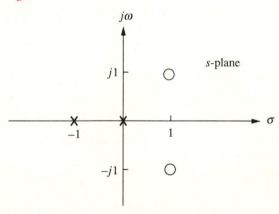

5. Sketch the root locus of the unity feedback system shown in Figure P8.3, where

$$G(s) = \frac{K(s + 4)(s + 6)}{s(s + 2)}$$

and find the break-in and breakaway points.

6. The characteristic polynomial of a feedback control system, which is the denominator of the closed-loop transfer function, is given by $s^3 + 3s^2 + (K + 2)s + 10K$. Sketch the root locus for this system.

7. Figure P8.5 shows open-loop poles and zeros. There are two possibilities for the sketch of the root locus. Sketch each of the two possibilities. Be aware that only one can be the *real* locus.

Figure P8.5

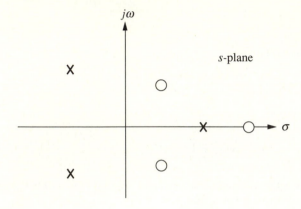

8. Plot the root locus for the unity feedback system shown in Figure P8.3, where

$$G(s) = \frac{K(s + 1)(s - 1)}{(s^2 + 1)}$$

For what range of K will the poles be in the right half-plane?

9. For the unity feedback system shown in Figure P8.3, where

$$G(s) = \frac{K(s^2 + 1)}{(s^2 - 1)}$$

sketch the root locus and tell for what values of K the system is stable and unstable.

10. Sketch the root locus for the unity feedback system shown in Figure P8.3, where

$$G(s) = \frac{K(s^2 + 1)}{(s + 1)(s + 2)}$$

Give the values for all critical points of interest. Is the system ever unstable? If so, for what range of K?

11. For each system shown in Figure P8.6, make an accurate plot of the root locus and find the following:
    a. The breakaway and break-in points
    b. The range of K to keep the system stable
    c. The value of K that yields a stable system with critically damped second-order poles
    d. The value of K that yields a stable system with a pair of second-order poles that have a damping ratio of 0.707

Figure P8.6

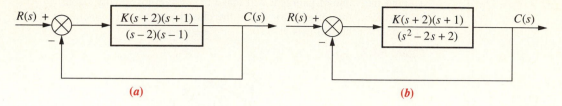

(a) *(b)*

12. Sketch the root locus and find the range of K for stability for the unity feedback system shown in Figure P8.3 for the following conditions.

   a. $G(s) = \dfrac{K(s^2 + 1)}{(s - 1)(s + 2)(s + 3)}$

   b. $G(s) = \dfrac{K(s - 1)(s + 2)(s + 3)}{(s^2 + 1)}$

   c. $G(s) = \dfrac{K(s^2 - 2s + 2)}{s(s + 1)(s + 2)}$

13. For the unity feedback system of Figure P8.3, where

$$G(s) = \frac{K(s + 2)}{(s^2 + 1)(s - 1)(s + 4)}$$

   sketch the root locus and find the range of K such that there will be only two right-half-plane poles for the closed-loop system.

14. For the unity feedback system of Figure P8.3, where

$$G(s) = \frac{K}{s(s + 4)(s + 8)}$$

   plot the root locus and calibrate your plot for gain. Find all the critical points such as breakaways, asymptotes, $j\omega$-axis crossing, and so forth.

15. Given the unity feedback system of Figure P8.3, make an accurate plot of the root locus for the following:

   a. $G(s) = \dfrac{K(s^2 - 2s + 2)}{(s + 1)(s + 2)}$

   b. $G(s) = \dfrac{K(s - 1)(s - 2)}{(s + 1)(s + 2)}$

   Calibrate the gain for at least four points for each case. Also find the breakaway points, the $j\omega$-axis crossing, and the range of gain for stability for each case.

16. Given the root locus shown in Figure P8.7,
  a. Find the value of gain that will make the system marginally stable.
  b. Find the value of gain for which the closed-loop transfer function will have a pole on the real axis at -10.

Figure P8.7

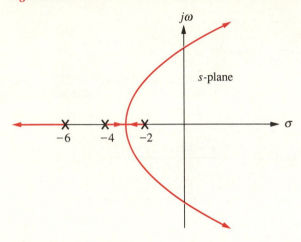

17. Given the unity feedback system of Figure P8.3, where

$$G(s) = \frac{K(s + 1)}{s(s + 2)(s + 3)(s + 4)}$$

do the following:
  a. Sketch the root locus.
  b. Find the asymptotes.
  c. Find the value of gain that will make the system marginally stable.
  d. Find the value of gain for which the closed-loop transfer function will have a pole on the real axis at -0.5.

18. For the unity feedback system of Figure P8.3, where

$$G(s) = \frac{K(s + \alpha)}{s(s + 3)(s + 6)}$$

find the values of α and K that will yield a second-order closed-loop pair of poles at $-1 \pm j10$.

19. For the unity feedback system of Figure P8.3, where

$$G(s) = \frac{K(s - 1)(s - 2)}{s(s + 1)}$$

sketch the root locus and find the following:
  a. The breakaway and break-in points
  b. The $j\omega$-axis crossing
  c. The range of gain to keep the system stable
  d. The value of K to yield a stable system with second-order complex poles, with a damping ratio of 0.5

20. For the system of Figure P8.8(*a*), sketch the root locus and find the following:
  a. Asymptotes
  b. Breakaway points
  c. The range of K for stability
  d. The value of K to yield a 0.7 damping ratio for the dominant second-order pair

Figure P8.8

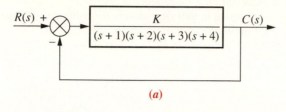

(*a*)

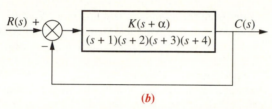

(*b*)

To improve stability, we desire the root locus to cross the $j\omega$-axis at $j5.5$. To accomplish this, the open-loop function is cascaded with a zero, as shown in Figure P8.8(*b*).
  e. Find the value of α and sketch the new root locus.
  f. Repeat part (**c**) for the new locus.
  g. Compare the results of parts (**c**) and (**f**). What improvement in transient response do you notice?

21. Sketch the root locus for the positive-feedback system shown in Figure P8.9.

Figure P8.9

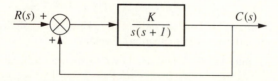

22. Root loci are usually plotted for variations in the gain. Sometimes we are interested in the variation of the closed-loop poles as other parameters are changed. For the system shown in Figure P8.10, sketch the root locus as α is varied.

Figure P8.10

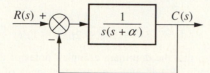

23. Given the unity feedback system shown in Figure P8.3, where

$$G(s) = \frac{K}{(s + 1)(s + 2)(s + 3)}$$

do the problem parts below by first making a second-order approximation. After you are finished with all of the parts, justify your second-order approximation.
a. Sketch the root locus.
b. Find K for 20% overshoot.
c. For K found in part (**b**), what is the settling time, and what is the peak time?
d. Find the range of K for stability.
e. Find the locations of higher-order poles.

24. The unity feedback system shown in Figure P8.3, where

$$G(s) = \frac{K(s + 2)(s + 3)}{s(s + 1)}$$

is to be designed for minimum damping ratio. Find the following:
a. The value of K that will yield minimum damping ratio
b. The estimated percent overshoot for the above case
c. The estimated settling time and peak time for the above case
d. The justification of a second-order approximation (discuss)
e. The expected steady-state error for a unit ramp input for the case of minimum damping ratio

25. Prove that a second-order approximation is valid for the following closed-loop system if the zero at $-\alpha$ is far enough away from the second-order underdamped poles.

$$T(s) = \frac{\left(\dfrac{\omega_n^2}{\alpha}\right)(s + \alpha)}{(s^2 + 2\zeta\omega_n s + \omega_n^2)}$$

26. For the unity feedback system shown in Figure P8.3, where

$$G(s) = \frac{K}{s(s + 1)(s + 2)}$$

find K to yield a damping ratio of 0.7. Does your solution require a justification of a second-order approximation? Explain.

27. For the unity feedback system shown in Figure P8.3, where

$$G(s) = \frac{K(s + \alpha)}{s(s + 1)(s + 10)}$$

find the value of α so that the system will have a settling time of 4 seconds for large values of K. Sketch the resulting root locus.

28. For the unity feedback system shown in Figure P8.3, where

$$G(s) = \frac{K(s + 6)}{(s^2 + 10s + 26)(s + 1)^2(s + \alpha)}$$

design K and α so that the dominant complex poles of the closed-loop function have a damping ratio of 0.45 and a natural frequency of $9/8$.

29. For the unity feedback system shown in Figure P8.3, where

$$G(s) = \frac{K}{s(s+1)(s+5)(s+6)}$$

do the following:
a. Sketch the root locus.
b. Find the value of K that will yield a 10% overshoot.
c. Locate all nondominant poles. What can you say about the second-order approximation that led to your answer in part **(b)**?
d. Find the range of K that yields a stable system.

30. Sketch the root locus for the system of Figure P8.11 and find the following:
a. The range of gain to yield stability
b. The value of gain that will yield a damping ratio of 0.707 for the system's dominant poles
c. The value of gain that will yield closed-loop poles that are critically damped

Figure P8.11

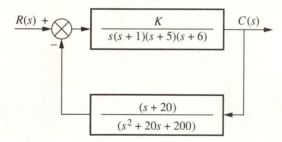

31. Given the unity feedback system shown in Figure P8.3, where

$$G(s) = \frac{K(s+z)}{s^2(s+20)}$$

do the following:
a. If $z = 6$, find K so that the damped frequency of oscillation of the transient response is 10 rad/s.
b. For the system of part **(a)**, what static error constant (finite) can be specified? What is its value?
c. The system is to be redesigned by changing the values of z and K. If the new specifications are $\%OS = 4.32\%$ and $T_s = 0.4$ s, find the new values of z and K.

32. Given the unity feedback system shown in Figure P8.3, where

$$G(s) = \frac{K}{(s+1)(s+3)(s+6)^2}$$

find the following:
a. The value of gain, K, that will yield a settling time of 4 seconds
b. The value of gain, K, that will yield a critically damped system

33. Given the unity feedback system shown in Figure P8.3, where

$$G(s) = \frac{K}{s(s+1)(s+5)}$$

evaluate the pole sensitivity of the closed-loop system if the second-order, underdamped closed-loop poles are set for

a. $\zeta = 0.591$

b. $\zeta = 0.456$

c. Which of the two previous cases has the most desirable sensitivity?

34. *Chapter-objective problem:* Given the schematic for the antenna azimuth position control system shown in Figure P2.36, find the preamplifier gain, K, required for an 8-second settling time.

35. A floppy disk drive is a position control system in which a read/write head is positioned over a magnetic disk. The system responds to a command from a computer to position itself at a particular track on the disk. A physical representation of the system and a block diagram are shown in Figure P8.12.

a. Find K to yield a settling time of 0.1 seconds.

b. What is the resulting percent overshoot?

c. What is the range of K that keeps the system stable?

Figure P8.12

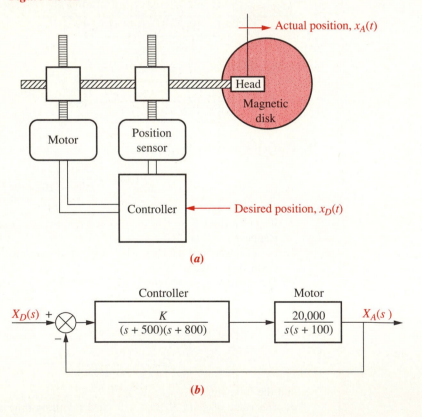

(a)

(b)

36. Figure P8.13(a) shows a robot equipped to perform heavy-duty spot welding. This device can also be configured as a six-degrees-of-freedom industrial robot that can transfer objects according to a desired program. Assume the block diagram of the

Figure P8.13 (*a*) A Robot Equipped to Perform Heavy-Duty Spot Welding (Courtesy of Prab Robots); (*b*) Block Diagram for Swing Motion System (Adapted from Hardy, H.L. Multi-Loop Servo Controls Programmed Robot. *Instruments and Control Systems,* June 1967, p. 107)

(*a*)

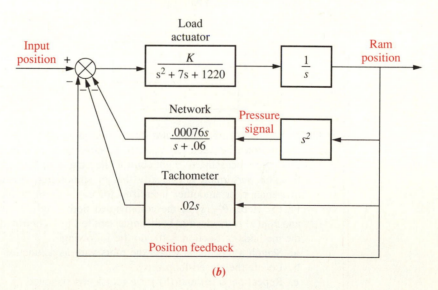

(*b*)

swing motion system shown in Figure P8.13(*b*) (Hardy, 1967). If $K = 64,512.2$, make a second-order approximation and estimate the following:

a. Damping ratio
b. Percent overshoot

c. Natural frequency
d. Settling time
e. Peak time

What can you say about your original second-order approximation?

37. A simplified block diagram of a human pupil servomechanism is shown in Figure P8.14. The term $e^{-0.18s}$ represents a time delay. This function can be approximated by what is known as a *Padé approximation*. This approximation can take on many increasingly complicated forms, depending upon the degree of accuracy required. If we use the Padé approximation

$$e^{-x} = \frac{1}{1 + x + \dfrac{x^2}{2!}}$$

then

$$e^{-0.18s} = \frac{61.73}{s^2 + 11.11s + 61.73}$$

Since the retinal light flux is a function of the opening of the iris, oscillations in the amount of retinal light flux imply oscillations of the iris (Guy, 1976). Find the following:
a. The value of K that will yield oscillations
b. The frequency of these oscillations
c. The settling time for the iris if K is such that the eye is operating with 20% overshoot

Figure P8.14 Simplified Block Diagram of the Pupil Servomechanism

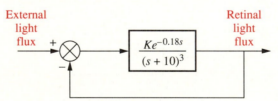

38. During ascent, the automatic steering program aboard the space shuttle provides the interface between the low-rate processing of guidance (commands) and the high-rate processing of flight control (steering in response to the commands). The function performed is basically that of smoothing. A simplified representation of a maneuver smoother linearized for coplaner maneuvers is shown in Figure P8.15. Here, $\theta_{CB}(s)$ is the commanded body angle as calculated by guidance, and $\theta_{DB}(s)$ is the desired body angle sent to flight control after smoothing.[4] Using the methods of Section 8.10, do the following:
a. Sketch a root locus where the roots vary as a function of K_3.
b. Locate the closed-loop zeros.
c. Repeat parts **(a)** and **(b)** for a root locus sketched as a function of K_2.

---

[4] Source: Rockwell International.

Figure P8.15 Block Diagram of Smoother (Adapted from Rockwell International. *GNC FSSR FC Ascent,* vol. 1. Downey, Calif., 30 June 1985)

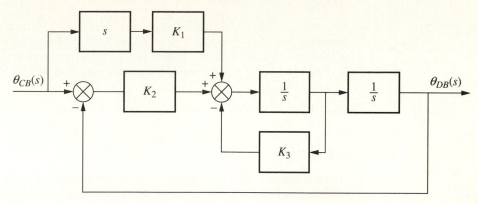

39. Repeat Problem 3, but sketch your root loci for negative values of K.

40. A possible active suspension system for AMTRAK trains has been proposed. The system uses a pneumatic actuator in parallel with the passive suspension system, as shown in Figure P8.16. The force of the actuator subtracts from the force applied by the ground, as represented by displacement, $y_g(t)$. Acceleration is sensed by an accelerometer, and signals proportional to acceleration and velocity are fed back to the force actuator. The transfer function relating acceleration to

Figure P8.16 Active Suspension System (Adapted from Cho, D., and Hedrick, J. K. Pneumatic Actuators for Vehicle Active Suspension Applications. *ASME, Journal of Dynamic Systems, Measurement, and Control,* March 1985, p. 68)

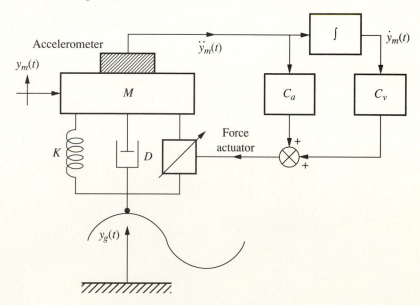

ground displacement is

$$\frac{\ddot{Y}_m(s)}{Y_g(s)} = \frac{s^2\,(Ds + K)}{(C_a + M)s^2 + (C_v + D)s + K}$$

Assuming that $M = 1$ and $D = K = C_v = 2$, do the following (Cho, 1985):

a. Sketch a root locus for this system as C_a varies from zero to infinity.

b. Find the value of C_a that would yield a damping ratio of 0.69 for the closed-loop poles.

BIBLIOGRAPHY

Cho, D., and Hedrick, J. K. Pneumatic Actuators for Vehicle Active Suspension Applications. *Journal of Dynamic Systems, Measurement, and Control,* March 1985, pp. 67–72.

Dorf, R. C. *Modern Control Systems.* 5th ed. Addison-Wesley, Reading, Mass., 1989.

Evans, W. R. Control System Synthesis by Root Locus Method. *AIEE Transactions*, vol. 69, 1950, pp. 66–69.

Evans, W. R. Graphical Analysis of Control Systems. *AIEE Transactions*, vol. 67, 1948, pp. 547–551.

Guy, W., *The Human Pupil Servomechanism.* Computers in Education Division of ASEE, Application Note No. 45, 1976.

Hardy, H. L. Multi-Loop Servo Controls Programmed Robot. *Instruments and Control Systems*, June 1967, pp. 105–111.

Johnson, H., et al. *Unmanned Free-Swimming Submersible (UFSS) System Description.* NRL Memorandum Report 4393. Naval Research Laboratory, Washington, D.C., 1980.

Kuo, B. C. *Automatic Control Systems.* 5th ed. Prentice Hall, Englewood Cliffs, N.J., 1987.

9

DESIGN OF COMPENSATORS
VIA THE ROOT LOCUS

9.1 Introduction and Definition of the Problem

In Chapter 8 we saw that the root locus graphically displayed both transient response and stability information. The locus can be sketched quickly to get a general idea of the changes in transient response generated by changes in gain. Specific points on the locus also can be found accurately to give quantitative design information.

The root locus typically allows us to choose the proper loop gain to meet a transient response specification. As the gain is varied, we move through different regions of response. Setting the gain at a particular value yields the transient response dictated by the poles at that point on the root locus. Thus, *we are limited to those responses that exist along the root locus.*

Improving Transient Response

Flexibility in the design of a desired transient response can be increased if we can design for transient responses that are not on the root locus. Figure 9.1(*a*) illustrates the concept. Assume that the desired transient response, defined by percent overshoot and settling time, is represented by point *B*. Unfortunately, on the current root locus at the specified percent overshoot, we only can obtain the settling time represented by point *A* after a simple gain adjustment. Thus, our goal is to speed up the response at *A* to that of *B*, without affecting the percent overshoot. This increase in speed cannot be accomplished by a simple gain adjustment since point *B* does not lie on the root locus. Figure 9.1(*b*) illustrates the improvement in the transient response we seek: the faster response has the same percent overshoot as the slower response.

One way of solving our problem is to replace the existing system with a system whose root locus intersects the desired design point, *B*. Unfortunately this replacement is expensive and counterproductive. Most systems are chosen for characteristics other than transient response. For example, an elevator cage and motor are chosen for speed and power. Components chosen for their transient response may not necessarily meet, for example, power requirements.

Rather than change the existing system, we augment, or *compensate,* the system with *additional* poles and zeros, so that the compensated system has a

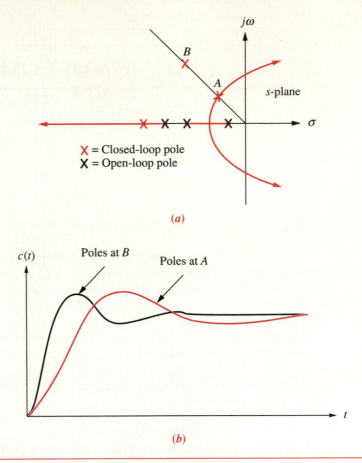

(a)

(b)

Figure 9.1 (**a**) Sample Root Locus, Showing Possible Design Point Arrived at through Gain Adjustment (*A*) and Desired Design Point that Cannot Be Met through Simple Gain Adjustment (*B*); (**b**) Comparison of Responses from Poles at Point *A* and Point *B*

root locus that goes through the desired pole location for some value of gain. One of the advantages of compensating a system in this way is that additional poles and zeros can be added at the low-power end of the system before the plant. Addition of compensating poles and zeros need not interfere with the power output requirements of the system nor present additional load or design problems. The compensating poles and zeros can be generated with a passive or active network.

One method of compensating for transient response that will be discussed later is to insert a differentiator in the forward path in parallel with the gain. We can visualize the operation of the differentiator with the following example. Assuming a position control with a step input, we note that the error undergoes an initial large change. Differentiating this rapid change yields a large signal that drives the plant. The output from the differentiator is much larger than the output from the pure gain. This large, initial input to the plant produces a faster response. As the error approaches its final value, its derivative approaches zero, and the output from the differentiator becomes negligible compared to the output from the gain.

Improving Steady-State Errors

Compensators are not only used to improve the transient response of a system; they are also used *independently* to improve the steady-state error characteristics. Previously, when the system gain was adjusted to meet the transient response specification, steady-state error performance deteriorated, since both the transient response and the static error constant were related to the gain. The higher the gain, the smaller the steady-state error, but the larger the percent overshoot. On the other hand, reducing gain to reduce overshoot increased the steady-state error. If we use dynamic compensators, compensating networks can be designed that will allow us to meet transient and steady-state error specifications *simultaneously*.[1] We no longer need to compromise between transient response and steady-state error.

In Chapter 7 we learned that steady-state error can be improved by adding an open-loop pole at the origin in the forward path, thus increasing the system type and driving the associated steady-state error to zero. This additional pole at the origin requires an integrator for its realization.

In summary then, transient response is improved with the addition of differentiation, and steady-state error is improved with the addition of integration in the forward path.

Configurations

Two configurations of compensation will be covered in this chapter: (1) cascade compensation and (2) feedback compensation. These methods are modeled in Figure 9.2. With cascade compensation, the compensating network, $G_1(s)$, is placed at the low-power end of the forward path in cascade with the plant. If feedback compensation is used, the compensator, $H_1(s)$, is placed in the feedback path. Both methods change the open-loop poles and zeros, thereby creating a new root locus that goes through the desired closed-loop pole location.

Compensators

Compensators that use pure integration for improving steady-state error or pure differentiation for improving transient response are defined as *ideal compensators*. Ideal compensators must be implemented with active networks, which, in the case of electric networks, require the use of active amplifiers and possible additional power sources. An advantage of ideal compensators is that steady-state error is reduced to zero. Electromechanical ideal compensators are often used to improve transient response since these compensators can be conveniently interfaced with the plant.

---

[1]The word *dynamic* means compensators with noninstantaneous transient response. The transfer functions of such compensators are functions of the Laplace variable, s, rather than pure gain.

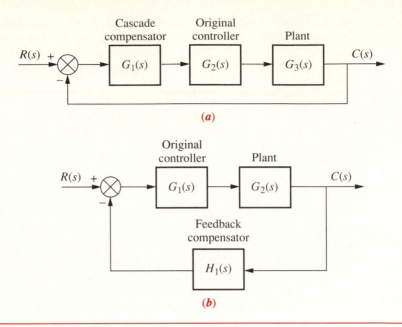

Figure 9.2 Compensation Techniques: (*a*) Cascade Compensation; (*b*) Feedback Compensation

Other design techniques that preclude the use of active devices for compensation can be adopted. These compensators, which can be implemented with passive elements such as resistors and capacitors, do not use pure integration and differentiation and are not ideal compensators. An advantage of passive networks is that they are less expensive and do not require additional power sources for their operation. The disadvantage of passive networks is that the steady-state error is not driven to zero.

Thus, the choice of an active or passive compensator revolves around cost, weight, desired performance, transfer function, and the interface between the compensator and other hardware. In Sections 9.2, 9.3, and 9.4, we first discuss cascade compensator design using ideal compensation and then discuss cascade compensation using compensators that are not implemented with pure integration and differentiation.

Chapter Objective

Given the azimuth position control system of Figure 2.2, the student will be able to design cascade compensation to meet percent overshoot, settling time, and steady-state error requirements. In Section 9.2 we discuss the design of cascade compensation to improve the steady-state error. In Section 9.3 we design cascade compensators to improve the transient response. Finally, in Section 9.4, cascade compensators are designed to improve both the steady-state error and the transient response independently. Following our discussion of cascade compensators, we cover feedback compensation in Section 9.5.

9.2 Cascade Compensation to Improve Steady-State Error

In this section we discuss two ways of improving the steady-state error of a feedback control system using cascade compensation. One objective of this design is to improve the steady-state error without appreciably affecting the transient response.

The first technique is *ideal integral compensation,* which uses a pure integrator to place an open-loop, forward-path pole at the origin, thus increasing the system type and reducing the error to zero. The second technique does not use pure integration. This compensation technique places the pole near the origin, and although it does not drive the steady-state error to zero, it does yield a measurable reduction in steady-state error.

While the first technique reduces the steady-state error to zero, the compensator must be implemented with active networks, such as amplifiers. The second technique, although it does not reduce the error to zero, does have the advantage that it can be implemented with a less-expensive passive network that does not require additional power sources.

The names associated with the compensators come either from the method of implementing the compensator or the compensator's characteristics. For example, in this section we will call the ideal integral compensator a *Proportional-plus-Integral (PI) controller* since the implementation, as we will see, consists of feeding the error (proportional) plus the integral of the error forward to the plant. The second technique uses what we call a *lag compensator.* The name of this compensator comes from its frequency response characteristics, which will be discussed in Chapter 11. Thus, we will use the name *PI controller* interchangeably with *ideal integral compensator,* and we will use the name *lag compensator* when the cascade compensator does not employ pure integration.

Ideal Integral Compensation (PI)

Steady-state error can be improved by placing an open-loop pole at the origin since this action increases the system type by one. For example, a Type 0 system responding to a step input with a finite error responds with zero error if the system type is increased by one. Active circuits can be used to place poles at the origin. Later in this chapter we show how to implement an integrator with active electronic circuits.

To see how to improve the steady-state error without affecting the transient response, look at Figure 9.3(*a*). Here we have a system operating with a desirable

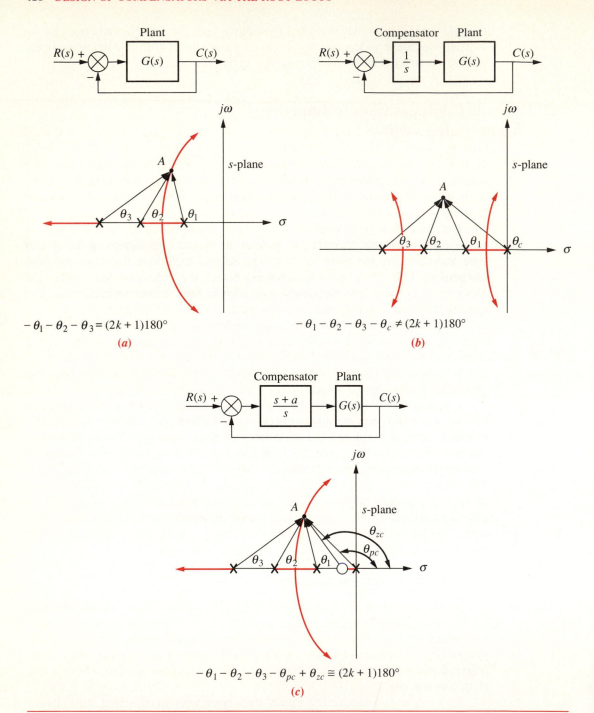

$$-\theta_1 - \theta_2 - \theta_3 = (2k+1)180°$$

(a)

$$-\theta_1 - \theta_2 - \theta_3 - \theta_c \neq (2k+1)180°$$

(b)

$$-\theta_1 - \theta_2 - \theta_3 - \theta_{pc} + \theta_{zc} \cong (2k+1)180°$$

(c)

Figure 9.3 **(a)** Without Compensator—Pole at *A* Is on the Root Locus; **(b)** With Compensator Pole—Pole at *A* Is Not on the Root Locus; **(c)** With Compensator Pole and Zero—Pole at *A* Is Approximately on the Root Locus

transient response generated by the closed-loop poles at A. If we add a pole at the origin to increase the system type, the angular contribution of the open-loop poles at point A is no longer $180°$, and the root locus no longer goes through point A, as shown in Figure 9.3(b).

To solve the problem, we also add a zero close to the pole at the origin, as shown in Figure 9.3(c). Now the angular contribution of the compensator zero and compensator pole cancel out, point A is still on the root locus, and the system type has been increased. Furthermore, the required gain at the dominant pole is about the same as before compensation, since the ratio of lengths from the compensator pole and the compensator zero is approximately unity. Thus, we have improved the steady-state error without appreciably affecting the transient response. A compensator with a pole at the origin and a zero close to the pole is called an *ideal integral compensator*.

In the example that follows, we demonstrate the design of ideal integral compensation. An open-loop pole will be placed at the origin to increase the system type and drive the steady-state error to zero. An open-loop zero will be placed very close to the open-loop pole at the origin so that the original closed-loop poles on the original root locus still remain at approximately the same points on the compensated root locus.

***

Example 9.1

Analyze an ideal integral compensator.

Problem   Given the system of Figure 9.4(a), operating with a damping ratio of 0.174, show that the addition of the ideal integral compensating network shown in Figure 9.4(b) will reduce the steady-state error to zero for a step input without appreciably affecting the transient response. The compensating network is chosen with a pole at the origin to increase the system type and a zero at -0.1, close to the compensator pole, so that the angular contribution of the compensator evaluated at the original, dominant,

Figure 9.4 Closed-Loop System for Example 9.1: (**a**) Before Compensation; (**b**) After Ideal Integral Compensation

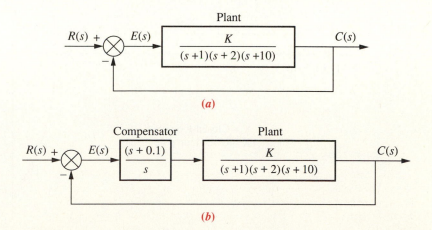

(a)

(b)

second-order poles is approximately zero. Thus the original, dominant, second-order closed-loop poles are still approximately on the new root locus.

Solution We first analyze the uncompensated system and determine the location of the dominant, second-order poles. Next, we evaluate the uncompensated steady-state error for a unit step input. The root locus for the uncompensated system is shown in Figure 9.5.

A damping ratio of 0.174 is represented by a radial line drawn on the s-plane at 100.02°. Searching along this line with the root locus program in Appendix C, we find that the dominant poles are $-0.694 \pm j3.926$ for a gain, K, of 164.565. Now look for the third pole on the root locus beyond -10 on the real axis. Using the root locus program and searching for the same gain as that of the dominant pair, $K = 164.565$, we find that the third pole is approximately at -11.613.

This gain yields $K_p = 8.238$. Hence, the steady-state error is

$$e(\infty) = \frac{1}{1 + K_p} = \frac{1}{1 + 8.228} = 0.108 \tag{9.1}$$

Adding an ideal integral compensator with a zero at -0.1, as shown in Figure 9.4 (*b*), we obtain the root locus shown in Figure 9.6. The dominant second-order poles, the third pole beyond -10, and the gain are approximately the same as for

Figure 9.5 Sketch of the Root Locus for the Uncompensated System of Figure 9.4(*a*)

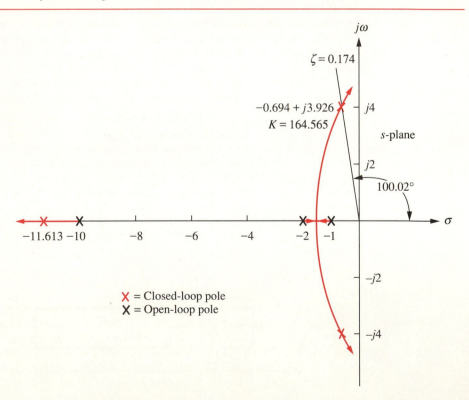

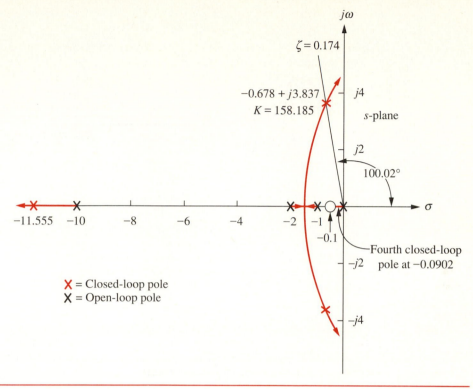

Figure 9.6 Sketch of the Root Locus for the Compensated
System of Figure 9.4(b)

the uncompensated system. Another section of the compensated root locus is between
the origin and −0.1. Searching this region for the same gain at the dominant pair,
$K = 158.185$, the fourth closed-loop pole is found at −0.0902, close enough to
the zero to cause pole-zero cancellation. Thus, the compensated system's closed-loop
poles and gain are approximately the same as the uncompensated system's closed-loop
poles and gain, which indicates that the transient response of the compensated system
is about the same as that of the uncompensated system. However, the compensated
system, with its pole at the origin, is a Type 1 system; unlike the uncompensated
system, it will respond to a step input with zero error.

Figure 9.7 compares the uncompensated response with the ideal integral compensated response. The step response of the ideal integral compensated system approaches
unity in the steady-state, while the uncompensated system approaches 0.892. Thus,
the ideal integral compensated system responds with zero steady-state error. The transient response of both the uncompensated and ideal integral compensated systems is
the same up to approximately 3 seconds. After that time the integrator in the compensator, shown in Figure 9.4(b), slowly compensates for the error until zero error is
finally reached. The simulation shows that it takes 18 seconds for the compensated
system to reach to within ±2% of the final value of unity, while the uncompensated
system takes about 6 seconds to settle to within ±2% of its final value of 0.892.

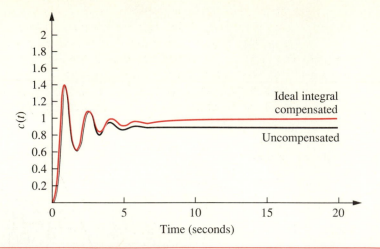

Figure 9.7 Comparison of the Ideal Integral Compensated System Response and the Uncompensated System Response of Example 9.1

The compensation at first may appear to yield deterioration in the settling time. However, notice that the compensated system reaches the uncompensated system's final value in about the same time. The remaining time is used to improve the steady-state error over that of the uncompensated system.

A method of implementing an ideal integral compensator is shown in Figure 9.8. The compensating network precedes $G(s)$ and is an ideal integral compensator since

$$G_c(s) = K_1 + \frac{K_2}{s} = \frac{K_1\left(s + \dfrac{K_2}{K_1}\right)}{s} \tag{9.2}$$

Figure 9.8 A PI Controller

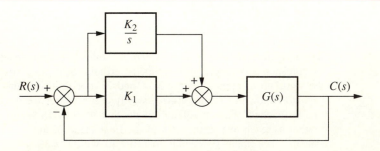

The value of the zero can be adjusted by varying K_2/K_1. In this implementation, the error and the integral of the error are fed forward to the plant, $G(s)$. Systems that feed the error forward to the plant are called *proportional control systems*. Systems that feed the integral of the error to the plant are called *integral control systems*. The system of Figure 9.8 has both proportional and integral control. The controller, or compensator, is thus given the name *PI controller,* which stands for "proportional plus integral." Later in the chapter we will see how to implement each block, K_1 and K_2/s.

Lag Compensation

Ideal integral compensation, with its pole on the origin, requires an active integrator. If we use passive networks, the pole and zero are moved to the left, close to the origin, as shown in Figure 9.9(c). One may guess that this placement of the pole, although it does not increase the system type, does yield an improvement in the static error constant over an uncompensated system. Without loss of generality, we demonstrate that this improvement is indeed realized for a Type 1 system.

Figure 9.9 (**a**) Type 1 Uncompensated System; (**b**) Type 1 Compensated System; (**c**) Compensator Pole-Zero Plot

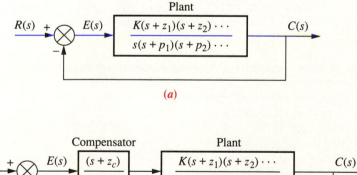

(a)

(b)

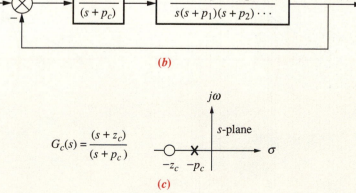

$$G_c(s) = \frac{(s + z_c)}{(s + p_c)}$$

(c)

Assume the uncompensated system shown in Figure 9.9(a). The static error constant, K_{v_O}, for the system is

$$K_{v_O} = \frac{K z_1 z_2 \cdots}{p_1 p_2 \cdots} \tag{9.3}$$

Assuming the lag compensator shown in Figure 9.9(b) and (c), the new static error constant is

$$K_{v_N} = \frac{(K z_1 z_2 \cdots)(z_c)}{(p_1 p_2 \cdots)(p_c)} \tag{9.4}$$

What is the effect upon the transient response? Figure 9.10 shows the effect on the root locus of adding the lag compensator. The uncompensated system's root locus is shown in Figure 9.10(a), where point P is assumed to be the dominant pole. If the lag compensator pole and zero are close together, the angular contribution of the compensator to point P is approximately zero degrees. Thus, in Figure 9.10(b), where the compensator has been added, point P is still approximately at the same location on the compensated root locus.

What is the effect upon the required gain, K? After inserting the compensator, we find that the value of K is virtually the same for both the uncompensated and compensated systems since the lengths of the vectors drawn from the lag compensator are approximately equal, and all other vectors have not changed appreciably.

Now, what improvement can we expect in the steady-state error? Since we established that the gain, K, is about the same for both the uncompensated and

Figure 9.10 (**a**) Root Locus prior to Lag Compensation; (**b**) Root Locus after Lag Compensation

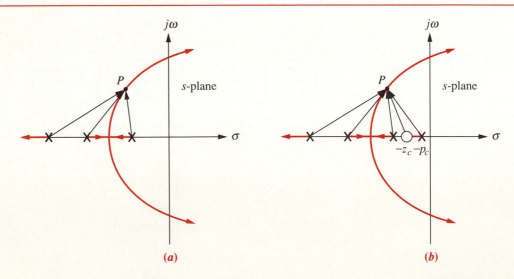

compensated systems, we can substitute Equation 9.3 into 9.4 and obtain

$$K_{v_N} = K_{v_0} \frac{z_c}{p_c} \qquad (9.5)$$

Equation 9.5 shows that the improvement in the compensated system's K_v over the uncompensated system's K_v is equal to the ratio of the magnitude of the compensator zero to the compensator pole. In order to keep the transient response unchanged, we know the compensator pole and zero must be close to each other. The only way the ratio of z_c to p_c can be large in order to yield an appreciable improvement in steady-state error and simultaneously have the compensator's pole and zero close to each other to minimize the angular contribution is to place the compensator's pole-zero pair close to the origin. For example, the ratio of z_c to p_c can be equal to 10 if the pole is at -0.001 and the zero is at -0.01. Thus, the ratio is 10, yet the pole and zero are very close, and the angular contribution of the compensator is small.

In conclusion, although the ideal compensator drives the steady-state error to zero, a lag compensator with a pole that is not at the origin will improve the static error constant by a factor equal to z_c/p_c. There also will be a minimal effect upon the transient response if the pole-zero pair of the compensator is placed close to the origin. Later in the chapter we will show circuit configurations for the lag compensator. These circuit configurations can be obtained with passive networks and thus do not require the active amplifiers and possible additional power supplies that are required by the ideal integral (PI) compensator. In the following example, we design a lag compensator to yield a specified improvement in steady-state error.

Example 9.2   Design a lag compensator.

Problem   Compensate the system of Figure 9.4(a), whose root locus is shown in Figure 9.5, to improve the steady-state error by a factor of 10 if the system is operating with a damping ratio of 0.174.

Solution   The uncompensated system error from Example 9.1 was 0.108 with $K_p = 8.228$. A tenfold improvement means a steady-state error of

$$e(\infty) = \frac{0.108}{10} = 0.0108 \qquad (9.6)$$

Since

$$e(\infty) = \frac{1}{1 + K_p} = 0.0108 \qquad (9.7)$$

rearranging and solving for the required K_p yields

$$K_p = \frac{1 - e(\infty)}{e(\infty)} = \frac{1 - 0.0108}{0.0108} = 91.593 \qquad (9.8)$$

The improvement in K_p from the uncompensated system to the compensated system is the required ratio of the compensator zero to the compensator pole, or

$$\frac{z_c}{p_c} = \frac{K_{p_N}}{K_{p_O}} = \frac{91.593}{8.228} = 11.132 \tag{9.9}$$

Arbitrarily selecting

$$p_c = 0.01 \tag{9.10}$$

we use Equation 9.9 and find

$$z_c = 11.132 p_c \approx 0.111 \tag{9.11}$$

Let us now compare the compensated and uncompensated systems. First, sketch the root locus, as shown in Figure 9.12, of the compensated system shown in Figure 9.11. Next, we search along the $\zeta = 0.174$ line for a multiple of 180° and find that the second-order dominant poles are at $-0.678 \pm j3.836$ with a gain, K, of 158.097. The third and fourth closed-loop poles are at -11.554 and -0.101, respectively, and are found by searching the real axis for a gain equal to that of the dominant poles. All transient and steady-state results for both the uncompensated and compensated systems are shown in Table 9.1.

The fourth pole of the compensated system cancels its zero. This cancellation leaves the remaining three closed-loop poles of the compensated system very close in value to the three closed-loop poles of the uncompensated system. Hence the transient response of both systems is approximately the same as is the system gain, but notice that the steady-state error of the compensated system is 9.818 times less than that of the uncompensated system and is close to the design specification of a tenfold improvement.

Figure 9.13 shows the effect of the lag compensator in the time domain. Even though the transient responses of both the uncompensated and lag-compensated systems are the same, the lag-compensated system exhibits less steady-state error by approaching unity more closely than the uncompensated system.

We now examine another design possibility for the lag compensator and compare the response to Figure 9.13. Let us assume a lag compensator whose pole and zero are 10 times closer to the origin than in the previous design. The results are compared in Figure 9.14. Even though both responses will eventually reach approximately the same steady-state value, the lag compensator previously designed, $G_c(s) = \dfrac{(s + 0.111)}{(s + 0.01)}$,

Figure 9.11 Compensated System for Example 9.2

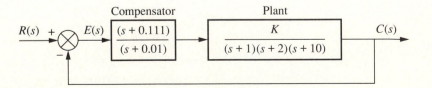

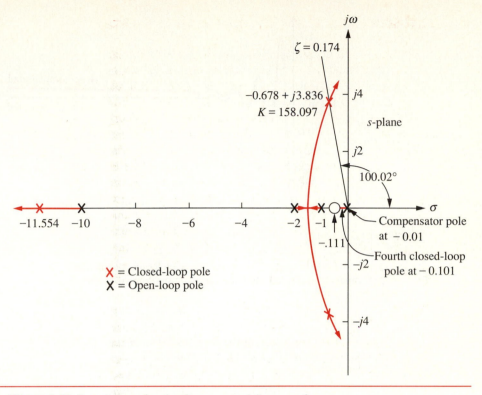

Figure 9.12 Root Locus for the Compensated System of Figure 9.11

Table 9.1 Comparison of Uncompensated and Lag-Compensated Systems' Predicted Characteristics for Example 9.2

Parameter	Uncompensated	Lag-compensated
Plant	$\dfrac{K}{(s + 1)(s + 2)(s + 10)}$	$\dfrac{K(s + 0.111)}{(s + 1)(s + 2)(s + 10)(s + 0.01)}$
K	164.565	158.097
K_p	8.228	87.744
$e(\infty)$	0.108	0.011
Dominant second-order poles	$-0.694 \pm j3.926$	$-0.678 \pm j3.836$
Third pole	-11.613	-11.554
Fourth pole	None	-0.101
Zero	None	-0.111

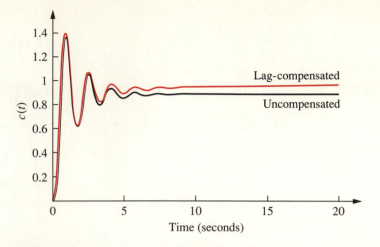

Figure 9.13 Comparison between the Step Response of the
Uncompensated System and the Lag-Compensated System
for Example 9.2

Figure 9.14 Comparison of the Step Response of the
System for Example 9.2 Using Two Different Lag
Compensators

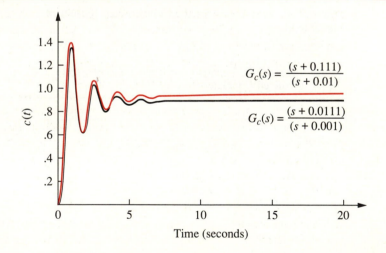

approaches the final value faster than the proposed lag compensator, $G_c(s) = \dfrac{(s + 0.0111)}{(s + 0.001)}$.

We can explain this phenomenon as follows. From Table 9.1, the previously designed lag compensator has a fourth closed-loop pole at -0.101. Using the same analysis for the new lag compensator with its open-loop pole 10 times closer to the imaginary axis, we find its fourth closed-loop pole at -0.01. Thus, the new lag compensator has a closed-loop pole closer to the imaginary axis than the original lag compensator. This pole at -0.01 will produce a slower transient response than the original pole at -0.101, and the steady-state value will not be reached as quickly.

Since we have solved the problem of improving the steady-state error without affecting the transient response, let us now improve the transient response itself.

9.3 Cascade Compensation to Improve Transient Response

In this section we discuss two ways of improving the transient response of a feedback control system using cascade compensation. Typically, the objective is to design a response that has a desirable percent overshoot and a faster settling time than the uncompensated system.

The first technique for improving transient response that we will discuss is *ideal derivative compensation*. With ideal derivative compensation, a pure differentiator is added to the forward path of the feedback control system. We will see that the result of adding differentiation is the addition of a zero to the forward-path transfer function. This type of compensation requires an active network for its realization. Further, differentiation is a noisy process; although the level of the noise is low, the frequency of the noise is high compared to the signal. Thus, differentiating high-frequency noise yields a large, unwanted signal.

The second technique does not use pure differentiation. Instead, it approximates differentiation with a passive network by adding a zero and a more distant pole to the forward-path transfer function. The zero approximates pure differentiation as described above.

As with compensation to improve steady-state error, we introduce names associated with the implementation of the compensators. We call an ideal derivative compensator a <u>P</u>roportional-plus-<u>D</u>erivative (PD) controller since the implementation, as we will see, consists of feeding the error (proportional) plus the derivative of the error forward to the plant. The second technique uses a passive network called a *lead compensator*. As with the lag compensator, the name comes from its frequency response, which will be discussed in Chapter 11. Thus, we use the name *PD controller* interchangeably with *ideal derivative compensator,* and we use the name *lead compensator* when the cascade compensator does not employ pure differentiation.

Ideal Derivative Compensation (PD)

The transient response of a system can be selected by choosing an appropriate closed-loop pole location on the s-plane. If this point is on the root locus, then a simple gain adjustment is all that is required in order to meet the transient response specification. If the closed-loop pole location is not on the root locus, then the root locus must be reshaped so that the compensated (new) root locus goes through the selected closed-loop pole location. In order to accomplish the latter task, poles and zeros can be added in the forward path to produce a new open-loop function whose root locus goes through the design point on the s-plane. One method that generally works to speed up the original system is the addition of a single zero to the forward path.

This zero can be represented by a compensator whose transfer function, $G_c(s)$, is

$$G_c(s) = s + z_c \tag{9.12}$$

This function, the sum of a differentiator and a pure gain, is called an *ideal derivative,* or *PD controller.* Judicious choice of the position of the compensator zero can quicken the response over the uncompensated system. In summary, transient responses unattainable by a simple gain adjustment can be obtained by augmenting the system's poles and zeros with an ideal derivative compensator.

We will now show that ideal derivative compensation speeds up the response of a system. Several simple examples are shown in Figure 9.15, where the uncompensated system of Figure 9.15(*a*), operating with a damping ratio of 0.4, becomes a compensated system by the addition of a compensating zero at -2, -3, and -4 in Figures 9.15(*b*), (*c*), and (*d*), respectively. In each design, the zero is moved to a different position, and the root locus is shown. For each compensated case, we moved the dominant, second-order poles further out along the 0.4 damping ratio line.

Each of the compensated cases has dominant poles with the same damping ratio as the uncompensated case. Thus, we predict that the percent overshoot will be the same for each case.

Also, the compensated, dominant, closed-loop poles have more negative real parts than the uncompensated, dominant, closed-loop poles. Hence, we predict that the settling times for the compensated cases will be shorter than for the uncompensated case. The compensated, dominant, closed-loop poles with the more negative real parts will have the shorter settling times. The system in Figure 9.15(*b*) will have the shortest settling time.

All of the compensated systems will have smaller peak times than the uncompensated system since the imaginary parts of the compensated systems are larger. The system of Figure 9.15(*b*) will have the smallest peak time.

Also notice that as the zero is placed further from the dominant poles, the closed-loop, compensated dominant poles move closer to the origin and to the uncompensated, dominant closed-loop poles. Table 9.2 summarizes the results obtained from the root locus of each of the design cases shown in Figure 9.15.

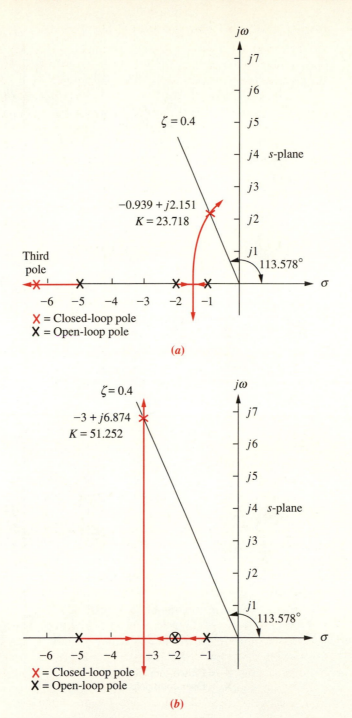

Figure 9.15 Using Ideal Derivative Compensation:
(**a**) Uncompensated; (**b**) Compensator Zero at −2

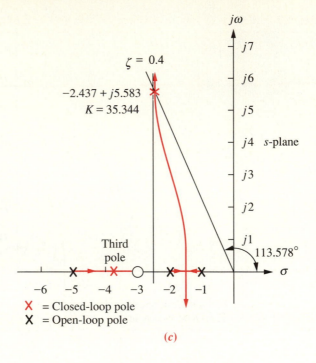

(c)

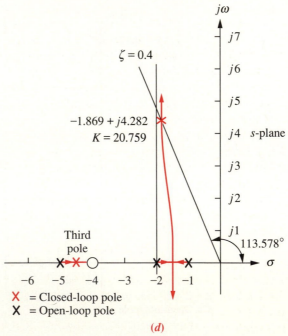

(d)

Figure 9.15 (continued) Using Ideal Derivative Compensation: (c) Compensator Zero at −3; (d) Compensator Zero at −4

Table 9.2 Summary of Predicted Characteristics for the Systems of Figure 9.15

	Uncompensated	Compensation b	Compensation c	Compensation d
Plant	$\dfrac{K}{(s+1)(s+2)(s+5)}$	$\dfrac{K(s+2)}{(s+1)(s+2)(s+5)}$	$\dfrac{K(s+3)}{(s+1)(s+2)(s+5)}$	$\dfrac{K(s+4)}{(s+1)(s+2)(s+5)}$
Dom. poles	$-0.939 \pm j2.151$	$-3 \pm j6.874$	$-2.437 \pm j5.583$	$-1.869 \pm j4.282$
K	23.718	51.252	35.344	20.759
ζ	0.4	0.4	0.4	0.4
ω_n	2.347	7.5	6.091	4.673
%OS	25.38	25.38	25.38	25.38
T_s	4.26	1.33	1.64	2.14
T_p	1.46	0.46	0.56	0.733
K_p	2.372	10.25	10.6	8.304
$e(\infty)$	0.297	0.089	0.086	0.107
Third pole	-6.123	None	-3.127	-4.263
Zero	None	None	-3	-4
Comments	Second-order approx. OK	Pure second-order	Second-order approx. OK	Second-order approx. OK

In summary, although compensation methods c and d yield slower responses than method b, the addition of ideal derivative compensation shortened the response time in each case while still keeping the percent overshoot the same. This change can best be seen in the settling time and peak time, where there is at least a doubling of speed across all of the cases of compensation. An added benefit is the improvement in the steady-state error, even though lag compensation was not used. Here, the steady-state error of the compensated system is at least one-third that of the uncompensated system, as seen from $e(\infty)$ and K_p. All systems in Table 9.2 are Type 0, and some steady-state error is expected. The reader must not assume that, in general, improvement in transient response always yields an improvement in steady-state error.

The time response of each case shown in Table 9.2 is shown in Figure 9.16. We see that the compensated responses are faster and exhibit less error than the uncompensated response.

Now that we have seen what ideal derivative compensation can do, we are ready to design our own ideal derivative compensator to meet a transient response specification. Basically, we will evaluate the sum of angles from the open-loop poles and zeros to a design point that is the closed-loop pole that yields the desired transient response. The difference between 180° and the calculated angle must be the angular contribution of the compensator zero. Trigonometry is then used to locate the position of the zero to yield the required difference in angle.

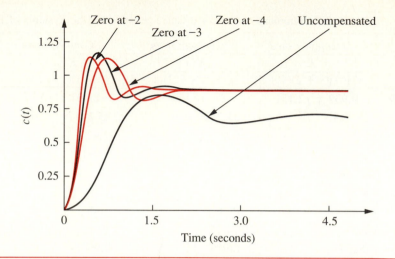

Figure 9.16 Comparison of the Uncompensated System and Three Ideal Derivative Compensation Solutions from Table 9.2

Example 9.3

Design an ideal derivative compensator.

Problem Given the system of Figure 9.17, design an ideal derivative compensator to yield a 16% overshoot, with a threefold reduction in settling time.

Solution Let us first evaluate the performance of the uncompensated system operating with 16% overshoot. The root locus for the uncompensated system is shown in Figure 9.18. Since 16% overshoot is equivalent to $\zeta = 0.504$, we search along that damping ratio line for an odd multiple of 180° and find that the dominant, second-order pair of poles is at $-1.205 \pm j2.064$. Thus, the settling time of the uncompensated system is

$$T_s = \frac{4}{\zeta\omega_n} = \frac{4}{1.205} = 3.320 \tag{9.13}$$

Since our evaluation of percent overshoot and settling time is based upon a second-order approximation, we must check the assumption by finding the third pole and justifying the second-order approximation. Searching beyond -6 on the real axis for

Figure 9.17 Feedback Control System for Example 9.3

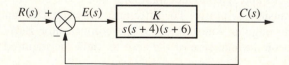

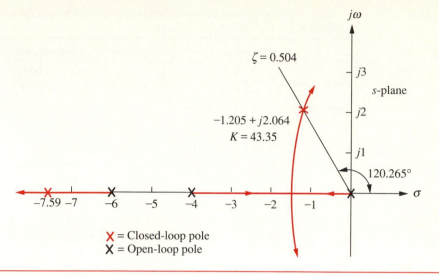

Figure 9.18 The Root Locus for the Uncompensated System Shown in Figure 9.17

a gain equal to the gain of the dominant, second-order pair, 43.35, we find a third pole at -7.59, which is over six times further from the $j\omega$-axis than the dominant, second-order pair. We conclude that our approximation is valid. The transient and steady-state error characteristics of the uncompensated system are summarized in Table 9.3.

Now we proceed to compensate the system. First, we find the location of the compensated system's dominant poles. In order to have a threefold reduction in the settling time, the compensated system's settling time will be one-third of Equation 9.13. The new settling time will be 1.107. Therefore, the real part of the compensated system's dominant, second-order pole is

$$\sigma = \frac{4}{T_s} = \frac{4}{1.107} = 3.613 \tag{9.14}$$

Figure 9.19 shows the designed dominant, second-order pole, with a real part equal to -3.613 and an imaginary part of

$$\omega_d = 3.613 \tan(180° - 120.265°) = 6.192 \tag{9.15}$$

Next, we design the location of the compensator zero. Input the uncompensated system's poles and zeros in the root locus program as well as the design point $-3.613 + j6.192$ as a test point. The result is the sum of the angles to the design point of all the poles and zeros of the compensated system except for those of the compensator zero itself. The difference between the result obtained and 180° is the angular contribution required of the compensator zero. Using the open-loop poles shown in Figure 9.19 and the test point, $-3.613 + j6.192$, which is the desired dominant second-order pole, we obtain the sum of the angles as $-275.607°$. Hence, the angular contribution required from the compensator zero for the test point to be on the root locus is $+275.607° - 180° = 95.607°$. The geometry is shown in Figure 9.20, where we now must solve for σ, which is the location of the compensator zero.

Table 9.3 Comparison of the Uncompensated System and the Compensated System for Example 9.3

	Uncompensated	Simulation	Compensated	Simulation
Plant	$\dfrac{K}{s(s+4)(s+6)}$		$\dfrac{K(s+3.005)}{s(s+4)(s+6)}$	
Dom. poles	$-1.205 \pm j2.064$		$-3.613 \pm j6.192$	
K	43.35		47.439	
ζ	0.504		0.504	
ω_n	2.39		7.169	
%OS	16	14.8	16	11.8
T_s	3.320	3.6	1.107	1.2
T_p	1.522	1.7	0.507	0.5
K_v	1.806		5.94	
$e(\infty)$	0.554		0.168	
Third pole	-7.591		-2.774	
Zero	None		-3.005	
Comments	Second-order approx. OK		Pole-zero not canceling	

Figure 9.19 Compensated Dominant Pole Superimposed over the Uncompensated Root Locus for Example 9.3

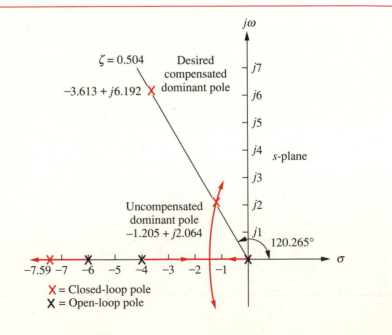

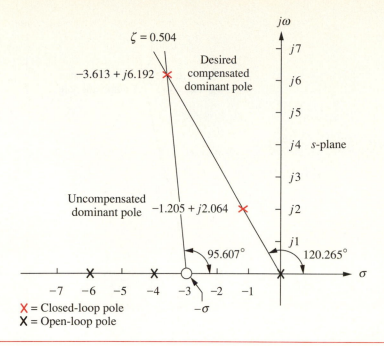

Figure 9.20 Evaluating the Location of the Compensating Zero for Example 9.3

From the figure,

$$\frac{6.192}{3.613 - \sigma} = \tan(180° - 95.607°) \tag{9.16}$$

Thus, $\sigma = 3.005$. The complete root locus for the compensated system is shown in Figure 9.21. Table 9.3 summarizes the results for both the uncompensated system and the compensated system.

For the uncompensated system, the estimate of the transient response is accurate since the third pole is at least five times the real part of the dominant, second-order pair. The second-order approximation for the compensated system, however, may be invalid because there is no approximate closed-loop third-pole and zero cancellation. A simulation is required, and the results are shown in the second column for the uncompensated system and the fourth column for the compensated system. The simulation results can be obtained using a program like the state-space step response program described in Appendix C. The percent overshoot differs by 3% between the uncompensated and compensated systems, while there is approximately a threefold improvement in speed as evaluated from the settling time.

The final results are displayed in Figure 9.22, which compares the uncompensated system and the faster compensated system.

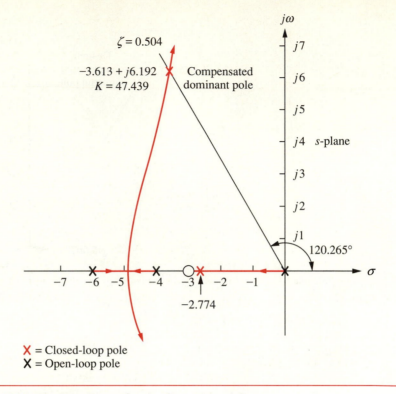

X = Closed-loop pole
X = Open-loop pole

Figure 9.21 The Root Locus for the Compensated System of Example 9.3

Figure 9.22 Comparison of the Uncompensated and Compensated Systems of Example 9.3

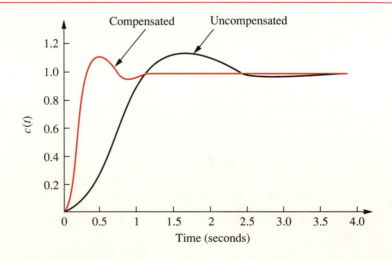

Once we decide on the location of the compensating zero, how do we implement the ideal derivative, or PD controller? The ideal integral compensator that improved steady-state error was implemented with a proportional-plus-integral (PI) controller. The ideal derivative compensator used to improve the transient response is implemented with a proportional-plus-derivative (PD) controller. For example, in Figure 9.23 the transfer function of the controller is

$$G_c(s) = K_2 s + K_1 = K_2 \left(s + \frac{K_1}{K_2} \right) \tag{9.17}$$

Hence, K_1/K_2 is chosen to equal the negative of the compensator zero, and K_2 is chosen to contribute to the required loop-gain value. Later in the chapter, we will study circuits that can be used to create differentiation and gain.

While the ideal derivative compensator can improve the transient response of the system, it has two drawbacks. First, it requires an active circuit to perform the differentiation. Second, differentiation is a noisy process; the level of the noise is low, but the frequency of the noise is high compared to the signal. The lead compensator is a passive network used to overcome the disadvantages of ideal differentiation and still retain the ability to improve the transient response.

Lead Compensation

Just as the active ideal integral compensator can be approximated with a passive lag network, an active ideal derivative compensator can be approximated with a passive lead compensator. When passive networks are used, a single zero cannot be produced; rather, a compensator zero and a pole result. However, if the pole is further from the imaginary axis than the zero, the angular contribution of the compensator is still positive and thus approximates an equivalent single zero. In other words, the angular contribution of the compensator pole subtracts from the angular contribution of the zero but does not preclude the use of the compensator to improve transient response since the net angular contribution is positive, just as for a single PD controller zero.

The advantages of a passive lead network over an active PD controller are that (1) no additional power supplies are required, and (2) noise due to differentiation is reduced. The disadvantage is that the additional pole does not reduce the number of branches of the root locus that cross the imaginary axis into the right

Figure 9.23 A PD Controller

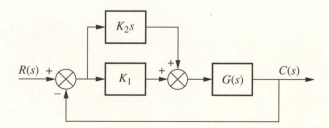

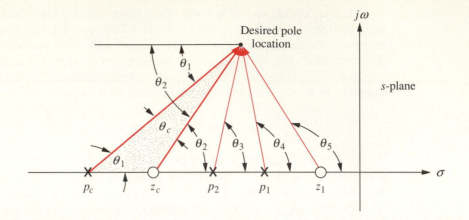

Figure 9.24 Geometry of Lead Compensation

half-plane, while the addition of the single zero of the PD controller tends to reduce the number of branches of the root locus that cross into the right half-plane.

Let us first look at the concept behind lead compensation. If we select a desired dominant, second-order pole on the s-plane, the sum of the angles from the uncompensated system's poles and zeros to the design point can be found. The difference between $180°$ and the sum of the angles must be the angular contribution required of the compensator.

For example, looking at Figure 9.24, we see that

$$\theta_2 - \theta_1 - \theta_3 - \theta_4 + \theta_5 = (2k + 1)180° \qquad (9.18)$$

where $(\theta_2 - \theta_1) = \theta_c$ is the angular contribution of the lead compensator. From Figure 9.24 we see that θ_c is the angle of a ray extending from the design point and intersecting the real axis at the pole value and zero value of the compensator. Now, visualize this ray rotating about the desired closed-loop pole location and

Figure 9.25 Three of the Infinite Possible Lead Compensator Solutions

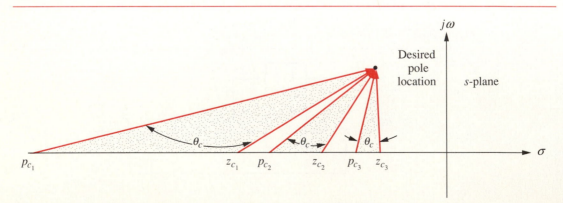

intersecting the real axis at the compensator pole and zero, as illustrated in Figure 9.25. We realize that an infinite number of lead compensators could be used to meet the transient response requirement.

How do the possible lead compensators differ? The differences are in the values of static error constants, in the gain required to reach the design point on the compensated root locus, the difficulty in justifying a second-order approximation when the design is complete, and the ensuing transient response.

For design, we arbitrarily select either a lead compensator pole or zero and find the angular contribution at the design point of this pole or zero along with the system's open-loop poles and zeros. The difference between this angle and 180° is the required contribution of the remaining compensator pole or zero. Let us look at an example.

Example 9.4   Design a system with lead compensation.

Problem   Design three lead compensators for the system of Figure 9.17 that will reduce the settling time by a factor of 2 while maintaining 30% overshoot. Compare the system specifications among the three designs.

Solution   First, determine the characteristics of the uncompensated system operating at 30% overshoot to see what the uncompensated settling time is. Since 30% overshoot is equivalent to a damping ratio of 0.358, we search along the $\zeta = 0.358$ line for the uncompensated dominant poles on the root locus, as shown in Figure 9.26. From the pole's real part, we calculate the uncompensated settling time as

Figure 9.26 Design of Lead Compensator, Showing Evaluation of Uncompensated and Compensated Dominant Poles for Example 9.4

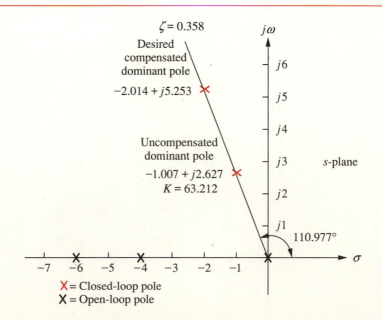

Table 9.4 Comparison of Three Lead Compensation Designs for Example 9.4

	Uncompensated	Compensation a	Compensation b	Compensation c
Plant	$\dfrac{K}{s(s+4)(s+6)}$	$\dfrac{K(s+5)}{s(s+4)(s+6)(s+42.969)}$	$\dfrac{K(s+4)}{s(s+4)(s+6)(s+20.089)}$	$\dfrac{K(s+2)}{s(s+4)(s+6)(s+8.971)}$
Dom. poles	$-1.007 \pm j2.627$	$-2.014 \pm j5.253$	$-2.014 \pm j5.254$	$-2.014 \pm j5.254$
K	63.212	1423.658	698.490	345.849
ζ	0.358	0.358	0.358	0.358
ω_n	2.813	5.626	5.627	5.627
$\%OS$	30 (28)	30 (30.7)	30 (28.2)	30 (14.5)
T_s	3.972 (4)	1.986 (2)	1.986 (2)	1.986 (1.7)
T_p	1.196 (1.3)	0.598 (0.6)	0.598 (0.6)	0.598 (0.7)
K_v	2.634	6.903	5.795	3.213
$e(\infty)$	0.380	0.145	0.173	0.311
Other poles	-7.986	$-43.807, -5.134$	-22.06	$-13.3, -1.642$
Zero	None	-5	None	-2
Comments	Second-order approx. OK	Second-order approx. OK	Second-order approx. OK	No pole-zero cancellation

Note: Simulation results are shown in parentheses.

$T_s = 4/1.007 = 3.972$ seconds. The remaining characteristics of the uncompensated system are summarized in Table 9.4.

Next, we find the design point. A twofold reduction in settling time yields $T_s = 3.972/2 = 1.986$ seconds, from which the real part of the desired pole location is $-\zeta\omega_n = -4/T_s = -2.014$. The imaginary part is $\omega_d = 2.014 \tan(110.977°) = 5.253$.

We continue by designing the lead compensator. Arbitrarily assume a compensator zero at -5 on the real axis as a possible solution. Using the root locus program, sum the angles from both this zero and the uncompensated system's poles and zeros, using the design point as a test point. The resulting angle is $-172.691°$. The difference between this angle and $180°$ is the angular contribution required from the compensator pole in order to place the design point on the root locus. Hence, an angular contribution of $-7.309°$ is required from the compensator pole.

The geometry shown in Figure 9.27 is used to calculate the location of the compensator pole. From the figure,

$$\frac{5.253}{p_c - 2.014} = \tan 7.309° \tag{9.19}$$

from which the compensator pole is found to be

$$p_c = 42.969 \tag{9.20}$$

The compensated system root locus is sketched in Figure 9.28.

In order to justify our estimates of percent overshoot and settling time, we must show that the second-order approximation is valid. To perform this validity check, we search for the third and fourth closed-loop poles found beyond -42.969 and between -5 and -6 in Figure 9.28. Searching these regions for the gain equal to that of the compensated dominant pole, 1423.658, we find that the third and fourth poles are at -43.807 and -5.134, respectively. Since -43.807 is more than 20 times the real part of the dominant pole, the effect of the third closed-loop pole is negligible. Since the closed-loop pole at -5.134 is close to the zero at -5, we have pole-zero cancellation, and the second-order approximation is valid.

All results for this design and two other designs, which place the compensator zero arbitrarily at -2 and -4 and follow similar design techniques, are summarized in Table 9.4. Each design should be verified by a simulation, which could consist

Figure 9.27 *s*-Plane Picture Used to Calculate the Location of the Compensator Pole for Example 9.4

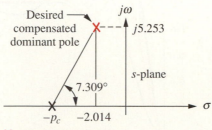

X = Closed-loop pole
X = Open-loop pole

Note: This figure is not drawn to scale.

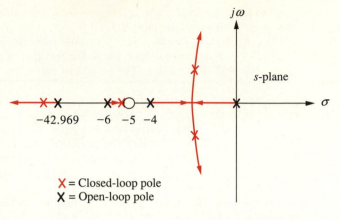

X = Closed-loop pole
X = Open-loop pole

Note: This figure is not drawn to scale.

Figure 9.28 Compensated System Root Locus

of using the state-space model and the step-response program covered in Chapter 4 and Appendix C. We have performed a simulation for this design problem, and the results are shown by parenthetical entries next to the estimated values in the table. The only design that disagrees with the simulation is the case where the compensator zero is at −2. For this case, the closed-loop pole and zero do not cancel.

Let us now discuss and compare the results shown in Table 9.4. First, we notice differences in the following:

1. The position of the arbitrarily selected zero
2. The amount of improvement in the steady-state error

Figure 9.29 Comparison of the Uncompensated System and the Three Lead Compensation Solutions for Example 9.4

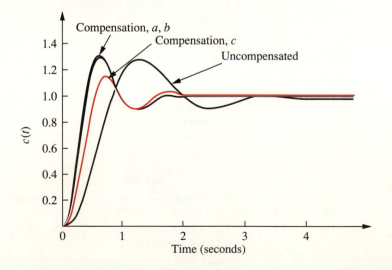

3. The amount of required gain, K
4. The position of the third and fourth poles and their relative effect upon the second-order approximation. This effect is measured by their distance from the dominant poles or the degree of cancellation with the closed-loop zero.

Once a simulation verifies desired performance, the choice of compensation can be based upon the amount of gain required or the improvement in steady-state error that can be obtained without a lag compensator.

The results of Table 9.4 are supported by simulations of the step response shown in Figure 9.29 for the uncompensated system and the three lead compensation solutions.

In this section we improved transient response or steady-state error through the use of cascade compensators. In the next section, we improve both steady-state errors and transient response independently.

9.4 Cascade Compensation to Improve Both Steady-State Error and Transient Response

We now combine the design techniques covered in Sections 9.2 and 9.3 to obtain improvement in steady-state error and transient response *independently*. Basically, we first improve the transient response by using the methods of Section 9.3. Then we improve the steady-state error of this compensated system by applying the methods of Section 9.2. A disadvantage of this approach is the slight decrease in the speed of the response when the steady-state error is improved.

As an alternative, we can improve the steady-state error first and then follow with the design to improve the transient response. A disadvantage of this approach is that the improvement in transient response in some cases yields deterioration in the improvement of the steady-state error, which was designed first. In other cases, the improvement in transient response yields further improvement in steady-state errors. Thus, a system can be overdesigned with respect to steady-state errors. Overdesign is usually not a problem unless it impacts cost or produces other design problems. In this textbook, we first design for transient response and then design for steady-state error.

The design can use either active or passive compensators, as previously described. If we design an active PD controller followed by an active PI controller, the resulting compensator is called a *Proportional-Plus-Integral-Plus-Derivative (PID) controller*. If we first design a passive lead compensator and then design a passive lag compensator, the resulting compensator is called a *lag-lead compensator*.

PID Controller Design

A PID controller is shown in Figure 9.30. The transfer function of the PID controller is

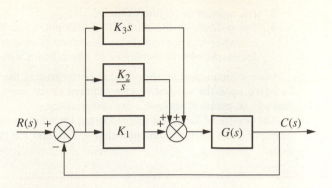

Figure 9.30 A PID Controller

$$G_c(s) = K_1 + \frac{K_2}{s} + K_3 s = \frac{K_1 s + K_2 + K_3 s^2}{s} = \frac{K_3\left(s^2 + \dfrac{K_1}{K_3}s + \dfrac{K_2}{K_3}\right)}{s} \quad (9.21)$$

which has two zeros plus a pole at the origin. One zero and the pole at the origin can be designed as the ideal integral compensator; the other zero can be designed as the ideal derivative compensator.

The design technique, which is demonstrated in Example 9.5, consists of the following steps:

1. Evaluate the performance of the uncompensated system in order to determine how much improvement in transient response is required.
2. Design the PD controller to meet the transient response specifications. The design includes the zero location and the loop gain.
3. Simulate the system to be sure all requirements have been met.
4. Redesign if the simulation shows that requirements have not been met.
5. Design the PI controller to yield the required steady-state error.
6. Determine the gains, K_1, K_2, and K_3, in Figure 9.30.
7. Simulate the system to be sure all requirements have been met.
8. Redesign if simulation shows that requirements have not been met.

Example 9.5   Design a PID controller.

Problem   Given the system of Figure 9.31, design a PID controller so that the system can operate with a peak time that is two-thirds that of the uncompensated system at 20% overshoot and with zero steady-state error for a step input.

Solution   Note that our solution follows the eight-step procedure described above.

Step 1   Let us first evaluate the uncompensated system operating at 20% overshoot. Searching along the 20% overshoot line ($\zeta = 0.456$) in Figure 9.32, we find the dominant poles to be $-5.415 \pm j10.569$ with a gain of 121.510. A third pole, which exists at -8.169, is found by searching the region between -8 and -10 for a gain equivalent to that at the dominant poles. The complete performance of the

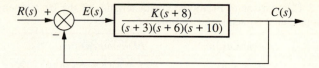

Figure 9.31 Uncompensated Feedback Control System for Example 9.5

uncompensated system is shown in the first column of Table 9.5, where we compare the calculated values to those obtained through simulation (Figure 9.35). We estimate that the uncompensated system has a peak time of 0.297 seconds at 20% overshoot.

Step 2 To compensate the system to reduce the peak time to two-thirds of that of the uncompensated system, we must first find the compensated system's dominant pole location. The imaginary part of the compensated dominant pole is

$$\omega_d = \frac{\pi}{T_p} = \frac{\pi}{(2/3)(0.297)} = 15.867 \tag{9.22}$$

Figure 9.32 Root Locus for the Uncompensated System of Example 9.5

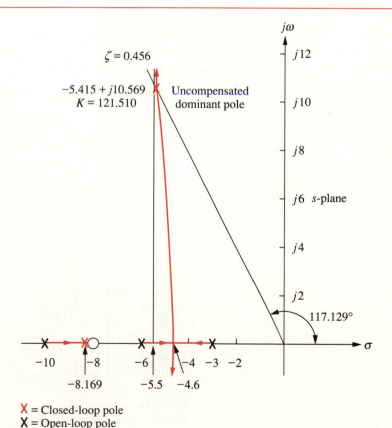

X = Closed-loop pole
X = Open-loop pole

Table 9.5 Comparison of Predicted Characteristics of the Uncompensated, PD- , and PID-Compensated Systems of Example 9.5

	Uncompensated	PD-compensated	PID-compensated
Plant	$\dfrac{K(s+8)}{(s+3)(s+6)(s+10)}$	$\dfrac{K(s+8)(s+55.909)}{(s+3)(s+6)(s+10)}$	$\dfrac{K(s+8)(s+55.909)(s+0.5)}{(s+3)(s+6)(s+10)s}$
Dom. poles	$-5.415 \pm j10.569$	$-8.13 \pm j15.867$	$-7.519 \pm j14.676$
K	121.510	5.339	4.604
ζ	0.456	0.456	0.456
ω_n	11.875	17.829	16.49
%OS	20	20	20
T_s	0.739	0.492	0.532
T_p	0.297	0.198	0.214
K_p	5.4	13.267	∞
$e(\infty)$	0.156	0.070	0
Other poles	-8.169	-8.079	$-7.906 \; -0.468$
Zeros	-8	$-8, -55.909$	$-8, -55.909, -0.5$
Comments	Second-order approx. OK	Zero at -55.909 not canceled	Zero at -55.909 and -0.5 not canceled

Thus, the real part of the compensated dominant pole is

$$\sigma = \frac{\omega_d}{\tan 117.129°} = -8.13 \tag{9.23}$$

Next, we design the compensator. Using the geometry shown in Figure 9.33, we calculate the compensating zero's location. Using the root locus program, we find the sum of angles from the uncompensated system's poles and zeros to the desired compensated dominant pole to be $-198.371°$. Thus, the contribution required from the compensator zero is $180° - 198.371° = 18.371°$. Assume that the compensator

Figure 9.33 Calculating the PD Compensator Zero for Example 9.5

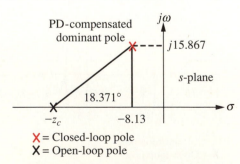

X = Closed-loop pole
X = Open-loop pole

Note: This figure is not drawn to scale.

zero is located at $-z_c$, as shown in Figure 9.33. Since

$$\frac{15.867}{z_c - 8.13} = \tan 18.371° \tag{9.24}$$

then

$$z_c = 55.909 \tag{9.25}$$

Thus, the PD controller is

$$G_{PD}(s) = (s + 55.909) \tag{9.26}$$

The complete root locus for the PD-compensated system is sketched in Figure 9.34. Using a root locus program, the gain at the design point is 5.339. Complete specifications for ideal derivative compensation are shown in the third column of Table 9.5.

Steps 3 and 4   We simulate the PD-compensated system, as shown in Figure 9.35. We see the reduction in peak time and the improvement in steady-state error over the uncompensated system.

Step 5   After we design the PD controller, we design the ideal integral compensator to reduce the steady-state error to zero for a step input. Any ideal integral compensator zero will work, as long as the zero is placed close to the origin. Choosing the ideal

Figure 9.34 Root Locus for the PD-Compensated System of Example 9.5

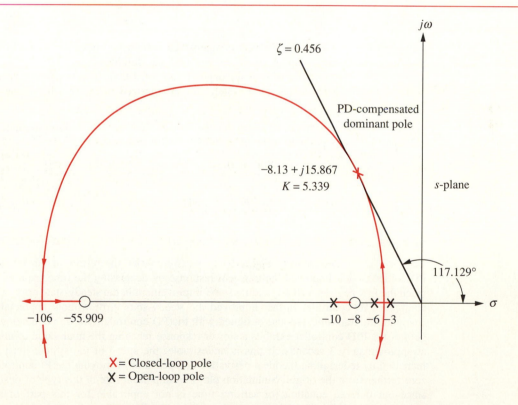

$\zeta = 0.456$

PD-compensated dominant pole

$-8.13 + j15.867$
$K = 5.339$

s-plane

$117.129°$

-106 -55.909 -10 -8 -6 -3

$j\omega$

σ

X = Closed-loop pole
X = Open-loop pole

Note: This figure is not drawn to scale.

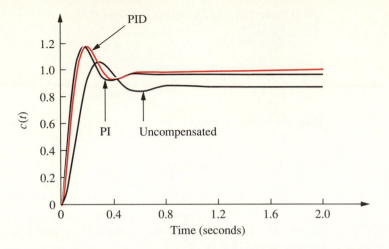

Figure 9.35 Comparison of the Step Response for the Uncompensated, PD-Compensated, and PID-Compensated Systems of Example 9.5

integral compensator to be

$$G_{PI}(s) = \frac{s + 0.5}{s} \tag{9.27}$$

we sketch the root locus for the PID-compensated system, as shown in Figure 9.36. Searching the 0.456 damping ratio line, we find the dominant, second-order poles to be $-7.519 \pm j14.676$, with an associated gain of 4.604. The remaining specifications for the PID-compensated system are summarized in the fourth column of Table 9.5.

Step 6 Now we determine the gains, K_1, K_2, and K_3 in Figure 9.30. From Equations 9.26 and 9.27, the product of the gain and the PID controller is

$$G_{PID}(s) = \frac{K(s + 55.909)(s + 0.5)}{s} = \frac{4.604(s + 55.909)(s + 0.5)}{s}$$

$$= \frac{4.604(s^2 + 56.409 + 27.954)}{s} \tag{9.28}$$

Matching Equations 9.21 and 9.28, $K_1 = 259.707$, $K_2 = 128.7$, and $K_3 = 4.604$.

Steps 7 and 8 Returning to Figure 9.35, we summarize the results of our design. PD compensation improved the transient response by decreasing the time required to reach the first peak as well as yielding some improvement in the steady-state error. The complete PID controller further improved the steady-state error without appreciably changing the transient response designed with the PD controller. As we've mentioned before, the PID controller exhibits a slower response reaching the final value of unity at approximately 3 seconds. If this is undesirable, the speed of the system must be increased by redesigning the ideal derivative compensator or moving the PI controller zero further from the origin. Simulation plays an important role in this type of design since our derived equation for settling time is not applicable for this part of the response, where there is a slow correction of the steady-state error.

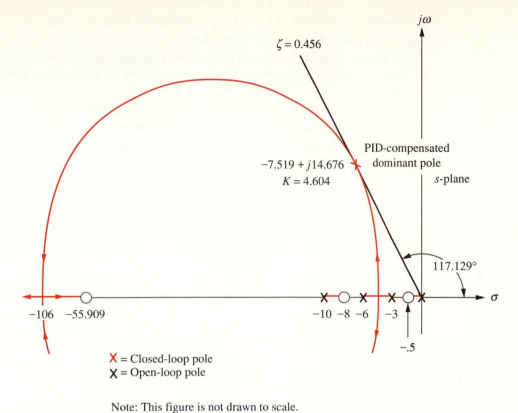

Figure 9.36 Root Locus for the PID-Compensated System of Example 9.5

Lag-Lead Compensator Design

In the previous example, we serially combined the concepts of ideal derivative and ideal integral compensation to arrive at the design of a PID controller that improved both the transient and steady-state error performance. In the next example, we improve transient response and steady-state error by using a lead compensator and a lag compensator rather than the ideal PID. Our compensator is called a *lag-lead compensator.*

We first design the lead compensator to improve the transient response. Next, we evaluate the improvement in steady-state error still required. Finally, we design the lag compensator to meet the steady-state error requirement. Later in the chapter, we show circuit designs for the passive network. The following steps summarize the design procedure.

1. Evaluate the performance of the uncompensated system in order to determine how much improvement in transient response is required.

2. Design the lead compensator to meet the transient response specifications. The design includes the zero location, pole location, and the loop gain.

3. Simulate the system to be sure all requirements have been met.

4. Redesign if the simulation shows that requirements have not been met.

5. Evaluate the steady-state error performance for the lead-compensated system to determine how much more improvement in steady-state error is required.

6. Design the lag compensator to yield the required steady-state error.

7. Simulate the system to be sure all requirements have been met.

8. Redesign if the simulation shows that requirements have not been met.

Example 9.6

Design a lag-lead compensator.

Problem   Design a lag-lead compensator for the system of Figure 9.37 so that the system will operate with 20% overshoot, and a twofold reduction in settling time. Further, the compensated system will exhibit a tenfold improvement in steady-state error for a ramp input.

Solution   Again, our solution follows the steps described above.

Step 1   First, we evaluate the performance of the uncompensated system. Searching along the 20% overshoot line ($\zeta = 0.456$) in Figure 9.38, we find the dominant poles at $-1.794 \pm j3.501$, with a gain of 192.064. The performance of the uncompensated system is summarized in Table 9.6.

Step 2   Next, we begin the lead compensator design by selecting the location of the compensated system's dominant poles. In order to realize a twofold reduction in settling time, the real part of the dominant pole must be increased by a factor of 2, since the settling time is inversely proportional to the real part. Thus,

$$-\zeta\omega_n = -2(1.794) = -3.588 \tag{9.29}$$

The imaginary part of the design point is

$$\omega_d = \zeta\omega_n \tan 117.129° = 3.588 \tan 117.129° = 7.003 \tag{9.30}$$

Now we design the lead compensator. Arbitrarily select a location for the lead compensator zero. For this example we select the location of the compensator zero coincident with the open-loop pole at -6. This choice will eliminate a zero and leave the lead-compensated system with three poles, the same number that the uncompensated system has.

We complete the design by finding the location of the compensator pole. Using the root locus program, sum the angles to the design point from the uncompensated

Figure 9.37 Uncompensated System for Example 9.6

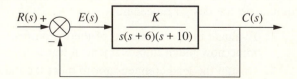

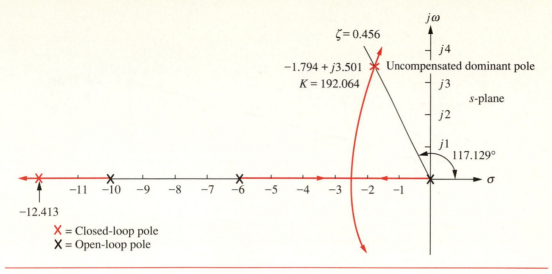

Figure 9.38 Root Locus for Uncompensated System of Example 9.6

Table 9.6 Comparison of Predicted Characteristics of the Uncompensated, Lead-Compensated, and Lag-Lead-Compensated Systems of Example 9.6

	Uncompensated	Lead-compensated	Lag-lead-compensated
Plant	$\dfrac{K}{s(s+6)(s+10)}$	$\dfrac{K}{s(s+10)(s+29.101)}$	$\dfrac{K(s+0.04713)}{s(s+10)(s+29.101)(s+0.01)}$
Dom. poles	$-1.794 \pm j3.501$	$-3.588 \pm j7.003$	$-3.574 \pm j6.976$
K	192.064	1976.662	1971.569
ζ	0.456	0.456	0.456
ω_n	3.934	7.869	7.838
%OS	20	20	20
T_s	2.230	1.115	1.119
T_p	0.897	0.449	0.450
K_v	3.201	6.792	31.930
$e(\infty)$	0.312	0.147	0.0313
Third pole	-12.413	-31.920	$-31.916,\ -0.04739$
Zero	None	None	-0.04713
Comments	Second-order approx. OK	Second-order approx. OK	Second-order approx. OK

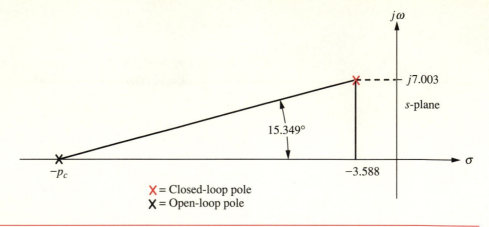

Figure 9.39 Evaluating the Compensator Pole for Example 9.6

system's poles and zeros and the compensator zero and get $-164.651°$. The difference between $180°$ and this quantity is the angular contribution required from the compensator pole, or $-15.349°$. Using the geometry shown in Figure 9.39,

$$\frac{7.003}{p_c - 3.588} = \tan 15.349° \tag{9.31}$$

from which the location of the compensator pole, p_c, is found to be -29.101.

The complete root locus for the lead-compensated system is sketched in Figure 9.40. The gain setting at the design point is found to be 1976.662.

Figure 9.40 Root Locus for the Lead-Compensated System of Example 9.6

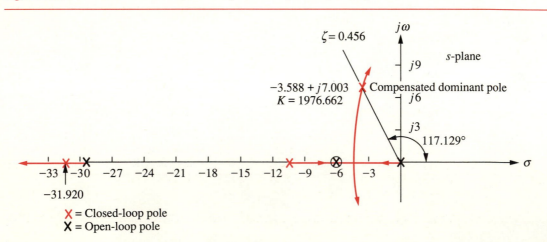

Step 3 and 4 Check the design with a simulation. The result for the lead-compensated system is shown in Figure 9.42 and is satisfactory.

Step 5 Continue by designing the lag compensator to improve the steady-state error. Since the uncompensated system's open-loop transfer function is

$$G(s) = \frac{192.064}{s(s+6)(s+10)} \tag{9.32}$$

the static error constant, K_v, which is inversely proportional to the steady-state error, is 3.201. Since the open-loop transfer function of the lead-compensated system is

$$G_{LC}(s) = \frac{1976.662}{s(s+10)(s+29.101)} \tag{9.33}$$

the static error constant, K_v, which is inversely proportional to the steady-state error, is 6.792. Thus, the addition of lead compensation has improved the steady-state error by a factor of 2.122. Since the requirements of the problem specified a tenfold improvement, the lag compensator must be designed to improve the steady-state error by a factor of 4.713 ($10/2.122 = 4.713$) over the lead-compensated system.

Step 6 We arbitrarily choose the lag compensator pole at 0.01, which then places the lag compensator zero at 0.04713, yielding

$$G_{Lag}(s) = \frac{(s+0.04713)}{(s+0.01)} \tag{9.34}$$

as the lag compensator. The lag-lead-compensated system's open-loop transfer function is

$$G_{LLC}(s) = \frac{K(s+0.04713)}{s(s+10)(s+29.101)(s+0.01)} \tag{9.35}$$

where the uncompensated system pole at -6 canceled the lead compensator zero at -6. By drawing the complete root locus for the lag-lead-compensated system and by searching along the 0.456 damping ratio line, we find the dominant, closed-loop poles to be at $-3.574 \pm j6.976$, with a gain of 1971.57. The lag-lead-compensated root locus is shown in Figure 9.41.

A summary of our design is shown in Table 9.6. Notice that the lag-lead compensation has indeed increased the speed of the system, as witnessed by the settling time or the peak time. The steady-state error for a ramp input has also decreased by about 10 times, as seen from $e(\infty)$.

Step 7 The final proof of our designs is shown by the simulations of Figures 9.42 and 9.43. The improvement in the transient response is shown in Figure 9.42, where we see the peak time occurring sooner in the lag-lead-compensated system. Improvement in the steady-state error for a ramp input is seen in Figure 9.43, where each step of our design yields more improvement. The improvement for the lead-compensated system is shown in Figure 9.43(a), and the final improvement due to the addition of the lag is shown in Figure 9.43(b).

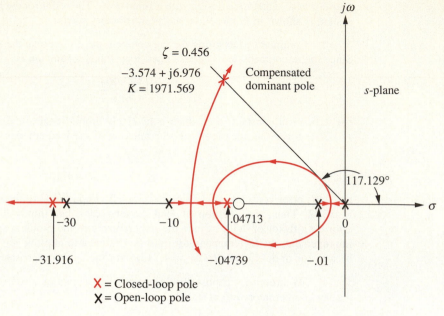

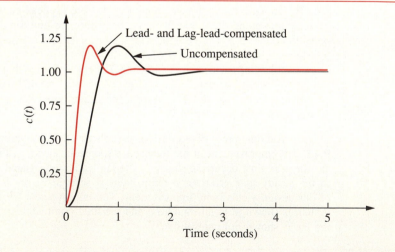

Figure 9.41 Root Locus for the Lag-Lead-Compensated
System of Example 9.6

Figure 9.42 Improvement in Step Response for the
Lag-Lead-Compensated System of Example 9.6

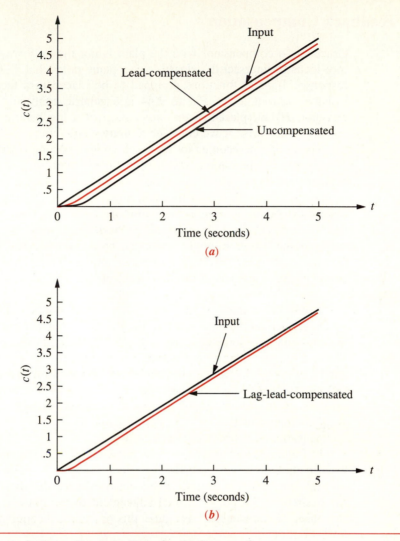

Figure 9.43 Improvement in Ramp Response Error for the System of Example 9.6: (**a**) Lead-Compensated; (**b**) Lag-Lead-Compensated

In the previous example, we canceled the system pole at −6 with the lead compensator zero. The design technique is the same if you place the lead compensator zero at a different location. Placing a zero at a different location and not canceling the open-loop pole yields a system with one more pole than the example. This increased complexity could make it more difficult to justify a second-order approximation. In any case, simulations should be used at each step to verify performance.

In Section 9.4 we used cascade compensation as a way to improve both transient response and steady-state response independently of one another. In the next section, we examine another compensation technique to meet the specifications of a system.

9.5 Feedback Compensation

Cascading a compensator with the plant is not the only way we can reshape the root locus to intersect the closed-loop s-plane poles that yield a desired transient response. Transfer functions designed to be placed in a feedback path can also reshape the root locus. Figure 9.44 is a generic configuration showing a compensator, $H_c(s)$, placed in the *minor loop* of a feedback control system. Other configurations arise if we consider K unity, $G_2(s)$ unity, or both unity.

The design procedures for feedback compensation are more complicated than for cascade compensation. On the other hand, with feedback compensation, the designer has the luxury of isolating portions of the control loop to design desired responses. For example, the transient response of the attitude control system of an aircraft can be designed separately from the response of the entire aircraft steering control loop. Another advantage of feedback compensation over cascade compensation is that faster response times can be acquired. Also, feedback compensation may not require additional amplification since the signal passing through the compensator originates at the high-level output and is delivered to the low-level input.

A popular control system feedback compensator is a differentiator. In many systems, such as the one shown in Figure 9.45, this differentiator may be implemented with a tachometer or a rate gyro. A tachometer is a voltage generator that yields a voltage output proportional to rotational input speed. This compensator can be easily geared to the position output of a system. In aircraft and ship applications, the feedback component can be a rate gyro that responds to an angular position input with an output voltage proportional to angular velocity. The block diagram representation of a tachometer and two configurations for tachometer compensation are shown in Figure 9.46.

In Figure 9.46(*b*) we show a tachometer used for *major-loop compensation*. This method adds a zero to the open-loop transfer function that can be adjusted with the tachometer gain, K_t. This zero reshapes the root locus to go through the desired design point. A final adjustment of the gain, K_1, yields the desired response. The reader may recognize this procedure as equivalent to the design of a PD controller.

The second method, *minor-loop compensation*, which is shown in Figure

Figure 9.44 Generic Control System with Feedback Compensation

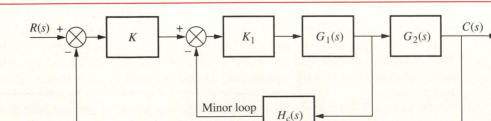

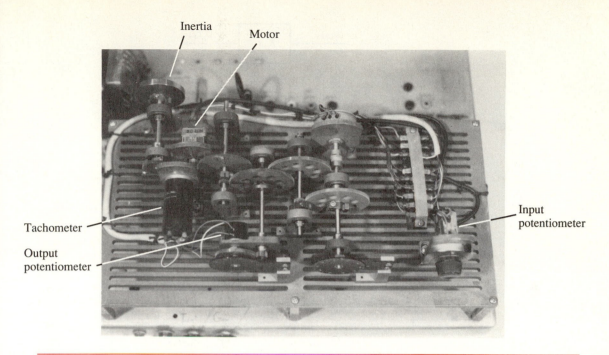

Figure 9.45 A Position Control System That Uses a Tachometer as a Differentiator in the Feedback Path. Can You See the Similarity between This System and Figure 2.36? (Photo by Mark E. Van Dusen)

9.46(*c*), uses the feedback compensator to design desired pole locations for a minor closed loop. Typically the minor loop is around the plant or around the controller. Thus, minor-loop compensation allows the engineer to design the appropriate response for the plant or controller prior to closing the major loop. For example, in an aircraft, proper design of the transient response of the aerosurfaces is implemented prior to the design of the entire aircraft's transient response to roll, pitch, or yaw commands. Another way of looking at minor-loop compensation is to visualize the feedback compensator as a device that moves the plant's or controller's open-loop poles. Thus, with minor-loop compensation, we are literally changing the open-loop poles of the system and reshaping the root locus to yield the desired closed-loop poles.

In summary then, major-loop compensation *adds a compensating zero* to reshape the root locus. Minor-loop compensation on the other hand reshapes the root locus by *changing the system's open-loop poles*. Minor-loop compensation has the added advantage that the desired response of the minor loop can be designed at the same time. Let us take a detailed look at each design procedure.

Major-Loop Compensation

Major-loop compensation imitates the design of a PD controller. If the compensator is in parallel with the major-loop unity feedback path, as shown in Figure

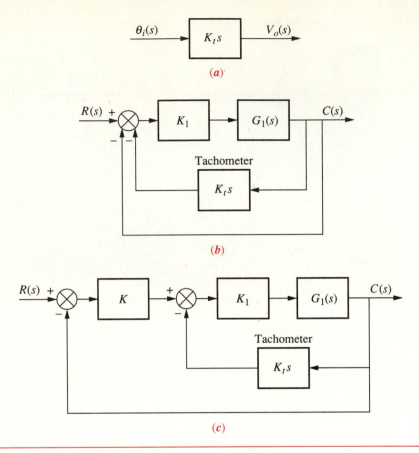

Figure 9.46 (**a**) Transfer Function of a Tachometer;
(**b**) Major-Loop Feedback Compensation; (**c**) Minor-Loop
Feedback Compensation

9.46(*b*), the equivalent feedback is

$$H(s) = 1 + K_t s = K_t \left(s + \frac{1}{K_t} \right) \tag{9.36}$$

The equivalent open-loop function is $K_1 G_1(s) H(s) = K_1 K_t G_1(s) \left(s + \frac{1}{K_t} \right)$.

Hence, the feedback compensator adds an adjustable zero to the open-loop function. The root locus can now be reshaped by a judicious choice of the zero, just as in cascade compensation, and followed by a gain adjustment. In contrast to cascade compensation, however, remember that the compensator zero is not a zero of the closed-loop transfer function.[2] We illustrate the design with an example.

---

[2]See Section 8.1.

Example 9.7   Design a compensating zero using tachometer compensation.

Problem   Given the system of Figure 9.47(*a*), design tachometer compensation to reduce the settling time by a factor of 4 while continuing to operate the system with 20% overshoot.

Figure 9.47 (*a*) System for Example 9.7; (*b*) System with Tachometer Compensation; (*c*) Equivalent Compensated System; (*d*) Equivalent Compensated System, Showing Unity Feedback

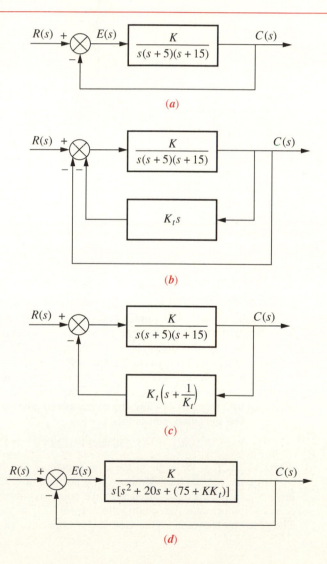

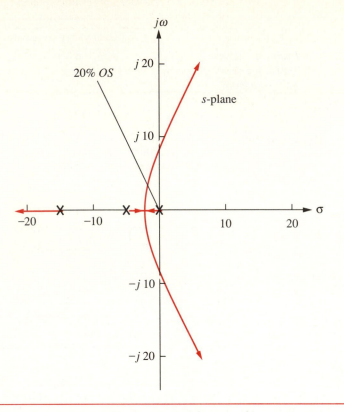

Figure 9.48 Root Locus for the Uncompensated System of Example 9.7

Solution For the uncompensated system, search along the 20% overshoot line (ζ = 0.456) and find that the dominant poles are at $-1.809 \pm j3.531$, as shown in Figure 9.48. The estimated specifications for the uncompensated system are shown in Table 9.7, and the step response is shown in Figure 9.49. The settling time is 2.21 seconds and must be reduced by a factor of 4 to 0.55.

Next, determine the location of the dominant poles for the compensated system. To achieve a fourfold decrease in the settling time, the real part of the pole must be increased by a factor of 4. Thus, the compensated pole has a real part of $4(-1.809) = -7.236$. The imaginary part is then

$$\omega_d = -7.236 \tan 117.129° = 14.123 \tag{9.37}$$

where $117.129°$ is the angle of the 20% overshoot line.

Using the compensated dominant pole position of $-7.236 \pm j14.123$, we sum the angles from the uncompensated system's poles and obtain $-277.33°$. This angle requires a compensator zero contribution of $+97.33°$ to yield $180°$ at the design point. The geometry shown in Figure 9.50 leads to the calculation of the compensator's zero

Table 9.7 Comparison of Predicted Characteristics of the Uncompensated and Compensated Systems of Example 9.7

	Uncompensated	Compensated
Plant	$\dfrac{K}{s(s + 5)(s + 15)}$	$\dfrac{K}{s(s + 5)(s + 15)}$
Feedback	1	$0.185(s + 5.42)$
Dom. poles	$-1.809 \pm j3.531$	$-7.236 \pm j14.123$
K	257.841	1388.211
ζ	0.456	0.456
ω_n	3.97	15.88
%OS	20	20
T_s	2.21	0.55
T_p	0.89	0.22
K_v	3.44	4.18
$e(\infty)$ (ramp)	0.29	0.24
Other poles	-16.4	-5.51
Zero	None	None
Comments	Second-order approx. OK	Simulate

Figure 9.49 Step Response for the Uncompensated System of Example 9.7

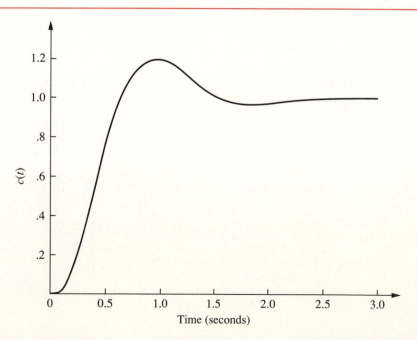

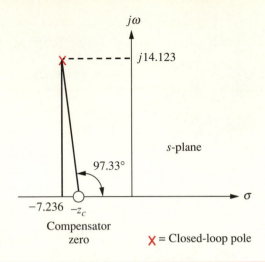

Figure 9.50 Geometry for Finding the Compensator Zero in Example 9.7

Figure 9.51 Root Locus for the Compensated System of Example 9.7

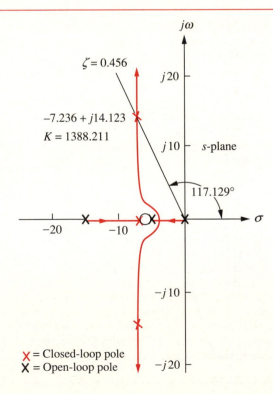

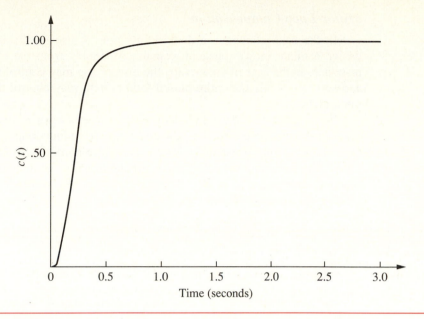

Figure 9.52 Step Response for the Compensated System of
Example 9.7

location. Hence,

$$\frac{14.123}{7.236 - z_c} = \tan(180° - 97.33°) \tag{9.38}$$

from which $z_c = 5.42$.

The root locus for the compensated system is shown in Figure 9.51. The gain at the design point, which is KK_t from Figure 9.47(c), is found to be 256.819. Since K_t is the reciprocal of the compensator zero, $K_t = 0.185$. Thus, $K = 1388.211$.

In order to evaluate the steady-state error characteristic, K_v is found from Figure 9.47(d) to be

$$K_v = \frac{K}{75 + KK_t} = 4.18 \tag{9.39}$$

Predicted performance for the compensated system is shown in Table 9.7. Notice that the higher-order pole is not far enough away from the dominant poles and thus cannot be neglected. There is no pole-zero cancellation since the open-loop zero is not a closed-loop zero. Hence, the design must be checked by simulation.

The results of the simulation are shown in Figure 9.52 and show an overdamped response with a settling time of 0.75 seconds compared to the uncompensated system's settling time of approximately 2.2 seconds. Although not meeting the design requirements, the response still represents an improvement over the uncompensated system of Figure 9.49. Typically, less overshoot is acceptable. The system should be redesigned for further reduction in settling time.

Minor-Loop Compensation

In the next example we demonstrate how to use feedback compensation to design a minor loop's transient response separately from the closed-loop system response. In the case of an aircraft, the minor loop may control the position of the aerosurfaces, while the entire closed-loop system may control the entire aircraft's pitch angle.

We will see that the minor loop basically represents a forward-path transfer function whose poles can be adjusted with the minor-loop gain. These poles, then, become the open-loop poles for the entire control system. In other words, rather than reshaping the root locus with additional poles and zeros, as in cascade compensation, we can actually change the plant's poles through a gain adjustment. Finally, the closed-loop poles are set by the loop gain, as in cascade compensation.

Example 9.8   Design minor-loop feedback compensation.

Problem   For the system of Figure 9.53(*a*), design minor-loop feedback compensation, as shown in Figure 9.53(*b*), to yield a damping ratio of 0.8 for the minor loop and a damping ratio of 0.6 for the closed-loop system.

Solution   The minor loop is defined as the loop containing the plant, $\dfrac{1}{s(s + 5)(s + 15)}$, and the feedback compensator, $K_t s$. The value of K_t will be adjusted to set the location of the minor-loop poles, and then K will be adjusted to yield the desired closed-loop response.

Figure 9.53 (***a***) Uncompensated System for Example 9.8; (***b***) Feedback-Compensated System for Example 9.8

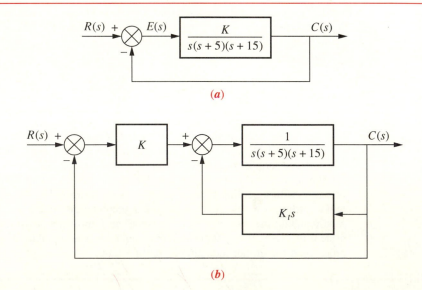

(*a*)

(*b*)

The transfer function of the minor loop, $G_{ML}(s)$, is

$$G_{ML}(s) = \frac{1}{s\left[s^2 + 20s + (75 + Kt)\right]}$$ (9.40)

The poles of $G_{ML}(s)$ can be found analytically or via the root locus. The root locus for the minor loop, where $\dfrac{K_t s}{s(s + 5)(s + 15)}$, is the open-loop transfer function is shown in Figure 9.54. Since the zero at the origin comes from the feedback function of the minor loop, this zero is not a zero of the closed-loop function of the minor loop. Hence, the pole at the origin appears to remain stationary and there is no pole-zero cancellation at the origin. Equation 9.40 also shows this phenomenon. We can see a stationary pole at the origin and two complex poles that change with gain. Drawing the $\zeta = 0.8$ line in Figure 9.54 yields the complex poles at $-10 \pm j7.5$. The gain, K_t, which equals 81.25, places the minor-loop poles in a position to meet the specifications. The poles just found, $-10 \pm j7.5$, as well as the pole at the origin (Equation 9.40) act as open-loop poles that generate a root locus for variations of the gain, K.

Figure 9.54 Root Locus for Minor Loop of Example 9.8

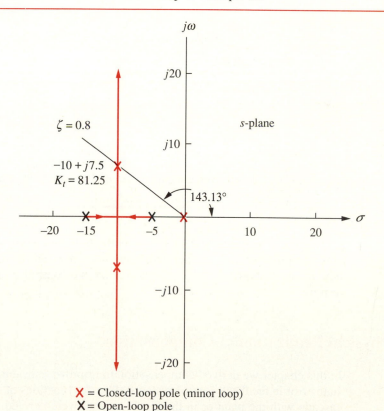

X = Closed-loop pole (minor loop)
X = Open-loop pole

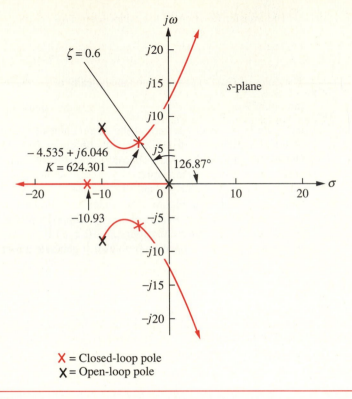

Figure 9.55 Root Locus for Closed-Loop System of Example 9.8

The final root locus for the system is shown in Figure 9.55. The $\zeta = 0.6$ damping ratio line is drawn and searched. The closed-loop complex poles are found to be $-4.535 \pm j6.046$, with a required gain of 624.301. A third pole is at -10.93.

Our discussion of compensation methods is now complete. We studied both cascade and feedback compensation and compared and contrasted them. We are now ready to show how to physically realize the controllers and compensators we designed.

9.6 Physical Realization of Compensation

In this chapter we derived compensation to improve transient response and steady-state error in feedback control systems. Transfer functions of compensators used in cascade with the plant or in the feedback path were derived. These compensators were defined by their pole-zero configuration. They were either active PI, PD, or PID controllers or passive lag, lead, or lag-lead compensators. In this section we show how to implement the active controllers and the passive compensators. Let us begin by paving the way toward the development of active controllers by discussing a basic component for their realization, the *operational amplifier*. After discussing the operational amplifier, we can use it as a building block for active controllers and compensators.

The Operational Amplifier

An operational amplifier, pictured in Figure 9.56(*a*), is an electronic amplifier used as a basic building block to implement transfer functions. It has the following characteristics:

1. Differential input, $v_2 - v_1$
2. High input impedance
3. Low output impedance
4. High gain, A

The output, $v_o(t)$, is given by

$$v_o(t) = A[v_2(t) - v_1(t)] \tag{9.41}$$

If $v_2(t)$ is grounded, as shown in Figure 9.56(*b*), the amplifier is called an *inverting operational amplifier*. An alternate representation of the inverting operational amplifier is shown in Figure 9.56(*c*). For the inverting operational amplifier,

$$v_o(t) = -Av_1(t) \tag{9.42}$$

Figure 9.56 (*a*) The Operational Amplifier; (*b*) Configured as an Inverting Operational Amplifier; (*c*) Alternate Schematic for an Inverting Operational Amplifier; (*d*) Operational Amplifier Configured for Transfer Function Realization

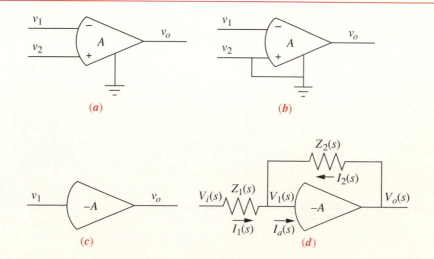

Table 9.8 Active Realization of Controllers and Compensators, Using an Operational Amplifier

Function	$Z_1(s)$	$Z_2(s)$	$G_c(s) = -\dfrac{Z_2(s)}{Z_1(s)}$
Gain	R_1	R_2	$-\dfrac{R_2}{R_1}$
Integration	R	C	$-\dfrac{\frac{1}{RC}}{s}$
Differentiation	C	R	$-RC\,s$
PI controller	R_1	$R_2 \quad C$	$-\dfrac{R_2}{R_1}\dfrac{\left(s + \frac{1}{R_2 C}\right)}{s}$
PD controller	C / R_1	R_2	$-R_2 C\left(s + \dfrac{1}{R_1 C}\right)$
PID controller	C_1 / R_1	$R_2 \quad C_2$	$-\left[\left(\dfrac{R_2}{R_1} + \dfrac{C_1}{C_2}\right) + R_2 C_1 s + \dfrac{\frac{1}{R_1 C_2}}{s}\right]$
Lag compensation	C_1 / R_1	C_2 / R_2	$-\dfrac{C_1}{C_2}\dfrac{\left(s + \frac{1}{R_1 C_1}\right)}{\left(s + \frac{1}{R_2 C_2}\right)}$ where $R_2 C_2 > R_1 C_1$
Lead compensation	C_1 / R_1	C_2 / R_2	$-\dfrac{C_1}{C_2}\dfrac{\left(s + \frac{1}{R_1 C_1}\right)}{\left(s + \frac{1}{R_2 C_2}\right)}$ where $R_1 C_1 > R_2 C_2$

If two impedances are connected to the inverting operational amplifier, as shown in Figure 9.56(d), we can derive an interesting result if the amplifier has the characteristics mentioned in the beginning of this section. If the input impedance to the amplifier is high, then by Kirchhoff's current law, $I_a(s) = 0$ and $I_1(s) = -I_2(s)$. Also, since the gain, A, is large, $v_1 \approx 0$. Thus, $I_1(s) = V_i(s)/Z_1(s)$, and $I_2 = V_o(s)/Z_2(s)$. Equating the two currents, $V_o(s)/Z_2(s) = -V_i(s)/Z_1(s)$, or the transfer function of the inverting operational amplifier configured as shown in Figure 9.56(d) is

$$\frac{V_o(s)}{V_i(s)} = -\frac{Z_2(s)}{Z_1(s)} \tag{9.43}$$

Active-Circuit Realization of Controllers and Compensators

Using Equation 9.43, we can implement many transfer functions, including PID controllers, by judicious choice of $Z_1(s)$ and $Z_2(s)$ and the use of the active operational amplifier, as shown in Figure 9.56(d). Table 9.8 summarizes the realization of PI, PD, and PID controllers as well as lag, lead, and lag-lead compensators using operational amplifiers. The reader can verify the table using the methods of Chapter 2 to find the impedances.

Other compensators can be realized by cascading compensators shown in the table. For example, a lag-lead compensator can be formed by cascading the lag compensator with the lead compensator, as shown in Figure 9.57. As an example, let us implement one of the controllers we designed earlier in the chapter.

Figure 9.57 Lag-Lead Compensator Implemented with Operational Amplifiers

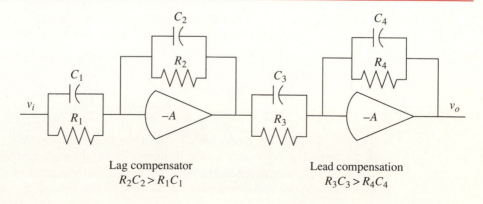

Lag compensator
$R_2 C_2 > R_1 C_1$

Lead compensation
$R_3 C_3 > R_4 C_4$

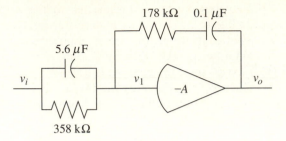

Figure 9.58 PID Controller

Example 9.9

Implement a PID controller.

Problem Implement the PID controller of Example 9.5.

Solution The transfer function of the PID controller is $G_c(s) = \dfrac{(s + 55.909)(s + 0.5)}{s}$,

which can be put in the form, $G_c(s) = s + 56.409 + \dfrac{27.954}{s}$. From the PID controller

in Table 9.8, $\left[\dfrac{R_2}{R_1} + \dfrac{C_1}{C_2} \right] = 56.409$, $R_2 C_1 = 1$, and $\dfrac{1}{R_1 C_2} = 27.954$.

Since there are four unknowns and three equations, we arbitrarily select a practical value for one of the elements. Selecting $C_2 = 0.1\ \mu F$, the remaining values are found to be $R_1 = 357.731\ k\Omega$, $R_2 = 178.862\ k\Omega$, and $C_1 = 5.591\ \mu F$.

The complete circuit is shown in Figure 9.58, where the circuit element values have been rounded off.

Passive-Circuit Realization of Compensators

Lag, lead, and lag-lead compensators can also be implemented with passive networks. Table 9.9 summarizes the networks and their transfer functions. The transfer functions can be derived with the methods of Chapter 2.

The lag-lead transfer function can be put in the following form:

$$G_c(s) = \frac{\left(s + \dfrac{1}{T_1}\right)\left(s + \dfrac{1}{T_2}\right)}{\left(s + \dfrac{1}{\alpha T_1}\right)\left(s + \dfrac{\alpha}{T_2}\right)} \tag{9.44}$$

where $\alpha < 1$. Thus the terms with T_1 form the lead compensator, and the terms with T_2 form the lag compensator. Equation 9.44 shows a restriction inherent in using this passive realization. What we see is that the ratio of the lead compensator zero to the lead compensator pole must be the same as the ratio of the lag

Table 9.9 Passive Realization of Compensators

Function	Network	Transfer function, $\dfrac{V_o(s)}{V_i(s)}$
Lag compensation		$\dfrac{R_2}{R_1 + R_2} \; \dfrac{s + \dfrac{1}{R_2 C}}{s + \dfrac{1}{(R_1 + R_2)C}}$
Lead compensation		$\dfrac{s + \dfrac{1}{R_1 C}}{s + \dfrac{1}{R_1 C} + \dfrac{1}{R_2 C}}$
Lag-lead compensation		$\dfrac{\left(s + \dfrac{1}{R_1 C_1}\right)\left(s + \dfrac{1}{R_2 C_2}\right)}{s^2 + \left(\dfrac{1}{R_1 C_1} + \dfrac{1}{R_2 C_2} + \dfrac{1}{R_2 C_1}\right)s + \dfrac{1}{R_1 R_2 C_1 C_2}}$

compensator pole to the lag compensator zero. In Chapter 11 we will design a lag-lead compensator with this restriction.

A lag-lead compensator without this restriction can be realized with an active network as previously shown or with passive networks by cascading the lead and lag networks shown in Table 9.9. Remember, though, that isolation must be provided between the two networks to ensure that one network does not load the other. If the networks load each other, the transfer function will not be the product of the individual transfer functions. A possible realization using the passive networks incorporates an operational amplifier to provide isolation. The circuit is shown in Figure 9.59. The following example demonstrates the design of a passive compensator.

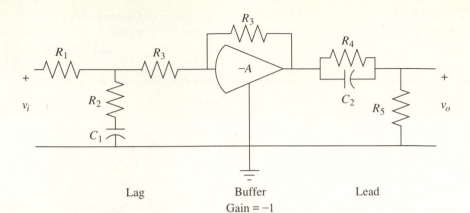

Figure 9.59 Lag-Lead Compensator Implemented with Cascaded Lag and Lead Networks with Isolation

Example 9.10   Realize a lead compensator.

Problem   Realize the lead compensator designed in Example 9.4 (Compensator *b*).

Solution   The transfer function of the lead compensator is $G_c(s) = \dfrac{s + 4}{s + 20.089}$.

Thus, from the transfer function of a lead network shown in Table 9.9, $\dfrac{1}{R_1 C} = 4$, and $\dfrac{1}{R_1 C} + \dfrac{1}{R_2 C} = 20.089$. Hence $R_1 C = 0.25$, and $R_2 C = 0.0622$. Since there are three network elements and two equations, we may select one of the element values arbitrarily. Letting $C = 1\,\mu\text{F}$, $R_1 = 250\ \text{k}\Omega$, and $R_2 = 62.2\ \text{k}\Omega$.

Our discussion has led us through the design and implementation of controllers and compensators. A review of the chapter follows.

9.7 Chapter-Objective Demonstration Problems

For the chapter-objective demonstration problem of Chapter 8, we obtained a 25% overshoot using a simple gain adjustment. Once this percent overshoot was obtained, the settling time was determined. If we try to improve the settling time by increasing the gain, the percent overshoot also increases. In this section, we continue with the design of the antenna azimuth position control by designing a cascade compensator that yields 25% overshoot at a reduced settling time. Further, we effect an improvement in the steady-state error performance of the system.

Example 9.11

Design cascade compensation to improve transient and steady-state error characteristics.

Problem   Given the schematic of the antenna azimuth position control system shown in Figure 2.36, design cascade compensation to meet the following requirements: (1) 25% overshoot, (2) 2-second settling time, and (3) $K_v = 20$.

Solution   In the demonstration problem (Section 8.13) in Chapter 8, a preamplifier gain of 64.3 yielded 25% overshoot, with the dominant, second-order poles at $-0.833 \pm j1.887$. The settling time is thus $4/\zeta\omega_n = 4/.833 = 4.8$ seconds. The open-loop function for the system as derived in the demonstration problem in Section 5.9 is $G(s) = \dfrac{6.63K}{s(s + 1.71)(s + 100)}$. Hence $K_v = \dfrac{6.63K}{1.71 \times 100} = 2.49$. Comparing these values to this example's problem statement, we want to improve the settling time by a factor of 2.4 and we want approximately an eightfold improvement in K_v.

Lead Compensator Design to Improve Transient Response   First, locate the dominant second-order pole. To obtain a settling time, T_s, of 2 seconds and a percent overshoot of 25%, the real part of the dominant second-order pole should be at $-4/T_s = -2$. Locating the pole on the 113.828° line ($\zeta = 0.404$ corresponding to 25% overshoot) yields an imaginary part of 4.529 (see Figure 9.60).

Second, assume a lead compensator zero and find the compensator pole. Assuming a compensator zero at -2, along with the uncompensated system's open-loop poles and zeros, use the root locus program in Appendix C to find that there is an angular contribution of $-120.14°$ at the design point of $-2 \pm j4.529$. Therefore, the compensator's pole must contribute $120.14° - 180° = -59.86°$ for the design point to be on the compensated system's root locus. The geometry is shown in Figure 9.60.

To calculate the compensator pole, we use $\dfrac{4.529}{p_c - 2} = \tan 59.86°$ or $p_c = 4.63$.

Figure 9.60 Geometry for Locating Compensator Pole

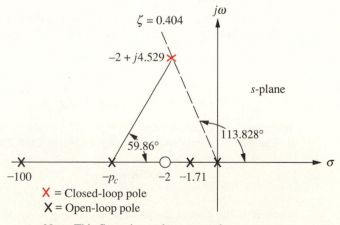

$\zeta = 0.404$

$-2 + j4.529$

$j\omega$

s-plane

113.828°

59.86°

-100 $-p_c$ -2 -1.71 σ

X = Closed-loop pole
X = Open-loop pole

Note: This figure is not drawn to scale.

Now determine the gain. Using the lead-compensated system's open-loop function,

$$G(s) = \frac{6.63K(s + 2)}{s(s + 1.71)(s + 100)(s + 4.63)} \tag{9.45}$$

and the design point $-2 + j4.529$ as the test point in the root locus program, the gain, $6.63K$, is found to be 2548.993.

Lag Compensator Design to Improve the Steady-State Error K_v for the lead-compensated system is found using Equation 9.45. Hence,

$$K_v = \frac{2548.993(2)}{(1.71)(100)(4.63)} = 6.44 \tag{9.46}$$

Since we want $K_v = 20$, the amount of improvement required over the lead-compensated system is $20/6.44 = 3.1$. Choose $p_c = -0.01$ and calculate $z_c = .031$, which is 3.1 times larger.

Determine Gain   The complete lag-lead-compensated open-loop function, $G_{LLC}(s)$, is

$$G_{LLC}(s) = \frac{6.63K(s + 2)(s + .031)}{s(s + .01)(s + 1.71)(s + 4.63)(s + 100)} \tag{9.47}$$

Using the root locus program in Appendix C and the poles and zeros of Equation 9.47, search along the 25% overshoot line (113.82°) for the design point. We find that this point has moved slightly with the addition of the lag compensator to $-1.99 \pm j4.51$. The gain at this point equals 2533.259, which is $6.63K$. Solving for K yields $K = 382.09$.

Realization of the Compensator   A realization of the lag-lead compensator is shown in Figure 9.59. From Table 9.9, the lag portion has the following transfer function:

$$G_{lag}(s) = \frac{R_2}{R_1 + R_2} \frac{s + \dfrac{1}{R_2C}}{s + \dfrac{1}{(R_1 + R_2)C}} = \frac{R_2}{R_1 + R_2} \frac{(s + 0.031)}{(s + 0.01)} \tag{9.48}$$

Selecting $C = 10 \, \mu F$, we find $R_2 = 3.2 \, M\Omega$ and $R_1 = 6.8 \, M\Omega$.

From Table 9.9, the lead compensator portion has the following transfer function:

$$G_{lead}(s) = \frac{s + \dfrac{1}{R_1C}}{s + \dfrac{1}{R_1C} + \dfrac{1}{R_2C}} = \frac{(s + 2)}{(s + 4.63)} \tag{9.49}$$

Selecting $C = 10 \, \mu F$, we find $R_1 = 50 \, k\Omega$ and $R_2 = 38 \, k\Omega$.

The total loop gain required by the system is 2533.259. Hence,

$$6.63K \frac{R_2}{R_1 + R_2} = 2533.259 \tag{9.50}$$

where K is the gain of the preamplifier, and $\dfrac{R_2}{R_1 + R_2}$ is the gain of the lag portion.

Using the values of R_1 and R_2 found during the realization of the lag portion, we find $K = 1194.032$.

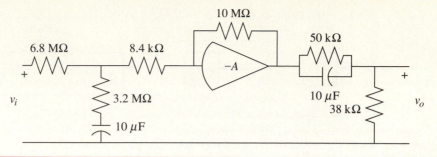

Figure 9.61 Realization of Lag-Lead Compensator

The final circuit is shown in Figure 9.61, where the preamplifier is implemented with an operational amplifier whose feedback and input resistor ratio approximately equals 1194.032, the required preamplifier gain. The preamplifier is placed to act as a buffer isolating the lag and lead portions of the compensator.

Summary of the Design Results   Using Equation 9.47 along with $K = 382.09$ yields the compensated value of K_v. Thus,

$$K_v = \lim_{s \to 0} s G_{LLC}(s) = \frac{2533.259(2)(0.031)}{(0.01)(1.71)(4.63)(100)} = 19.84 \tag{9.51}$$

which is an improvement over the gain-compensated system of Example 8.9, where $K_v = 2.49$.

Finally, checking the second-order approximation via simulation, we see in Figure 9.62 the actual transient response. Compare this response to the gain-compensated

Figure 9.62 Step Response of Lag-Lead-Compensated Antenna Azimuth Control System

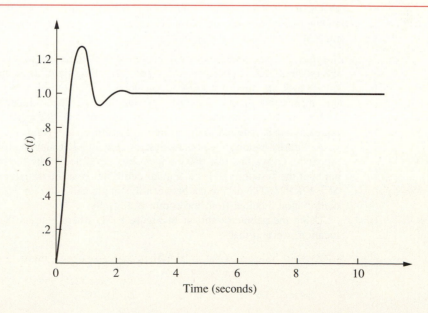

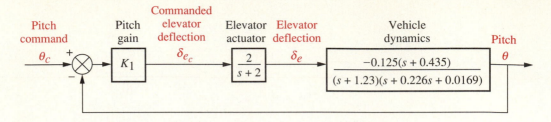

Figure 9.63 Pitch Control Loop for the UFSS Vehicle (Adapted from Johnson, H., et al. *Unmanned Free-Swimming Submersible (UFSS) System Description*. NRL Memorandum Report 4393. Naval Research Laboratory, Washington, D.C.)

system response of Figure 8.27 to see the improvement effected by cascade compensation over simple gain adjustment. The gain-compensated system yielded 25%, with a settling time of about 4 seconds. The lag-lead-compensated system yields 28% overshoot with a settling time of about 2 seconds. If the results are not adequate for the application, the system should be redesigned to reduce the percent overshoot.

As a final example, we will redesign the pitch control loop for the Unmanned Free-Swimming Submersible (UFSS) vehicle. In Example 8.10, we saw that rate feedback improved the transient response. In the next example we will replace the rate feedback with a cascade compensator.

Example 9.12   Design a lead compensator.

Problem   Given the pitch control loop without rate feedback for the UFSS vehicle shown in Figure 9.63, design a compensator to yield 20% overshoot and a settling time of 4 seconds.

Solution   First, determine the location of the dominant closed-loop poles. Using the required 20% overshoot and a 4-second settling time, a second-order approximation shows the dominant closed-loop poles are located at $-1 \pm j1.951$. From the uncompensated system analyzed in Example 8.10, the estimated settling time was 19.8 seconds for dominant closed-loop poles of $-0.202 \pm j0.394$. Hence, a lead compensator is required to speed up the system.

Arbitrarily assume a lead compensator zero at -1. Using the root locus program in Appendix C, we find that this compensator zero, along with the open-loop poles and zeros of the system, yields an angular contribution at the design point, $-1 + j1.951$, of $-178.92°$. The difference between this angle and $180°$, or $-1.08°$, is the angular contribution required from the compensator pole.

Using the geometry shown in Figure 9.64, where $-p_c$ is the compensator pole location, we find that

$$\frac{1.951}{p_c - 1} = \tan 1.08° \tag{9.52}$$

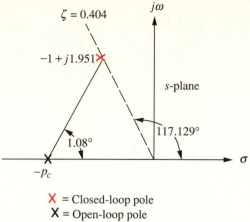

$\zeta = 0.404$

$-1 + j1.951$

s-plane

117.129°

1.08°

$-p_c$

X = Closed-loop pole
X = Open-loop pole
Note: This figure is not drawn to scale.

Figure 9.64 Geometry for Locating Compensator Pole

from which $p_c = 104.49$. The compensated open-loop transfer function is thus

$$G(s) = \frac{-0.25K_1(s + 0.435)(s + 1)}{(s + 1.23)(s + 2)(s^2 + 0.226s + 0.0169)(s + 104.49)} \tag{9.53}$$

where the compensator is

$$G_c(s) = \frac{(s + 1)}{(s + 104.49)} \tag{9.54}$$

Using all poles and zeros shown in Equation 9.53, the root locus program shows that a gain of 516.41 is required at the design point, $-1 \pm j1.951$. The root locus of the compensated system is shown in Figure 9.65.

A test of the second-order approximation shows three more closed-loop poles at -0.5, -0.9, and -104.54. Since the open-loop zeros are at -0.435 and -1, simulation is required to see if there is effectively closed-loop pole-zero cancellation with the closed-loop poles at -0.5 and -0.9, respectively. Further, the closed-loop pole at -104.54 is more than five times the real part of the dominant closed-loop pole, $-1 \pm j1.951$ and its effect upon the transient response is therefore negligible.

The step response of the closed-loop system is shown in Figure 9.66, where we see a 26% overshoot and a settling time of about 4.5 seconds. Comparing this response with Figure 8.30, the response of the uncompensated system, we see considerable improvement in the settling time and steady-state error. However, the transient response performance does not meet the design requirements. Thus, a re-design of the system to reduce the percent overshoot is suggested if required by the application.

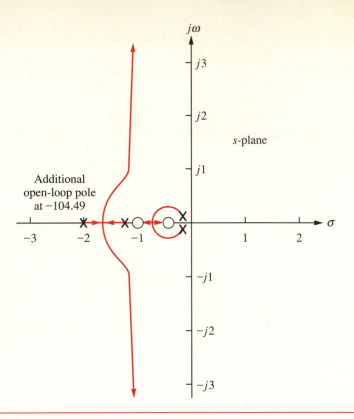

Figure 9.65 Root Locus for Lead-Compensated System

Figure 9.66 Step Response of Lead-Compensated UFSS Vehicle

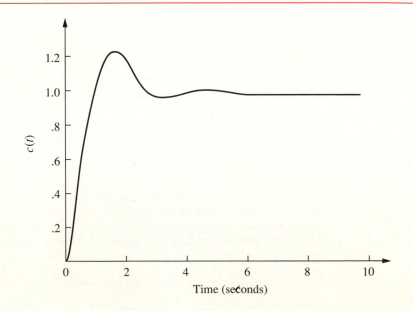

9.8 Summary

In this chapter we learned how to design a system to meet both transient and steady-state specifications. These design techniques overcame limitations in the design methodology covered in Chapter 8, where a transient response could be created only if the poles generating that response were on the root locus. Subsequent gain adjustment yielded the desired response. Since this value of gain dictated the amount of steady-state error present in the response, a trade-off was required between the desired transient response and the desired steady-state error.

Cascade or feedback compensation are used to overcome the disadvantages of gain adjustment as a compensating technique. In this chapter we saw that the transient response and the steady-state error can be designed separately from each other. No longer was a trade-off between these two specifications required. Further, we were able to design for a transient response that was not represented on the original root locus.

The transient response design technique covered in this chapter is based upon reshaping the root locus to go through a desired transient response point, followed by a gain adjustment. Typically, the resulting gain is much higher than the original if the compensated system response is faster than the uncompensated response. The root locus is reshaped by adding additional poles and zeros via a cascade or feedback compensator. The additional poles and zeros must be checked to see that any second-order approximations used in the design are valid. All poles beside the dominant second-order pair must yield a response that is much faster than the designed response. Thus, nondominant poles must be at least five times further from the imaginary axis than the dominant pair. Further, any zeros of the system must be close to a nondominant pole for pole-zero cancellation, or far from the dominant pole-pair. The resulting system can then be approximated by two dominant poles.

The steady-state response design technique is based upon placing a pole on or near the origin in order to increase or nearly increase the system type and then placing a zero near this pole so that the effect upon the transient response is negligible. The same arguments about other poles yielding fast responses and about zeros being canceled in order to validate a second-order approximation also hold true for this technique.

If the second-order approximations cannot be justified, then a simulation is required to make sure the design is within tolerance.

Steady-state design compensators are implemented via PI controllers or lag compensators. PI controllers add a pole at the origin, thereby increasing the system type. Lag compensators, usually implemented with passive networks, place the pole off of but near the origin. Both methods add a zero very close to the pole in order not to affect the transient response.

The transient response design compensators are implemented through PD controllers or lead compensators. PD controllers add a zero to compensate the transient response; they are considered ideal. Lead compensators, on the other hand, are not ideal since they add a pole along with the zero. Lead compensators are usually passive networks.

We can correct both transient response and steady-state error with a PID or lag-lead compensator. Both of these are simply combinations of the previously described compensators.

Feedback compensation can also be used to improve the transient response. Here, the compensator is placed in the feedback path. The feedback gain is used to change the compensator zero or the system's open-loop poles, giving the designer a wide choice of various root loci. The system gain is then varied to move along the selected root locus to the design point. An advantage of feedback compensation is the ability to design a particular response into a subsystem independently of the system's total response.

In the next chapter we look at another method of design, frequency response, which is an alternate method to the root locus.

REVIEW QUESTIONS

1. Briefly distinguish between the design techniques in Chapter 8 and Chapter 9.
2. Name two major advantages of the design techniques of Chapter 9 over the design techniques of Chapter 8.
3. What kind of compensation improves the steady-state error?
4. What kind of compensation improves transient response?
5. What kind of compensation improves both steady-state error and transient response?
6. Cascade compensation to improve the steady-state error is based upon what pole-zero placement of the compensator? Also, state the reasons for this placement.
7. Cascade compensation to improve the transient response is based upon what pole-zero placement of the compensator? Also, state the reasons for this placement.
8. What difference on the s-plane is noted between using a PD controller or using a lead network to improve the transient response?
9. In order to speed up a system without changing the percent overshoot, where must the compensated system's poles on the s-plane be located in comparison to the uncompensated system's poles?
10. Why is there more improvement in steady-state error if a PI controller is used instead of a lag network?
11. When compensating for steady-state error, what effect is sometimes noted in the transient response?
12. A lag compensator with the zero 25 times further from the imaginary axis than the compensator pole will yield approximately how much improvement in steady-state error?
13. If the zero of a feedback compensator is at -3, and a closed-loop system pole is at -3.001, can you say there will be pole-zero cancellation? Why?
14. Name two advantages of feedback compensation.

PROBLEMS

1. Design a PI controller to drive the step response error to zero for the unity feedback system shown in Figure P9.1, where

$$G(s) = \frac{K}{(s + 3)(s + 6)}$$

The system operates with a damping ratio of 0.707. Compare the specifications of the uncompensated and compensated systems.

Figure P9.1

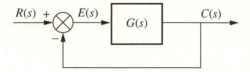

2. The unity feedback system shown in Figure P9.1 with

$$G(s) = \frac{K}{(s + 1)(s + 3)(s + 5)}$$

is operating with 10% overshoot.
   a. What is the value of the appropriate static error constant?
   b. Find the transfer function of a lag network so that the appropriate static error constant equals 4 without appreciably changing the dominant poles of the uncompensated system.
   c. Simulate the system to see the effect of your compensator.

3. Consider the unity feedback system shown in Figure P9.1 with

$$G(s) = \frac{K}{(s + 2)(s + 4)(s + 6)}$$

   a. Design a compensator that will yield $K_p = 20$ without appreciably changing the dominant pole location that yields a 10% overshoot for the uncompensated system.
   *b. Simulate the uncompensated and compensated systems.
   *c. How much time does it take the slow response of the lag compensator to bring the output to within 2% of its final compensated value?

4. The unity feedback system shown in Figure P9.1 with

$$G(s) = \frac{K}{(s + 1)(s + 2)(s + 3)(s + 6)}$$

is operating with a dominant pole damping ratio of 0.707. Design a PD controller so that the settling time is reduced by a factor of 2. Compare the transient and steady-state performance of the uncompensated and compensated systems. Describe any problems with your design.

5. Consider the unity feedback system shown in Figure P9.1 with

$$G(s) = \frac{K}{(s + 2)^2(s + 3)}$$

a. Find the location of the dominant poles to yield a 1.6-second settling time and an overshoot of 25%.
b. If a compensator with a zero at -1 is used to achieve the conditions of part (**a**), what must the angular contribution of the compensator pole be?
c. Find the location of the compensator pole.
d. Find the gain required to meet the requirements of part (**a**).
e. Find the location of other closed-loop poles for the compensated system.
f. Discuss the validity of your second-order approximation.
g. Simulate the compensated system to check your design.

6. The unity feedback system shown in Figure P9.1 with

$$G(s) = \frac{K}{s^2}$$

is to be designed for a settling time of 1.667 seconds and a 16.3% overshoot. If the compensator zero is placed at -1, do the following:
a. Find the coordinates of the dominant poles.
b. Find the compensator pole.
c. Find the system gain.
d. Find the location of all nondominant poles.
e. Estimate the accuracy of your second-order approximation.
f. Evaluate the steady-state error characteristics.
*g. Simulate the system and evaluate the actual transient response characteristics for a step input.

7. Given the unity feedback system of Figure P9.1, with

$$G(s) = \frac{K(s + 3)}{(s + 1)(s + 2)(s + 4)(s + 5)}$$

do the following:
a. Sketch the root locus.
b. Find the coordinates of the dominant poles for which $\zeta = 0.8$.
c. Find the gain for which $\zeta = 0.8$.
d. If the system is to be cascade-compensated so that $T_s = 4/3$ seconds and $\zeta = 0.8$, find the compensator pole if the compensator zero is at -2.5.
e. Discuss the validity of your second-order approximation.
*f. Simulate the compensated and uncompensated systems and compare the results to those expected.

8. Consider the unity feedback system of Figure P9.1 with

$$G(s) = \frac{K}{s(s + 10)(s + 30)}$$

The system is operating at 15% overshoot. Design a compensator to decrease the settling time by a factor of 3 without affecting the percent overshoot and do the following:

a. Evaluate the uncompensated system's dominant poles, gain, and settling time.

b. Evaluate the compensated system's dominant poles, gain, and settling time.

c. Evaluate the compensator's pole and zero.

*d. Simulate the compensated and uncompensated systems' step response.

9. The unity feedback system shown in Figure P9.1 with

$$G(s) = \frac{K}{(s + 10)(s^2 + 4s + 5)}$$

is operating with 37.23% overshoot, and the compensator zero is at -5.

a. Find the transfer function of a cascade compensator, the system gain, and the dominant pole location that will cut the settling time in half.

b. Find other poles and zeros and discuss your second-order approximation.

*c. Simulate both the uncompensated and compensated systems to see the effect of your compensator.

10. Given the unity feedback system of Figure P9.1, with

$$G(s) = \frac{K}{(s + 2)(s + 4)(s + 6)(s + 8)}$$

find the transfer function of a lead compensator that will yield a settling time 0.5 seconds shorter than that of the uncompensated system, with a damping ratio of 0.5. The compensator zero is at -5. Also, find the compensated system's gain. Justify any second-order approximations or verify the design through simulation.

11. A unity feedback control system has the following forward transfer function:

$$G(s) = \frac{K}{s^2(s + 10)}$$

a. Design a compensator to yield dominant poles with a damping ratio of 0.357 and a natural frequency of 1.6 rad/s. Be sure to specify the value of K.

b. Estimate the expected percent overshoot and settling time.

c. Is your second-order approximation valid?

*d. Simulate and compare the transient response of the compensated system to the predicted transient response.

12. For the unity feedback system of Figure P9.1, with

$$G(s) = \frac{K}{(s^2 + 20s + 101)(s + 20)}$$

the damping ratio for the dominant poles is to be 0.4, and the settling time is to be 0.5 seconds.

a. Find the coordinates of the dominant poles.

b. Find the location of the compensator zero if the compensator pole is at -15.

c. Find the required system gain.

d. Compare the performance of the uncompensated and compensated systems.

*e. Simulate the system to check your design.

13. Given the uncompensated unity feedback system of Figure P9.1, with

$$G(s) = \frac{K}{s(s + 1)(s + 3)}$$

do the following:

a. Design a compensator to yield the following specifications: settling time = 2.86 seconds, percent overshoot = 4.32%; the steady-state error is to be improved by a factor of 2 over the uncompensated system.

b. Compare the transient and steady-state error specifications of the uncompensated and compensated systems.

c. Compare the gains of the uncompensated and compensated systems.

d. Discuss the validity of your second-order approximation.

*e. Simulate the uncompensated and compensated systems and verify the specifications.

14. Consider the unity feedback system of Figure P9.1, with

$$G(s) = \frac{K}{(s + 3)(s + 5)}$$

a. Show that the system cannot operate with a settling time of $2/3$ second and a percent overshoot of 1.5% with a simple gain adjustment.

b. Design a lead-compensator so that the system meets the transient response characteristics of part (**a**).

c. Specify the compensator's pole, zero, and the required gain.

15. For the unity feedback system in Figure P9.1, with

$$G(s) = \frac{K}{(s + 1)(s + 4)}$$

design a PID controller that will yield a peak time of 1.047 seconds and a damping ratio of 0.8, with zero error for a step input.

16. Consider the unity feedback system in Figure P9.1, with

$$G(s) = \frac{K}{(s + 2)(s + 4)}$$

The system is operated with 4.32% overshoot. In order to improve the steady-state error, K_p is to be increased by at least a factor of 5. A lag compensator of the form

$$G_c(s) = \frac{(s + 0.5)}{(s + 0.1)}$$

is to be used.

a. Find the gain required for both the compensated and uncompensated systems.

b. Find the value of K_p for both the compensated and uncompensated systems.

c. Estimate the percent overshoot and settling time for both the compensated and uncompensated systems.

d. Discuss the validity of the second-order approximation used for your results in (**c**).

*e. Simulate the step response for the uncompensated and compensated systems. What do you notice about the compensated system's response?

*f. Design a lead compensator that will correct the objection you notice in part (e).

17. a. Find the transfer function of a motor whose torque-speed curve and load are given in Figure P9.2.

b. Design a tachometer compensator to yield a damping ratio of 0.5 for a position control employing a power amplifier of gain 1, and a preamplifier of gain 5000.

c. Compare the transient and steady-state characteristics of the uncompensated system and the compensated system.

Figure P9.2

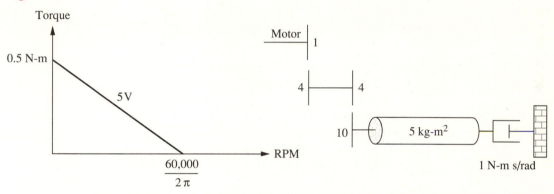

18. If the system of Figure P9.3 operates with a damping ratio of 0.517 for the dominant second-order poles, find the location of all closed-loop poles and zeros.

Figure P9.3

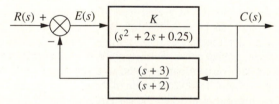

19. You are given the motor whose transfer function is shown in Figure P9.4(a).

a. If this motor were the forward transfer function of a unity feedback system, calculate the percent overshoot and settling time that could be expected.

b. You want to improve the closed-loop response. Since the motor constants cannot be changed, and since you cannot use a different motor, an amplifier and tachometer are inserted into the loop as shown in Figure P9.4(b). Find the values of K_1 and K_2 to yield a percent overshoot of 25% and a settling time of 0.2 seconds.

Figure P9.4

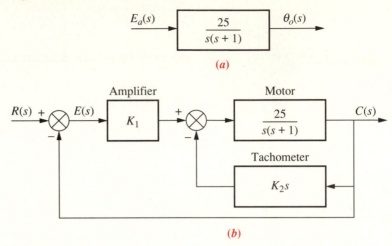

(a)

(b)

c. Evaluate the steady-state error specifications for both the uncompensated and compensated systems.

20. A position control is to be designed with a 20% overshoot and a settling time of 2 seconds. You have on hand an amplifier with a variable gain of K_1 and a pole at -20, and a power amplifier of unity gain with which to drive the motor. Two ten-turn pots are available to convert shaft position into voltage. A voltage of $\pm 5\pi$ volts is placed across the pots. A dc motor whose transfer function is of the form

$$\frac{\theta_o(s)}{E_a(s)} = \frac{K}{s(s + a)}$$

is also available. The transfer function of the motor is found experimentally as follows. The motor and geared load are driven open-loop by applying a large, short, rectangular pulse to the armature. An oscillogram of the response shows that the motor reached 63% of its final output value at $1/2$ second after the application of the pulse. Further, with a constant 10 volts dc applied to the armature, the constant output speed was 100 rad/s.

a. Draw a complete block diagram of the system, specifying the transfer function of each component when the system is operating with 20% overshoot.

b. What will the steady-state error be for a unit ramp input?

c. Determine the transient response characteristics.

d. If tachometer feedback is used around the motor, as shown in Figure P9.5, find the tachometer and the amplifier gain to meet the original specifications. Summarize the transient and steady-state characteristics.

21. Given the system shown in Figure P9.6, find the values of K and K_t so that the closed-loop dominant poles will have a damping ratio of 0.5, and the underdamped poles of the minor loop will have a damping ratio of 0.8.

Figure P9.5

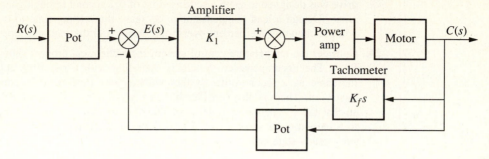

Figure P9.6

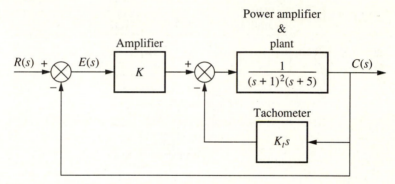

22. Given the system in Figure P9.7, find the values of K and K_t so that the closed-loop system will have a 4.32% overshoot, and the minor loop will have a damping ratio of 0.8. Compare the specifications of the system without tachometer compensation to the performance with tachometer compensation.

Figure P9.7

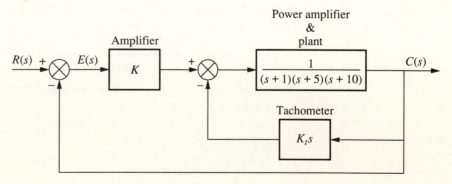

23. In Problem 35 of Chapter 8, a head-position control system for a floppy disk drive was designed to yield a settling time of 0.1 second through gain adjustment alone. Design a lead compensator to decrease the settling time to 0.05 seconds without changing the percent overshoot. Also, find the required loop gain.

24. Consider the temperature control system for a chemical process shown in Figure P9.8. The uncompensated system is operating with a rise time approximately the same as a second-order system with a peak time of 16 seconds and 5% overshoot. There is also considerable steady-state error. Design a PID controller so that the compensated system will have a rise time approximately equivalent to a second-order system with a peak time of 8 seconds and 5% overshoot and zero steady-state error for a step input.

Figure P9.8 Chemical Process Temperature Control System

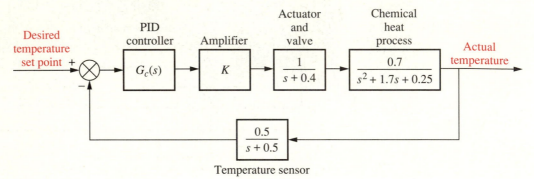

Temperature sensor

25. *Chapter-objective problem:* Consider the schematic for the antenna azimuth position control system shown in Figure P2.36. In Chapter 8, Problem 34, you were asked to design, via gain adjustment, an 8-second settling time.
   a. For your solution to Problem 34 in Chapter 8, evaluate the percent overshoot and the value of the appropriate static error constant.
   b. Design a cascade compensator to reduce the percent overshoot by a factor of 4 and the settling time by a factor of 2. Also, improve the appropriate static error constant by a factor of 2.

26. The heading control system for the UFSS vehicle is shown in Figure P9.9. The minor loop contains the rudder and vehicle dynamics, and the major loop relates output and input heading (Johnson, 1980). Find the values of K and K_t so that the minor-loop dominant poles have a damping ratio of 0.6 and the major-loop dominant poles have a damping ratio of 0.5.

27. Identify and realize the following controllers with operational amplifiers.

   a. $\dfrac{s + 0.01}{s}$

   b. $s + 2$

   c. $\dfrac{(s + 0.01)(s + 2)}{s}$

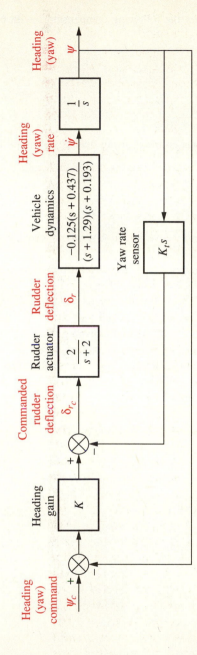

Figure P9.9 Simplified Block Diagram of the Heading Control System for the UFSS Vehicle (Adapted from Johnson, H., et al. *Unmanned Free-Swimming Submersible (UFSS) System Description*. NRL Memorandum Report 4393. Naval Research Laboratory, Washington, D.C.)

28. Identify and realize the following compensators with passive networks.

    a. $\dfrac{s + 0.1}{s + 0.01}$

    b. $\dfrac{s + 2}{s + 5}$

    c. $\left(\dfrac{s + 0.1}{s + 0.01}\right)\left(\dfrac{s + 1}{s + 10}\right)$

29. Repeat the previous problem, using operational amplifiers.

BIBLIOGRAPHY

Budak, A. *Passive and Active Network Analysis and Synthesis*. Houghton Mifflin, Boston, 1974.

D'Azzo, J. J., and Houpis, C. H. *Feedback Control System Analysis and Synthesis*. 2d ed. McGraw-Hill, New York, 1966.

Dorf, R. C. *Modern Control Systems*. 5th ed. Addison-Wesley, Reading, Mass., 1989.

Hostetter, G. H.; Savant, C. J., Jr.; and Stefani, R. T. *Design of Feedback Control Systems*. 2d ed. Saunders College Publishing, New York, 1989.

Johnson, H., et al. *Unmanned Free-Swimming Submersible (UFSS) System Description*. NRL Memorandum Report 4393. Naval Research Laboratory, Washington, D.C., 1980.

Ogata, K. *Modern Control Engineering*. 2d ed. Prentice Hall, Englewood Cliffs, N.J., 1990.

Van de Vegte, J. *Feedback Control Systems*. 2d ed. Prentice Hall, Englewood Cliffs, N.J., 1990.

10

FREQUENCY RESPONSE
METHODS

10.1 Introduction

The root locus method for transient design, steady-state design, and stability was covered in Chapters 8 and 9. In Chapter 8 we covered the simple case of design through gain adjustment, where a trade-off was made between a desired transient response and a desired steady-state error. In Chapter 9, the need for this trade-off was eliminated by using compensation networks, so that transient and steady-state errors could be separately specified and designed. Further, a desired transient response no longer had to be on the original system's root locus.

This chapter and Chapter 11 present the design of feedback control systems through gain adjustment and compensation networks from another perspective — that of frequency response. The results of frequency response compensation techniques are not new or different from the results of root locus techniques.

Frequency response methods, developed by Nyquist and Bode in the 1930s, are older than the root locus method, which was discovered by Evans in 1948 (Nyquist, 1932; Bode, 1945). The older method, which is covered in this chapter, is not as intuitive as the root locus. However, frequency response yields a new vantage point from which to view feedback control systems. This technique has distinct advantages in the following situations: (1) when modeling transfer functions from physical data, as shown in Figure 10.1; (2) when designing lead compensators to meet a steady-state error requirement and a transient response requirement; (3) when finding the stability of nonlinear systems; and (4) in settling ambiguities when sketching a root locus.

Chapter Objective

Given the antenna azimuth position control system shown in Figure 2.2, the student, using frequency response techniques, will be able to find the range of system gain that will keep the system stable. Further, given a system with a specified gain, the student will be able to find the percent overshoot, settling time, peak time, and rise time by using frequency response techniques.

We first discuss the concept of frequency response, define frequency response, derive analytical expressions for the frequency response, plot the frequency re-

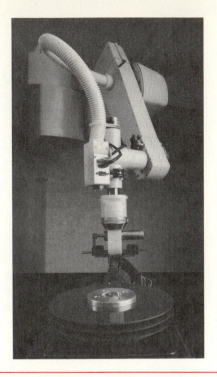

Figure 10.1 An Intelledex Model 660 Robot Holds a Disk Drive's Magnetic Head in Place During Frequency Response Testing to Obtain the Transfer Function (Courtesy of Intelledex)

sponse, develop ways of sketching the frequency response, and then apply the concept to control system design and analysis.

The Concept and Definition of Frequency Response

In the steady-state, sinusoidal inputs to a linear system generate sinusoidal responses of the same frequency. Even though these responses are of the same frequency as the input, they differ in amplitude and phase angle from the input. These differences are functions of frequency.

Before defining frequency response, let us look at a convenient representation of sinusoids. Sinusoids can be represented as complex numbers called *phasors*. The magnitude of the complex number is the amplitude of the sinusoid and the angle of the complex number is the phase angle of the sinusoid. Thus, $M_1 \cos(\omega t + \phi_1)$ can be represented as $M_1 \angle \phi_1$ where the frequency, ω, is implicit.

Since a system causes both the amplitude and phase angle of the input to be changed, we can therefore think of the system itself represented by a complex number, defined so that the product of the input phasor and the system function yields the phasor representation of the output.

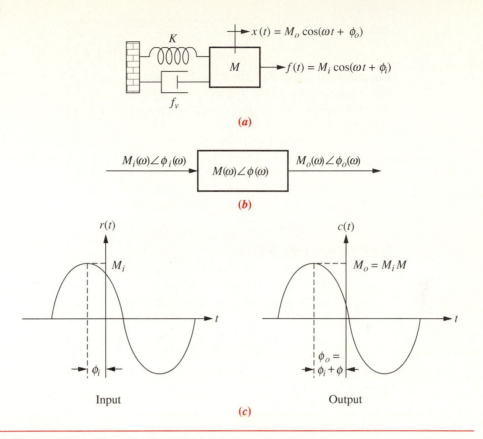

Figure 10.2 A View of a Sinusoidal Frequency Response Function: (*a*) System; (*b*) Transfer Function; (*c*) Input and Output Waveforms

Consider the mechanical system of Figure 10.2(*a*). If the input force, $f(t)$, is sinusoidal, the steady-state output response, $x(t)$, of the system is also sinusoidal and at the same frequency as the input. In Figure 10.2(*b*) the input and output sinusoids are represented by complex numbers, or phasors, $M_i(\omega)\angle\phi_i(\omega)$ and $M_o(\omega)\angle\phi_o(\omega)$, respectively. Here, the M's are the amplitudes of the sinusoids and the ϕ's are the phase angles of the sinusoids as shown in Figure 10.2(*c*). Assume that the system is represented by the complex number $M(\omega)\angle\phi(\omega)$. The output steady-state sinusoid is found by multiplying the complex number representation of the input times the complex number representation of the system. Thus, the steady-state output sinusoid is

$$M_o(\omega)\angle\phi_o(\omega) = M_i(\omega)M(\omega)\angle[\phi_i(\omega) + \phi(\omega)] \qquad (10.1)$$

From Equation 10.1 we see that the system function is given by

$$M(\omega) = \frac{M_o(\omega)}{M_i(\omega)} \qquad (10.2)$$

and

$$\phi(\omega) = \phi_o(\omega) - \phi_i(\omega) \tag{10.3}$$

Equations 10.2 and 10.3 form our definition of frequency response. We call $M(\omega)$ the *magnitude frequency response* and $\phi(\omega)$ the *phase frequency response*. The combination of the magnitude and phase frequency response is called the *frequency response* and is $M(\omega)\angle\phi(\omega)$.

In other words, we define the magnitude frequency response to be the ratio of the output sinusoid's magnitude to the input sinusoid's magnitude. We define the phase response to be the difference in phase angle between the output and the input sinusoids. Both responses are functions of frequency and apply only to the steady-state sinusoidal response of the system.

Development of Analytical Expressions for Frequency Response

Now that we have defined frequency response, let us obtain the analytical expression for it (Nilsson, 1986). Later in the chapter we will use this analytical expression to determine stability, transient response, and steady-state error. Figure 10.3 shows a system, $G(s)$, with the Laplace transform of a general sinusoid, $r(t) = A\cos\omega t + B\sin\omega t = \sqrt{A^2 + B^2}\cos\left(\omega t - \tan^{-1}\dfrac{B}{A}\right)$ as the input. We can represent the input as a phasor in three ways: (1) in polar form, $M_i\angle\phi_i$, where $M_i = \sqrt{A^2 + B^2}$ and $\phi_i = -\tan^{-1}\dfrac{B}{A}$; (2) in rectangular form, $A - jB$; and (3) using Euler's formula, $M_i e^{j\phi_i}$.

We now solve for the forced response portion of $C(s)$, from which we evaluate the frequency response. From Figure 10.3,

$$C(s) = \frac{As + B\omega}{(s^2 + \omega^2)}\,G(s) \tag{10.4}$$

We separate the forced solution from the transient solution by performing a partial fraction expansion on Equation 10.4. Thus,

$$
\begin{aligned}
C(s) &= \frac{As + B\omega}{(s + j\omega)(s - j\omega)}G(s) \\
&= \frac{K_1}{s + j\omega} + \frac{K_2}{s - j\omega} + \text{Partial fraction terms from } G(s) \tag{10.5}
\end{aligned}
$$

Figure 10.3 A System with a Sinusoidal Input

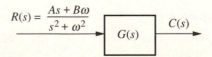

where

$$K_1 = \frac{As + B\omega}{s - j\omega}G(s)\bigg|_{s \to -j\omega} = \frac{1}{2}(A + jB)G(-j\omega) = \frac{1}{2}M_i e^{-j\phi_i}M_G e^{-j\phi_G}$$

$$= \frac{M_i M_G}{2}e^{-j(\phi_1 + \phi_G)} \tag{10.6a}$$

$$K_2 = \frac{As + B\omega}{s + j\omega}G(s)\bigg|_{s \to +j\omega} = \frac{1}{2}(A - jB)G(j\omega) = \frac{1}{2}M_i e^{j\phi_i}M_G e^{j\phi_G}$$

$$= \frac{M_i M_G}{2}e^{j(\phi_i + \phi_G)} = K_1^* \tag{10.6b}$$

For Equations 10.6, K_1^* is the complex conjugate of K_1,

$$M_G = |G(j\omega)| \tag{10.7}$$

and

$$\phi_G = \text{angle of } G(j\omega) \tag{10.8}$$

The steady-state response is that portion of the partial fraction expansion that comes from the input waveform's poles, or just the first two terms of Equation 10.5. Hence the sinusoidal steady-state output, $C_{ss}(s)$, is

$$C_{ss}(s) = \frac{K_1}{s + j\omega} + \frac{K_2}{s - j\omega} \tag{10.9}$$

Substituting Equations 10.6 into Equation 10.9, we obtain

$$C_{ss}(s) = \frac{\dfrac{M_i M_G}{2}e^{-j(\phi_i + \phi_G)}}{s + j\omega} + \frac{\dfrac{M_i M_G}{2}e^{j(\phi_i + \phi_G)}}{s - j\omega} \tag{10.10}$$

Taking the inverse Laplace transformation, we obtain

$$c(t) = M_i M_G \left(\frac{e^{-j(\omega t + \phi_i + \phi_G)} + e^{j(\omega t + \phi_i + \phi_G)}}{2}\right)$$

$$= M_i M_G \cos(\omega t + \phi_i + \phi_G) \tag{10.11}$$

which can be represented in phasor form as $M_o \angle \phi_o = (M_1 \angle \phi_1)(M_G \angle \phi_G)$, where $M_G \angle \phi_G$ is the frequency response function. But, from Equations 10.7 and 10.8, $M_G \angle \phi_G = G(j\omega)$. In other words, the frequency response of a system whose transfer function is $G(s)$ is

$$G(j\omega) = G(s)\big|_{s \to j\omega} \tag{10.12}$$

Plotting Frequency Response

$G(j\omega) = M_G(\omega)\angle \phi_G(\omega)$ can be plotted in several ways; two of them are (1) as a function of frequency, with separate magnitude and phase plots; and (2) as a polar plot, where the phasor length is the magnitude, and the phasor angle is the phase. When plotting separate magnitude and phase plots, the magnitude curve can be plotted in decibels (dB) vs $\log \omega$, where dB $= 20 \log M$.[1] The phase curve is plotted as phase angle vs $\log \omega$. The motivation for these plots will be shown in Section 10.2.

Using the concepts covered in Section 8.1, data for the plots also can be obtained using vectors on the s-plane drawn from the poles and zeros of $G(s)$ to the imaginary axis. Here, the magnitude response at a particular frequency is the product of the vector lengths from the zeros of $G(s)$ divided by the product of the vector lengths from the poles of $G(s)$ drawn to points on the imaginary axis. The phase response is the sum of the angles from the zeros of $G(s)$ minus the sum of the angles from the poles of $G(s)$ drawn to points on the imaginary axis. Performing this operation for successive points along the imaginary axis yields the data for the frequency response. Remember, each point is equivalent to substituting that point, $s = j\omega_1$, into $G(s)$ and evaluating its value.

The plots also can be made from a computer program that calculates the frequency response. For example, the root locus program in Appendix C can be used with test points that are on the imaginary axis. The calculated K value at each frequency is the reciprocal of the scaled magnitude response, and the calculated angle is, directly, the phase angle response at that frequency.

The following example demonstrates how to obtain an analytical expression for frequency response and make a plot of the result.

Example 10.1 Find and plot the analytical expression for the frequency response of a system.

Problem Find the analytical expression for the magnitude frequency response and the phase frequency response for the system $G(s) = \dfrac{1}{(s + 2)}$. Also, plot both the separate magnitude and phase diagrams and the polar plot.

Solution First, substitute $s = j\omega$ in the system function and obtain $G(j\omega) = \dfrac{1}{(j\omega + 2)} = \dfrac{(2 - j\omega)}{(\omega^2 + 4)}$. The magnitude of this complex number, $|G(j\omega)| = M(\omega) = \dfrac{1}{\sqrt{(\omega^2 + 4)}}$, is the magnitude frequency response. The phase angle of $G(j\omega)$, $\phi(\omega) = -\tan^{-1}(\omega/2)$, is the phase frequency response.

$G(j\omega)$ can be plotted in two ways: (1) in separate magnitude and phase plots, and (2) in a polar plot. Figure 10.4 shows separate magnitude and phase diagrams, where the magnitude diagram is $20 \log M(\omega) = 20 \log \dfrac{1}{\sqrt{\omega^2 + 4}}$ vs $\log \omega$, and the phase diagram is $\phi(\omega) = -\tan^{-1}(\omega/2)$ vs $\log \omega$. The polar plot, shown in Figure 10.5, is a plot of $M(\omega)\angle\phi(\omega) = \dfrac{1}{\sqrt{\omega^2 + 4}}\angle - \tan^{-1}(\omega/2)$ for different ω.

---

[1] Throughout this book, log is used to mean $\log_{10}$, or logarithm to the base 10.

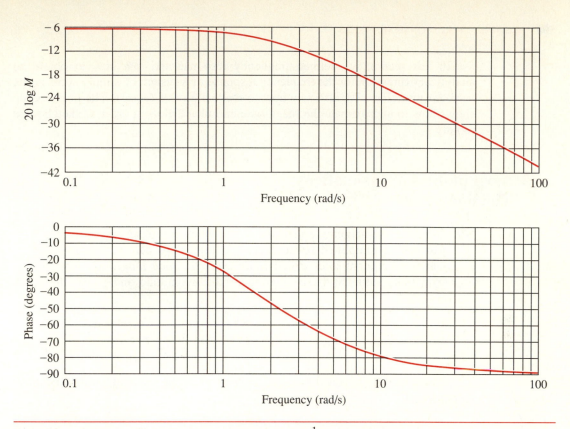

Figure 10.4 Frequency Response Plots for $G(s) = \dfrac{1}{(s + 2)}$: Separate Magnitude and Phase

Figure 10.5 Frequency Response Plot for $G(s) = \dfrac{1}{(s + 2)}$: Polar Plot

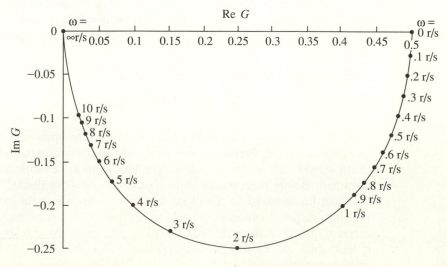

Note: r/s = rad/s

In this section we defined frequency response and saw how to obtain an analytical expression for the frequency response of a system by simply substituting $s = j\omega$ into $G(s)$. We also saw how to make a plot of $G(j\omega)$. The next section shows how to approximate the magnitude and phase plots in order to sketch them rapidly.

10.2 Asymptotic Approximations: The Bode Plots

Sketching the log-magnitude and phase frequency response curves as functions of $\log \omega$ can be simplified because these relationships can be approximated as a sequence of straight lines called *Bode plots* or *Bode diagrams*. Straight-line approximations simplify the evaluation of the magnitude and phase frequency response.

Consider the following transfer function:

$$G(s) = \frac{K(s + z_1)(s + z_2) \cdots (s + z_n)}{s^m(s + p_1)(s + p_2) \cdots (s + p_n)} \tag{10.13}$$

The magnitude frequency response is the product of the magnitude frequency responses of each term, or

$$|G(j\omega)| = \left. \frac{K|(s + z_1)||(s + z_2)| \cdots |(s + z_n)|}{|s^m||(s + p_1)||(s + p_2)| \cdots |(s + p_n)|} \right|_{s \to j\omega} \tag{10.14}$$

Thus, if we know the magnitude response of each pole and zero term, we can find the total magnitude response. The process can be simplified by working with the logarithm of the magnitude since the zero terms' magnitude responses would be added and the pole terms' magnitude responses would be subtracted, rather than respectively multiplied or divided, to yield the logarithm of the total magnitude response. Converting the magnitude response into dB, we obtain

$$20 \log |G(j\omega)| = 20 \log K + 20 \log |(s + z_1)| + 20 \log |(s + z_2)|$$
$$+ \cdots - 20 \log |s^m| - 20 \log |(s + p_1)| - \cdots \Big|_{s \to j\omega} \tag{10.15}$$

Thus, if we knew the response of each term, the algebraic sum would yield the total response in dB. Further, if we could make an approximation of each term that would consist only of straight lines, graphic addition of terms would be greatly simplified. Before proceeding, let us look at the phase response.

From Equation 10.13 the phase frequency response is the *sum* of the phase frequency response curves of the zero terms minus the *sum* of the phase frequency response curves of the pole terms. Again, since the phase response is the sum of individual terms, straight-line approximations to these individual responses simplify graphic addition.

Let us now show how to approximate the frequency response of simple pole and zero terms by straight-line approximations. Later, we will show how to combine these responses to sketch the frequency response of more complicated functions. In subsequent sections, after a discussion of the Nyquist stability criterion, we will learn how to use the Bode plots for the analysis and design of stability and transient response.

Bode Plots for $G(s) = (s + a)$

Consider a function, $G(s) = (s + a)$, for which we want to sketch separate logarithmic magnitude and phase response plots. Letting $s = j\omega$, we have

$$G(j\omega) = (j\omega + a) = a\left(1 + j\frac{\omega}{a}\right) \tag{10.16}$$

At low frequencies, when ω approaches zero,

$$G(j\omega) \approx a \tag{10.17}$$

The magnitude response in dB is

$$20 \log M = 20 \log a \tag{10.18}$$

where $M = |G(j\omega)|$ and is a constant. Equation 10.18 is shown plotted in Figure 10.6(a) from $\omega = 0.01a$ to a.

At high frequencies, where $\omega \gg a$, Equation 10.16 becomes

$$G(j\omega) \approx a\left(\frac{j\omega}{a}\right) = a\left(\frac{\omega}{a}\right)\angle 90° = \omega\angle 90° \tag{10.19}$$

The magnitude response in dB is

$$20 \log M = 20 \log a + 20 \log \frac{\omega}{a} = 20 \log \omega \tag{10.20}$$

where $a < \omega < \infty$. Notice from the middle term that the high-frequency approximation is equal to the low-frequency approximation when $\omega = a$ and increases for $\omega > a$.

If we plot dB, $20 \log M$, against $\log \omega$, then Equation 10.20 becomes a straight line:

$$y = 20x \tag{10.21}$$

where $y = 20 \log M$, and $x = \log \omega$. The line has a slope of 20 when plotted as dB vs $\log \omega$.

Since each doubling of frequency causes $20 \log \omega$ to increase by 6 dB, the line rises at an equivalent slope of 6 dB/octave, where an *octave* is a doubling of frequency. This rise begins at $\omega = a$, where the low-frequency approximation equals the high-frequency approximation.

We call the straight-line approximations *asymptotes*. The low-frequency approximation is called the *low-frequency asymptote,* and the high-frequency approximation is called the *high-frequency asymptote.* The frequency, a, is called the

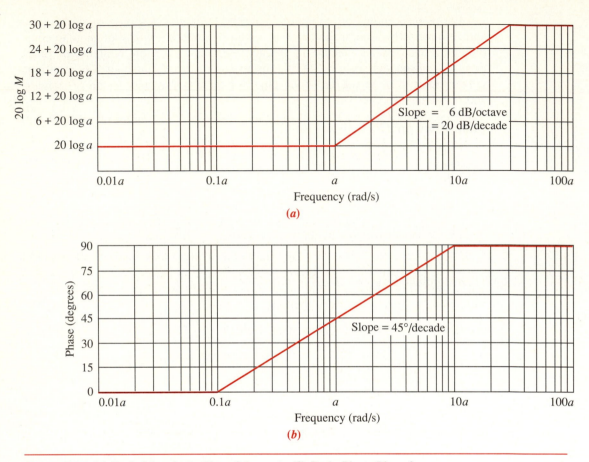

Figure 10.6 (**a**) Bode Magnitude Plot of $(s + a)$; (**b**) Bode Phase Plot of $(s + a)$

break frequency because it is the break between the low- and the high-frequency asymptotes.

Sometimes it is convenient to draw the line over a decade rather than an octave, where a *decade* is 10 times the initial frequency. Over one decade, $20 \log \omega$ increases by 20 dB. Thus a slope of 6 dB/octave is equivalent to a slope of 20 dB/decade. The plot is shown in Figure 10.6(a) from $\omega = 0.01a$ to $100a$.

Let us now turn to the phase response, which can be drawn as follows. At the break frequency, a, Equation 10.16 shows the phase to be $45°$. At low frequencies, Equation 10.17 shows that the phase is $0°$. At high frequencies, Equation 10.19 shows that the phase is $90°$. To draw the curve, start one decade (1/10) below the break frequency, $0.1a$, with $0°$ phase, and draw a line of slope $+45°$/decade passing through $45°$ at the break frequency and continuing to $90°$ one decade above the break frequency, $10a$. The resulting phase diagram is shown in Figure 10.6(b).

It is often convenient to *normalize* the magnitude and *scale* the frequency so that the log-magnitude plot will be 0 dB at a break frequency of unity. Normalizing and scaling helps when comparing different frequency response plots. To normalize $(s + a)$, we factor out the quantity a and form $a\left(\dfrac{s}{a} + 1\right)$. The frequency is scaled by defining a new frequency variable, $s_1 = \dfrac{s}{a}$. Then the magnitude is divided by the quantity a to yield 0 dB at the break frequency. Hence, the normalized and scaled function is $(s_1 + 1)$. To obtain the original frequency response, the magnitude and frequency are multiplied by the quantity a.

We now use the concepts of normalization and scaling to compare the asymptotic approximation to the actual magnitude and phase plot for $(s + a)$. Table 10.1 shows the comparison for the normalized and scaled frequency response of $(s + a)$. Notice that the actual magnitude curve is never greater than 3.01 dB from the asymptotes. This maximum difference occurs at the break frequency. The maximum difference for the phase curve is 5.71°, which occurs at the decades above and below the break frequency. For convenience, the data in Table 10.1 is plotted in Figures 10.7 and 10.8.

Table 10.1 Comparison of the Asymptotic and Actual Normalized and Scaled Frequency Response for $(s + a)$

Frequency $\dfrac{}{a}$	$20 \log \dfrac{M}{a}$ (dB)		Phase (degrees)	
(rad/s)	Asymptotic	Actual	Asymptotic	Actual
0.02	0	0.00	0.00	1.15
0.04	0	0.01	0.00	2.29
0.06	0	0.02	0.00	3.43
0.08	0	0.03	0.00	4.57
0.1	0	0.04	0.00	5.71
0.2	0	0.17	13.55	11.31
0.4	0	0.64	27.09	21.80
0.6	0	1.34	35.02	30.96
0.8	0	2.15	40.64	38.66
1	0	3.01	45.00	45.00
2	6	6.99	58.55	63.43
4	12	12.30	72.09	75.96
6	15.56	15.68	80.02	80.54
8	18	18.13	85.64	82.87
10	20	20.04	90.00	84.29
20	26.02	26.03	90.00	87.14
40	32.04	32.04	90.00	88.57
60	35.56	35.56	90.00	89.05
80	38.06	38.06	90.00	89.28
100	40	40.00	90.00	89.43

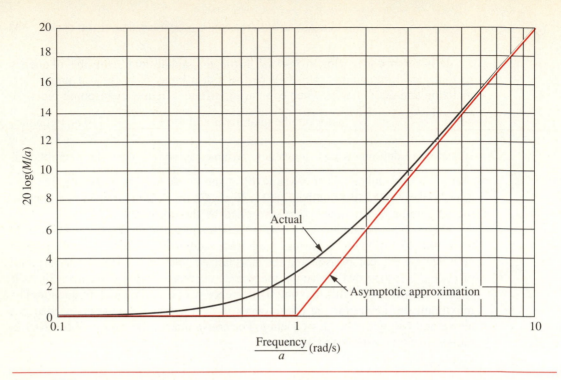

Figure 10.7 Comparison of the Asymptotic and Actual Normalized and
Scaled Magnitude Response of $(s + a)$

Figure 10.8 Comparison of the Asymptotic and Actual Normalized and
Scaled Phase Response of $(s + a)$

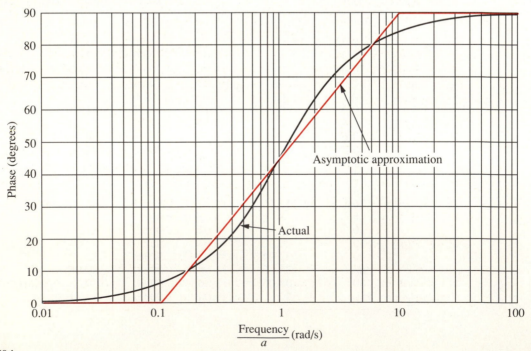

We now find the Bode plots for other common transfer functions.

Bode Plots for $G(s) = 1/(s + a)$

Let us find the Bode plots for the following transfer function:

$$G(s) = \frac{1}{(s + a)} = \frac{1}{a\left(\dfrac{s}{a} + 1\right)} \tag{10.22}$$

This function has a low-frequency asymptote of $20 \log(1/a)$ which is found by letting the frequency, s, approach zero. The Bode plot is constant until the break frequency, a radians/second, is reached. The plot is then approximated by the high-frequency asymptote found by letting s approach ∞. Thus, at high frequencies,

$$G(j\omega) = \frac{1}{a\left(\dfrac{s}{a}\right)}\Bigg|_{s \to j\omega} = \frac{1}{a\left(\dfrac{j\omega}{a}\right)} = \frac{\dfrac{1}{a}}{\dfrac{\omega}{a}} \angle -90° = \frac{1}{\omega} \angle -90° \tag{10.23}$$

or, in dB,

$$20 \log M = 20 \log \frac{1}{a} - 20 \log \frac{\omega}{a} = -20 \log \omega \tag{10.24}$$

Notice from the middle term that the high-frequency approximation equals the low-frequency approximation when $\omega = a$, and decreases for $\omega > a$. This result is similar to Equation 10.20, except the slope is negative rather than positive. The Bode log-magnitude diagram will decrease at a rate of 6 dB/octave rather than increase at a rate of 6 dB/octave after the break frequency.

The phase plot is the negative of the previous example since the function is the inverse. The phase begins at $0°$ and reaches $-90°$ at high frequencies, going through $-45°$ at the break frequency. Both the Bode normalized and scaled log-magnitude and phase plots are shown in Figure 10.9(d).

Bode Plots for $G(s) = s$

Our next function, $G(s) = s$, has only a high-frequency asymptote. Letting $s = j\omega$, the magnitude is $20 \log \omega$, which is the same as Equation 10.20. Hence, the Bode magnitude plot is a straight line drawn with a $+6$ dB/octave slope passing through zero dB when $\omega = 1$. The phase plot, which is a constant $+90°$, is shown with the magnitude plot in Figure 10.9(a).

Bode Plots for $G(s) = 1/s$

The frequency response of the inverse of the preceding function, $G(s) = 1/s$, is shown in Figure 10.9(b) and is a straight line with a -6 dB/octave slope passing through zero dB at $\omega = 1$. The Bode phase plot is equal to a constant $-90°$.

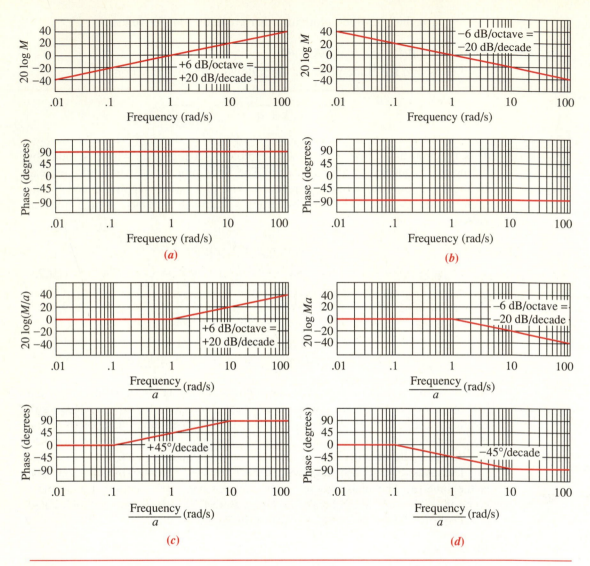

Figure 10.9 Normalized and Scaled Bode Plots for (**a**) $G(s) = s$; (**b**) $G(s) = 1/s$; (**c**) $G(s) = (s + a)$; (**d**) $G(s) = 1/(s + a)$

We have covered four functions that have first-order polynomials in s in the numerator or denominator. Before proceeding to second-order polynomials, let us look at an example of drawing the Bode plots for a function that consists of the product of first-order polynomials in the numerator and denominator. The plots will be made by adding together the individual frequency response curves.

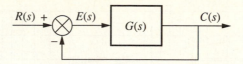

Figure 10.10 Closed-Loop Unity Feedback System

Example 10.2

Make a Bode plot of a function that consists of the product of first-order polynomials in the numerator and denominator.

Problem Draw the Bode plots for the system shown in Figure 10.10, where $G(s) = \dfrac{K(s + 3)}{s(s + 1)(s + 2)}$.

Solution We will make a Bode plot for the open-loop function $G(s) = \dfrac{K(s + 3)}{s(s + 1)(s + 2)}$. The Bode plot is the sum of the Bode plots for each first-order term.

First, determine that the break frequencies are at -1, -2, and -3. The magnitude plot should begin a decade below the lowest break frequency and extend a decade above the highest break frequency. Hence, we choose 0.1 radians to 100 radians, or three decades, as the extent of our plot.

At $\omega = 0.1$, the low-frequency value of the function is found using the low-frequency values for all of the $(s + a)$ terms (i.e., $s = 0$) and the actual value for the s term in the denominator. Thus,

$$G(j0.1) \approx \frac{3K}{(0.1)(1)(2)} = 15K \tag{10.25}$$

The effect of K is to move the magnitude curve up (increasing K) or down (decreasing K) by an amount $20 \log K$. K has no effect upon the phase curve. If we choose $K = 1$, the magnitude plot can be denormalized later for any value of K that is calculated or known.[2]

Figure 10.11(a) shows each component part of the Bode log-magnitude frequency response. Summing the component parts yields the composite plot shown in Figure 10.11(b). The Bode magnitude plot starts at $\omega = 0.1$ with a value of $20 \log 15 = 23.52$ dB and decreases immediately at a rate of -6 dB/octave due to the s term in the denominator. At $\omega = 1$, the $(s+1)$ term in the denominator begins its 6 dB/octave

---

[2]An alternate approach to plotting the log-magnitude curve is to let $G(s) = \dfrac{\dfrac{3}{2}K\left(\dfrac{s}{3} + 1\right)}{s(s + 1)\left(\dfrac{s}{2} + 1\right)}$.

This form yields $G(s) = \dfrac{1}{s}$ as the low-frequency asymptote if the magnitude is divided by $\dfrac{3}{2}K$. Using this form, the plot is normalized to go through 0 dB at $\omega = 1$ rad/s. The normalized plot is then denormalized by adding $20 \log \dfrac{3}{2}K$.

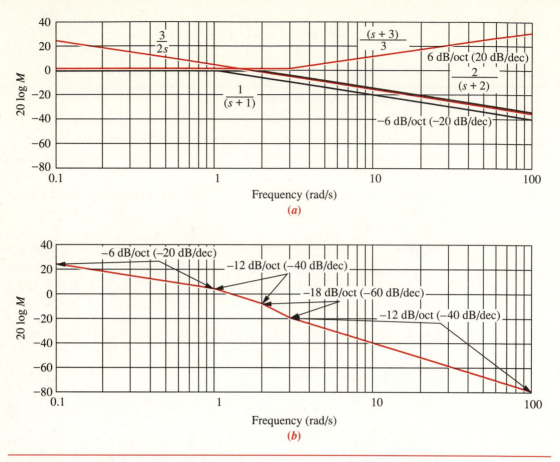

Figure 10.11 Bode Log-Magnitude Plot for Example 10.2: (*a*) Component Parts; (*b*) Composite

downward slope and causes an additional 6 dB/octave negative slope, or a total of -12 dB/octave. At $\omega = 2$, the term $(s + 2)$ begins its -6 dB/octave slope, adding yet another -6 dB/octave to the resultant plot, or a total of -18 dB/octave slope that continues until $\omega = 3$. At this frequency, the $(s + 3)$ term in the numerator begins its positive 6 dB/octave slope. The resultant magnitude plot, therefore, changes from a slope of -18 dB/octave to -12 dB/octave at $\omega = 3$, and continues at that slope since there are no more break frequencies.

The slopes are easily drawn by sketching straight-line segments decreasing by 6 dB over an octave (or 20 dB over a decade). For example, the initial -6 dB/octave (-20 dB/decade) slope is drawn from 23.52 dB at $\omega = 0.1$ to 3.52 dB (a 20 dB decrease) at $\omega = 1$. The -12 dB/octave (-40 dB/decade) slope starting at $\omega = 1$ is drawn by sketching a line segment from 3.52 dB at $\omega = 1$ to -36.48 dB (a 40 dB decrease) at $\omega = 10$ and using only the portion from $\omega = 1$ to $\omega = 2$. The next slope of -18 dB/octave (-60 dB/decade) is drawn by first sketching a line segment from $\omega = 2$ to $\omega = 20$ (1 decade) that drops down by 60 dB, and using only that

Table 10.2 Bode Phase Plot: Slope Contribution from Each Pole and Zero in Example 10.2

	Start: pole at -1	Start: pole at -2	Start: zero at -3	End: pole at -1	End: pole at -2	End: zero at -3
Frequency (rad/s)	0.1	0.2	0.3	10	20	30
Pole at -1	-45	-45	-45	0		
Pole at -2		-45	-45	-45	0	
Zero at -3			45	45	45	0
Total slope (deg/dec)	-45	-90	-45	0	45	0

portion of the line from $\omega = 2$ to $\omega = 3$. The final slope is drawn by sketching a line segment from $\omega = 3$ to $\omega = 30$ (1 decade) that drops by 40 dB. This slope continues to the end of the plot.

Phase is handled similarly. However, the existence of breaks a decade below and a decade above the break frequency requires a little more bookkeeping. Table 10.2 shows the starting and stopping frequencies of the $45°$/decade slope for each of the poles and zeros. For example, reading across for the pole at -2, we see that the $-45°$ slope starts at a frequency of 0.2 and ends at 20. Filling in the rows for each pole and then summing the columns yields the slope portrait of the resulting phase plot. Looking at the row marked *Total slope*, we see that the phase plot will have a slope of $-45°$/decade from a frequency of 0.1 to 0.2. The slope will then increase to $-90°$/decade from 0.2 to 0.3. The slope will return to $-45°$/decade from 0.3 to 10 rad/s. A slope of 0 ensues from 10 to 20 rad/s, followed by a slope of $+45°$/decade from 20 to 30 rad/s. Finally, from 30 rad/s to infinity, the slope is $0°$/decade.

The resulting component and composite phase plots are shown in Figure 10.12. Since the pole at the origin yields a constant $-90°$ phase shift, the plot begins at $-90°$ and follows the slope portrait just described.

Bode Plots for $G(s) = s^2 + 2\zeta\omega_n s + \omega_n^2$

Now that we have covered Bode plots for first-order systems, we turn to the Bode log-magnitude and phase plots for second-order polynomials in s. The second-order polynomial is of the form

$$G(s) = s^2 + 2\zeta\omega_n s + \omega_n^2 \tag{10.26}$$

or its reciprocal. Unlike the first-order frequency response approximation, the difference between the asymptotic approximation and the actual frequency response can be great for some values of ζ. A correction to the Bode diagrams can be made to improve the accuracy. We first derive the asymptotic approximation and then show the difference between the asymptotic approximation and the actual frequency response curves.

At low frequencies, Equation 10.26 becomes

$$G(s) \approx \omega_n^2 = \omega_n^2 \angle 0° \tag{10.27}$$

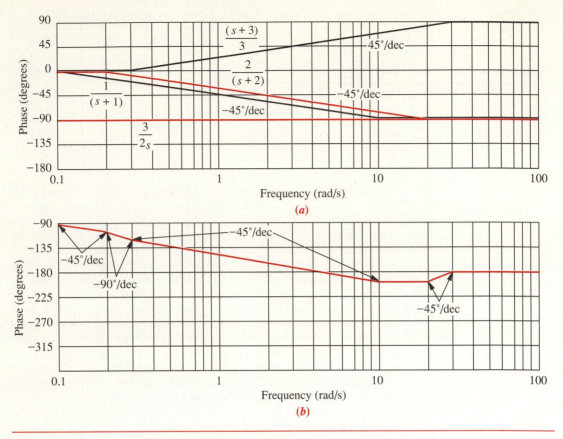

Figure 10.12 Bode Phase Plot for Example 10.2: (*a*) Component Parts;
(*b*) Composite

The magnitude, M, in dB at low frequencies is therefore

$$20 \log M = 20 \log |G(j\omega)| = 20 \log \omega_n^2 \qquad (10.28)$$

At high frequencies,

$$G(s) \approx s^2 \qquad (10.29)$$

or

$$G(j\omega) \approx -\omega^2 = \omega^2 \angle 180° \qquad (10.30)$$

The log-magnitude is

$$20 \log M = 20 \log |G(j\omega)| = 20 \log \omega^2 = 40 \log \omega \qquad (10.31)$$

Equation 10.31 is a straight line with twice the slope of a first-order term (Equation 10.20). Its slope is 12 dB/octave, or 40 dB/decade.

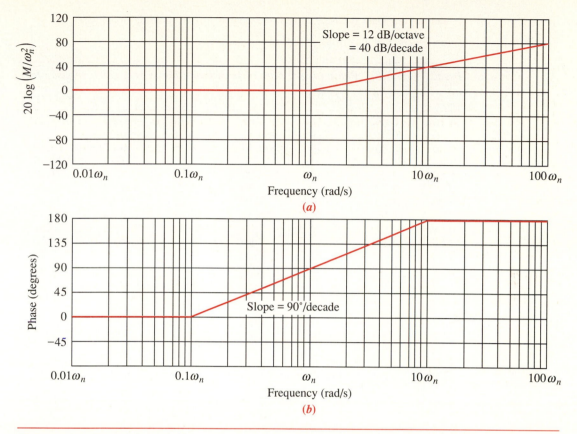

Figure 10.13 Bode Asymptotes for Normalized $G(s) = s^2 + 2\zeta\omega_n s + \omega_n^2$: **(a)** Magnitude; **(b)** Phase

The low-frequency asymptote (Equation 10.27) and the high-frequency asymptote (Equation 10.31) are equal when $\omega = \omega_n$. Thus, ω_n is the break frequency for the second-order polynomial. Figure 10.13(a) shows the asymptotes for the magnitude plot.

We now draw the phase plot shown in Figure 10.13(b). It is $0°$ at low frequencies (Equation 10.27) and $180°$ at high frequencies (Equation 10.30). To find the phase at the natural frequency, first evaluate $G(j\omega)$,

$$G(j\omega) = s^2 + 2\zeta\omega_n s + \omega_n^2 \big|_{s \to j\omega} = (\omega_n^2 - \omega^2) + j2\zeta\omega_n\omega \qquad (10.32)$$

and then find the function value at the natural frequency by substituting $\omega = \omega_n$. Since the result is $j2\zeta\omega_n^2$, the phase at the natural frequency is $+90°$. Thus, the phase plot increases at a rate of $90°/$decade from $0.1\omega_n$ to $10\omega_n$ and passes through $90°$ at ω_n.

Table 10.3 Data for Normalized and Scaled Log-Magnitude and Phase Plot for $(s^2 + 2\zeta\omega_n s + \omega_n^2)$. Mag $= 20 \log \dfrac{M}{\omega_n^2}$

Freq. $\dfrac{\omega}{\omega_n}$	Mag (dB) $\zeta = 0.1$	Phase (deg) $\zeta = 0.1$	Mag (dB) $\zeta = 0.2$	Phase (deg) $\zeta = 0.2$	Mag (dB) $\zeta = 0.3$	Phase (deg) $\zeta = 0.3$
0.10	−0.09	1.16	−0.08	2.31	−0.07	3.47
0.20	−0.35	2.39	−0.32	4.76	−0.29	7.13
0.30	−0.80	3.77	−0.74	7.51	−0.65	11.19
0.40	−1.48	5.44	−1.36	10.78	−1.17	15.95
0.50	−2.42	7.59	−2.20	14.93	−1.85	21.80
0.60	−3.73	10.62	−3.30	20.56	−2.68	29.36
0.70	−5.53	15.35	−4.70	28.77	−3.60	39.47
0.80	−8.09	23.96	−6.35	41.63	−4.44	53.13
0.90	−11.64	43.45	−7.81	62.18	−4.85	70.62
1.00	−13.98	90.00	−7.96	90.00	−4.44	90.00
1.10	−10.34	133.67	−6.24	115.51	−3.19	107.65
1.20	−6.00	151.39	−3.73	132.51	−1.48	121.43
1.30	−2.65	159.35	−1.27	143.00	0.35	131.50
1.40	0.00	163.74	0.92	149.74	2.11	138.81
1.50	2.18	166.50	2.84	154.36	3.75	144.25
1.60	4.04	168.41	4.54	157.69	5.26	148.39
1.70	5.67	169.80	6.06	160.21	6.64	151.65
1.80	7.12	170.87	7.43	162.18	7.91	154.26
1.90	8.42	171.72	8.69	163.77	9.09	156.41
2.00	9.62	172.41	9.84	165.07	10.19	158.20
3.00	18.09	175.71	18.16	171.47	18.28	167.32
4.00	23.53	176.95	23.57	173.91	23.63	170.91
5.00	27.61	177.61	27.63	175.24	27.67	172.87
6.00	30.89	178.04	30.90	176.08	30.93	174.13
7.00	33.63	178.33	33.64	176.66	33.66	175.00
8.00	35.99	178.55	36.00	177.09	36.01	175.64
9.00	38.06	178.71	38.07	177.42	38.08	176.14
10.00	39.91	178.84	39.92	177.69	39.93	176.53

Bode Plots for $G(s) = \dfrac{\omega_n^2}{(s^2 + 2\zeta\omega_n s + \omega_n^2)}$

The reciprocal function, $1/G(s)$, for the second-order system is similarly found. The magnitude curve breaks at the natural frequency and decreases at a rate of -12 dB/octave (-40 dB/decade). The phase plot is $0°$ at low frequencies. At $0.1\omega_n$ it begins a decrease of $-90°$/decade and continues until $\omega = 10\omega_n$, where it levels off at $-180°$.

Corrections to Second-Order Bode Plots

Let us now examine the error between the actual response and the asymptotic approximation of the second-order polynomial. Whereas the first-order polynomial has a disparity of no more than 3.01 dB magnitude and 5.71° phase, the second-order function may have a greater disparity, which depends upon the value of ζ.

From Equation 10.32, the actual magnitude and phase for $G(s) = s^2 + 2\zeta\omega_n s + \omega_n^2$ are, respectively,

Freq. $\frac{\omega}{\omega_n}$	Mag (dB) $\zeta = 0.5$	Phase (deg) $\zeta = 0.5$	Mag (dB) $\zeta = 0.7$	Phase (deg) $\zeta = 0.7$	Mag (dB) $\zeta = 1$	Phase (deg) $\zeta = 1$
0.10	−0.04	5.77	0.00	8.05	0.09	11.42
0.20	−0.17	11.77	0.00	16.26	0.34	22.62
0.30	−0.37	18.25	0.02	24.78	0.75	33.40
0.40	−0.63	25.46	0.08	33.69	1.29	43.60
0.50	−0.90	33.69	0.22	43.03	1.94	53.13
0.60	−1.14	43.15	0.47	52.70	2.67	61.93
0.70	−1.25	53.92	0.87	62.51	3.46	69.98
0.80	−1.14	65.77	1.41	72.18	4.30	77.32
0.90	−0.73	78.08	2.11	81.42	5.15	83.97
1.00	0.00	90.00	2.92	90.00	6.02	90.00
1.10	0.98	100.81	3.83	97.77	6.89	95.45
1.20	2.13	110.14	4.79	104.68	7.75	100.39
1.30	3.36	117.96	5.78	110.76	8.60	104.86
1.40	4.60	124.44	6.78	116.10	9.43	108.92
1.50	5.81	129.81	7.76	120.76	10.24	112.62
1.60	6.98	134.27	8.72	124.85	11.03	115.99
1.70	8.10	138.03	9.66	128.45	11.80	119.07
1.80	9.17	141.22	10.56	131.63	12.55	121.89
1.90	10.18	143.95	11.43	134.46	13.27	124.48
2.00	11.14	146.31	12.26	136.97	13.98	126.87
3.00	18.63	159.44	19.12	152.30	20.00	143.13
4.00	23.82	165.07	24.09	159.53	24.61	151.93
5.00	27.79	168.23	27.96	163.74	28.30	157.38
6.00	31.01	170.27	31.12	166.50	31.36	161.08
7.00	33.72	171.70	33.80	168.46	33.98	163.74
8.00	36.06	172.76	36.12	169.92	36.26	165.75
9.00	38.12	173.58	38.17	171.05	38.28	167.32
10.00	39.96	174.23	40.00	171.95	40.09	168.58

$$M = \sqrt{(\omega_n^2 - \omega^2)^2 + (2\zeta\omega_n\omega)^2} \tag{10.33}$$

$$\text{Phase} = \tan^{-1}\frac{2\zeta\omega_n\omega}{\omega_n^2 - \omega^2} \tag{10.34}$$

These relationships are tabulated in Table 10.3 for a range of values of ζ and plotted in Figures 10.14 and 10.15 along with the asymptotic approximations for normalized magnitude and scaled frequency. In Figure 10.14, which is scaled to the square of the natural frequency, the normalized log-magnitude at the scaled natural frequency is $-20\log 2\zeta$. The reader should verify that the actual magnitude at the unscaled natural frequency is $-20\log 2\zeta\omega_n^2$. Table 10.3 and Figures 10.14 and 10.15 can be used to improve accuracy when drawing the Bode diagrams.

Before proceeding with an example, let us mention that the frequency response of the inverse function, $1/(s^2 + 2\zeta\omega_n s + \omega_n^2)$ follows the same derivation. The magnitude correction at the scaled natural frequency is $+20\log 2\zeta$.

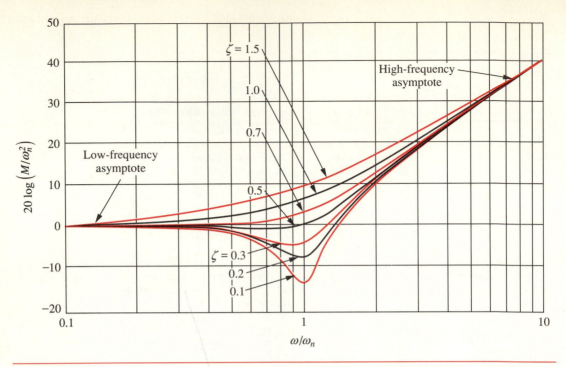

Figure 10.14 Normalized Log-Magnitude Response for $(s^2 + 2\zeta\omega_n s + \omega_n^2)$

Figure 10.15 Phase Response for $(s^2 + 2\zeta\omega_n s + \omega_n^2)$

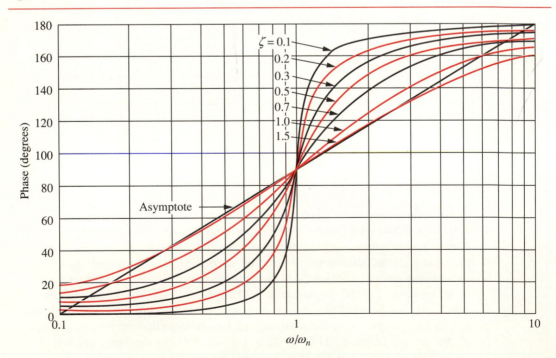

Example 10.3   Draw Bode plots for transfer functions with first- and second-order terms.

Problem   Draw the Bode log-magnitude and phase plots of $G(s)$ for the unity feedback system shown in Figure 10.10, where $G(s) = \dfrac{(s + 3)}{(s + 2)(s^2 + 2s + 25)}$.

Solution   The Bode log-magnitude diagram is shown in Figure 10.16(b) and is the sum of the individual first- and second-order terms of $G(s)$ shown in Figure 10.16(a). We will solve this problem by adding the slopes of these component parts, beginning and ending at the appropriate frequencies. The low-frequency value for $G(s)$, found by letting $s = 0$, is $3/50$, or -24.44 dB. The Bode magnitude plot starts out at this value and continues until the first break frequency at 2 rad/s. Here, the pole at -2 yields a -20 dB/decade slope until the next break at 3 rad/s. The zero at -3 causes an upward slope of $+20$ dB/decade, which, when added to the previous -20 dB/decade curve, gives a net slope of 0. At a frequency of 5 rad/s, the second-order term initiates a 40 dB/decade downward slope, which continues to infinity.

Figure 10.16 Bode Magnitude Plot for $G(s) = \dfrac{(s + 3)}{(s + 2)(s^2 + 2s + 25)}$:
(***a***) Components; (***b***) Composite

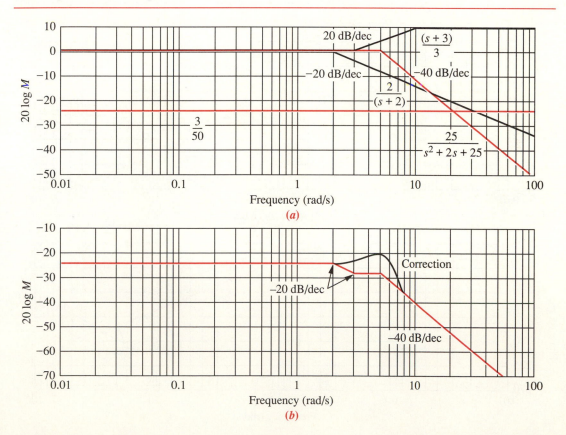

Table 10.4 Phase Diagram Slopes for Example 10.3

	Start: pole at -2	Start: zero at -3	Start: ω_n at -5	End: pole at -2	End: zero at -3	End: ω_n at -5
Frequency (rad/s)	0.2	0.3	0.5	20	30	50
Pole at -2	-45	-45	-45	0		
Zero at -3		45	45	45	0	
ω_n at -5			-90	-90	-90	0
Total slope (deg/dec)	-45	0	-90	-45	-90	0

The correction to the log-magnitude curve due to the underdamped second-order term can be found by plotting a point $20 \log 2\zeta$ above the asymptotes at the natural frequency. Since $\zeta = 0.2$ for the second-order term in the denominator of $G(s)$, the correction is 7.96 dB. Points close to the natural frequency can be corrected by taking the values from the inverse of the curves of Figure 10.14.

We now turn to the phase plot. Table 10.4 is formed to determine the progression of slopes on the phase diagram. The first-order pole at -2 yields a phase angle

Figure 10.17 Bode Phase Plot for $G(s) = \dfrac{(s + 3)}{(s + 2)(s^2 + 2s + 25)}$: **(a)** Components; **(b)** Composite

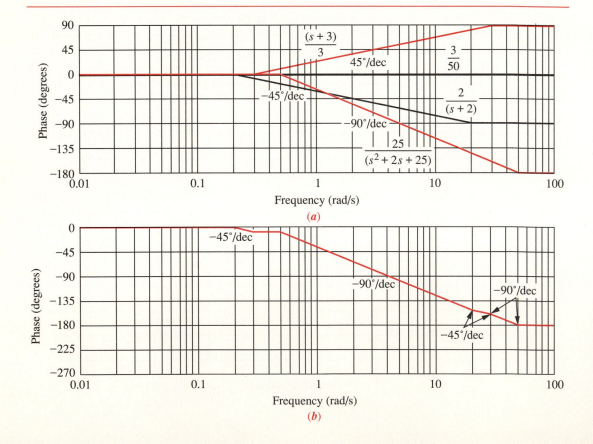

that starts at $0°$ and ends at $-90°$ via a $-45°$/decade slope starting a decade below its break frequency and ending a decade above its break frequency. The first-order zero yields a phase angle that starts at $0°$ and ends at $+90°$ via a $+45°$/decade slope starting a decade below its break frequency and ending a decade above its break frequency. The second-order poles yield a phase angle that starts at $0°$ and ends at $-180°$ via a $-90°$/decade slope starting a decade below their natural frequency ($\omega_n = 5$) and ending a decade above their natural frequency. The slopes, shown in Figure 10.17(a), are summed over each frequency range, and the final Bode phase plot is shown in Figure 10.17(b).

In this section, we learned how to construct Bode log-magnitude and Bode phase plots. The Bode plots are separate magnitude and phase frequency response curves for a system, $G(s)$. In the next section, we develop the Nyquist criterion for stability, which makes use of the frequency response of a system. The Bode plots can then be used to determine the stability of a system.

10.3 Derivation of the Nyquist Criterion for Stability

The Nyquist criterion relates the stability of a closed-loop system to the open-loop frequency response and open-loop pole location. Thus, knowledge of the open-loop system's frequency response yields information about the stability of the closed-loop system. This concept is similar to the root locus, where we began with information about the open-loop system, its poles and zeros, and developed transient and stability information about the closed-loop system.

Although the Nyquist criterion will yield stability information at first, we will extend the concept to transient response and steady-state errors. Thus, frequency response techniques are an alternate approach to the root locus.

Derivation of the Nyquist Criterion

Consider the system of Figure 10.18. The Nyquist criterion can tell us how many closed-loop poles are in the right half-plane. Before deriving the criterion, let us establish four important concepts that will be used during the derivation: (1)

Figure 10.18 Closed-Loop Control System

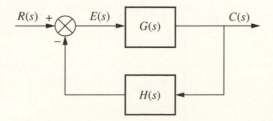

the relationship between the poles of $1 + G(s)H(s)$ and the poles of $G(s)H(s)$; (2) the relationship between the zeros of $1 + G(s)H(s)$ and the poles of the closed-loop transfer function, $T(s)$; (3) the concept of *mapping* points; and (4) the concept of mapping *contours*.

Letting

$$G(s) = \frac{N_G}{D_G} \tag{10.35a}$$

$$H(s) = \frac{N_H}{D_H} \tag{10.35b}$$

we find

$$G(s)H(s) = \frac{N_G N_H}{D_G D_H} \tag{10.36a}$$

$$1 + G(s)H(s) = 1 + \frac{N_G N_H}{D_G D_H} = \frac{D_G D_H + N_G N_H}{D_G D_H} \tag{10.36b}$$

$$T(s) = \frac{G(s)}{1 + G(s)H(s)} = \frac{N_G D_H}{D_G D_H + N_G N_H} \tag{10.36c}$$

From Equations 10.36, we conclude that (1) the poles of $1 + G(s)H(s)$ are the same as the poles of $G(s)H(s)$, the open-loop system, and (2) the zeros of $1 + G(s)H(s)$ are the same as the poles of $T(s)$, the closed-loop system.

Next, let us define the term *mapping*. If we take a complex number on the s-plane and substitute it into a function, $F(s)$, another complex number will result. This process is called *mapping*. For example, substituting $s = 4 + j3$ into the function $(s^2 + 2s + 1)$ yields $16 + j30$. We say that $4 + j3$ maps into $16 + j30$ through the function $(s^2 + 2s + 1)$.

Finally, we discuss the concept of mapping *contours*. Consider the collection of points, called a *contour,* shown in Figure 10.19 as contour A. Also, assume that

$$F(s) = \frac{(s - z_1)(s - z_2) \cdots}{(s - p_1)(s - p_2) \cdots} \tag{10.37}$$

Figure 10.19 Mapping Contour A Through Function $F(s)$ to Contour B

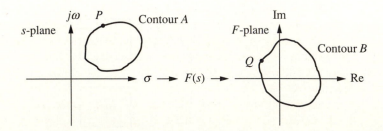

Contour A can be mapped through $F(s)$ into contour B by substituting each point of contour A into the function $F(s)$ and plotting the resulting complex numbers. For example, point P in Figure 10.19 maps into point Q through the function $F(s)$.

The vector approach to performing the calculation, covered in Section 8.1, can be used as an alternative. Some examples of contour mapping are shown in Figure 10.20 for some simple $F(s)$. The mapping of each point is defined by complex arithmetic, where the resulting complex number, R, is evaluated from the complex numbers represented by V, as shown in the last column of Figure 10.20. The student should verify that if we assume a clockwise direction for mapping the points on contour A, then contour B maps in a clockwise direction if $F(s)$ in Figure 10.20 has just zeros, and in a counterclockwise direction if $F(s)$ has just poles. Also, the student should verify that if the pole or zero of $F(s)$ is enclosed by contour A, the mapping encircles the origin. In the last case of Figure 10.20, the pole and zero rotation cancel, and the mapping does not encircle the origin.

Let us now begin the derivation of the criterion for stability. We will show that a unique relationship exists between the number of poles of $F(s)$ contained inside contour A, the number of zeros of $F(s)$ contained inside contour A, and the number of counterclockwise encirclements of the origin for the mapping of contour B. We will then show how this interrelationship can be used to determine the stability of closed-loop systems. This method of determining stability is called the *Nyquist criterion*.

Let us first assume that $F(s) = 1 + G(s)H(s)$, with the picture of the poles and zeros of $1 + G(s)H(s)$ as shown in Figure 10.21 near contour A. Hence, $R = \dfrac{V_1 V_2}{V_3 V_4 V_5}$. As each point P of the contour A is substituted into $1 + G(s)H(s)$, a mapped point will result on contour B. Assuming that $F(s) = 1 + G(s)H(s)$ has two zeros and three poles, each parenthetical term of Equation 10.37 is a vector in Figure 10.21. As we move around contour A in a clockwise direction, each vector of Equation 10.37 that lies inside contour A will appear to undergo a complete rotation, or a change in angle of $360°$. On the other hand, each vector drawn from the poles and zeros of $1 + G(s)H(s)$ that exist outside contour A will appear to oscillate and return to its previous position, undergoing a net angular change of $0°$.

Each pole or zero factor of $1 + G(s)H(s)$ whose vector undergoes a complete rotation around contour A must yield a change of $360°$ in the resultant, R, or a complete rotation of the mapping of contour B. If we move in a clockwise direction along contour A, each zero inside contour A yields a rotation in the clockwise direction, while each pole inside contour A yields a rotation in the counterclockwise direction since poles are in the denominator of Equation 10.37.

Thus, $N = P - Z$, where N equals the number of counterclockwise rotations of contour B about the origin; P equals the number of poles of $1 + G(s)H(s)$ inside contour A, and Z equals the number of zeros of $1 + G(s)H(s)$ inside contour A.

Since the poles shown in Figure 10.21 are poles of $1 + G(s)H(s)$, we know from Equations 10.36 that they are also the poles of $G(s)H(s)$ and are known.

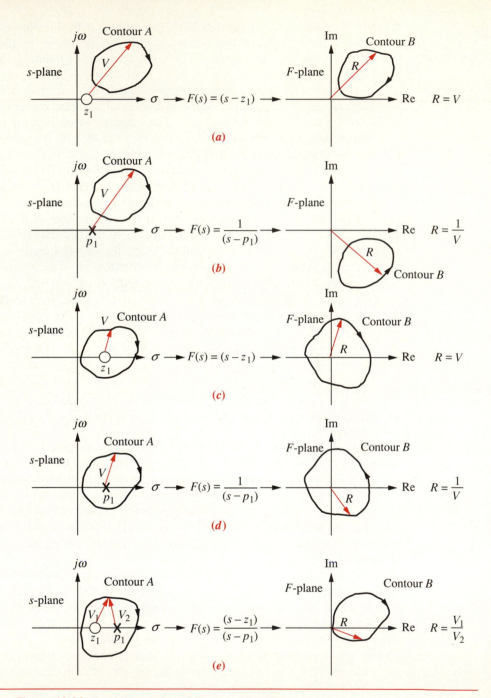

Figure 10.20 Some Examples of Contour Mapping

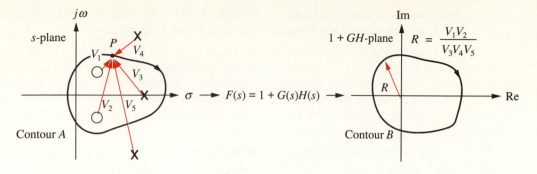

Figure 10.21 Vector Representation of Mapping

But, since *the zeros shown in Figure 10.21 are the zeros of* $1 + G(s)H(s)$, we know from Equations 10.36 that *they are also the poles of the closed-loop system and are not known.*

Hence, $N = P - Z$, or alternately $Z = P - N$, tells us that the number of closed-loop poles inside the contour (which is the same as the zeros inside the contour) equals the number of open-loop poles of $G(s)H(s)$ inside the contour minus the number of counterclockwise rotations of the mapping about the origin.

If we extend the contour to include the entire right half-plane, as shown in Figure 10.22, we can then count the number of right-half-plane, closed-loop poles inside contour A and determine a system's stability. Since we can count the number of open-loop poles, P, inside the contour, which are the same as the right-half-plane poles of $G(s)H(s)$, the only problem remaining is how to obtain the mapping and find N.

Since all of the poles and zeros of $G(s)H(s)$ are known, what if we map through $G(s)H(s)$ instead of $1 + G(s)H(s)$? The resulting contour is the same as a mapping through $1 + G(s)H(s)$, except that it is translated one unit to the left;

Figure 10.22 Contour Enclosing Right Half-Plane to Determine Stability

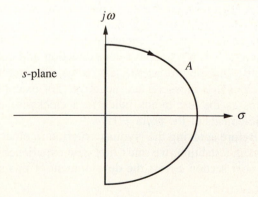

thus, we count rotations about -1 instead of rotations about the origin. Hence, the final statement of the Nyquist stability criterion is as follows:

If a contour, A, that encircles the entire right half-plane is mapped through $G(s)H(s)$, then the number of closed-loop poles, Z, in the right half-plane equals the number of open-loop poles, P, that are in the right half-plane minus the number of counterclockwise revolutions, N, around -1 of the mapping; that is, $Z = P - N$. The mapping is called the Nyquist diagram, or Nyquist plot, of $G(s)H(s)$.

We can now see why this method is classified as a frequency response technique. Around contour A in Figure 10.22, the mapping of the points on the $j\omega$-axis through the function $G(s)H(s)$ is the same as substituting $s = j\omega$ into $G(s)H(s)$ to form the frequency response function $G(j\omega)H(j\omega)$. We are thus finding the frequency response of $G(s)H(s)$ over that part of contour A on the positive $j\omega$-axis. In other words, part of the Nyquist diagram is the polar plot of the frequency response of $G(s)H(s)$.

Introduction to the Application of the Nyquist Criterion to Determine Stability

Before describing how to sketch a Nyquist diagram, let us look at some typical examples that use the Nyquist criterion to determine the stability of a system. These examples give us perspective prior to engaging in the details of mapping. Figure 10.23(a) shows a contour A that does not enclose closed-loop poles, that is, the zeros of $1 + G(s)H(s)$. The contour thus maps through $G(s)H(s)$ into a Nyquist diagram that does not encircle -1. Hence, $P = 0$, $N = 0$, and $Z = P - N = 0$. Since Z is the number of closed-loop poles inside contour A, which encircles the right half-plane, this system has no right-half-plane poles and is stable.

On the other hand, Figure 10.23(b) shows a contour A that, while it does not enclose open-loop poles, does generate two clockwise encirclements of -1. Thus, $P = 0$, $N = -2$, and the system is unstable; it has two closed-loop poles in the right half-plane since $Z = P - N = 2$. The two closed-loop poles are shown inside contour A in Figure 10.23(b) as zeros of $1 + G(s)H(s)$. The reader should keep in mind that the existence of these poles is not known a priori.

In this example notice that clockwise encirclements imply a negative value for N. The number of encirclements can be determined by drawing a test radius from -1 in any convenient direction and counting the number of times the Nyquist diagram crosses the test radius. Counterclockwise crossings are positive, and clockwise crossings are negative. For example, in Figure 10.23(b), contour B crosses the test radius twice in a clockwise direction. Hence, there are -2 encirclements of the point -1.

Before applying the Nyquist criterion to other examples in order to determine a system's stability, we must first gain experience in sketching Nyquist diagrams. The next section covers the development of this skill.

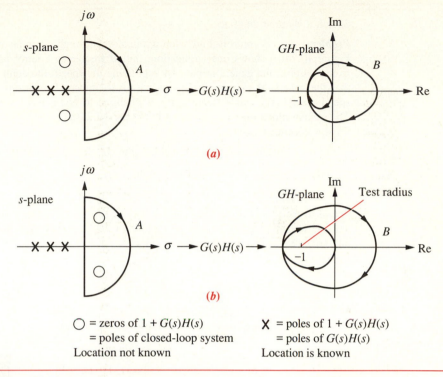

(a)

(b)

○ = zeros of $1 + G(s)H(s)$ **X** = poles of $1 + G(s)H(s)$
   = poles of closed-loop system   = poles of $G(s)H(s)$
   Location not known   Location is known

Figure 10.23 Examples of Mapping for the Nyquist
Criterion: (*a*) Contour Does Not Enclose Closed-Loop
Poles; (*b*) Contour Encloses Closed-Loop Poles

10.4 Sketching the Nyquist Diagram

The contour that encloses the right half-plane can be mapped through the function $G(s)H(s)$ by substituting points along the contour into $G(s)H(s)$. The points along the positive extension of the imaginary axis yield the polar frequency response of $G(s)H(s)$. Approximations can be made to $G(s)$ for points around the infinite semicircle by assuming that the vectors originate at the origin. Thus their length is infinite, and their angles are easily evaluated.

However, most of the time, a simple sketch of the Nyquist plot is all that is needed. A sketch can be obtained rapidly by looking at the vectors of $G(s)H(s)$ and their motion along the contour. In the examples that follow, we will stress this rapid method for sketching the Nyquist diagram. However, the examples will also include analytical expressions for $G(s)H(s)$ for each section of the contour to aid the reader in determining the shape of the Nyquist diagram.

Example 10.4   Sketch a Nyquist diagram.

Problem   Speed controls find wide application throughout industry and the home. Figure 10.24(*a*) shows one application: output frequency control of electrical power from a turbine and generator pair. By regulating the speed, the control system ensures that the generated frequency remains within tolerance. Deviations from the desired speed are sensed, and a steam valve is changed to compensate for the speed error. The system block diagram is shown in Figure 10.24(*b*). Sketch the Nyquist diagram for the system of Figure 10.24.

Solution   Conceptually, the Nyquist diagram is plotted by substituting the points of the contour shown in Figure 10.25(*a*) into $G(s) = \dfrac{500}{(s + 1)(s + 3)(s + 10)}$. This process is equivalent to performing complex arithmetic using the vectors of $G(s)$ drawn to the points of the contour, as shown in Figure 10.25(*a*) and (*b*). Each pole and zero term of $G(s)$ shown in Figure 10.24(*b*) is a vector in Figure 10.25(*a*) and (*b*). The resultant vector, R, found at any point along the contour is in general the product of the zero vectors divided by the product of the pole vectors (see Figure 10.25(*c*)). Thus, the magnitude of the resultant is the product of the zero lengths

Figure 10.24 (***a***) Turbine and Generator; (***b***) Block Diagram of Speed Control System for Example 10.4

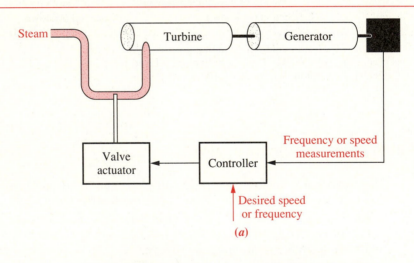

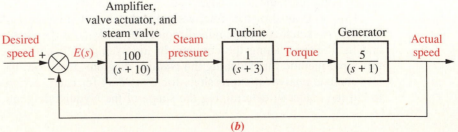

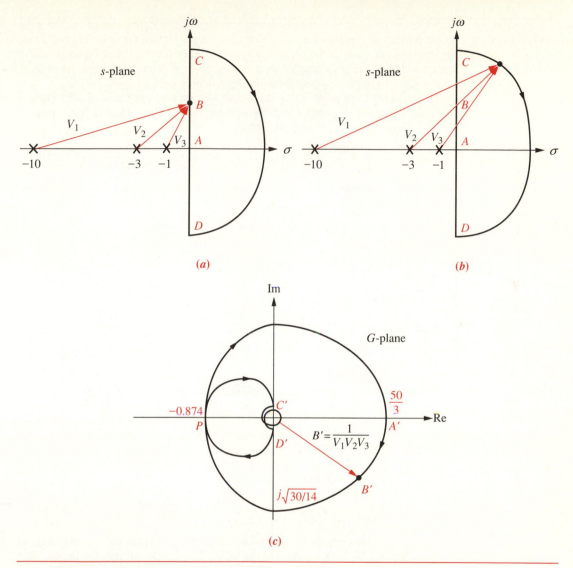

Figure 10.25 Complex Arithmetic Using Vectors for Evaluating the Nyquist Diagram for Example 10.4: (***a***) Vectors on Contour at Low Frequency; (***b***) Vectors on Contour around Infinity; (***c***) Nyquist Diagram

divided by the product of the pole lengths, and the angle of the resultant is the sum of the zero angles minus the sum of the pole angles.

As we move in a clockwise direction around the contour from point A to point C in Figure 10.25(a), the resultant angle goes from $0°$ to $-3 \times 90° = -270°$, or from A' to C' in Figure 10.25(c). Since the angles emanate from poles in the denominator of $G(s)$, the rotation or increase in angle is really a decrease in angle of the function $G(s)$; the poles gain $270°$ in a counterclockwise direction, which explains why the function loses $270°$.

While the resultant moves from A' to C' in Figure 10.25(c), its magnitude changes as the product of the zero lengths divided by the product of the pole lengths. Thus, the resultant goes from a finite value at zero frequency (at point A of Figure 10.25(a) there are three finite pole lengths) to zero magnitude at infinite frequency at point C (at point C of Figure 10.25(a) there are three infinite pole lengths).

The mapping from point A to point C can also be explained analytically. From A to C the collection of points along the contour is imaginary. Hence, from A to C, $G(s) = G(j\omega)$, or from Figure 10.24(b),

$$G(j\omega) = \frac{500}{(s+1)(s+3)(s+10)}\bigg|_{s \to j\omega} = \frac{500}{(-14\omega^2 + 30) + j(43\omega - \omega^3)} \quad (10.38)$$

Multiplying the numerator and denominator by the complex conjugate of the denominator, we obtain

$$G(j\omega) = 500\frac{(-14\omega^2 + 30) - j(43\omega - \omega^3)}{(-14\omega^2 + 30)^2 + (43\omega - \omega^3)^2} \quad (10.39)$$

At zero frequency, $G(j\omega) = 500/30 = 50/3$. Thus, the Nyquist diagram starts at $50/3$ at an angle of $0°$. As ω increases, the real part remains positive, and the imaginary part remains negative. At $\omega = \sqrt{30/14}$, the real part becomes negative. At $\omega = \sqrt{43}$, the Nyquist diagram crosses the negative real axis since the imaginary term goes to zero. The real value at the axis crossing, point P in Figure 10.25(c), found by substituting $\omega = \sqrt{43}$ into Equation 10.39, is -0.874. Continuing toward $\omega = \infty$, the real part is negative, and the imaginary part is positive. At infinite frequency, $G(j\omega) \approx 500j/\omega^3$, or approximately zero at $90°$.

Around the infinite semicircle from point C to point D, shown in Figure 10.25(b), the vectors rotate clockwise, each by $180°$. Hence, the resultant undergoes a counterclockwise rotation of $3 \times 180°$, starting at point C' and ending at point D' of Figure 10.25(c). Analytically, we can see this by assuming that around the infinite semicircle, the vectors originate approximately at the origin and have infinite length. For any point on the s-plane, the value of $G(s)$ can be found by representing each complex number in polar form, as follows:

$$G(s) = \frac{500}{(R_{-1}e^{j\theta_{-1}})(R_{-3}e^{j\theta_{-3}})(R_{-10}e^{j\theta_{-10}})} \quad (10.40)$$

where R_{-i} is the magnitude of the complex number $(s+i)$, and θ_{-i} is the angle of the complex number $(s+i)$. Around the infinite semicircle, all R_{-i} are infinite, and we can use our assumption to approximate the angles as if the vectors originated at the origin. Thus, around the infinite semicircle,

$$G(s) = \frac{500}{\infty \angle(\theta_{-1} + \theta_{-3} + \theta_{-10})} = 0\angle -(\theta_{-1} + \theta_{-3} + \theta_{-10}) \quad (10.41)$$

At point C in Figure 10.25(b), the angles are all $90°$. Hence the resultant is $0\angle -270°$, shown as point C' in Figure 10.25(c). Similarly, at point D, $G(s) = 0\angle +270°$, and maps into point D'. The reader can select intermediate points to verify the spiral whose radius vector approaches zero at the origin, as shown in Figure 10.25(c).

The negative imaginary axis can be mapped by realizing that the real part of $G(j\omega)H(j\omega)$ is always an even function, whereas the imaginary part of $G(j\omega)H(j\omega)$ is an odd function. That is, the real part will not change sign when negative values of ω are used, whereas the imaginary part will change sign. Thus, the mapping of the

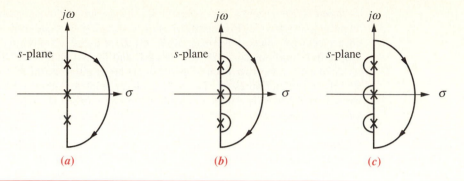

Figure 10.26 Selecting a Contour That Detours around Open-Loop Poles: (**a**) Pole on Contour; (**b**) Detour Right; (**c**) Detour Left

negative imaginary axis is a mirror image of the mapping of the positive imaginary axis. The mapping of the section of the contour from points D to A is drawn as a mirror image about the real axis of the mapping of points A to C.

In the previous example there were no open-loop poles situated along the contour enclosing the right half-plane. If such poles exist, then a detour around the poles on the contour is required; otherwise the mapping would go to infinity in an undetermined way, without angular information. Subsequently, a complete sketch of the Nyquist diagram could not be made, and the number of encirclements of -1 could not be found.

Let us assume a $G(s) = N(s)/sD(s)$ where $D(s)$ has imaginary roots. The s term in the denominator and the imaginary roots of $D(s)$ are poles of $G(s)$ that lie on the contour, as shown in Figure 10.26(*a*). To sketch the Nyquist diagram, the contour must detour around each open-loop pole lying on its path. The detour can be to the right of the pole, as shown in Figure 10.26(*b*), which makes it clear that each pole's vector rotates through $+180°$ as we move around the contour near that pole. This knowledge of the angular rotation of the poles on the contour permits us to complete the Nyquist diagram. Of course, our detour must carry us only an infinitesimal distance into the right half-plane, or else some closed-loop, right-half-plane poles will be excluded in the count.

We can also detour to the left of the open-loop poles. In this case each pole rotates through an angle of $-180°$ as we detour around it. Again, the detour must be infinitesimally small, or else we might include some left-half-plane poles in the count. Let us look at an example.

Example 10.5   Find the Nyquist diagram for an open-loop function with poles on the contour.

Problem   Find the Nyquist diagram of the unity feedback system of Figure 10.10, where $G(s) = \dfrac{s+2}{s^2}$.

Solution The system's two poles at the origin are on the contour and must be bypassed, as shown in Figure 10.27(*a*). The mapping starts at point *A* and continues in a clockwise direction. Points *A*, *B*, *C*, *D*, *E*, and *F* of Figure 10.27(*a*) map respectively into points *A'*, *B'*, *C'*, *D'*, *E'*, and *F'* of Figure 10.27(*b*).

At point *A*, the two open-loop poles at the origin contribute $2 \times 90° = 180°$ and the zero contributes $0°$. The total angle at point *A* is thus $-180°$. Close to the origin, the function is infinite in magnitude because of the close proximity to the two open-loop poles. Thus, point *A* maps into point *A'*, located at infinity at an angle of $-180°$.

Moving from point *A* to point *B* along the contour yields a net change in angle of $+90°$ from the zero alone. The angles of the poles remain the same. Thus, the mapping changes by $+90°$ in the counterclockwise direction. The mapped vector goes from $-180°$ at *A'* to $-90°$ at *B'*. At the same time, the magnitude changes from infinity to zero since at point *B*, there is one infinite length from the zero divided by two infinite lengths from the poles.

Alternately, the frequency response can be determined analytically from $G(j\omega) = \dfrac{2 + j\omega}{-\omega^2}$ and considering ω going from 0 to ∞. At low frequencies, $G(j\omega) \approx \dfrac{2}{-\omega^2}$, or $\infty \angle 180°$. At high frequencies, $G(j\omega) \approx \dfrac{j}{-\omega}$, or $0 \angle -90°$. Also, the real and imaginary parts are always negative.

As we travel along the contour *BCD*, the function magnitude stays at zero (one infinite zero length divided by two infinite pole lengths). As the vectors move through *BCD*, the zero's vector and the two poles' vectors undergo changes of $-180°$ each. Thus, the mapped vector undergoes a net change of $+180°$ which is the angular change of the zero minus the sum of the angular changes of the poles $\{-180 - [2(-180)] = +180\}$. The mapping is shown as *B'C'D'*, where the resultant vector changes by $+180°$ with a magnitude of ϵ that approaches zero.

From the analytical point of view,

$$G(s) = \frac{R_{-2} \angle \theta_{-2}}{(R_0 \angle \theta_0)(R_0 \angle \theta_0)} \tag{10.42}$$

Figure 10.27 (*a*) Contour for Example 10.5; (*b*) Nyquist Diagram for Example 10.5

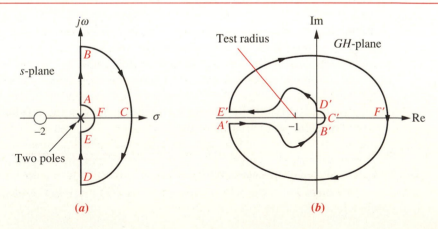

(a)   *(b)*

anywhere on the s-plane where $R_{-2}\angle\theta_{-2}$ is the vector from the zero at -2 to any point on the s-plane, and $R_0\angle\theta_0$ is the vector from a pole at the origin to any point on the s-plane. Around the infinite semicircle, all $R_{-i} = \infty$, and all angles can be approximated as if the vectors originated at the origin. Thus at point B, $G(s) = 0\angle - 90°$ since all $\theta_{-i} = 90°$ in Equation 10.42. At point C, all $R_{-i} = \infty$, and all $\theta_{-i} = 0°$ in Equation 10.42. Thus, $G(s) = 0\angle 0°$. At point D, all $R_{-i} = \infty$, and all $\theta_{-i} = -90°$ in Equation 10.42. Thus, $G(s) = 0\angle 90°$.

The mapping of the section of the contour from D to E is a mirror image of the mapping of A to B. The result is D' to E'.

Finally, over the section EFA, the resultant magnitude approaches infinity. The angle of the zero does not change, but each pole changes by $+180°$. This change yields a change in the function of $-2 \times 180° = -360°$. Thus, the mapping from E' to A' is shown as infinite in length and rotating $-360°$. Analytically, we can use Equation 10.42 for the points along the contour EFA. At E, $G(s) = \dfrac{(2\angle 0°)}{(\epsilon\angle - 90°)(\epsilon\angle - 90°)} = \infty\angle 180°$. At F, $G(s) = \dfrac{(2\angle 0°)}{(\epsilon\angle 0°)(\epsilon\angle 0°)} = \infty\angle 0°$. At A, $G(s) = \dfrac{(2\angle 0°)}{(\epsilon\angle 90°)(\epsilon\angle 90°)} = \infty\angle - 180°$.

The Nyquist diagram is now complete, and a test radius drawn from -1 in Figure 10.27(b) shows one counterclockwise revolution and one clockwise revolution yielding zero encirclements.

In this section we learned how to sketch a Nyquist diagram. We saw how to calculate the value of the intersection of the Nyquist diagram with the negative real axis. This intersection is important in determining the number of encirclements of -1. Also, we showed how to sketch the Nyquist diagram when open-loop poles exist on the contour; this case required detours around the poles. In the next section we apply the Nyquist criterion to determine the stability of feedback control systems.

10.5 Determining Stability via the Nyquist Diagram

We now use the Nyquist diagram to determine a system's stability, using the simple equation $Z = P - N$. The values of P, the number of open-loop poles of $G(s)H(s)$ enclosed by the contour, and N, the number of encirclements the Nyquist diagram makes about -1, are used to determine Z, the number of right-half-plane poles of the closed-loop system.

If the closed-loop system has a variable gain in the loop, one question we would like to ask is, "For what range of gain is the system stable?" This question, previously answered by the root locus method and the Routh-Hurwitz criterion, is now answered via the Nyquist criterion. The general approach is to set the loop gain equal to unity and draw the Nyquist diagram. Since gain is simply a multiplying factor, the effect of the gain is to multiply the resultant by a constant anywhere along the Nyquist diagram.

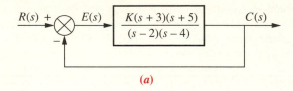

(a)

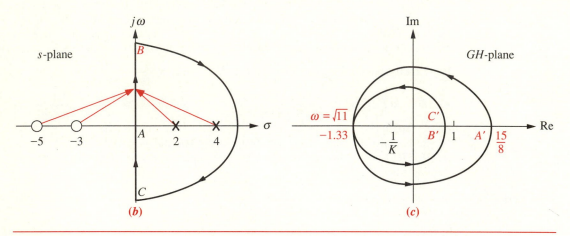

(b)

(c)

Figure 10.28 Feedback Control System to Demonstrate Nyquist Stability:
(**a**) System; (**b**) Contour; (**c**) Nyquist Diagram

For example, consider Figure 10.28, which summarizes the Nyquist approach for a system with variable gain, K. As the gain is varied, we can visualize the Nyquist diagram in Figure 10.28(c) expanding (increased gain) or shrinking (decreased gain) like a balloon. This motion could move the Nyquist diagram past the -1 point, changing the stability picture. For this system, since $P = 2$, the critical point must be encircled by the Nyquist diagram to yield $N = 2$ and a stable system. A reduction in gain would place the critical point outside the Nyquist diagram where $N = 0$, yielding $Z = 2$, an unstable system.

From another perspective, we can think of the Nyquist diagram as remaining stationary and the -1 point moving along the real axis. In order to do this, we set the gain to unity and position the critical point at $-1/K$ rather than -1. Thus, the critical point appears to move closer to the origin as K increases.

Finally, if the Nyquist diagram intersects the real axis at -1, then $G(j\omega)H(j\omega) = -1$. From the root locus concepts, when $G(s)H(s) = -1$, the variable s is a closed-loop pole of the system. Thus, the frequency at which the Nyquist diagram intersects -1 is the same frequency at which the root locus crosses the $j\omega$-axis. Hence, the system is marginally stable if the Nyquist diagram intersects the real axis at -1.

In summary, then, if the open-loop system contains a variable gain, K, set $K = 1$ and sketch the Nyquist diagram. Consider the critical point to be at $-1/K$ rather than at -1. Adjust the value of K to yield stability, based upon the Nyquist criterion.

Example 10.6   Find the range of gain for stability, instability, and marginal stability via the Nyquist criterion.

Problem   For the unity feedback system of Figure 10.10, where $G(s) = \dfrac{K}{s(s + 3)(s + 5)}$, find the range of gain, K, for stability and for instability, and the value of gain for marginal stability. For marginal stability, also find the frequency of oscillation. Use the Nyquist criterion.

Solution   First, set $K = 1$ and sketch the Nyquist diagram for the system, using the contour shown in Figure 10.29(a). For all points on the imaginary axis,

$$G(j\omega)H(j\omega) = \left.\frac{K}{s(s + 3)(s + 5)}\right|_{\substack{K=1 \\ s=j\omega}} = \frac{-8\omega^2 - j(15\omega - \omega^3)}{64\omega^4 + \omega^2(15 - \omega^2)^2} \tag{10.43}$$

At $\omega = 0$, $G(j\omega)H(j\omega) = -0.0356 - j\infty$.

Next, find the point where the Nyquist diagram intersects the negative real axis. Setting the imaginary part of Equation 10.43 equal to zero, we find $\omega = \sqrt{15}$. Substituting this value of ω back into Equation 10.43 yields the real part of -0.0083.

Finally, at $\omega = \infty$, $G(j\omega)H(j\omega) = G(s)H(s)\big|_{s\to j\infty} = \dfrac{1}{(j\infty)^3} = 0\angle-270°$.

From the contour of Figure 10.29(a), $P = 0$; for stability, N must then be equal to zero. From Figure 10.29(b), the system is stable if the critical point lies outside the contour ($N = 0$), so that $Z = P - N = 0$. Thus, K can be increased by

Figure 10.29 (**a**) Contour for Example 10.6; (**b**) Nyquist Diagram

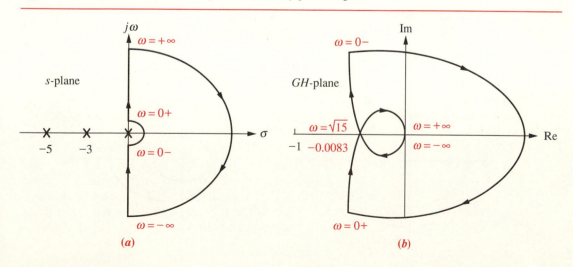

$1/0.0083 = 120.48$ before the Nyquist diagram encircles -1. Hence, for stability, $K < 120.48$. For marginal stability, $K = 120.48$. At this gain, the Nyquist diagram intersects -1, and the frequency of oscillation is $\sqrt{15}$ rad/s.

Now that we have used the Nyquist diagram to determine stability, we can develop a simplified approach that uses only the mapping of the positive $j\omega$-axis.

Stability via the Nyquist Criterion, Using Mapping of the Positive $j\omega$-Axis Alone

Once the stability of a system is determined by the Nyquist criterion, continued evaluation of a system can be simplified by using just the mapping of the positive $j\omega$-axis. This concept plays a major role in the next two sections, where we discuss stability margin and the implementation of the Nyquist criterion with Bode plots.

Consider the system shown in Figure 10.30, which is stable at low values of gain and unstable at high values of gain. Since the contour does not encircle open-loop poles, the Nyquist criterion tells us that we must have no encirclements of -1 for the system to be stable. We can see from the Nyquist diagram that the encirclements of the critical point can be determined from the mapping of the positive $j\omega$-axis alone. If the gain is small, the mapping will pass to the right of -1, and the system will be stable. If the gain is high, the mapping will pass to the left of -1, and the system will be unstable. Thus, this system is stable for the range of loop gain, K, that ensures that the *open-loop magnitude is less than unity at that frequency where the phase angle is 180° (or, equivalently, $-180°$).* This statement is thus an alternative to the Nyquist criterion for this system.

Now consider the system shown in Figure 10.31, which is unstable at low values of gain and stable at high values of gain. Since the contour encloses two open-loop poles, two counterclockwise encirclements of the critical point are

Figure 10.30 (**a**) Contour and Root Locus of a System That Is Stable for Small Gain and Unstable for Large Gain; (**b**) Nyquist Diagram

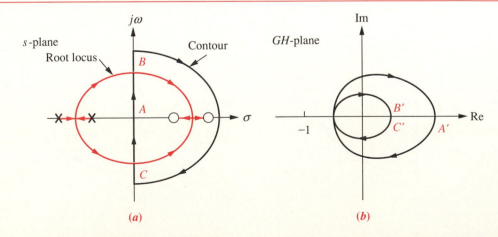

(a)   (b)

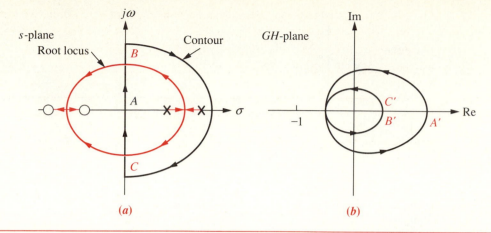

Figure 10.31 (**a**) Contour and Root Locus of a System That Is Unstable for Small Gain and Stable for Large Gain; (**b**) Nyquist Diagram

required for stability. Thus, for this case, the system is stable if the *open-loop magnitude is greater than unity at that frequency where the phase angle is* 180° *(or, equivalently,* −180°*).*

In summary, first determine stability from the Nyquist criterion and the Nyquist diagram. Next, interpret the Nyquist criterion and determine whether the mapping of just the positive imaginary axis should have a gain of less than or greater than unity at 180°. If the Nyquist diagram crosses ±180° at multiple frequencies, determine the interpretation from the Nyquist criterion.

Example 10.7

Find the range of gain for stability and instability, and the gain and frequency of oscillation for marginal stability via the mapping of only the positive imaginary axis.

Problem Find the range of gain for stability and instability, and the gain for marginal stability for the unity feedback system shown in Figure 10.10, where $G(s) = \dfrac{K}{(s^2 + 2s + 2)(s + 2)}$ For marginal stability, find the radian frequency of oscillation. Use the Nyquist criterion and the mapping of only the positive imaginary axis.

Solution Since the open-loop poles are only in the left half-plane, the Nyquist criterion tells us that we want no encirclements of −1 for stability. Hence, a gain less than unity at ±180° is required. Begin by letting $K = 1$ and draw the portion of the contour along the positive imaginary axis, as shown in Figure 10.32(*a*). In Figure 10.32(*b*), the intersection with the negative real axis is found by letting $s = j\omega$ in $G(s)H(s)$, setting the imaginary part equal to zero to find the frequency, and then substituting the frequency into the real part of $G(j\omega)H(j\omega)$. Thus, for any point on the positive imaginary axis,

$$
\begin{aligned}
G(j\omega)H(j\omega) &= \frac{1}{(s^2 + 2s + 2)(s + 2)}\bigg|_{s \to j\omega} \\
&= \frac{4(1 - \omega^2) - j\omega(6 - \omega^2)}{16(1 - \omega^2)^2 + \omega^2(6 - \omega^2)^2}
\end{aligned}
$$

(10.44)

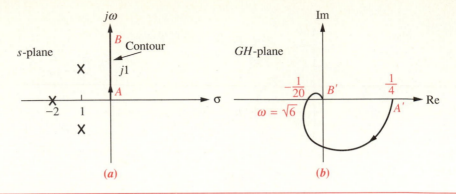

Figure 10.32 (*a*) Portion of Contour to be Mapped for Example 10.7; (*b*) Nyquist Diagram of Mapping of Positive Imaginary Axis

Setting the imaginary part equal to zero, we find $\omega = \sqrt{6}$. Substituting this value back into Equation 10.44 yields the real part, $-\dfrac{1}{20} = \dfrac{1}{20}\angle 180°$.

This closed-loop system is stable if the magnitude of the frequency response is less than unity at 180°. Hence, the system is stable for $K < 20$, unstable for $K > 20$, and marginally stable for $K = 20$. When the system is marginally stable, the radian frequency of oscillation is $\sqrt{6}$.

Now that we know how to sketch and interpret a Nyquist diagram to determine a closed-loop system's stability, let us extend our discussion to the concepts of gain margin and phase margin. These concepts will eventually lead us to the design of transient response characteristics via frequency response techniques.

10.6 Gain Margin and Phase Margin

Using the Nyquist diagram, we now define two quantitative measures of how stable a system is. These quantities are called *gain margin* and *phase margin*. Systems with greater gain and phase margins can withstand greater changes in system parameters before becoming unstable. In a sense, gain and phase margins can be qualitatively related to the root locus, in that systems whose poles are further from the imaginary axis have a greater degree of stability.

In the last section we discussed stability from the point of view of gain at 180° phase shift. This concept leads to the following definitions of gain margin and phase margin:

Gain margin, G_M: The gain margin is the change in open-loop gain, K', expressed in decibels (dB), required at 180° of phase shift to make the closed-loop system unstable.

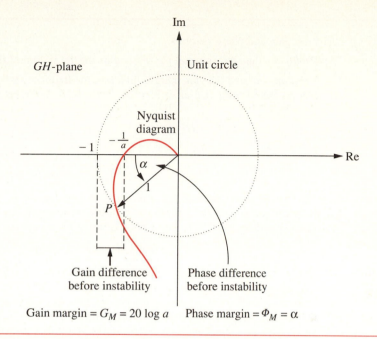

Gain margin = G_M = 20 log a Phase margin = $\Phi_M = \alpha$

Figure 10.33 Nyquist Diagram Showing Definitions of Gain and Phase Margins

Phase margin, Φ_M : The phase margin is the change in open-loop phase shift, Φ_M, required at unity gain to make the closed-loop system unstable.

These two definitions are shown graphically on the Nyquist diagram of Figure 10.33.

Assume a system that is stable if there are no encirclements of -1. Using Figure 10.33, let us focus on the definition of gain margin. Here, a gain difference between the Nyquist diagram's crossing of the real axis at $-1/a$ and the -1 critical point determines the proximity of the system to instability. Thus, if the gain of the system were multiplied by a units, the Nyquist diagram would intersect the critical point. We then say that the gain margin is a units, or, expressed in dB, $G_M = 20 \log a$. Notice that the gain margin is the reciprocal of the real-axis crossing expressed in dB.

In Figure 10.33, we also see the phase margin graphically displayed. At point P, where the gain is unity, α represents the system's proximity to instability. That is, at unity gain, if a phase shift of α degrees occurs, the system becomes unstable. Hence, the amount of phase margin is α. Later in the chapter, we show that phase margin can be related to the damping ratio. Thus, we will be able to relate frequency response characteristics to transient response characteristics as well as stability. We will also show that the calculations of gain and phase margins are more convenient if Bode plots are used rather than a Nyquist diagram such as that shown in Figure 10.33.

For now, let us look at an example that shows the calculation of the gain and phase margins.

Example 10.8   Find gain and phase margins.

Problem   Find the gain and phase margins for the system of Example 10.7 if $K = 6$.

Solution   To find the gain margin, first find the frequency where the Nyquist diagram crosses the negative real axis. Finding $G(j\omega)H(j\omega)$, we have

$$G(j\omega)H(j\omega) = \frac{6}{(s^2 + 2s + 2)(s + 2)}\Bigg|_{s \to j\omega}$$

$$= \frac{6[4(1 - \omega^2) - j\omega(6 - \omega^2)]}{16(1 - \omega^2)^2 + \omega^2(6 - \omega^2)^2} \qquad (10.45)$$

The Nyquist diagram crosses the real axis at a frequency of $\sqrt{6}$ rad/s. The real part is calculated to be -0.3. Thus, the gain can be increased by $(1/0.3) = 3.33$ before the real part becomes -1. The gain margin is

$$G_M = 20 \log 3.33 = 10.45 \text{ dB} \qquad (10.46)$$

To find the phase margin, find the frequency in Equation 10.45 for which the magnitude is unity. As the problem stands, this calculation requires computational tools such as a function solver or the program described in Appendix C. Later in the chapter we will simplify the process by using Bode plots. Equation 10.45 has unity gain at a frequency of 1.253 rad/s. At this frequency the phase angle is $-112.33°$. The difference between this angle and $-180°$ is $67.67°$, which is the phase margin.

In this section we defined gain margin and phase margin and calculated them via the Nyquist diagram. In the next section we show how to use Bode diagrams to implement the stability calculations performed in Sections 10.5 and 10.6 using the Nyquist diagram. We will see that the Bode plots reduce the time and simplify the calculations required to obtain results.

10.7 Stability Calculations Using Bode Plots

In this section, we determine stability, gain and phase margins, and the range of gain required for stability. All of these topics were covered previously in this chapter, using Nyquist diagrams as the tool. Now, we use Bode plots to determine these characteristics. Bode plots are subsets of the complete Nyquist diagram, but in another form. They are a viable alternative to Nyquist plots since they are easily drawn without the aid of the computational devices or long calculations required for the Nyquist diagram and root locus. The student should remember that all calculations applied to stability were derived from and based upon the Nyquist stability criterion. The Bode plots are an alternate way of visualizing and implementing the theoretical concepts.

Determining Stability via Bode Plots

Let us look at an example and determine the stability of a system, implementing the Nyquist stability criterion by using Bode plots. We will draw a Bode log-magnitude plot and then determine the value of gain that ensures that the magnitude is less than 0 dB (unity gain) at that frequency where the phase is $\pm180°$.

Example 10.9

Determine the range of K for the stability of a system via the Bode plots.

Problem Use Bode plots to determine the range of K within which the unity feedback system of Figure 10.10 is stable. Let $G(s) = \dfrac{K}{(s + 2)(s + 4)(s + 5)}$.

Solution Since this system has all of its open-loop poles in the left half-plane, the open-loop system is stable. Hence, from the discussion of Section 10.5, the closed-loop system will be stable if the frequency response has a gain less than unity when the phase is 180°.

Begin by sketching the Bode magnitude and phase diagrams shown in Figure 10.34. The low-frequency gain of $G(s)H(s)$ is found by setting s to zero. Thus, the

Figure 10.34 Bode Log-Magnitude and Phase Diagrams for the System of Example 10.9

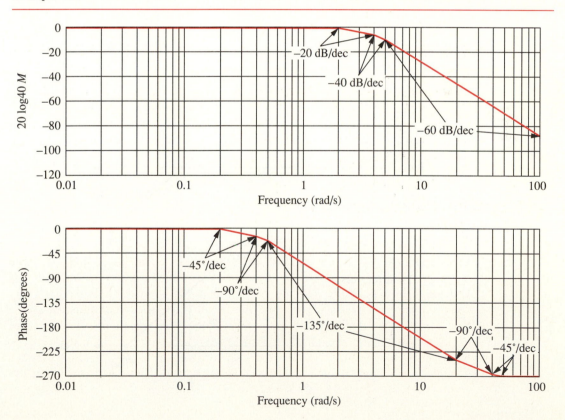

Bode magnitude plot starts at $K/40$. For convenience, let $K = 40$ so that the log-magnitude plot starts at 0 dB. At each break frequency, 2, 4, and 5, a 20 dB/decade increase in negative slope is drawn, yielding the log-magnitude plot shown in Figure 10.34.

The phase diagram begins at $0°$ until a decade below the first break frequency of 2 rad/s. At 0.2 rad/s the curve decreases at a rate of $-45°$/decade, decreasing an additional 45°/decade at each subsequent frequency (0.4 and 0.5 rad/s) a decade below each break. At a decade above each break frequency, the slopes are reduced by 45°/decade at each frequency.

The Nyquist criterion for this example tells us that we want zero encirclements of -1 for stability. Thus, we recognize that the Bode log-magnitude plot must be less than unity when the Bode phase plot is 180°. Accordingly, we see that at a frequency of approximately 7 rad/s, when the phase plot is $-180°$, the magnitude plot is -20 dB. Therefore, an increase in gain of $+20$ dB is possible before the system becomes unstable. Since the gain plot was scaled for a gain of 40, $+20$ dB (a gain of 10) represents the required increase in gain above 40. Hence, the gain for instability is $40 \times 10 = 400$. The final result is $0 < K < 400$ for stability.

This result, obtained by approximating the frequency response by Bode asymptotes, can be compared to the result obtained from the actual frequency response, which yields a gain of 378 at a frequency of 6.16 rad/s.

Figure 10.35 Gain and Phase Margins on the Bode Diagrams

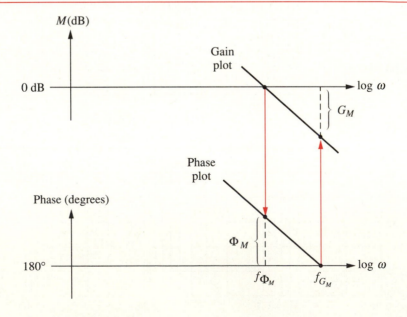

Evaluating Gain and Phase Margins via the Bode Plots

Next, we show how to evaluate the gain and phase margins using Bode plots (see Figure 10.35). The gain margin is found by using the phase plot to find the frequency, f_{G_M}, where the phase angle is 180°. At this frequency look at the magnitude plot to determine the gain margin, G_M, which is the gain required to raise the magnitude curve to 0 dB. To illustrate, in the previous example with $K = 40$, the gain margin was found to be 20 dB.

The phase margin is found by using the magnitude curve to find the frequency, f_{Φ_M}, where the gain is 0 dB. On the phase curve at that frequency, the phase margin, Φ_M, is the difference between the phase value and 180°.

Example 10.10   Find gain and phase margins from the Bode plots.

Problem   If $K = 200$ in the system of Example 10.9, find the gain margin and the phase margin.

Solution   The Bode plot in Figure 10.34 is scaled to a gain of 40. If $K = 200$ (five times greater), the magnitude plot would be $20 \log 5 = 13.98$ dB higher.

To find the gain margin, look at the phase plot and find the frequency where the phase is 180°. At this frequency, determine from the magnitude plot how much the gain can be increased before reaching 0 dB. In Figure 10.34, the phase angle is 180° at approximately 7 rad/s. On the magnitude plot, the gain is $-20 + 13.98 = -6.02$ dB. Thus, the gain margin is 6.02 dB.

To find the phase margin, we look on the magnitude plot for the frequency where the gain is 0 dB. At this frequency, we look on the phase plot to find the difference between the phase and 180°. This difference is the phase margin. Again, remembering that the magnitude plot of Figure 10.34 is 13.98 dB lower than the actual plot, the 0 dB crossing (-13.98 dB for the normalized plot shown in Figure 10.34) occurs at 5.5 rad/s. At this frequency, the phase angle is $-165°$. Thus, the phase margin is $180° - 165° = 15°$.

We have seen that the open-loop frequency response curves can be used not only to determine whether a system is stable, but to calculate the range of loop gain that will ensure stability. We have also seen how to calculate the gain margin and the phase margin from the Bode diagrams. Is it then possible to parallel the root locus technique and analyze and design systems for transient response using frequency response methods? We begin to explore the answer in the next section.

10.8 Relation between Closed-Loop Transient and Closed-Loop Frequency Responses

Relationship between Damping Ratio (Percent Overshoot) and Closed-Loop Frequency Response for a Two-Pole System

In this section we show that a relationship exists between a system's transient response and its closed-loop frequency response. Consider the second-order feedback control system shown in Figure 10.36, whose closed-loop transfer function is the standard second-order system that we studied in Chapter 4. Thus,

$$\frac{C(s)}{R(s)} = T(s) = \frac{\omega_n^2}{s^2 + 2\zeta\omega_n s + \omega_n^2} \tag{10.47}$$

Substituting $s = j\omega$ into Equation 10.47, we evaluate the magnitude closed-loop frequency response as

$$M = |T(j\omega)| = \frac{\omega_n^2}{\sqrt{(\omega_n^2 - \omega^2)^2 + 4\zeta^2\omega_n^2\omega^2}} \tag{10.48}$$

A representative sketch of the log plot of Equation 10.48 is shown in Figure 10.37.

We now show that a relationship exists between the peak value of the closed-loop magnitude response and the damping ratio. Squaring Equation 10.48, differentiating with respect to ω^2, and setting the derivative equal to zero yields the maximum value of M, M_p, where

$$M_p = \frac{1}{2\zeta\sqrt{1 - \zeta^2}} \tag{10.49}$$

at a frequency, ω_p, of

$$\omega_p = \omega_n\sqrt{1 - 2\zeta^2} \tag{10.50}$$

Since ζ is related to percent overshoot, we can plot M_p vs percent overshoot. The result is shown in Figure 10.38.

Equation 10.49 shows that the maximum magnitude on the frequency response curve is directly related to the damping ratio and, hence, the percent overshoot. Also notice from Equation 10.50 that the peak frequency, ω_p, is not the

Figure 10.36 A Second-Order Closed-Loop System

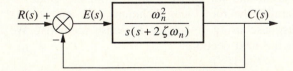

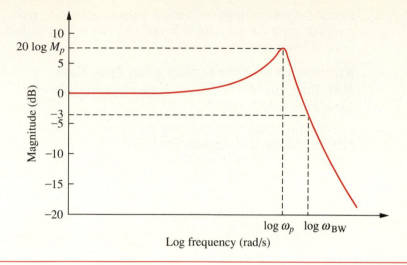

Figure 10.37 Representative Log-Magnitude Plot of the Closed-Loop Transfer Function of a Two-Pole System

Figure 10.38 Relationship between the Closed-Loop Frequency Response Peak and Percent Overshoot for a Two-Pole System

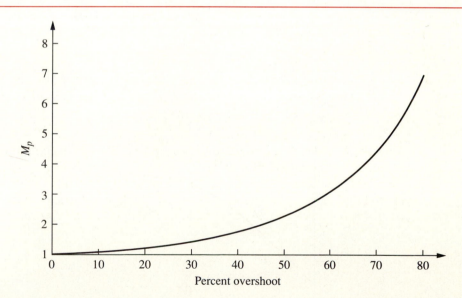

natural frequency. However, for low values of damping ratio, we can assume that the peak occurs at the natural frequency. Finally, notice that there will not be a peak at frequencies above zero if $\zeta > 0.707$.

Relationship between Settling Time, Peak Time, Rise Time, and Closed-Loop Frequency Response for a Two-Pole System

Another relationship between the frequency response and time response is between the speed of the time response (as measured by settling time, peak time, and rise time) and the *bandwidth* of the closed-loop frequency response, which is defined here to be that frequency, ω_{BW}, at which the magnitude response curve is 3 dB down from its value at zero frequency (see Figure 10.37).

Figure 10.39 Bandwidth Normalized by (*a*) Settling Time, (*b*) Peak Time, and (*c*) Rise Time as Functions of Damping Ratio

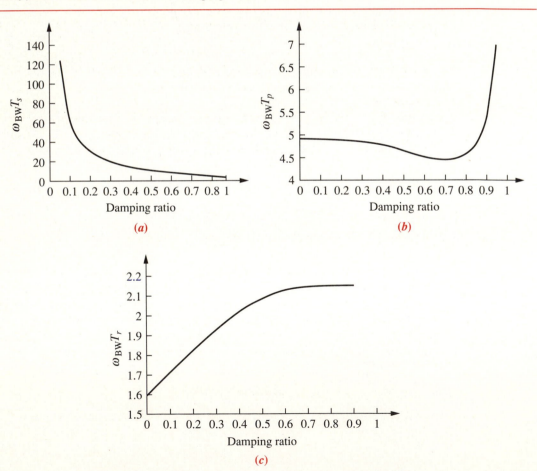

The bandwidth of a two-pole system can be found by finding that frequency for which $M = 1/\sqrt{2}$ (i.e., -3 dB) in Equation 10.48. The derivation is left as an exercise for the student. The result is

$$\omega_{BW} = \omega_n \sqrt{(1 - 2\zeta^2) + \sqrt{4\zeta^4 - 4\zeta^2 + 2}} \qquad (10.51)$$

To relate ω_{BW} to settling time, we substitute $\omega_n = 4/T_s\zeta$ into Equation 10.51 and obtain

$$\omega_{BW} = \frac{4}{T_s\zeta} \sqrt{(1 - 2\zeta^2) + \sqrt{4\zeta^4 - 4\zeta^2 + 2}} \qquad (10.52)$$

Similarly, since $\omega_n = \dfrac{\pi}{T_p \sqrt{1 - \zeta^2}}$,

$$\omega_{BW} = \frac{\pi}{T_p \sqrt{1 - \zeta^2}} \sqrt{(1 - 2\zeta^2) + \sqrt{4\zeta^4 - 4\zeta^2 + 2}} \qquad (10.53)$$

To relate the bandwidth to rise time, T_r, we use Figure 4.15, knowing the desired ζ and T_r. For example, assume $\zeta = 0.4$ and $T_r = 0.2$ seconds. Using Figure 4.15, the ordinate $T_r\omega_n = 1.463$, from which $\omega_n = 1.463/0.2 = 7.315$ rad/s. Using Equation 10.51, $\omega_{BW} = 10.05$ rad/s. Normalized plots of Equations 10.52 and 10.53 and the relationship between bandwidth normalized by rise time and damping ratio are shown in Figure 10.39.

In this section we related the closed-loop transient response to the closed-loop frequency response via bandwidth. We continue by relating the closed-loop frequency response to the open-loop frequency response and explaining the impetus.

10.9 Relation between Closed-Loop Frequency and Open-Loop Frequency Responses

At this point, we do not have an easy way of finding the closed-loop frequency response from which we could determine M_p and thus the transient response. As we have seen, we are equipped to rapidly sketch the open-loop frequency response, but not the closed-loop frequency response. However, if the open-loop response is related to the closed-loop response, we can combine the ease of sketching the open-loop response with the transient response information contained in the closed-loop response.

Constant M Circles and Constant N Circles

Consider a unity feedback system whose closed-loop transfer function is

$$T(s) = \frac{G(s)}{1 + G(s)} \qquad (10.54)$$

The frequency response of this closed-loop function is

$$T(j\omega) = \frac{G(j\omega)}{1 + G(j\omega)} \tag{10.55}$$

Since $G(j\omega)$ is a complex number, let $G(j\omega) = P(\omega) + jQ(\omega)$ in Equation 10.55, which yields

$$T(j\omega) = \frac{P(\omega) + jQ(\omega)}{[(P(\omega) + 1) + jQ(\omega)]} \tag{10.56}$$

Therefore,

$$M^2 = |T^2(j\omega)| = \frac{P^2(\omega) + Q^2(\omega)}{[(P(\omega) + 1)^2 + Q^2(\omega)]} \tag{10.57}$$

Equation 10.57 can be put into the form

$$\left(P + \frac{M^2}{M^2 - 1}\right)^2 + Q^2 = \frac{M^2}{(M^2 - 1)^2} \tag{10.58}$$

which is the equation of a circle of radius $\dfrac{M}{(M^2 - 1)}$ centered at $\left(\dfrac{-M^2}{M^2 - 1}, 0\right)$.
These circles, shown plotted in Figure 10.40 for various values of M, are called *constant M circles* and are the locus of the closed-loop magnitude frequency response for unity feedback systems. Thus, if the polar frequency response of an open-loop function, $G(s)$, is plotted and superimposed on top of the constant M circles, the closed-loop magnitude frequency response is determined by each intersection of this polar plot with the constant M circles.

Before demonstrating the use of the constant M circles with an example, let us go through a similar development for the closed-loop phase plot, the constant N circles. From Equation 10.56, the phase angle, ϕ, of the closed-loop response is

$$\phi = \tan^{-1} \frac{Q(\omega)}{P(\omega)} - \tan^{-1} \frac{Q(\omega)}{P(\omega) + 1}$$

$$= \tan^{-1} \frac{\dfrac{Q(\omega)}{P(\omega)} - \dfrac{Q(\omega)}{P(\omega) + 1}}{1 + \dfrac{Q(\omega)}{P(\omega)}\left(\dfrac{Q(\omega)}{P(\omega) + 1}\right)} \tag{10.59}$$

after using $\tan(\alpha - \beta) = (\tan\alpha - \tan\beta)/(1 + \tan\alpha\tan\beta)$. Dropping the functional notation,

$$\tan\phi = N = \frac{Q}{P^2 + P + Q^2} \tag{10.60}$$

Equation 10.60 can be put into the form of a circle,

$$\left(P + \frac{1}{2}\right)^2 + \left(Q - \frac{1}{2N}\right)^2 = \frac{N^2 + 1}{4N^2} \tag{10.61}$$

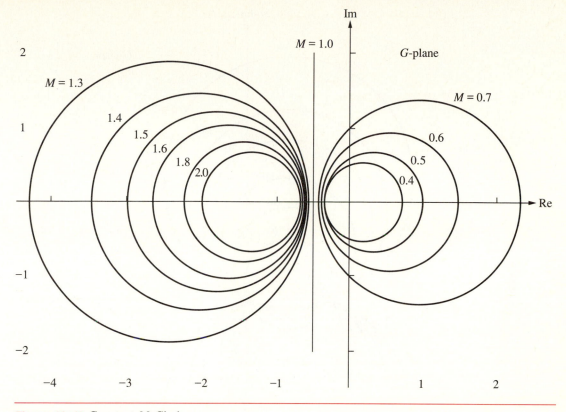

Figure 10.40 Constant M Circles

which is plotted in Figure 10.41 for various values of N. The circles of this plot are called *constant N circles*. Superimposing a unity feedback, open-loop frequency response over the constant N circles yields the closed-loop phase response of the system. Let us now look at an example that demonstrates the use of the constant M and N circles.

Example 10.11   Find the closed-loop frequency response of a unity feedback control system using the open-loop frequency response and the constant M and N circles.

Problem   Find the closed-loop frequency response of the unity feedback system shown in Figure 10.10, where $G(s) = \dfrac{50}{s(s+3)(s+6)}$, using the constant M circles, N circles, and the open-loop polar frequency response curve.

Solution   First, evaluate the open-loop frequency function and make a polar frequency response plot superimposed over the constant M and N circles. The open-loop

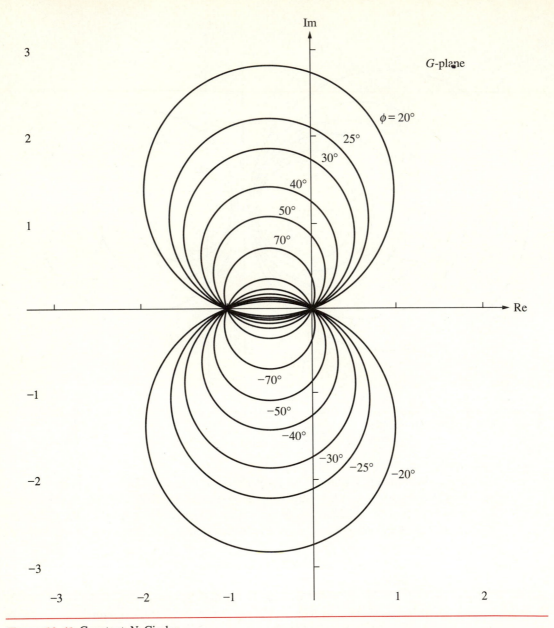

Figure 10.41 Constant N Circles

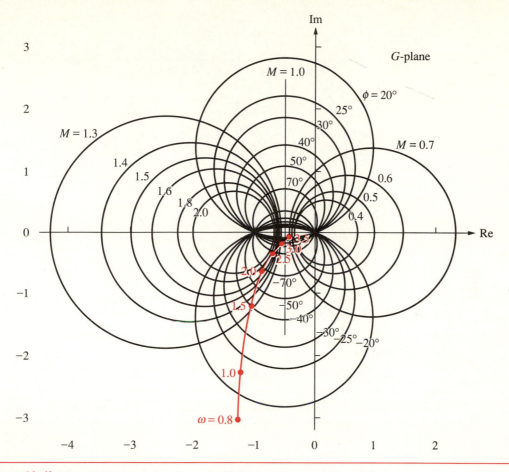

Figure 10.42 Nyquist Diagram for Example 10.11 and Constant M and N
Circles

frequency function is

$$G(j\omega) = \frac{50}{-9\omega^2 + j(18\omega - \omega^3)} \tag{10.62}$$

from which the magnitude, $|G(j\omega)|$, and phase, $\angle G(j\omega)$, can be found and plot-
ted. The polar plot of the open-loop frequency response (Nyquist diagram) is shown
superimposed over the M and N circles in Figure 10.42.

The closed-loop magnitude frequency response can now be obtained by finding the
intersection of each point of the Nyquist plot with the M circles, while the closed-
loop phase response can be obtained by finding the intersection of each point of the
Nyquist plot with the N circles. The result is shown in Figure 10.43.[3]

---

[3]The student is cautioned not to use the *closed-loop* polar plot for the Nyquist criterion. The
closed-loop frequency response, however, can be used to determine the closed-loop transient response,
as discussed in Section 10.8.

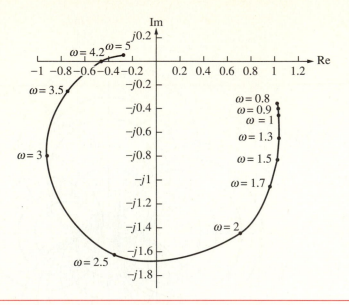

Figure 10.43 Closed-Loop Frequency Response for Example 10.11

Nichols Charts

A disadvantage of using the M and N circles is that changes of gain in the open-loop transfer function $G(s)$ cannot be handled easily. For example, in the Bode plot, a gain change is handled by moving the Bode magnitude curve up or down an amount equal to the gain change in dB. Since the M and N circles are not dB plots, changes in gain require each point of $G(j\omega)$ to be multiplied in length by the increase or decrease in gain.

Another presentation of the M and N circles, called a *Nichols chart,* displays the constant M circles in dB, so that changes in gain are as simple to handle as in the Bode plot. A Nichols chart is shown in Figure 10.44. The chart is a plot of open-loop magnitude in dB vs open-loop phase angle in degrees. Every point on the M circles can be transferred to the Nichols chart. Each point on the constant M circles is represented by magnitude and angle (polar coordinates). Converting the magnitude to dB, we can transfer the point to the Nichols chart, using the polar coordinates with magnitude in dB plotted as the ordinate and the phase angle plotted as the abscissa. Similarly, the N circles can also be transferred to the Nichols chart.

For example, assume the function

$$G(s) = \frac{K}{s(s + 1)(s + 2)} \tag{10.63}$$

Superimposing the frequency response of $G(s)$ on the Nichols chart by plotting magnitude in dB vs phase angle for a range of frequencies from 0.1 to 1 rad/s, we obtain the plot in Figure 10.45 for $K = 1$. If the gain is increased

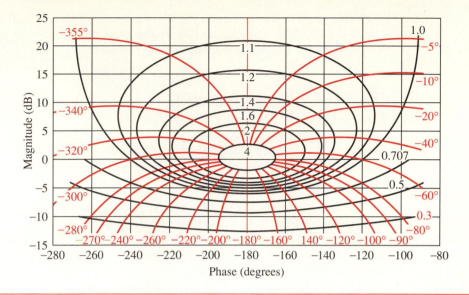

Figure 10.44 Nichols Chart

Figure 10.45 Nichols Chart with Frequency Response for
$$G(s) = \frac{K}{s(s + 1)(s + 2)}$$ Superimposed. Values of $K = 1$
and $K = 3.16$ Are Shown

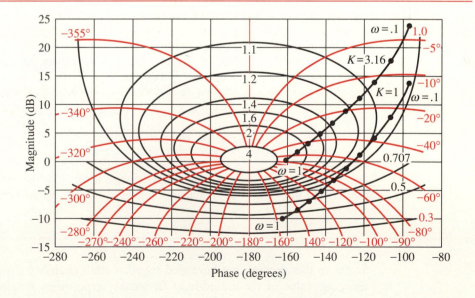

by 10 dB, simply raise the curve for $K = 1$ by 10 dB and obtain the curve for $K = 3.16$ (10 dB). The intersection of the plots of $G(j\omega)$ with the M and N circles yields the frequency response of the closed-loop system.

10.10 Relation between Closed-Loop Transient and Open-Loop Frequency Responses

Finding the Damping Ratio (Percent Overshoot) from M_P Evaluated from the Open-Loop Frequency Response and the M Circles

We can use the results of Example 10.11 to estimate the transient response characteristics of the system. We can find the peak of the closed-loop frequency response by finding the maximum M curve tangent to the open-loop frequency response. Then we can find the damping ratio, ζ, and subsequently the percent overshoot, via Equation 10.49. The following example demonstrates the use of the open-loop frequency response and the M circles to find the damping ratio, or, equivalently, the percent overshoot.

Example 10.12   Find the damping ratio and percent overshoot of a system via the open-loop frequency response.

Problem   Find the damping ratio and the percent overshoot expected from the system of Example 10.11 using the open-loop frequency response and the M circles.

Solution   Equation 10.49 shows that there is a unique relationship between the closed-loop system's damping ratio and the peak value, M_P, of the closed-loop system's magnitude frequency plot. From Figure 10.42, we see that the Nyquist diagram is tangent to the 1.6 M circle. We see that this is the maximum value for the closed-loop frequency response. Thus, $M_P = 1.6$.

We can solve for ζ by rearranging Equation 10.49 into the following form:

$$\zeta^4 - \zeta^2 + (1/4M_P^2) = 0 \tag{10.64}$$

Since $M_P = 1.8$, then $\zeta = 0.29$ and 0.96. From Equation 10.50, a damping ratio larger than 0.707 yields no peak above zero frequency. Thus, we select $\zeta = 0.29$, which is equivalent to 38.6% overshoot. Care must be taken, however, to be sure we can make a second-order approximation when associating the value of percent overshoot to the value of ζ. A computer simulation of the step response shows 36% overshoot.

So far in this section, we have tied together the system's transient response and the peak value of the closed-loop frequency response as obtained from the open-loop frequency response. We used the Nyquist plots and the M and N circles to obtain the closed-loop transient response. Another association exists between the open-loop frequency response and the closed-loop transient response that is easily implemented with the Bode plots, which are easier to draw than the Nyquist plots.

Damping Ratio (Percent Overshoot) from Phase Margin Obtained from the Open-Loop Frequency Response

Let us now derive the relationship between the phase margin and the damping ratio. This relationship will enable us to evaluate the percent overshoot from the phase margin found from the open-loop frequency response.

Consider a unity feedback system whose open-loop function,

$$G(s) = \frac{\omega_n^2}{s(s + 2\zeta\omega_n)} \tag{10.65}$$

yields the typical second-order, closed-loop transfer function

$$T(s) = \frac{\omega_n^2}{s^2 + 2\zeta\omega_n s + \omega_n^2} \tag{10.66}$$

In order to evaluate the phase margin, we first find the frequency for which $|G(j\omega)| = 1$. Hence,

$$|G(j\omega)| = \frac{\omega_n^2}{|-\omega^2 + j2\zeta\omega_n\omega|} = 1 \tag{10.67}$$

The frequency, ω_1, that satisfies Equation 10.67 is

$$\omega_1 = \omega_n \sqrt{-2\zeta^2 + \sqrt{1 + 4\zeta^4}} \tag{10.68}$$

The phase angle of $G(j\omega)$ at this frequency is

$$\angle G(j\omega) = -90 - \tan^{-1}\frac{\omega_1}{2\zeta\omega_n}$$

$$= -90 - \tan^{-1}\frac{\sqrt{-2\zeta^2 + \sqrt{4\zeta^4 + 1}}}{2\zeta} \tag{10.69}$$

The difference between the angle of Equation 10.69 and $-180°$ is the phase margin, Φ_M. Thus,

$$\Phi_M = 90 - \tan^{-1}\frac{\sqrt{-2\zeta^2 + \sqrt{1 + 4\zeta^4}}}{2\zeta}$$

$$= \tan^{-1}\frac{2\zeta}{\sqrt{-2\zeta^2 + \sqrt{1 + 4\zeta^4}}} \tag{10.70}$$

Equation 10.70, plotted in Figure 10.46, shows the relationship between phase margin and damping ratio.

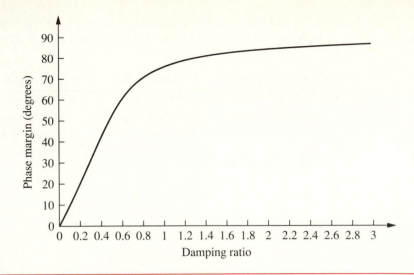

Figure 10.46 Relationship between Phase Margin
and Damping Ratio

As an example, Equation 10.50 tells us that there is no peak frequency if
$\zeta = 0.707$. Hence there is no peak to the closed-loop magnitude frequency
response curve for this value of damping ratio and larger. Thus, from Figure
10.46, a phase margin of 65.52° ($\zeta = 0.707$) or larger is required for the *open-
loop* frequency response to ensure there is no peaking in the *closed-loop* frequency
response.

Settling Time, Peak Time, and Rise Time
from the Open-Loop Frequency Response

Equations 10.52 and 10.53 relate the closed-loop bandwidth to the desired
settling or peak time and the damping ratio. We now show that the closed-loop
bandwidth can be estimated from the open-loop frequency response. From the
Nichols chart in Figure 10.44, we can see the relationship between the open-loop
gain and the closed-loop gain. The $M = 0.707$ (-3 dB) curve is replotted in
Figure 10.47 for clarity and shows the open-loop gain when the closed-loop gain
is -3 dB, which typically occurs at ω_{BW} if the low-frequency closed-loop gain is
0 dB. We can approximate Figure 10.47 by saying that the closed-loop bandwidth,
ω_{BW} (i.e., the frequency at which the closed-loop magnitude response is -3 dB),
equals the frequency at which the open-loop magnitude response is between -6
and -7.5 dB if the open-loop phase response is between $-135°$ and $225°$. Then,
using a second-order system approximation, Equations 10.52 and 10.53 can be
used, along with the desired damping ratio, ζ, to find settling time and peak time,
respectively. Let us look at an example.

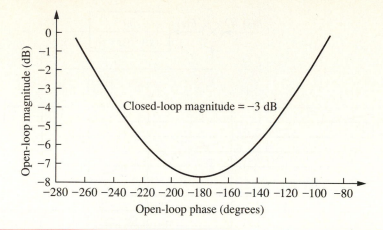

Figure 10.47 Open-Loop Gain When Closed-Loop Gain Is −3 dB, as a Function of Open-Loop Phase Angle

Example 10.13 Find the settling time and peak time from the open-loop frequency response.

Problem Given the system of Figure 10.48(*a*) and the Bode diagrams of Figure 10.48(*b*), estimate the settling time and peak time.

Solution Using Figure 10.48(*b*), we estimate the closed-loop bandwidth by finding the frequency where the open-loop magnitude response is in the range of −6 to −7.5 dB if the phase response is in the range of −135° to 225°. Since Figure 10.48(*b*) shows −6 to −7.5 dB at approximately 3.7 rad/s with a phase response in the stated region, $\omega_{BW} \cong 3.7$ rad/s.

Next, find ζ via the phase margin. From Figure 10.48(*b*), the phase margin is found by first finding the frequency at which the magnitude plot is 0 dB. At this frequency, 2.2 rad/s, the phase is about −145°. Hence the phase margin is approximately (180° − 145°) = 35°. Using Figure 10.46, $\zeta = 0.32$. Finally, using Equations 10.52 and 10.53, with the values of ω_{BW} and ζ just found, $T_s = 4.86$ seconds, and $T_p = 1.29$ seconds. Checking the analysis with a computer simulation shows $T_s = 5.5$ seconds, and $T_p = 1.43$ seconds.

In this section we showed how to obtain transient response data from the open-loop frequency response by using both the Nyquist diagram and the Bode plots. Section 10.11 shows how to obtain steady-state error data from the frequency response.

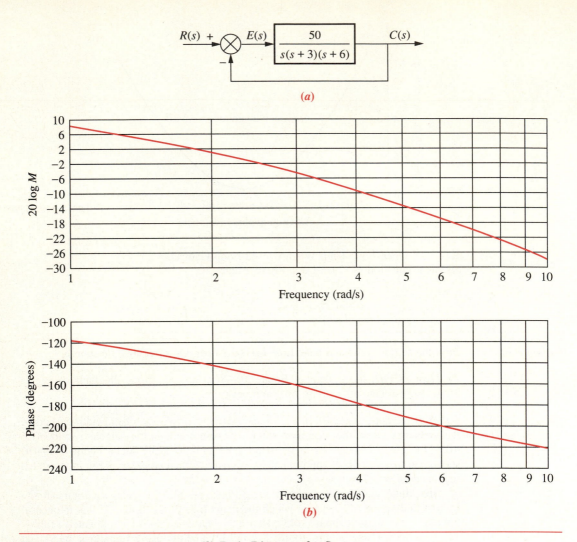

Figure 10.48 (**a**) Block Diagram; (**b**) Bode Diagrams for System of Example 10.13

10.11 Finding the Steady-State Error Specifications from the Frequency Response

In this section we show how to use Bode diagrams to find the values of the static error constants: K_p for a Type 0 system, K_v for a Type 1 system, and K_a for a Type 2 system. The results will be obtained from unscaled Bode log-magnitude plots.

Finding the Position Constant

To find K_p, consider the following Type 0 system:

$$G(s)H(s) = \frac{\prod\limits_{i=1}^{n}(s + z_i)}{\prod\limits_{i=1}^{m}(s + p_i)} \tag{10.71}$$

A typical, unscaled Bode log-magnitude plot is shown in Figure 10.49(a). The initial value is

$$20\log M = 20\log \frac{\prod\limits_{i=1}^{n} z_i}{\prod\limits_{i=1}^{m} p_i} \tag{10.72}$$

Figure 10.49 Typical, Unscaled Bode Log-Magnitude Plots Showing the Value of Static Error Constants: (a) Type 0; (b) Type 1; (c) Type 2

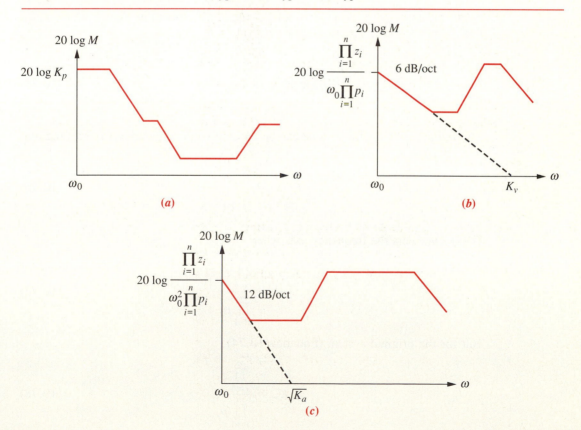

But for this system,

$$K_p = \frac{\prod\limits_{i=1}^{n} z_i}{\prod\limits_{i=1}^{m} p_i} \tag{10.73}$$

which is the same as the value of the low-frequency axis. Thus, for an unscaled, Bode log-magnitude plot, the low-frequency magnitude is $20 \log K_p$ for a Type 0 system.

Finding the Velocity Constant

To find K_v for a Type 1 system, consider the following open-loop transfer function of a Type 1 system:

$$G(s)H(s) = \frac{\prod\limits_{i=1}^{n} (s + z_i)}{s \prod\limits_{i=1}^{m} (s + p_i)} \tag{10.74}$$

A typical unscaled Bode log-magnitude diagram is shown in Figure 10.49(b) for this Type 1 system. The Bode plot starts at

$$20 \log M = 20 \log \frac{\prod\limits_{i=1}^{n} z_i}{\omega_0 \prod\limits_{i=1}^{m} p_i} \tag{10.75}$$

The initial -6 dB/octave slope can be thought of as originating from a function,

$$G'(s) = \frac{\prod\limits_{i=1}^{n} z_i}{s \prod\limits_{i=1}^{m} p_i} \tag{10.76}$$

$G'(s)$ intersects the frequency axis when

$$\omega = \frac{\prod\limits_{i=1}^{n} z_i}{\prod\limits_{i=1}^{m} p_i} \tag{10.77}$$

But for the original system (Equation 10.74),

$$K_v = \frac{\prod\limits_{i=1}^{n} z_i}{\prod\limits_{i=1}^{m} p_i} \tag{10.78}$$

Thus, we can find K_v by extending the initial -6 dB/octave slope to the frequency axis on an unscaled Bode diagram. The intersection with the frequency axis is K_v.

Finding the Acceleration Constant

To find K_a for a Type 2 system, consider the following:

$$G(s)H(s) = \frac{\prod\limits_{i=1}^{n} (s + z_i)}{s^2 \prod\limits_{i=1}^{m} (s + p_i)} \tag{10.79}$$

A typical, unscaled Bode plot for a Type 2 system is shown in Figure 10.49(c). The Bode plot starts at

$$20 \log M = 20 \log \frac{\prod\limits_{i=1}^{n} z_i}{\omega_0^2 \prod\limits_{i=1}^{m} p_i} \tag{10.80}$$

The initial -12 dB/octave slope can be thought of as coming from a function,

$$G'(s) = \frac{\prod\limits_{i=1}^{n} z_i}{s^2 \prod\limits_{i=1}^{m} p_i} \tag{10.81}$$

$G'(s)$ intersects the frequency axis when

$$\omega = \sqrt{\frac{\prod\limits_{i=1}^{n} z_i}{\prod\limits_{i=1}^{m} p_i}} \tag{10.82}$$

But for the original system (Equation 10.78),

$$K_a = \frac{\prod\limits_{i=1}^{n} z_i}{\prod\limits_{i=1}^{m} p_i} \tag{10.83}$$

Thus, the initial -12 dB/octave slope intersects the frequency axis at $\sqrt{K_a}$.

Example 10.14   Find the static error constants from the Bode log-magnitude plot.

Problem   For each unscaled Bode log-magnitude plot shown in Figure 10.50,
    a. Find the system type.
    b. Find the value of the appropriate static error constant.

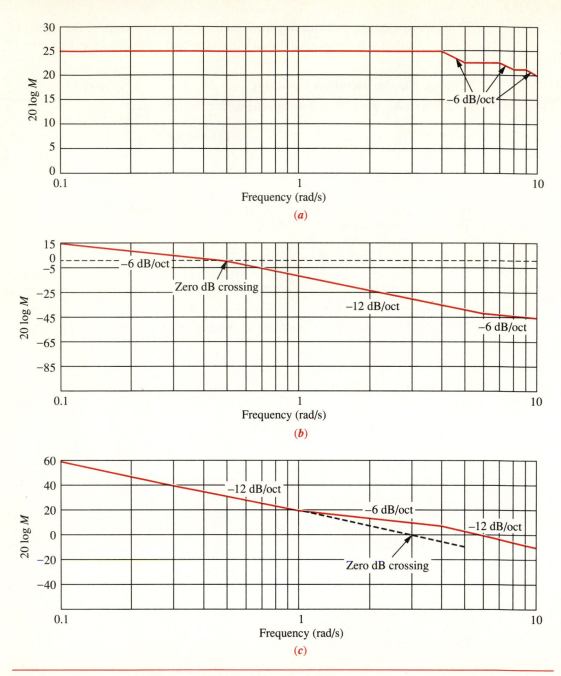

Figure 10.50 Bode Log-Magnitude Plots for Example 10.14

Solution   Figure 10.50(*a*) is a Type 0 system since the initial slope is zero. The value of K_p is given by the low-frequency asymptote value. Thus, $20 \log K_p = 25$, or $K_p = 17.78$.

Figure 10.50(*b*) is a Type 1 system since the initial slope is -6 dB/octave. The value of K_v is the value of the frequency that the initial slope intersects at the zero dB crossing of the frequency axis. Hence, $K_v = 0.45$.

Figure 10.50(*c*) is a Type 2 system since the initial slope is -12 dB/octave. The value of $\sqrt{K_a}$ is the value of the frequency that the initial slope intersects at the zero dB crossing of the frequency axis. Hence, $K_a = 3^2 = 9$.

10.12 Frequency Response Methods for Systems with Time Delay

Time delay occurs in control systems when there is a delay between the commanded response and the start of the output response. For example, consider a heating system that operates by heating water in a pipeline for distribution to radiators at distant locations. Since the heat diffuses through the water in the line, the radiators will not begin to get hot until after a specified time delay. In other words, the time between the command for more heat and the commencement of the rise in temperature at a distant location along the pipeline is the time delay. Notice that this is not the same as the transient response or the time it takes the temperature to rise to the desired level. During the time delay, nothing is occurring at the output.

Modeling Time Delay

Assume that an input, $R(s)$, to a system, $G(s)$, yields an output, $C(s)$. If another system, $G'(s)$, delays the output by T seconds, the output response is $c(t - T)$. From Table 2.2, Entry 5, the Laplace transform of $c(t-T)$ is $e^{-sT}C(s)$. Thus, for the system without delay, $C(s) = R(s)G(s)$, and for the system with delay, $e^{-sT}C(s) = R(s)G'(s)$. Dividing these two equations, $G'(s)/G(s) = e^{-sT}$. Thus, a system with time delay T can be represented in terms of an equivalent system without time delay as follows:

$$G'(s) = e^{-sT}G(s) \tag{10.84}$$

The effect of introducing time delay into a system can also be seen from the perspective of the frequency response by substituting $s = j\omega$ in Equation 10.84. Hence,

$$G'(j\omega) = e^{-j\omega T}G(j\omega) = |G(j\omega)|\angle\{-\omega T + \angle G(j\omega)\} \tag{10.85}$$

In other words, the time delay does not affect the magnitude frequency response curve of $G(j\omega)$, but it does subtract a linearly increasing phase shift, ωT, from the phase frequency response plot of $G(j\omega)$.

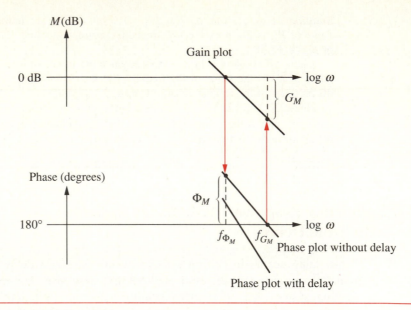

Figure 10.51 Effect of Delay upon Frequency Response

The typical effect of adding time delay can be seen in Figure 10.51. Assume that the gain and phase margins as well as the gain- and phase-margin frequencies shown in the figure apply to the system without delay. From the figure, we see that the reduction in phase shift caused by the delay reduces the phase margin. Using a second-order approximation, this reduction in phase margin yields a reduced damping ratio for the closed-loop system and a more oscillatory response. The reduction of phase also leads to a reduced gain-margin frequency. From the magnitude curve we can see that a reduced gain-margin frequency leads to reduced gain margin, thus moving the system closer to instability.

An example of plotting frequency response curves for systems with delay follows.

Example 10.15   Plot the frequency response of a system with time delay.

Problem   Plot the frequency response for the system $G(s) = \dfrac{1}{s(s+1)(s+2)}$ if there is a time delay of 1 second through the system. Use the Bode plots.

Solution   The magnitude curve can be plotted by the methods previously covered in the chapter. Since the magnitude curve is not affected by the delay, we will leave this part of the problem to the reader.

The phase plot, however, is affected by the delay. Figure 10.52 shows the result. First, draw the phase plot for the delay, $e^{-j\omega T} = 1\angle -\omega T = 1\angle -\omega$, since $T = 1$ from the problem statement. Next, draw the phase plot of the system, $G(j\omega)$, using

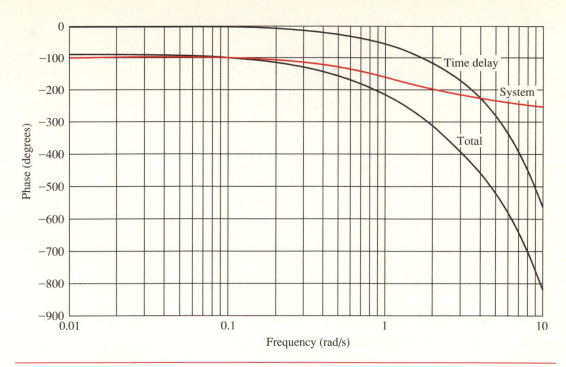

Figure 10.52 Phase Frequency Response Plot with a Delay of 1 Second for the System $G(s) = \dfrac{1}{s(s + 1)(s + 2)}$

the methods previously covered. Finally, add the two phase curves together to obtain the total phase response for $e^{-j\omega T} G(j\omega)$.

Notice that the delay yields a decreased phase margin since at any frequency the phase angle is more negative. Using a second-order approximation, this decrease in phase margin implies a lower damping ratio and a more oscillatory response for the closed-loop system.

Further, there is a decrease in the gain-margin frequency. On the magnitude curve, note that a reduction in the gain-margin frequency shows up as reduced gain margin, thus moving the system closer to instability.

In summary, then, systems with time delay can be handled using previously described frequency response techniques if the phase response is adjusted to reflect the time delay. Typically, time delay reduces gain and phase margins, resulting in increased percent overshoot or instability in the closed-loop response.

In this chapter we derived the relationships between time response performance and the frequency responses of the open and closed-loop systems. The methods derived, although yielding a different perspective, are simply alternatives to the root locus and steady-state error analysis previously covered.

10.13 Chapter-Objective Demonstration Problem

Our ongoing antenna position control system serves now as an example that summarizes the major objectives of the chapter. The problem demonstrates the use of frequency response methods to find the range of gain for stability and to design a value of gain to meet a percent overshoot requirement for the closed-loop step response.

Example 10.16

Design stability and percent overshoot using gain adjustment and frequency response methods for an antenna azimuth position control system.

Problem   Given the antenna azimuth position control system shown in Figure 2.36, use frequency response techniques to find the following:
 a. The range of preamplifier gain, K, required for stability
 b. Percent overshoot if the preamplifier gain is set to 30
 c. The estimated settling time
 d. The estimated peak time
 e. The estimated rise time

Solution   The schematic for the control system is shown in Figure 2.36. The resulting block diagram for the system as derived in Section 5.9 is shown in Figure 10.53. From Figure 10.53, the loop gain, $G(s)H(s)$, is

$$G(s)H(s) = \frac{6.63K}{s(s + 1.71)(s + 100)} \qquad (10.86)$$

Letting $K = 1$, we have the magnitude and phase frequency response plots shown in Figure 10.54.

 a. In order to find the range of K for stability, we notice from Figure 10.54 that the phase response is 180° at $\omega = 13.1$ rad/s. At this frequency, the magnitude plot is -68.41 dB. The gain, K, can be raised by 68.41 dB. Thus, $K = 2633.3$ will cause the system to be marginally stable. Hence, the system is stable if $0 < K < 2633.3$.

 b. To find the percent overshoot if $K = 30$, we first make a second-order approximation and assume that the second-order transient response equations relating percent overshoot, damping ratio, and phase margin are true for this system. In other words, we assume that Equation 10.70, which relates damping ratio to phase margin, is valid. If $K = 30$, the magnitude curve of Figure 10.54 is moved up by $20 \log 30 = 29.54$ dB. Therefore, the adjusted magnitude curve goes through zero dB at $\omega = 1$. At

Figure 10.53 Block Diagram for Antenna Azimuth Position Control System

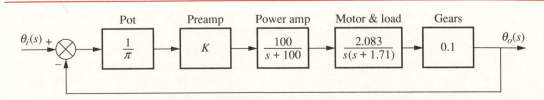

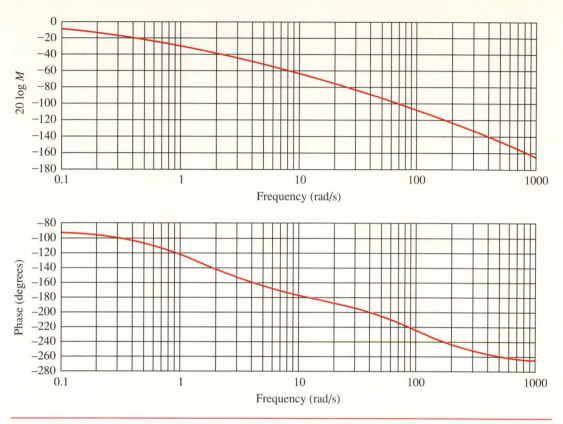

Figure 10.54 Open-Loop Frequency Response Plots for Antenna Azimuth
Position Control System ($K = 1$)

this frequency, the phase angle is $-120.89°$, yielding a phase margin of $59.11°$.
Using Equation 10.70 or Figure 10.46, $\zeta = 0.6$, or 9.48% overshoot. A computer
simulation shows 10%.

c. To estimate the settling time, we make a second-order approximation and use
Equation 10.52. Since $K = 30$ (29.54 dB), the open-loop magnitude response is -7
dB when the normalized magnitude response of Figure 10.54 is -36.54 dB. Thus,
the estimated bandwidth is 1.8 rad/s. Using Equation 10.52, $T_s = 4.25$ seconds. A
computer simulation shows a settling time of about 4.4 seconds.

d. Using the estimated bandwidth found in part **c**, along with Equation 10.53,
and the damping ratio found in part **a**, we estimate the peak time to be 2.5 seconds.
A computer simulation shows a peak time of 2.8 seconds.

e. To estimate the rise time, we use Figure 4.15 and find that the normalized
rise time for a damping ratio of 0.6 is 1.854. Using Equation 10.51, the estimated
bandwidth found in part **c**, and $\zeta = 0.6$, we find $\omega_n = 1.57$. Using the normalized
rise time and ω_n, we find $T_r = 1.854/1.57 = 1.18$ seconds. A simulation shows a
rise time of 1.2 seconds.

This chapter has dealt with the *analysis* of control systems via frequency response techniques. We have covered stability, transient response, and steady-state errors by showing the relationship of each to the frequency response. The next chapter covers the *design* of control systems, using the Bode plots.

10.14 Summary

Frequency response methods are an alternative to the root locus for analyzing and designing feedback control systems. Frequency response techniques can be used more effectively than the root locus to model physical systems in the laboratory. The input to a physical system can be sinusoidally varying with known frequency, amplitude, and phase angle. The system's output, which is also sinusoidal in the steady state, can then be measured for amplitude and phase angle at different frequencies. From this data the magnitude frequency response of the system, which is the ratio of the output amplitude to the input amplitude, can be plotted and used in place of an analytically obtained magnitude frequency response. Similarly, we can obtain the phase response by finding the difference between the output phase angle and the input phase angle at different frequencies. On the other hand, the root locus is more directly related to the time response.

The frequency response of a system can be represented either as a polar plot or as separate magnitude and phase diagrams. As a polar plot, the magnitude response is the length of a vector drawn from the origin to a point on the curve, whereas the phase response is the angle of that vector. In the polar plot, frequency is implicit and is represented by each point on the polar curve. The polar plot of $G(s)H(s)$ is known as a *Nyquist diagram*.

Separate magnitude and phase diagrams, sometimes referred to as *Bode plots,* present the data with frequency explicitly enumerated along the abscissa. The magnitude curve can be a plot of log-magnitude vs log-frequency. The other graph is a plot of phase angle vs log-frequency. An advantage of Bode plots over the Nyquist diagram is that they can be easily drawn using asymptotic approximations to the actual curve.

The Nyquist criterion sets forth the theoretical foundation from which the frequency response can be used to determine a system's stability. Using the Nyquist criterion and Nyquist diagram, or the Nyquist criterion and Bode plots, we can determine a system's stability.

Frequency response methods give us not only stability information, but also transient response information. By defining such frequency response quantities as gain margin and phase margin, the transient response can be analyzed or designed. Gain margin is the amount that the gain of a system can be increased before instability occurs if the phase angle is constant at 180°. Phase margin is the amount that the phase angle can be changed before instability occurs if the gain is held at unity.

While the open-loop frequency response leads to the results for stability and transient response described above, other design tools relate the closed-loop frequency response peak and bandwidth to the transient response. Since the closed-loop response is not as easy to obtain as the open-loop response because of the unavailability of the closed-loop poles, we use graphic aids in order to obtain the closed-loop frequency response from the open-loop frequency response. These graphic aids are the M and N circles and the Nichols chart. By superimposing the open-loop frequency response over the M and N circles or the Nichols chart, we are able to obtain the closed-loop frequency response and then analyze and design for transient response.

Today, with the availability of computers and appropriate software, frequency response plots can be obtained without relying on the graphic techniques described in this chapter. The program used for the root locus calculations and described in Appendix C is one such program.

This chapter has examined the analysis problem. We developed the relationships between frequency response and both stability and transient response. In the next chapter, we apply the concepts to the design of feedback control systems.

REVIEW QUESTIONS

1. Name four advantages of frequency response techniques over the root locus.
2. Define frequency response as applied to a physical system.
3. Name two ways to plot the frequency response.
4. Briefly describe how to obtain the frequency response analytically.
5. Define Bode plots.
6. Each pole of a system contributes how much of a slope to the Bode magnitude plot?
7. A system with only four poles and no zeros would exhibit what value of slope at high frequencies in a Bode magnitude plot?
8. A system with four poles and two zeros would exhibit what value of slope at high frequencies in a Bode magnitude plot?
9. Describe the asymptotic phase response of a system with a single pole at -2.
10. What is the major difference between Bode magnitude plots for first-order systems and for second-order systems?
11. For a system with three poles at -4, what is the maximum difference between the asymptotic approximation and the actual magnitude response?
12. Briefly state the Nyquist criterion.
13. What does the Nyquist criterion tell us?
14. What is a Nyquist diagram?
15. Why is the Nyquist criterion called a frequency response method?
16. When sketching a Nyquist diagram, what must be done with open-loop poles on the imaginary axis?

17. What simplification to the Nyquist criterion can we usually make for systems that are open-loop stable?

18. What simplification to the Nyquist criterion can we usually make for systems that are open-loop unstable?

19. Define gain margin.

20. Define phase margin.

21. Name two different frequency response characteristics that can be used to determine a system's transient response.

22. Name three different methods of finding the closed-loop frequency response from the open-loop transfer function.

23. Briefly explain how to find the static error constant from the Bode magnitude plot.

24. Describe the change in the open-loop frequency response magnitude plot if time delay is added to the plant.

25. If the phase response of a pure time delay were plotted on a linear phase vs linear frequency plot, what would be the shape of the curve?

PROBLEMS

1. Find analytical expressions for the magnitude and phase response for each $G(s)$ below.

   a. $G(s) = \dfrac{1}{(s + 2)(s + 4)}$

   b. $G(s) = \dfrac{1}{s(s + 2)(s + 4)}$

   c. $G(s) = \dfrac{(s + 5)}{(s + 2)(s + 4)}$

   d. $G(s) = \dfrac{(s + 3)(s + 5)}{s(s + 2)(s + 4)}$

2. For each function in Problem 1, make a plot of the log-magnitude and the phase, using log-frequency in rad/s as the ordinate. Do not use asymptotic approximations.

3. For each function in Problem 1, make a polar plot of the frequency response.

4. For each function in Problem 1, sketch the Bode asymptotic magnitude and asymptotic phase plots. Compare your results with your answers to Problem 1.

5. Sketch the Nyquist diagram for each of the systems in Figure P10.1.

6. Using the Nyquist criterion, find out whether or not each system of Problem 5 is stable or not.

7. Using the Nyquist criterion, find the range of K for stability for each of the systems in Figure P10.2.

Figure P10.1

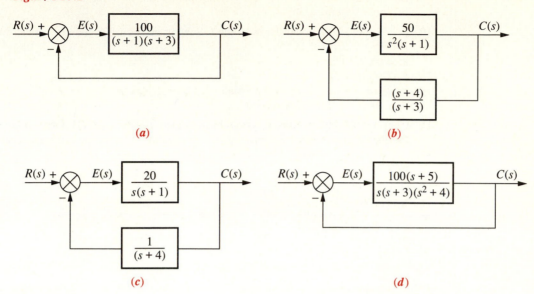

(a)

(b)

(c)

(d)

Figure P10.2

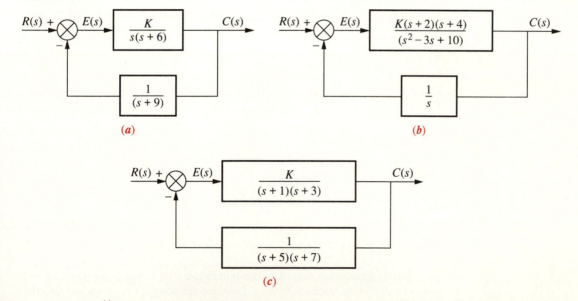

(a)

(b)

(c)

8. For each system of Problem 7, find the gain margin and phase margin if the value of K in each part of Problem 7 is
   a. $K = 500$
   b. $K = 600$
   c. $K = 10$

9. Derive Equation 10.51, the closed-loop bandwidth in terms of ζ and ω_n of a two-pole system.

10. For each closed-loop system with the following performance characteristics, find the closed-loop bandwidth.
    a. $\zeta = 0.3$, $T_s = 2$ seconds
    b. $\zeta = 0.3$, $T_p = 2$ seconds
    c. $T_s = 5$ seconds, $T_p = 3$ seconds
    d. $\zeta = 0.38$, $T_r = 2.8$ seconds

11. Consider the unity feedback system of Figure 10.10. For each $G(s)$ that follows, use the M and N circles to make a plot of the closed-loop frequency response.

    a. $G(s) = \dfrac{10}{s(s + 1)(s + 2)}$

    b. $G(s) = \dfrac{20}{(s + 3)(s + 4)(s + 5)(s + 6)}$

    c. $G(s) = \dfrac{50(s + 3)}{s(s + 2)(s + 4)}$

12. Repeat Problem 11 using the Nichols chart in place of the M and N circles.

13. Using the results of Problem 11, estimate the percent overshoot that can be expected in the step response for each system shown.

14. Use the results of Problem 12 to estimate the percent overshoot if the gain term in the numerator of the forward path of each part of the problem is respectively changed as follows:
    a. From 10 to 30
    b. From 20 to 50
    c. From 50 to 75

15. Using Bode diagrams, estimate the transient response of the system in Figure P10.3.

Figure P10.3

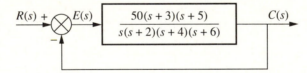

16. The open-loop frequency response shown in Figure P10.4, was experimentally obtained from a unity feedback system. Estimate the percent overshoot and steady-state error of the closed-loop system.

17. Consider the system in Figure P10.5.
    a. Find the phase margin if the system is stable for time delays of 0, 2, 3, 5, and 10 seconds.
    b. Find the gain margin if the system is stable for each of the time delays given in part a.

Figure P10.4

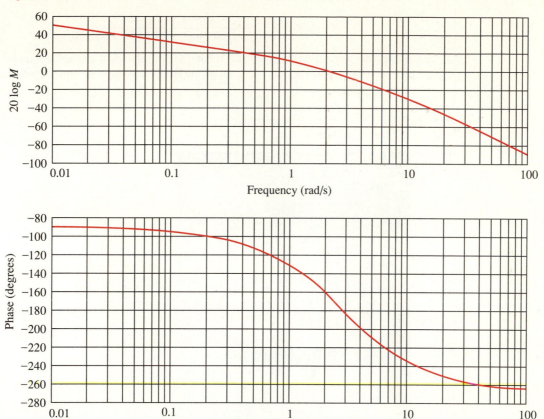

Figure P10.5

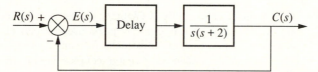

c. For what time delays mentioned in part **a**, is the system stable?

d. For each time delay that makes the system unstable, how much reduction in gain is required for the system to be stable?

18. Industrial robots, such as that shown in Figure P10.6, require accurate models for design of high performance. Many transfer function models for industrial robots assume interconnected rigid bodies with the drive-torque source modeled as a pure gain, or first-order system. Since the motions associated with the robot are connected to the drives through flexible linkages rather than rigid linkages, past modeling does not explain the resonances observed. An accurate, small-motion, linearized model has been developed that takes into consideration the flexible drive. The transfer function,

$$G(s) = 999.12 \frac{(s^2 + 8.94s + 44.7^2)}{(s + 20.7)(s^2 + 34.858s + 60.1^2)}$$

relates the angular velocity of the robot base to electrical current commands (Good, 1985). Make a Bode diagram of the frequency response and identify the resonant frequencies.

Figure P10.6 Robotic Manipulator (Courtesy of Odetics, Inc.)

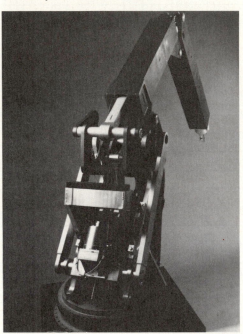

19. The charge-coupled device (CCD) that is used in video movie cameras to convert the image into electrical signals can be used as part of an automatic focusing system in 35-mm cameras. Automatic focusing can be implemented by focusing the center of the image on a charge-coupled device array through two lenses. The separation of the two images on the CCD is related to the focus. The camera senses the separation, and a computer drives the lens and focuses the image (see references to *Popular Photography* in this chapter's Bibliography). The automatic focus system is a position control, where the desired position of the lens is an input selected by pointing the camera at the subject. The output is the actual position of the lens. The Nikon N8008 in Figure P10.7(*a*) is an example of a camera that uses a CCD automatic focusing system. Figure P10.7(*b*) shows the automatic focusing feature represented as a position control system. Assuming the simplified model shown in Figure P10.7(*c*), draw the Bode diagrams and estimate the percent overshoot for a step input.

Figure P10.7 (*a*) A Cutaway View of the Nikon N8008 35-mm Camera Showing Parts of the CCD Automatic Focusing System (Courtesy of Nikon, Inc.) (*b*) Functional Block Diagram; (*c*) Block Diagram

(*a*)

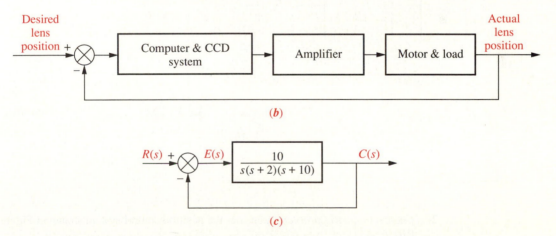

(*b*)

(*c*)

Figure P10.8 Block Diagram of a Ship's Roll Stabilizing System

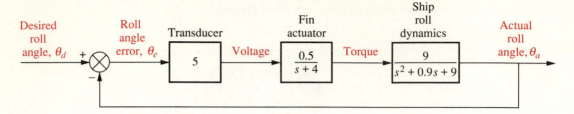

20. A ship's roll can be stabilized with a control system. A voltage applied to the fins' actuators creates a roll torque that is applied to the ship. The ship, in response to the roll torque, yields a roll angle. Assuming the block diagram for the roll control system shown in Figure P10.8, determine the gain and phase margins for the system and estimate the percent overshoot for a step change in the desired roll angle away from zero.

Figure P10.9 Position Control System

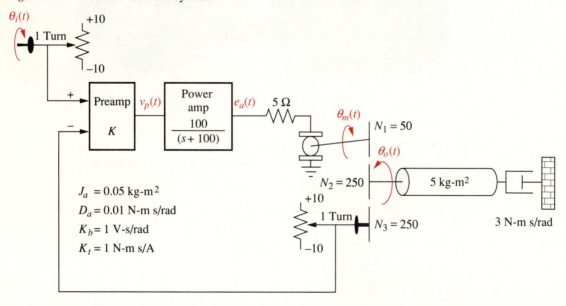

21. *Chapter-objective problem:* Consider the position control system shown in Figure P10.9.
    a. Find the range of gain, K, for stability.
    b. Find the percent overshoot for a step input if the gain, K, equals 3.

BIBLIOGRAPHY

Auto Focus SLR Update. *Popular Photography,* December 1987, pp. 72–75.

Bode, H. W. *Network Analysis and Feedback Amplifier Design.* Van Nostrand, Princeton, N.J., 1945.

Dorf, R. C. *Modern Control Systems.* 5th ed. Addison-Wesley, Reading, Mass., 1989.

Goldberg, N., and Frank, M. A. Pop Photo Camera Test of Minolta Maxxum 7000. *Popular Photography,* February 1986, pp. 44–48, 94.

Good, M. C.; Sweet, L. M.; and Strobel, K. L. Dynamic Models for Control System Design of Integrated Robot and Drive Systems. *Journal of Dynamic Systems, Measurement, and Control,* March 1985, pp. 53–59.

Hostetter, G. H.; Savant, C. J., Jr.; and Stefani, R. T. *Design of Feedback Control Systems.* 2d ed. Saunders College Publishing, New York, 1989.

Kuo, B. C. *Automatic Control Systems.* 5th ed. Prentice-Hall, Englewood Cliffs, N.J., 1987.

Kuo, F. F. *Network Analysis and Synthesis.* Wiley, New York, 1966.

Nikon N4004. *Popular Photography,* October 1987, pp. 52–56.

Nilsson, J. W. *Electric Circuits.* 2d ed. Addison-Wesley, Reading, Mass., 1986.

Nyquist, H. Regeneration Theory. *Bell Systems Technical Journal,* January 1932, pp. 126–147.

Ogata, K. *Modern Control Engineering.* 2d ed. Prentice Hall, Englewood Cliffs, N.J., 1990.

11

DESIGN VIA
FREQUENCY RESPONSE
METHODS

11.1 Introduction

In Chapter 8, we designed the transient response of a control system by adjusting the gain along the root locus. The design process consisted of finding the transient response specification on the root locus, setting the gain accordingly, and settling for the resulting steady-state error. The disadvantages of design by gain adjustment are that only the transient response and steady-state error represented by points along the root locus are available.

In order to meet other transient response specifications and, independently, steady-state error requirements, we designed cascade compensators in Chapter 9. In Chapter 11, we will use Bode plots to parallel the root locus design process from Chapters 8 and 9. Frequency response design methods, unlike the root locus method, can be implemented conveniently without a computer or other tool except for testing the design. On the other hand, frequency response methods are not as intuitive as the root locus.

When designing via frequency response methods, we use the concepts of stability, transient response, and steady-state error that we learned in Chapter 10. First, the Nyquist criterion tells us how to determine if a system is stable. Typically, an open-loop stable system is stable in closed-loop if the open-loop magnitude frequency response has a gain of less than 0 dB at the frequency where the phase frequency response is 180°. Second, percent overshoot is reduced by increasing the phase margin, and the speed of the response is increased by increasing the bandwidth. Finally, steady-state error is improved by increasing the low-frequency magnitude responses, even if the high-frequency magnitude response is attenuated.

These, then, are the basic facts underlying our design for stability, transient response, and steady-state error using frequency response methods, where the Nyquist criterion and the Nyquist diagram compose the underlying theory behind the design process. Thus, even though we use the Bode plots for ease in obtaining the frequency response, the design process can be verified with the Nyquist diagram when questions arise about interpreting the Bode plots. In particular, when the structure of the system is changed with additional compensator poles and zeros, the Nyquist diagram can offer a valuable perspective.

The emphasis in this chapter will be on the design of lag, lead, and lag-lead compensation. General design concepts are presented first, followed by step-by-step procedures. These procedures are only suggestions, and the student is encouraged to develop other procedures to arrive at the same goals. Although the concepts in general apply to the design of PI, PD, and PID controllers, in the interest of brevity, detailed procedures and examples will not be presented. The student is encouraged to extrapolate the concepts and designs covered and apply them to problems involving PI, PD, and PID compensation presented at the end of this chapter. Finally, the compensators developed in this chapter can be implemented with the realizations discussed in Section 9.6.

Chapter Objective

Given the antenna azimuth position control system shown in Figure 2.2, the student will be able to design the system gain to meet a percent overshoot requirement, using frequency response techniques. Further, the student will be able to design cascade compensation, using frequency response techniques, so that the system will meet a transient response requirement and, separately, a steady-state error requirement.

11.2 Gain Adjustment to Improve Transient Response

Let us begin our discussion of design via frequency response methods by talking about the simple gain-adjustment problem. Since the phase margin is linked to the system's transient response, we begin by designing the phase margin using gain adjustment.

Designing Phase Margin via Gain Adjustment on the Bode Plots

In Section 10.10 a relationship between damping ratio and phase margin was derived. Thus, percent overshoot can be designed by adjusting the gain on the open-loop frequency response to yield the required phase margin. This concept is shown in Figure 11.1.

The required phase margin is determined from the percent overshoot, using Equations 4.38 and 10.70. To design for the required phase margin using gain adjustment, (1) find the frequency, f_{Φ_M}, on the Bode phase diagram that yields the desired phase margin, CD, as shown in Figure 11.1, and (2) change the gain by an amount AB to force the magnitude curve to go through 0 dB at f_{Φ_M}. The amount of gain adjustment is the added gain required to produce the required phase margin.

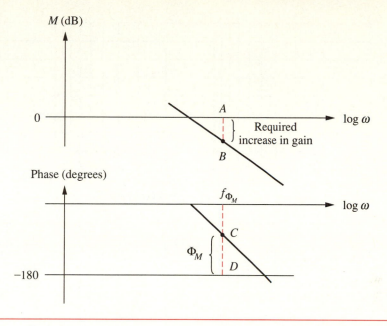

Figure 11.1 Bode Plots Showing Gain Adjustment for a
Specified Phase Margin

We now look at an example of designing the gain of a third-order system for
percent overshoot.

Example 11.1   Design a system for percent overshoot, using gain adjustment on the Bode plots.

Problem   For the position control system shown in Figure 11.2, find the value of
preamplifier gain, K, to yield a 9.48% overshoot in the transient response for a step
input. Use only frequency response methods.

Solution   Choose $K = 3.6$ in order to start the magnitude plot at 0 dB and at
$\omega = 0.1$ in Figure 11.3. For the first step, using Equation 4.38, we find that a
9.48% overshoot implies $\zeta = 0.6$. Equation 10.70 yields a 59.19° phase margin

Figure 11.2 System for Example 11.1

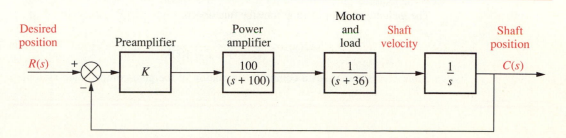

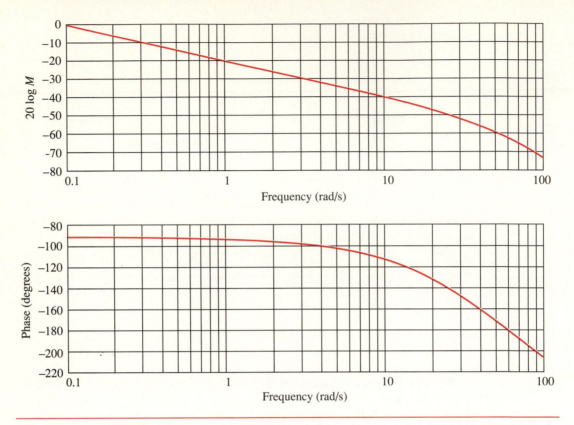

Figure 11.3 Bode Magnitude and Phase Plots for Example 11.1

for a damping ratio of 0.6. Now locate on the phase plot the frequency that yields a 59.19° phase margin. This frequency is found where the phase angle is the difference between −180° and 59.19°, or −120.81°. The value of the phase-margin frequency is 14.8 rad/s.

Now we proceed with step 2 of the design process. At a frequency of 14.8 rad/s on the magnitude plot, the gain is found to be −44.18 dB. This magnitude has to be raised to 0 dB to yield the required phase margin. Since the log-magnitude plot was drawn for $K = 3.6$, a 44.18 dB increase, or $K = 3.6 \times 161.808 = 582.51$, would yield the required phase margin for a 9.48% overshoot.

The gain-adjusted open-loop transfer function is

$$G(s) = \frac{58,251}{s(s + 36)(s + 100)} \tag{11.1}$$

Table 11.1 summarizes a computer simulation of the gain-compensated system.

Table 11.1 Characteristics of Gain-Compensated System of Example 11.1

Parameter	Value	Proposed Specification
K_v	16.18	—
Phase margin	59.23°	59.19°
Phase-margin frequency	14.8 rad/s	—
Percent overshoot	9	9.48
Peak time	0.18 seconds	—

In the next section we parallel the root locus compensator design in Chapter 9 and discuss the design of lag networks via the Bode diagrams.

11.3 Lag Compensation to Improve Steady-State Error

In Chapter 9 we used the root locus to design lag networks and PI controllers. Recall that these compensators permitted us to design for steady-state error without appreciably affecting the transient response. In this section we provide a parallel development using the Bode diagrams.

Visualizing the Effect of Lag Compensation on the Bode Plots

The function of the lag compensator as seen on the Bode diagrams is to (1) improve the static error constant by increasing only the low-frequency gain without any resulting instability, and (2) increase the phase margin of the system to yield the desired transient response. These concepts are illustrated in Figure 11.4.

The uncompensated system is unstable since the gain at 180° is greater than 0 dB. The lag compensator, while not changing the low-frequency gain, does reduce the high-frequency gain.[1] Thus, the low-frequency gain of the system can be made high to yield a large K_v without creating instability. This stabilizing effect of the lag network comes about because the gain at 180° of phase is reduced below 0 dB. Through judicious design, the magnitude curve can be reshaped, as shown in Figure 11.4, to go through 0 dB at the desired phase margin. Thus, both K_v and the desired transient response can be obtained.

---

[1] The name *lag compensator* comes from the fact that the typical phase angle response for the compensator, as shown in Figure 11.4, is always negative, or *lagging* in phase angle.

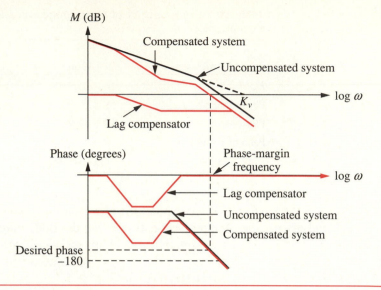

Figure 11.4 Concepts behind Lag Compensation

Lag Compensation Design Procedure

The plan of attack is as follows.

1. Set the gain, K, to that value that satisfies the steady-state error specification and plot the Bode magnitude and phase diagrams for this value of gain.

2. Find the frequency where the phase margin is 5° to 12° greater than the phase margin that yields the desired transient response (Ogata, 1990). This step compensates for the fact that the phase of the lag compensator may still contribute anywhere from −5° to −12° of phase at the phase-margin frequency.

3. Select a lag compensator whose magnitude response yields a composite Bode magnitude diagram that goes through 0 dB at the frequency found in step 2 above as follows: Draw the compensator's high-frequency asymptote to yield 0 dB at the frequency found in step 2; select the upper break frequency to be 1 decade below the frequency found in step 2; select the low-frequency asymptote to be at 0 dB; connect the compensator's high- and low-frequency asymptotes with a −20 dB/decade line to locate the lower break frequency.

4. Reset the system gain, K, to compensate for any attenuation in the lag network in order to keep the static error constant the same as that found in step 1.

From these steps, we see that we are relying upon the initial gain setting to meet the steady-state requirements and then relying upon the lag compensator's

−20 dB/decade slope to meet the transient response requirement by setting the zero dB crossing of the magnitude plot.

The transfer function of the lag compensator is

$$G_c(s) = \frac{s + \dfrac{1}{T}}{s + \dfrac{1}{\alpha T}} \qquad (11.2)$$

where $\alpha > 1$.

Figure 11.5 shows the frequency response curves for the lag compensator. The range of high frequencies shown in the phase plot is where we will design our phase margin. This region is after the second break frequency of the lag compensator, where we can rely on the attenuation characteristics of the lag network to reduce the total open-loop gain to unity at the phase-margin frequency. Further, in this region the phase response of the compensator will have minimal effect on our design of the phase margin. Since there is still some effect, approximately 5°

Figure 11.5 Frequency Response Plots of a Lag Compensator:
$$G_c(s) = \frac{(s + .1)}{(s + .01)}$$

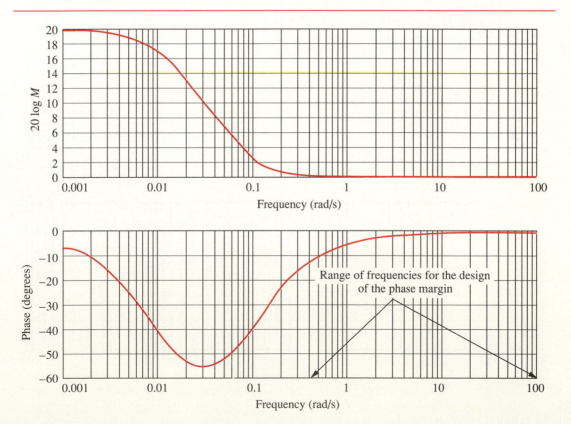

to 12°, we will add this amount to our phase margin to compensate for the phase response of the lag compensator (see step 2 above).

Example 11.2 Design lag compensation using Bode plots.

Problem Given the system of Figure 11.2, use Bode diagrams to design a lag compensator to yield a tenfold improvement in steady-state error over the gain-compensated system while keeping the percent overshoot at 9.48%

Solution We will follow the previously described lag compensation design procedure.

1. From Example 11.1, a gain, K, of 582.51 yields a 9.48% overshoot. Thus, for this system, $K_v = 16.181$. For a tenfold improvement in steady-state error, K_v must increase by a factor of 10, or $K_v = 161.81$. Therefore, the value of K in Figure 11.2 equals 5825.1, and the open-loop transfer function is

$$G(s)H(s) = \frac{582,510}{s(s + 36)(s + 100)} \tag{11.3}$$

Figure 11.6 Bode Plots for Example 11.2

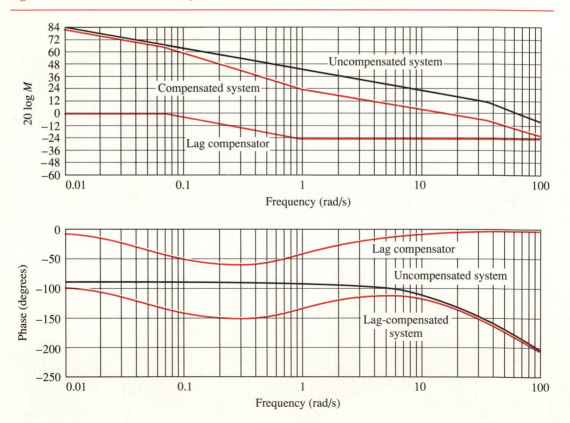

Table 11.2 Characteristics of the Lag-Compensated System of Example 11.2

Parameter	Value	Proposed Specification
K_v	161.74	161.81
Phase margin	62°	59.19°
Phase-margin frequency	11 rad/s	—
Percent overshoot	10	9.48
Peak time	0.25 seconds	—

The Bode plots for $K = 5825.1$ are shown in Figure 11.6.

2. The phase margin required for a 9.48% overshoot ($\zeta = 0.6$) is found from Equation 10.70 to be 59.19°. We increase this value of phase margin by 10° to 69.19° in order to compensate for the phase angle contribution of the lag compensator. Now find the frequency where the phase margin is 69.19°. This frequency occurs at a phase angle of $-180° + 69.19° = -110.81°$ and is 9.8 rad/s. At this frequency the magnitude plot must go through 0 dB. The magnitude at 9.8 rad/s is now $+24$ dB. Thus, the lag compensator must provide -24 dB attenuation at 9.8 rad/s.

3 & 4. We now design the compensator. First draw the high-frequency asymptote at -24 dB. Arbitrarily select the higher break frequency to be about one decade below the phase-margin frequency, or 0.98 rad/s. Starting at the intersection of this frequency with the lag compensator's high-frequency asymptote, draw a 20 dB/decade line until 0 dB is reached. The compensator must have a dc gain of unity to retain the value of K_v that we have already designed by setting $K = 5825.1$. The lower break frequency is found to be 0.07 rad/s. Hence the lag compensator's transfer function is

$$G_c(s) = \frac{0.0714(s + 0.98)}{(s + 0.07)} \tag{11.4}$$

where the gain of the compensator is 0.0714 to yield a dc gain of unity.

The compensated system's forward transfer function is thus

$$G(s)G_c(s) = \frac{41{,}591.21(s + 0.98)}{s(s + 36)(s + 100)(s + 0.07)} \tag{11.5}$$

The characteristics of the compensated system, found from a simulation and exact frequency response plots, are summarized in Table 11.2.

In this section we showed how to design a lag compensator to improve the steady-state error while keeping the transient response relatively unaffected. In the next section we discuss how to improve the transient response using frequency response methods.

11.4 Lead Compensation to Improve Transient Response

For second-order systems, we derived the relationship between phase margin and percent overshoot as well as the relationship between closed-loop bandwidth and other time-domain specifications, such as settling time, peak time, and rise time. When we designed the lag network to improve the steady-state error, we wanted a minimal effect on the phase diagram in order to yield an imperceptible change in the transient response. However, in designing lead compensators via Bode plots, we want to change the phase diagram, increasing the phase margin to reduce the percent overshoot, and increasing the gain crossover to realize a faster transient response.

Visualizing the Effect of Lead Compensation on the Bode Plots

The lead compensator increases the bandwidth by increasing the gain crossover frequency. At the same time, the phase diagram is raised at higher frequencies. The result is a larger phase margin and a higher phase-margin frequency. In the time domain, lower percent overshoots (larger phase margins) with smaller peak times (higher phase-margin frequency) are the results. The concepts are shown in Figure 11.7.

The uncompensated system has a small phase margin (B) and a low phase-margin frequency (A). Using a phase lead compensator, the phase angle plot (compensated system) is raised for higher frequencies.[2] At the same time the gain crossover frequency in the magnitude plot is increased from A rad/s to C rad/s. These effects yield a higher phase margin (D), a higher phase-margin frequency (C), and a larger bandwidth.

One advantage of the frequency response technique over the root locus is that we can implement a steady-state error requirement and then design a transient response. This specification of transient response with the constraint of a steady-state error is easier to implement with the frequency response technique than with the root locus. Notice that the initial slope, which determines the steady-state error, is not affected by the design for the transient response.

Lead Compensation Design Procedure

Let us first look at the frequency response characteristics of a lead network and derive some valuable relationships that will help us in the design process.

---

[2]The name *lead compensator* comes from the fact that the typical phase angle response shown in Figure 11.7 is always positive, or *leading* in phase angle.

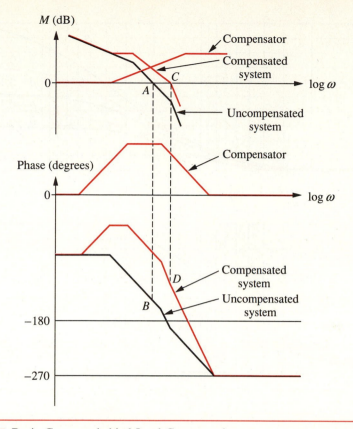

Figure 11.7 Basic Concepts behind Lead Compensation

Figure 11.8 shows plots of the lead network

$$G_c(s) = \frac{1}{\beta} \frac{s + \dfrac{1}{T}}{s + \dfrac{1}{\beta T}} \tag{11.6}$$

for various values of β, where $\beta < 1$. Notice that the peaks of the phase curve vary in maximum angle and in the frequency at which the maximum occurs. The dc gain of the compensator is set to unity with the coefficient $1/\beta$, in order not to change the dc gain designed for the static error constant when the compensator is inserted into the system.

In order to design a lead compensator and change both the phase margin and phase-margin frequency, it is helpful to have an analytical expression for the maximum value of phase and the frequency at which the maximum value of phase occurs, as shown in Figure 11.8.

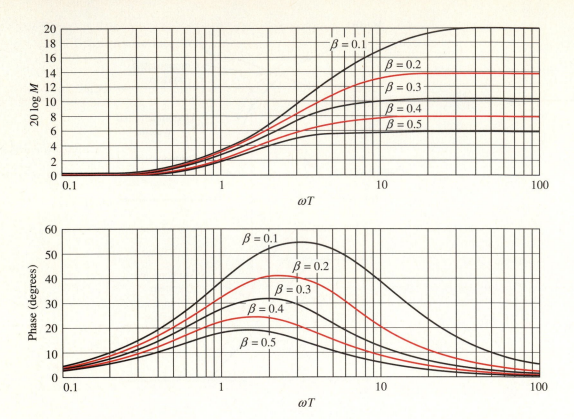

Figure 11.8 Frequency Response of a Lead Compensator:
$$G_c(s) = \frac{1}{\beta}\frac{(s + 1/T)}{(s + 1/\beta T)}$$

From Equation 11.6, the phase angle of the lead compensator, ϕ_c, is

$$\phi_c = \tan^{-1}\omega T - \tan^{-1}\omega\beta T \qquad (11.7)$$

Differentiating with respect to ω, we obtain

$$\frac{d\phi_c}{d\omega} = \frac{T}{1 + (\omega T)^2} - \frac{\beta T}{1 + (\omega\beta T)^2} \qquad (11.8)$$

Setting Equation 11.8 equal to zero, we find that the frequency, $\omega_{\max}$, at which the maximum phase angle, $\phi_{\max}$, occurs is

$$\omega_{\max} = \frac{1}{T\sqrt{\beta}} \qquad (11.9)$$

Substituting Equation 11.9 into Equation 11.6 with $s = j\omega_{max}$,

$$G_c(j\omega_{max}) = \frac{1}{\beta} \frac{j\omega_{max} + \dfrac{1}{T}}{j\omega_{max} + \dfrac{1}{\beta T}} = \frac{j\dfrac{1}{\sqrt{\beta}} + 1}{j\sqrt{\beta} + 1} \tag{11.10}$$

Making use of $\tan(\phi_1 - \phi_2) = (\tan\phi_1 - \tan\phi_2)/(1 + \tan\phi_1 \tan\phi_2)$, the maximum phase shift of the compensator, ϕ_{max}, is

$$\phi_{max} = \tan^{-1}\frac{1-\beta}{2\sqrt{\beta}} = \sin^{-1}\frac{1-\beta}{1+\beta} \tag{11.11}$$

and the compensator's magnitude at ω_{max} is

$$|G_c(j\omega_{max})| = \frac{1}{\sqrt{\beta}} \tag{11.12}$$

We are now ready to enumerate a design procedure. The plan of attack is as follows.

1. Find the closed-loop bandwidth required to meet the settling time, peak time, or rise time requirement (see Equations 10.51 through 10.53).
2. Since the lead compensator has negligible effect at low frequencies, set the gain, K, of the uncompensated system to the value that satisfies the steady-state error requirement.
3. Plot the Bode magnitude and phase diagrams for this value of gain and determine the uncompensated system's phase margin.
4. Find the phase margin to meet the damping ratio or percent overshoot requirement. Then evaluate the additional phase contribution required from the compensator.[3]
5. Determine the value of β (see Equations 11.6 and 11.11) from the lead compensator's required phase contribution.
6. Determine the compensator's magnitude at the peak of the phase curve (Equation 11.12).
7. Determine the new phase-margin frequency by finding where the uncompensated system's magnitude curve is the negative of the lead compensator's magnitude at the peak of the compensator's phase curve.
8. Design the lead compensator's break frequencies, using Equations 11.6 and 11.9 to find T and the break frequencies.
9. Reset the system gain to compensate for the lead compensator's gain.

---

[3]We know that the phase-margin frequency will be increased after the insertion of the compensator. At this new phase-margin frequency, the system's phase will be closer to $180°$ than originally estimated. Hence an additional phase should be added to that provided by the lead compensator to correct for the phase reduction caused by the original system.

10. Check the bandwidth to be sure the speed requirement in step 1 has been met.

11. Simulate to be sure all requirements are met.

12. Redesign if necessary to meet requirements.

From these steps, we see that we are increasing both the amount of phase margin (improving percent overshoot) and the gain-crossover frequency (increasing the speed). Using Equations 11.9, 11.11, 11.12, and steps 1 through 12 of this procedure, we can design a lead compensator to improve the transient response. Let us demonstrate.

Example 11.3   Design a lead compensator using Bode plots.

Problem   Given the system of Figure 11.2, design a lead compensator to yield a 20% overshoot and $K_v = 40$, with a peak time of 0.1 second.

Solution   The uncompensated system is $G(s) = \dfrac{100K}{s(s + 36)(s + 100)}$. We will follow the outlined procedure.

1. We first look at the closed-loop bandwidth needed to meet the speed requirement imposed by $T_p = 0.1$ second. From Equation 10.53, with $T_p = 0.1$ second and $\zeta = 0.456$ (i.e., 20% overshoot), a closed-loop bandwidth of 46.59 rad/s is required.

2. In order to meet the specification of $K_v = 40$, K must be set at 1440, yielding

$$G(s) = \frac{144,000}{s(s + 36)(s + 100)}.$$

3. The uncompensated system's frequency response plots for $K = 1440$ are shown in Figure 11.9.

4. A 20% overshoot implies a phase margin of 48.15°. The uncompensated system with $K = 1440$ has a phase margin of 33.97° at a phase-margin frequency of 29.7.[4] To increase the phase margin, we insert a lead network that adds enough phase to yield a 48.15° phase margin. Since we know that the lead network will also increase the phase-margin frequency, we add a correction factor to compensate for the lower uncompensated system's phase angle at this higher phase-margin frequency. Since we do not know the higher phase-margin frequency, we assume a correction factor of 10°. Thus, the total phase contribution required from the compensator is 48.15° − 33.97° + 10° = 24.18°. In summary, our compensated system should have a phase margin of 48.15° with a bandwidth of 46.59 rad/s. If the system's characteristics are not acceptable after the design, then a redesign with a different correction factor may be necessary.

5. Using Equation 11.11, $\beta = 0.42$ for $\phi_{max} = 24.18°$.

6. From Equation 11.12, the lead compensator's magnitude is 3.77 dB at ω_{max}.

7. If we select ω_{max} to be the new phase-margin frequency, the uncompensated system's magnitude at this frequency must be −3.77 dB to yield a 0 dB crossover at ω_{max} for the compensated system. The uncompensated system passes through −3.77 dB at $\omega_{max} = 39$ rad/s. This frequency is thus the new phase-margin frequency.

8. We now find the lead compensator's break frequencies. From Equation 11.9, $1/T = 25.27$ and $1/\beta T = 60.18$.

---

[4]Using actual (nonasymptotic) frequency response.

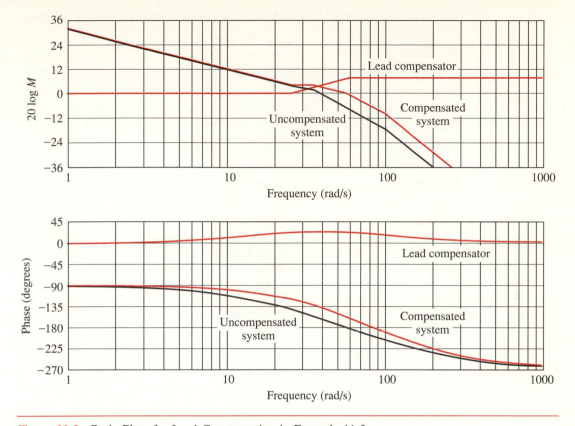

Figure 11.9 Bode Plots for Lead Compensation in Example 11.3

9. Hence, the compensator is given by

$$G_c(s) = \frac{1}{\beta} \frac{s + \dfrac{1}{T}}{s + \dfrac{1}{\beta T}} = 2.38 \frac{s + 25.27}{s + 60.18} \tag{11.13}$$

where 2.38 is the gain required to keep the dc gain of the compensator at unity so that $K_v = 40$ after the compensator is inserted.

The final, compensated open-loop transfer function is then given as

$$G_c(s)G(s) = \frac{342,720(s + 25.27)}{s(s + 36)(s + 100)(s + 60.18)} \tag{11.14}$$

10. From Figure 11.9, the lead-compensated open-loop magnitude response is -7 dB at approximately 85 rad/s. Thus, we estimate the closed-loop bandwidth to be 85 rad/s. Since this bandwidth exceeds the requirement of 46.59 rad/s, we assume the peak time specification is met. This conclusion about the peak time is based upon a second-order and asymptotic approximation that will be checked via simulation.

Figure 11.9 summarizes the design and shows the effect of the compensation.

Table 11.3 Characteristics of the Lead-Compensated System of Example 11.3

Parameter	Gain-Compensated System	Lead-Compensated System	Proposed Specification
K_v	40	40	40
Phase margin	33.97°	45.4°	48.15°
Phase-margin frequency	29.7 rad/s	39 rad/s	—
Closed-loop bandwidth	47 rad/s	68 rad/s	46.59 rad/s
Percent overshoot	37	21	20
Peak time	0.1 seconds	0.075 seconds	0.1 seconds

11. Final results, obtained from a simulation and the actual (nonasymptotic) frequency response, are shown in Table 11.3. Notice the increase in phase margin, phase-margin frequency, and closed-loop bandwidth after the lead compensator was added to the gain-adjusted system. The peak time and the steady-state error requirements have been met, although the phase margin is less than that proposed and the percent overshoot is 1% larger than proposed.

Finally, if the performance is not acceptable, a redesign is necessary.

Keep in mind that the previous examples were designs for third-order systems and must be simulated to ensure the desired transient results. In the next section we look at lag-lead compensation to improve steady-state error and transient response.

11.5 Lag-Lead Compensation to Improve Steady-State Error and Transient Response

In Section 9.4, using root locus, we designed lag-lead compensation to improve the transient response and the steady-state error of systems such as that shown in Figure 11.10. In this section we will repeat the design, using frequency response techniques. One method is to first design the lag compensation to lower the high-frequency gain, stabilize the system, and improve the steady-state error. Second, a lead compensator is designed to meet the phase-margin requirements. Let us look at another method.

Section 9.6 describes a passive lag-lead network that can be used in place of separate lag and lead networks. It may be more economical to use a single, passive network that performs both tasks since the buffer amplifier that separates the lag network from the lead network may be eliminated. In this section, we will emphasize lag-lead design, using a single, passive lag-lead network.

The transfer function of a single, passive lag-lead network is

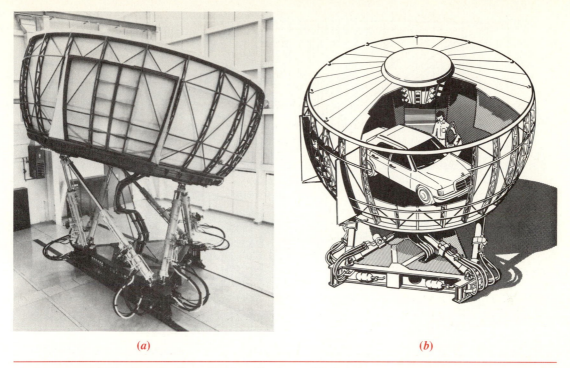

(*a*)   (*b*)

Figure 11.10 (*a*) The Daimler-Benz Driving Simulator; (*b*) Artist's
Conception Showing Vehicle Inside Simulator (Courtesy of
Mercedes-Benz of North America, Inc.)

$$G_c(s) = \left(\frac{s + \dfrac{1}{T_1}}{s + \dfrac{\gamma}{T_1}} \right) \left(\frac{s + \dfrac{1}{T_2}}{s + \dfrac{1}{\gamma T_2}} \right) \tag{11.15}$$

where $\gamma > 1$. The first term in parentheses produces the lead compensation, and
the second term in parentheses produces the lag compensation. The constraint that
we must follow here is that the single value γ replaces the quantity α for the lag
network in Equation 11.2 and the quantity β for the lead network in Equation
11.6. For our design, α and β must be reciprocals of each other. An example of
the frequency response of the passive lag-lead is shown in Figure 11.11.

Lag-Lead Design Procedure

1. Using a second-order approximation, find the closed-loop bandwidth required to meet the settling time, peak time, or rise time requirement (see Equations 10.52 and 10.53).

2. Set the gain, K, to the value required by the steady-state error specification.

3. Plot the Bode magnitude and phase diagrams for this value of gain.

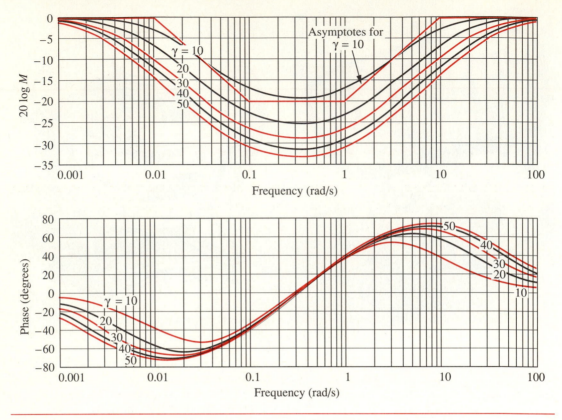

Figure 11.11 Sample Frequency Response Curves for a Lag-Lead

Compensator, $G_c(s) = \dfrac{(s + 1)(s + 0.1)}{(s + \gamma)(s + 0.1/\gamma)}$, where $\gamma = 10$

4. Using a second-order approximation, calculate the phase margin to meet the damping ratio or percent overshoot requirement, using equation 10.70.

5. Select a new phase-margin frequency near ω_{BW}.

6. At the new phase-margin frequency, determine the additional amount of phase lead required to meet the phase-margin requirement. Add a small contribution that will be required after the addition of the lag compensator.

7. Design the lag compensator by selecting the higher break frequency one decade below the new phase-margin frequency. The design of the lag compensator is not critical, and any design for the proper phase margin will be relegated to the lead compensator. The lag compensator simply provides stabilization of the system with the gain required for the steady-state error specification. Find the value of γ from the lead compensator's requirements. Using the phase required from the lead compensator, the phase response curve of Figure 11.8 can be used to find the value of

γ.[5] This value, along with the previously found lag's upper break frequency, allows us to find the lag's lower break frequency.

8. Design the lead compensator. Using the value of γ from the lag compensator design, and the value assumed for the new phase-margin frequency, find the lower and upper break frequency for the lead compensator, using Equation 11.9 and solving for T.

9. Check the bandwidth to be sure the speed requirement in step 1 has been met.

10. Redesign if phase-margin or transient specifications are not met.

Example 11.4

Design passive lag-lead compensation using Bode diagrams.

Problem   Given a unity feedback system where $G(s) = \dfrac{K}{s(s + 1)(s + 4)}$, design a passive lag-lead compensator using Bode diagrams to yield a 13.25% overshoot, a peak time of 2 seconds, and $K_v = 12$.

Solution   We will follow the steps previously mentioned in this section for lag-lead design.

1. The bandwidth required for a 2-second peak time is 2.29 rad/s.

2. In order to meet the steady-state error requirement, $K_v = 12$, the value of K is 48.

3. The Bode plots for the uncompensated system with $K = 48$ are shown in Figure 11.12. We can see that the system is unstable.

4. The required phase margin to yield a 13.25% overshoot is 55°.

5. Let us select $\omega = 1.8$ rad/s as the new phase-margin frequency.

6. At this frequency, the uncompensated phase is $-175°$ and would require, if we add a $-5°$ contribution from the lag compensator, a 55° contribution from the lead portion of the compensator.

7. The design of the lag compensator is next. The lag compensator allows us to keep the gain of 48 required for $K_v = 12$ and not have to lower the gain to stabilize the system. As long as the lag compensator stabilizes the system, the design parameters are not critical since the phase margin will be designed with the lead compensator. Thus, choose the lag compensator so that its phase response will have minimal effect at the new phase-margin frequency. Let us choose the lag compensator's higher break frequency to be 1 decade below the new phase-margin frequency, at 0.18 rad/s. Since we need to add 55° of phase shift with the lead compensator at $\omega = 1.8$ rad/s, Figure 11.8 tells us that if $\gamma = 10$ (since $\gamma = 1/\beta$, $\beta = 0.1$), we can obtain about 55° of phase shift from the lead compensator. Thus with $\gamma = 10$ and a new phase-margin frequency of $\omega = 1.8$ rad/s, the transfer function of the lag compensator is

$$G_{\text{Lag}}(s) = \frac{1}{\gamma} \frac{\left(s + \dfrac{1}{T_2}\right)}{\left(s + \dfrac{1}{\gamma T_2}\right)} = \frac{1}{10} \frac{(s + 0.18)}{(s + 0.018)} \tag{11.16}$$

---

[5] $\gamma = 1/\beta$

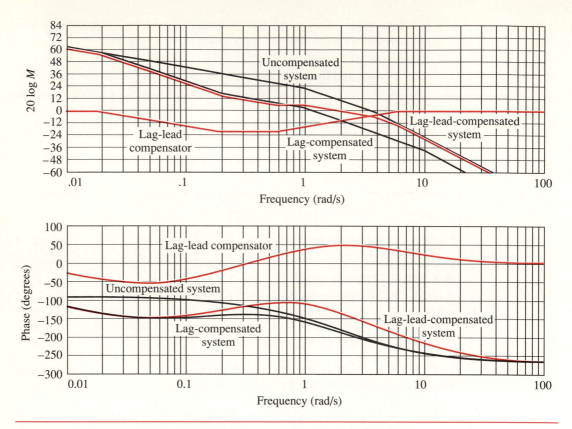

Figure 11.12 Bode Plots for Lag-Lead Compensation in Example 11.4

where the gain term, $1/\gamma$, keeps the dc gain of the lag compensator at 0 dB. The lag-compensated system's open-loop transfer function is

$$G_{\text{Lag-comp}}(s) = \frac{4.8(s + 0.18)}{s(s + 1)(s + 4)(s + 0.018)} \tag{11.17}$$

8. Now we design the lead compensator. At $\omega = 1.8$, the lag-compensated system has a phase angle of 180°. Using the values of $\omega_{\text{max}} = 1.8$ and $\beta = 0.1$, Equation 11.9 yields the lower break, $1/T_1 = 0.57$ rad/s. The higher break is then $1/\beta T_1 = 5.7$ rad/s. The lead compensator is

$$G_{\text{Lead}}(s) = \gamma \frac{\left(s + \dfrac{1}{T_1}\right)}{\left(s + \dfrac{\gamma}{T_2}\right)} = 10\frac{(s + 0.57)}{(s + 5.7)} \tag{11.18}$$

The lag-lead-compensated system's open-loop transfer function is

$$G_{\text{Lag-lead-comp}}(s) = \frac{48(s + 0.18)(s + 0.57)}{s(s + 1)(s + 4)(s + 0.018)(s + 5.7)} \tag{11.19}$$

Table 11.4 Characteristics of Gain-Compensated System of Example 11.4

Parameter	Value	Proposed Specifications
K_v	12	12
Phase margin	57°	55°
Phase-margin frequency	1.7 rad/s	—
Closed-loop bandwidth	3.2 rad/s	2.29 rad/s
Percent overshoot	12	13.25
Peak time	1.59 seconds	2.0 seconds

9. Now check the bandwidth. The closed-loop bandwidth is equal to that frequency where the open-loop magnitude response is approximately −7 dB. From Figure 11.12, the magnitude is −7 dB at approximately 4.5 rad/s (using the asymptotes). This bandwidth exceeds that required to meet the peak time requirement.

The design is now checked against a simulation and actual values. Table 11.4 summarizes the system's characteristics. The peak time requirement is also met. Again, if the requirements were not met, a redesign would be necessary.

Our discussion of design using the frequency response is now complete.

11.6 Chapter-Objective Demonstration Problems

Our ongoing antenna position control system serves now as an example to summarize the major objectives of the chapter. The problems demonstrate the use of frequency response methods to (1) design a value of gain to meet a percent overshoot requirement for the closed-loop step response and (2) design cascade compensation to meet both transient and steady-state error requirements.

Example 11.5   Design percent overshoot using gain-adjustment and frequency response methods for an antenna azimuth position control system.

Problem   Given the antenna azimuth position control system shown in Figure 2.36, use frequency response techniques to do the following:

  a. Find the preamplifier gain required for a closed-loop response of 20% overshoot for a step input.
  b. Estimate the settling time.

Solution The block diagram for the control system is shown in Figure 10.53, from which the loop gain is

$$G(s) = \frac{6.63K}{s(s + 1.71)(s + 100)} \qquad (11.20)$$

Letting $K = 1$, the magnitude and phase frequency response plots are shown in Figure 10.54.

a. To find K to yield a 20% overshoot, we first make a second-order approximation and assume that the second-order transient response equations relating percent overshoot, damping ratio, and phase margin are true for this system. Thus, a 20% overshoot implies a damping ratio of 0.456. Using Equation 10.70, this damping ratio implies a phase margin of 48.15°. The phase angle should therefore be $(-180° + 48.15°) = -131.85°$. The phase angle is $-131.85°$ at $\omega = 1.49$ rad/s, where the gain is -34.12 dB. Thus $K = 34.12$ dB $= 50.82$ for a 20% overshoot. Since the system is third-order, the second-order approximation should be checked. A computer simulation shows a 20% overshoot for the step response.

b. Letting $K = 50.82$ and unscaling the magnitude plot of Figure 10.54, we find -7 dB at $\omega = 2.5$ rad/s, which yields a closed-loop bandwidth of 2.5 rad/s. Using Equation 10.52 with $\zeta = 0.456$ and $\omega_{\text{BW}} = 2.5$, we find $T_s = 4.63$ seconds. A computer simulation shows a settling time of approximately 5 seconds.

We now demonstrate the design of cascade compensation, using frequency response techniques.

Example 11.6

Design transient response and steady-state error, using cascade compensation and frequency response techniques.

Problem Given the antenna azimuth position control system block diagram shown in Figure 10.53, use frequency response techniques and design cascade compensation for a closed-loop response of 20% overshoot for a step input, a fivefold improvement in steady-state error over the gain-compensated system operating at 20% overshoot, and a settling time of 2.5 seconds.

Solution Following the lag-lead design procedure, we first determine the value of gain, K, required to meet the steady-state error requirement.

1. Using Equation 10.52 with $\zeta = 0.456$, and $T_s = 2.5$ seconds, the required bandwidth is 4.63 rad/s.

2. From Example 11.5, the gain-compensated system's open-loop transfer function was, for $K = 50.82$,

$$G(s)H(s) = \frac{6.63K}{s(s + 1.71)(s + 100)} = \frac{336.94}{s(s + 1.71)(s + 100)} \qquad (11.21)$$

This function yields $K_v = 1.97$. If $K = 254.1$, then $K_v = 9.85$, a fivefold improvement.

3. The frequency response curves of Figure 10.54, which are plotted for $K = 1$, will be used for the solution.

4. Using a second-order approximation, a 20% overshoot requires a phase margin of 48.15°.

5. Select $\omega = 3$ rad/s to be the new phase-margin frequency.

6. The phase angle at the selected phase-margin frequency is $-152.04°$. This is a phase margin of 27.96°. Allowing for a 5° contribution from the lag compensator, the lead compensator must contribute $(48.15° - 27.96° + 5°) = 25.19°$.

7. The design of the lag compensator now follows. Choose the lag compensator upper break one decade below the new phase-margin frequency, or 0.3 rad/s. Figure 11.8 says that we can obtain 25.19° phase shift from the lead if $\beta = 0.4$ or $\gamma = 1/\beta = 2.5$. Thus, the lower break for the lag is at $1/(\gamma T) = 0.3/2.5 = 0.12$ rad/s. Hence,

$$G_{\text{Lag}}(s) = 0.4\frac{(s + 0.3)}{(s + 0.12)} \tag{11.22}$$

8. Finally, design the lead compensator. Using Equation 11.9, we have

$$T = \frac{1}{\omega_{\max}\sqrt{\beta}} = \frac{1}{3\sqrt{0.4}} = 0.527 \tag{11.23}$$

Therefore the lead compensator lower break frequency is $1/T = 1.897$ rad/s, and the upper break frequency is $1/(\beta T) = 4.74$ rad/s.

Thus, the lag-lead-compensated forward path is

$$G_{\text{Lag-lead-comp}}(s) = \frac{(6.63)(254.1)(s + 0.3)(s + 1.897)}{s(s + 1.71)(s + 100)(s + 0.12)(s + 4.74)} \tag{11.24}$$

9. A plot of the open-loop frequency response for the lag-lead-compensated system shows -7 dB at 5.3 rad/s. Thus, the bandwidth meets the design requirements for peak time.

Simulation of the compensated system shows a 21% overshoot and a settling time of approximately 2.8 seconds, compared to a 20% overshoot for the uncompensated system and a settling time of approximately 2.9 seconds. K_v for the compensated system is 9.86 compared to the uncompensated system value of 1.97.

11.7 Summary

This chapter covered the design of feedback control systems using frequency response techniques. In particular, we learned how to design by gain adjustment as well as cascaded lag, lead, and lag-lead compensation. Time response characteristics were related to the phase margin, phase-margin frequency, and bandwidth.

Design by gain adjustment consisted of adjusting the gain to meet a phase-margin specification. We located the phase-margin frequency and adjusted the gain to 0 dB.

A lag compensator is basically a low-pass filter. The low-frequency gain can be raised to improve the steady-state error, and the high-frequency gain is reduced to yield stability. Lag compensation consists first of setting the gain to meet the steady-state error requirement and then reducing the high-frequency gain to create stability and meet the phase-margin requirement for the transient response.

A lead compensator is basically a high-pass filter. The lead compensator increases the high-frequency gain while keeping the low-frequency gain the same. Thus, the steady-state error can be designed first. At the same time, the lead compensator increases the phase angle at high frequencies. The effect is to produce a faster, stable system since the uncompensated phase margin now occurs at a higher frequency.

A lag-lead compensator combines the advantages of both the lag and the lead compensator. First the lag compensator is designed to yield the proper steady-state error with improved stability. Next, the lead compensator is designed to speed up the transient response. If a single network is used as the lag-lead, additional design considerations are applied so that the ratio of the lag zero to the lag pole is the same as the ratio of the lead pole to the lead zero.

REVIEW QUESTIONS

1. What major advantage does compensator design by frequency response have over root locus design?
2. How is gain adjustment related to the transient response on the Bode diagrams?
3. Briefly explain how a lag network allows the low-frequency gain to be increased to improve steady-state error without having the system become unstable.
4. From the Bode plot perspective, briefly explain how the lag network does not appreciably affect the speed of the transient response.
5. Why is the phase margin increased above that desired when designing a lag compensator?
6. Compare the following for uncompensated and lag-compensated systems designed to yield the same transient response: low-frequency gain, phase-margin frequency, gain curve value around the phase-margin frequency, and phase curve values around the phase-margin frequency.
7. From the Bode diagram viewpoint, briefly explain how a lead network increases the speed of the transient response.
8. Based upon your answer to Question 7, explain why lead networks do not cause instability.
9. Why is a correction factor added to the phase margin required to meet the transient response?
10. When designing a lag-lead network, what difference is there in the design of the lag portion as compared to a separate lag compensator.

PROBLEMS

1. Design the value of gain, K, for a gain margin of 10 dB in the unity feedback system of Figure P11.1 if

   a. $G(s) = \dfrac{K}{(s + 2)(s + 5)(s + 8)}$

   b. $G(s) = \dfrac{K}{s(s + 2)(s + 5)}$

Figure P11.1

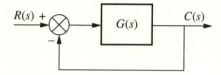

2. For each of the systems in Problem 1, design the gain, K, for a phase margin of 40°.

3. Given the unity feedback system of Figure P11.1, use frequency response methods to determine the value of gain, K, to yield a step response with a 20% overshoot if

   a. $G(s) = \dfrac{K}{s(s + 5)(s + 7)}$

   b. $G(s) = \dfrac{K(s + 2)}{s(s + 4)(s + 5)(s + 6)}.$

4. The unity feedback system of Figure P11.1 with

$$G(s) = \frac{K}{s(s + 4)}$$

is operating with a 15% overshoot and a 2-second settling time. Using frequency response techniques, design a compensator to yield $K_v = 50$ with the phase-margin frequency and phase margin remaining approximately the same as in the uncompensated system.

5. The unity feedback system shown in Figure P11.1 with

$$G(s) = \frac{K}{s(s + 5)(s + 8)}$$

is operating with a 20% overshoot. Using frequency response methods, design a compensator to yield a fivefold improvement in steady-state error without appreciably changing the transient response.

6. Design a PI controller for the system of Figure 11.2 that will yield zero error for a ramp input and a 9.48% overshoot for a step input.

7. Design a compensator for the unity feedback system of Figure P11.1 with

$$G(s) = \frac{K}{s(s + 2)(s + 4)(s + 6)}$$

to yield a $K_v = 2$ and a phase margin of $30°$.

8. The unity feedback system of Figure P11.1 with

$$G(s) = \frac{K(s + 6)}{s(s + 2)(s + 3)}$$

is operating with a 25% overshoot.
a. Find the settling time.
b. Find K_v.
c. Find the phase margin and the phase-margin frequency.
d. Using frequency response techniques, design a compensator that will yield a threefold improvement in K_v and a twofold reduction in settling time while keeping the overshoot at 25%.

9. Repeat the design of Example 11.3 using a PD controller.

10. An aircraft roll control system is shown in Figure P11.2. The torque on the aileron generates a roll rate. The resulting roll angle is then controlled through a feedback system as shown. Design a lead compensator for a $60°$ phase margin and $K_v = 5$.

Figure P11.2

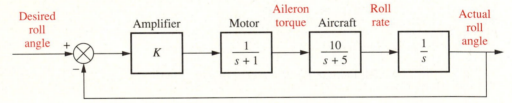

11. Self-guided vehicles, such as that shown in Figure P11.3(a), are used in factories to transport products from station to station. One method of construction is to embed a wire in the floor to provide guidance. Another method is to use an on-board computer and a laser scanning device. Bar-coded reflective devices at known locations allow the system to determine the vehicle's angular position. This system allows the vehicle to travel anywhere, including between buildings (Stefanides, 1987). Figure P11.3(b) shows a simplified block diagram of the vehicle's bearing control system. For 11% overshoot, K is set equal to 2. Design a lag compensator using frequency response techniques to improve the steady-state error by a factor of 30 over that of the uncompensated system.

12. Given a unity feedback system with

$$G(s) = \frac{K}{s(s + 1)(s + 4)}$$

design a PID controller to yield zero steady-state error for a ramp input, as well as a 12% overshoot, and a peak time less than 2 seconds for a step input. Use only frequency response methods.

13. *Chapter-objective problem:* Given the position control system shown in Figure P10.9, find the value of K to yield a 25% overshoot for a step input. Use only frequency response techniques.

Figure P11.3 (*a*) A Self-Guided Vehicle (Courtesy of Mannesman Demag Corp.); (*b*) Simplified Block Diagram

(a)

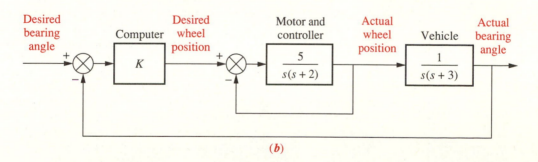

(b)

14. *Chapter-objective problem:* Given the position control system shown in Figure P10.9, design a lag-lead compensator to yield a 15% overshoot and $K_v = 20$. In order to speed up the system, the compensated system's phase-margin frequency will be set to 4.6 times the phase-margin frequency of the uncompensated system.

BIBLIOGRAPHY

D'Azzo, J. J. and Houpis, C. H. *Feedback Control System Analysis and Synthesis.* 2d ed. McGraw-Hill, New York, 1966.

Dorf, R. C. *Modern Control Systems.* 5th ed. Addison-Wesley, Reading, Mass., 1989.

Flower, T. L., and Son, M. Motor Drive Mechanics and Control Electronics for a High Performance Plotter. *HP Journal*, November 1981, pp. 12–15.

Hostetter, G. H.; Savant, C. J.; and Stefani, R. T. *Design of Feedback Control Systems.* 2d ed. Saunders College Publishing, New York, 1989.

Kuo, B. C. *Automatic Control Systems.* 5th ed. Prentice Hall, Englewood Cliffs, N.J., 1987.

Ogata, K. *Modern Control Engineering.* 2d ed. Prentice Hall, Englewood Cliffs, N.J., 1990.

Phillips, C. L., and Harbor, R. D. *Feedback Control Systems.* Prentice Hall, Englewood Cliffs, N.J., 1988.

Raven, F. H. *Automatic Control Engineering.* 4th ed. McGraw-Hill, New York, 1987.

Stefanides, E. J. Self-Guided Vehicles Upgrade Materials Handling. *Design News*, 7 December 1987, pp. 80–81.

*12

DESIGN VIA THE STATE-SPACE APPROACH

12.1 Introduction

In Chapter 3 we introduced the concepts of state-space analysis and system modeling. We showed that state-space methods, like transform methods, are simply tools for analyzing and designing feedback control systems. However, state-space techniques can be applied to a wider class of systems than transform methods. Systems with nonlinearities, such as that shown in Figure 12.1, and multiple-input, multiple-output systems are just two of the candidates for the state-space approach. In this book, however, we apply the approach only to linear systems.

In Chapters 9 and 11 we applied frequency domain methods to system design. The basic design technique is to create a compensator in cascade with the plant or in the feedback path that has the correct additional poles and zeros to yield a desired transient response and steady-state error.

One of the drawbacks of frequency domain methods of design, using either root locus or frequency response techniques, is that after designing the location of the dominant second-order pair of poles, we keep our fingers crossed, hoping that the higher-order poles do not affect the second-order approximation. What we would like to be able to do is specify *all* closed-loop poles of the higher-order system. Frequency domain methods of design prevent us from uniquely specifying all poles in systems of an order higher than two because the design procedure does not allow for a sufficient number of unknown parameters to uniquely place all of the closed-loop poles. One gain to adjust, or compensator pole and zero to select, does not yield a sufficient number of parameters to place all of the closed-loop poles at desired locations. Remember, to place n unknown quantities, you need n adjustable parameters. State-space methods solve this problem by introducing into the system (1) other adjustable parameters and (2) the technique for finding these parameter values, so that we can properly place all poles of the closed-loop system.[1]

---

[1] This is an advantage as long as we know where to place the higher-order poles. This is not always the case. One course of action is to place the higher-order poles far from the dominant second-order poles or near a closed-loop zero to keep the second-order system design valid. Another approach is to use optimal control concepts, which are beyond the scope of this text.

Figure 12.1 Robotic Arm Selects a Video Tape after Bar
Codes on the Tape Are Scanned for Identification. The Tape
Is Then Inserted into a Video Cassette Player. The System
Can Be Used by TV Stations (Courtesy of Odetics, Inc.)

On the other hand, state-space methods do not allow for the specification of closed-loop zero locations, which frequency domain methods do allow through placement of the lead compensator zero. This is a disadvantage of state-space methods since the location of the zero does indeed affect the transient response. Also, a state-space design may prove to be very sensitive to parameter changes.

Finally, there is a wide range of computational support for state-space methods; many software packages support the matrix algebra required by the design process. However, as mentioned before, the advantages of computer support are balanced by the loss of graphic insight into a design problem that the frequency domain methods yield.

This chapter should be considered only an introduction to state-space design; we introduce one state-space design technique and apply it only to linear systems. Advanced study is required to apply state-space techniques to the design of systems beyond the scope of this textbook.

Chapter Objective

Given the antenna azimuth position control system shown in Figure 2.2, the student will be able to design the desired transient response by (1) specifying all closed-loop poles, (2) introducing additional adjustable parameters through which the state variables are fed back to the input, and (3) finding the values of these parameters and placing the closed-loop poles at the desired locations. Further, the student will be able to design the system knowing how to estimate the state variables when all of the state variables of the plant cannot be measured.

12.2 Controller Design

In this section we show how to introduce additional parameters into a system so that we can control the location of all closed-loop poles. An nth-order feed back control system has an nth-order closed-loop characteristic equation of the form

$$s^n + a_{n-1}s^{n-1} + \cdots + a_1 s + a_0 = 0 \tag{12.1}$$

Since the coefficient of the highest power of s is unity, there are n coefficients whose values determine the system's closed-loop pole locations. Thus, if we can introduce n adjustable parameters into the system and relate them to the coefficients in Equation 12.1, all of the poles of the closed-loop system can be set to any desired location.

Topology for Pole Placement

In order to lay the groundwork for the approach, consider a plant represented in state space by

$$\dot{\mathbf{x}} = \mathbf{A}\mathbf{x} + \mathbf{B}u \tag{12.2a}$$

$$y = \mathbf{C}\mathbf{x} \tag{12.2b}$$

and shown pictorially in Figure 12.2(a), where light lines are scalars and the heavy lines are vectors.

In a typical feedback control system the output, y, is fed back to the summing junction. It is now that the topology of the design changes. Instead of feeding back y, what if we feed back all of the state variables? If each state variable is fed back to the control, u, through a gain, K_i, then there would be n gains, K_i, that could be adjusted to yield the required closed-loop pole values. The feedback through the gains, K_i, is represented in Figure 12.2(b) by the feedback vector $-\mathbf{K}$.

The state equations for the closed-loop system of Figure 12.2(b) can be written by inspection as

$$\dot{\mathbf{x}} = \mathbf{A}\mathbf{x} + \mathbf{B}u = \mathbf{A}\mathbf{x} + \mathbf{B}(-\mathbf{K}\mathbf{x} + r) = (\mathbf{A} - \mathbf{B}\mathbf{K})\mathbf{x} + \mathbf{B}r \tag{12.3a}$$

$$y = \mathbf{C}\mathbf{x} \tag{12.3b}$$

Before continuing, the reader should have a good idea of how the feedback system of Figure 12.2(b) is actually implemented. As an example, assume a plant signal-flow graph in phase-variable form, as shown in Figure 12.3(a). Each state variable is then fed back to the plant's input, u, through a gain, k_i, as shown in Figure 12.3(b). Although we will cover other representations later in the chapter,

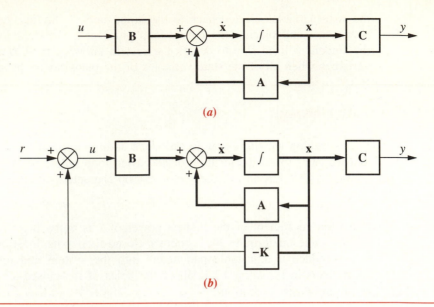

Figure 12.2 (*a*) Representation of a Plant Described in State Space; (*b*) Plant with State Feedback

the phase-variable representation yields the simplest evaluation of the feedback gains.

The design of state-variable feedback for closed-loop pole placement consists of equating the characteristic equation of a closed-loop system, such as that shown in Figure 12.3(*b*), to a desired characteristic equation and then finding the values of the feedback gains, k_i.

If a plant like that shown in Figure 12.3(*a*) is of high order and not represented in phase-variable form, the solution for the k_i's can be intricate. Thus, it is advisable to transform the system to phase-variable form, design the k_i's, and then transform the system back to its original representation. We perform this conversion in Section 12.4, where we develop a method for performing the transformations. Until then, let us direct our attention to plants represented in phase-variable form.

Development of Pole-Placement Methodology for Plants Represented in Phase-Variable Form

To apply pole-placement methodology to plants represented in phase-variable form, we take the following steps:

1. Represent the plant in phase-variable form.
2. Feedback each phase variable to the input of the plant through a gain, k_i.

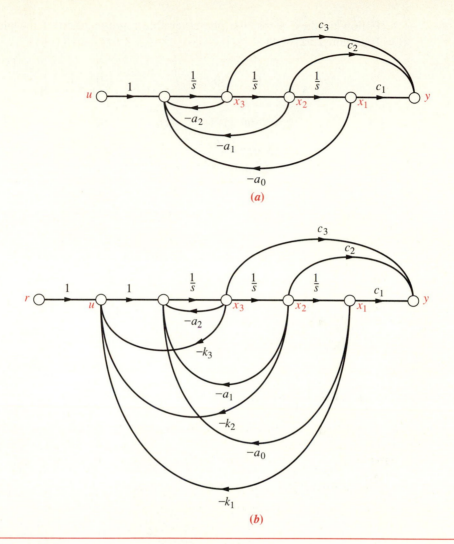

Figure 12.3 (*a*) Phase-Variable Representation for Plant;
(*b*) Plant with State-Variable Feedback

3. Find the characteristic equation for the closed-loop system represented in step 2.

4. Decide upon all closed-loop pole locations and determine an equivalent characteristic equation.

5. Equate like coefficients of the characteristic equations from steps 3 and 4 and solve for k_i.

Following these steps, the phase-variable representation of the plant is given by Equation 12.2, with

$$
\mathbf{A} = \begin{bmatrix} 0 & 1 & 0 & \cdots & 0 \\ 0 & 0 & 1 & \cdots & 0 \\ \vdots & \vdots & \vdots & \vdots & \vdots \\ -a_0 & -a_1 & -a_2 & \cdots & -a_{n-1} \end{bmatrix}; \quad \mathbf{B} = \begin{bmatrix} 0 \\ 0 \\ \vdots \\ 1 \end{bmatrix};
$$

$$
\mathbf{C} = \begin{bmatrix} c_1 & c_2 & \cdots & c_n \end{bmatrix} \tag{12.4}
$$

The characteristic equation of the plant is thus

$$
s^n + a_{n-1}s^{n-1} + \cdots + a_1 s + a_0 = 0 \tag{12.5}
$$

Now form the closed-loop system by feeding back each state variable to u, forming

$$
u = -\mathbf{Kx} \tag{12.6}
$$

where

$$
\mathbf{K} = \begin{bmatrix} k_1 & k_2 & \cdots & k_n \end{bmatrix} \tag{12.7}
$$

The k_i's are the phase variables' feedback gains.

Using Equation 12.3a with Equations 12.4 and 12.7, the system matrix, $\mathbf{A} - \mathbf{BK}$, for the closed-loop system is

$$
\mathbf{A} - \mathbf{BK} = \begin{bmatrix} 0 & 1 & 0 & \cdots & 0 \\ 0 & 0 & 1 & \cdots & 0 \\ \vdots & \vdots & \vdots & \vdots & \vdots \\ -(a_0 + k_1) & -(a_1 + k_2) & -(a_2 + k_3) & \cdots & -(a_{n-1} + k_n) \end{bmatrix}
$$
$$
\tag{12.8}
$$

Since Equation 12.8 is in phase-variable form, the characteristic equation of the closed-loop system can be written by inspection as

$$
\det(s\mathbf{I} - (\mathbf{A} - \mathbf{BK})) = s^n + (a_{n-1} + k_n)s^{n-1} + (a_{n-2} + k_{n-1})s^{n-2}
$$
$$
+ \cdots (a_1 + k_2)s + (a_0 + k_1) = 0 \tag{12.9}
$$

Notice the relationship between Equations 12.5 and 12.9. For plants represented in phase-variable form, we can write by inspection the closed-loop characteristic equation from the open-loop characteristic equation by adding the appropriate k_i to each coefficient.

Now assume that the desired characteristic equation for proper pole placement is

$$
s^n + d_{n-1}s^{n-1} + d_{n-2}s^{n-2} + \cdots + d_2 s^2 + d_1 s + d_0 = 0 \tag{12.10}
$$

where the d_i's are the desired coefficients. Equating Equations 12.9 and 12.10, we obtain

$$
d_i = a_i + k_{i+1} \qquad i = 0, 1, 2, \ldots, n - 1 \tag{12.11}
$$

from which

$$k_{i+1} = d_i - a_i \tag{12.12}$$

Now that we have found the denominator of the closed-loop transfer function, let us find the numerator. For systems represented in phase-variable form, we learned that the numerator polynomial is formed from the coefficients of the output coupling matrix, **C**. Since Figures 12.3(a) and (b) are both in phase-variable form and have the same output coupling matrix, we conclude that the numerators of their transfer functions are the same. Let us look at a design example.

Example 12.1

Find the feedback gains for proper closed-loop pole placement for a single-input, single-output system.

Problem Given the following plant,

$$G(s) = \frac{20(s + 5)}{s(s + 1)(s + 4)} \tag{12.13}$$

design the phase-variable feedback gains to yield a 9.48% overshoot and a settling time of 0.74 seconds.

Solution We begin by calculating the desired closed-loop characteristic equation. Using the transient response requirements, the closed-loop poles are $-5.4 \pm j7.2$. Since the system is third-order, we must select another closed-loop pole. The closed-loop system will have a zero at -5, the same as the open-loop system. We could select the third closed-loop pole to cancel the closed-loop zero. However, to demonstrate the effect of the third pole and the design process, including the need for simulation, let us choose -5.1 as the location of the third closed-loop pole.

Now, draw the signal-flow diagram for the plant. The result is shown in Figure 12.4(a). Next, feed back all state variables to the control, u, through gains k_i, as shown in Figure 12.4(b).

Writing the closed-loop system's state equations from Figure 12.4(b), we have

$$\dot{\mathbf{x}} = \begin{bmatrix} 0 & 1 & 0 \\ 0 & 0 & 1 \\ -k_1 & -(4 + k_2) & -(5 + k_3) \end{bmatrix} \mathbf{x} + \begin{bmatrix} 0 \\ 0 \\ 1 \end{bmatrix} r \tag{12.14a}$$

$$y = [\,100 \quad 20 \quad 0\,]\,\mathbf{x} \tag{12.14b}$$

Comparing Equation 12.14 to Equation 12.3, we identify the closed-loop system matrix as

$$\mathbf{A} - \mathbf{BK} = \begin{bmatrix} 0 & 1 & 0 \\ 0 & 0 & 1 \\ -k_1 & -(4 + k_2) & -(5 + k_3) \end{bmatrix} \tag{12.15}$$

To find the closed-loop system's characteristic equation, form

$$\det\,(s\mathbf{I} - (\mathbf{A} - \mathbf{BK})) = s^3 + (5 + k_3)s^2 + (4 + k_2)s + k_1 = 0 \tag{12.16}$$

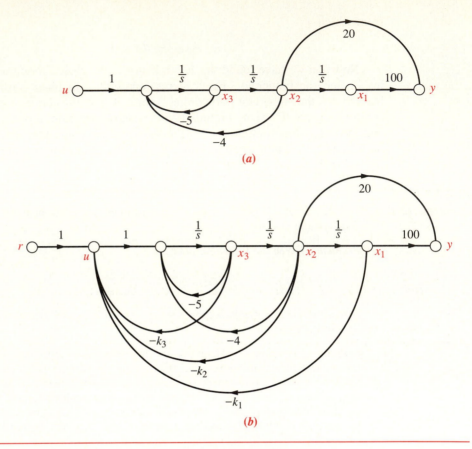

Figure 12.4 (**a**) Phase-Variable Representation for Plant of Example 12.1; (**b**) Plant with State-Variable Feedback

This equation must match the desired characteristic equation,

$$s^3 + 15.9s^2 + 136.08s + 413.1 = 0 \tag{12.17}$$

formed from the poles $-5.4 + j7.2$, $-5.4 - j7.2$, and -5.1, which were previously determined.

Equating the coefficients of Equations 12.16 and 12.17, we obtain

$$k_1 = 413.1; \quad k_2 = 132.08; \quad k_3 = 10.9 \tag{12.18}$$

Finally, the zero term of the closed-loop transfer function is the same as the zero term of the open-loop system, or $(s + 5)$.

Using Equation 12.14, we obtain the following state-space representation of the closed-loop system:

$$\dot{\mathbf{x}} = \begin{bmatrix} 0 & 1 & 0 \\ 0 & 0 & 1 \\ -413.1 & -136.08 & -15.9 \end{bmatrix} \mathbf{x} + \begin{bmatrix} 0 \\ 0 \\ 1 \end{bmatrix} r \tag{12.19a}$$

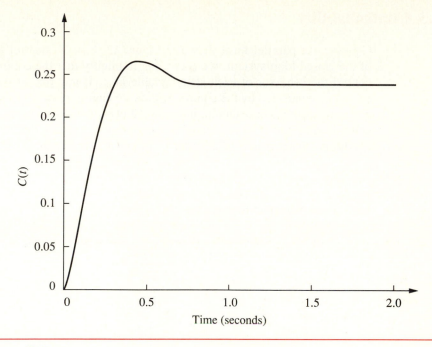

Figure 12.5 Simulation of Closed-Loop System of Example 12.1

$$y = [\,100 \quad 20 \quad 0\,]\mathbf{x} \qquad (12.19b)$$

The transfer function is

$$T(s) = \frac{20(s + 5)}{s^3 + 15.9s^2 + 136.08s + 413.1} \qquad (12.20)$$

Figure 12.5, a simulation of the closed-loop system, shows 11.5% overshoot and a settling time of 0.8 seconds. A redesign with the third pole canceling the zero at -5 will yield performance equal to the requirements.

Since the steady-state response approaches 0.24 instead of unity, there is a large steady-state error. Design techniques to reduce this error will be discussed in Section 12.8.

In this section we showed how to design feedback gains for plants represented in phase-variable form in order to place all of the closed-loop system's poles at desired locations on the s-plane. On the surface, it appears that the method should always work for any system. However, this is not the case. The conditions that must exist in order to be able to uniquely place the closed-loop poles where we want them is the topic of the next section.

12.3 Controllability

Consider the parallel form shown in Figure 12.6(a). To control the pole location of the closed-loop system, we are saying implicitly that the control signal, u, can control the behavior of each state variable in x. If any one of the state variables cannot be controlled by the control u, then we cannot place the poles of the system where we desire. For example, in Figure 12.6(b), if x_1 were not controllable by the control signal and if x_1 also exhibited an unstable response due to a nonzero initial condition, there would be no way to effect a state-feedback design to stabilize

Figure 12.6 Comparison of (**a**) Controllable and (**b**) Uncontrollable Systems

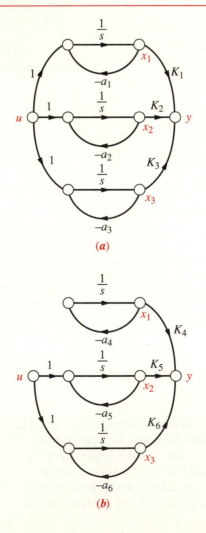

(a)

(b)

x_1; x_1 would perform in its own way regardless of the control signal, u. Thus, in some systems, a state-feedback design is not possible.

We now make the following definition based upon the previous discussion: If an input to a system can be found that takes every state variable from a desired initial state to a desired final state, then the system is said to be *controllable*; otherwise the system is said to be *uncontrollable*. Pole-placement is a viable design technique only for systems that are controllable. This section shows us how to determine, a priori, whether or not pole-placement is a viable design technique for a controller.

Controllability of Systems Represented in Parallel Form with Distinct Poles

We can explore controllability from another viewpoint: that of the state equation itself. When the system matrix is diagonal, as it is for the parallel form, it is apparent whether or not the system is controllable. For example, the state equation for Figure 12.6(a) is

$$\dot{\mathbf{x}} = \begin{bmatrix} -a_1 & 0 & 0 \\ 0 & -a_2 & 0 \\ 0 & 0 & -a_3 \end{bmatrix} \mathbf{x} + \begin{bmatrix} 1 \\ 1 \\ 1 \end{bmatrix} u \qquad (12.21)$$

or

$$\dot{x}_1 = -a_1 x_1 \qquad\qquad\qquad + u \qquad (12.22a)$$

$$\dot{x}_2 = \qquad\quad -a_2 x_2 \qquad\quad + u \qquad (12.22b)$$

$$\dot{x}_3 = \qquad\qquad\qquad -a_3 x_3 + u \qquad (12.22c)$$

Since each of Equations 12.22 is independent and decoupled from the rest, the control u affects each of the state variables. This is controllability from another perspective.

Now let us look at the state equations for the system of Figure 12.6(b):

$$\dot{\mathbf{x}} = \begin{bmatrix} -a_4 & 0 & 0 \\ 0 & -a_5 & 0 \\ 0 & 0 & -a_6 \end{bmatrix} \mathbf{x} + \begin{bmatrix} 0 \\ 1 \\ 1 \end{bmatrix} u \qquad (12.23)$$

or

$$\dot{x}_1 = -a_4 x_1 \qquad\qquad\qquad\qquad (12.24a)$$

$$\dot{x}_2 = \qquad\quad -a_5 x_2 \qquad\quad + u \qquad (12.24b)$$

$$\dot{x}_3 = \qquad\qquad\qquad -a_6 x_3 + u \qquad (12.24c)$$

From the state equations in 12.23 or 12.24, we see that state variable x_1 is not controlled by the control u. Thus the system is said to be uncontrollable.

In summary, a system with distinct eigenvalues and a diagonal system matrix is controllable if the input coupling matrix **B** does not have any rows that are zero.

Controllability of Systems with Multiple Poles and Systems Not Represented in Parallel Form: The Controllability Matrix

Tests for controllability that we have so far explored cannot be used for representations of the system other than the diagonal or parallel form with distinct eigenvalues. The problem of visualizing controllability gets more complicated if the system has multiple poles, even though it is represented in parallel form. Further, one cannot always determine controllability by inspection for systems that are not represented in parallel form. In other forms, the existence of paths from the input to the state variables is not a criterion for controllability since the equations are not decoupled.

In order to be able to determine controllability or, alternately, the ability to design state feedback for a plant under any representation or choice of state variables, a matrix can be derived that must have a particular property if all state variables are to be controlled by the plant input, u.[2] We now state the requirement for controllability, including the form, property, and name of this matrix.

An nth-order plant whose state equation is

$$\dot{\mathbf{x}} = \mathbf{Ax} + \mathbf{Bu} \qquad (12.25)$$

is controllable if the matrix

$$\mathbf{C_M} = [\,\mathbf{B} \quad \mathbf{AB} \quad \mathbf{A^2B} \quad \cdots \quad \mathbf{A}^{n-1}\mathbf{B}\,] \qquad (12.26)$$

is of rank n, where $\mathbf{C_M}$ is called the *controllability* matrix.[3] As an example, let us choose a system represented in parallel form with multiple roots.

Example 12.2   Determine the controllability of a system represented in parallel form with multiple roots.

Problem   Given the system of Figure 12.7, represented by a signal-flow diagram, determine its controllability.

Solution   The state equation for the system written from the signal-flow diagram is

$$\dot{\mathbf{x}} = \mathbf{Ax} + \mathbf{B}u = \begin{bmatrix} -1 & 1 & 0 \\ 0 & -1 & 0 \\ 0 & 0 & -2 \end{bmatrix} \mathbf{x} + \begin{bmatrix} 0 \\ 1 \\ 1 \end{bmatrix} u \qquad (12.27)$$

At first, it would appear that the system is not controllable because of the zero in the **B** matrix. Remember, though, that this configuration leads to uncontrollability only if the poles are real and distinct. In this case, we have multiple poles at -1.

---

[2] See the work listed in the Bibliography by Ogata (1990; 699–702) for the derivation.

[3] See Appendix B for the definition of rank. Instead of specifying rank n, we can alternately say that $\mathbf{C_M}$ must be nonsingular, possess an inverse (for single-input, single-output), or have linearly independent rows and columns.

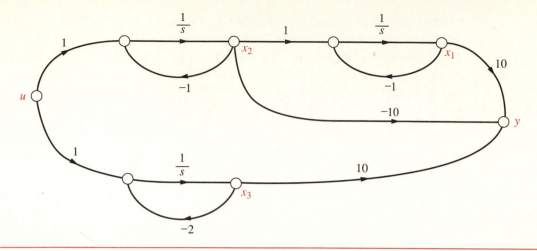

Figure 12.7 System for Example 12.2

The controllability matrix is

$$\mathbf{C_M} = [\,\mathbf{B} \quad \mathbf{AB} \quad \mathbf{A^2B}\,] = \begin{bmatrix} 0 & 1 & -2 \\ 1 & -1 & 1 \\ 1 & -2 & 4 \end{bmatrix} \tag{12.28}$$

The rank of $\mathbf{C_M}$ equals the number of linearly independent rows or columns. The rank can be found by finding the highest-order square sub matrix that is nonsingular. The determinant of $\mathbf{C_M} = -1$. Since the determinant is not zero, the 3×3 matrix is nonsingular, and the rank of $\mathbf{C_M}$ is 3. We conclude that the system is controllable since the rank of $\mathbf{C_M}$ equals the system order. Thus, the poles of the system can be placed using state-variable feedback design.

In the previous example we found that even though an element of the input coupling matrix was zero, the system was controllable. If we look at Figure 12.7 we can see why. In this figure, all of the state variables are driven by the input u.

On the other hand, if we disconnect the input at either dx_1/dt, dx_2/dt, or dx_3/dt, at least one state variable would not be controllable. To see the effect, let us disconnect the input at dx_2/dt. This causes the $\mathbf{B}$ matrix to become

$$\mathbf{B} = \begin{bmatrix} 0 \\ 0 \\ 1 \end{bmatrix} \tag{12.29}$$

We can see that the system is now uncontrollable since x_1 and x_2 are no longer controlled by the input. This conclusion is borne out by the controllability matrix, which is now

$$C_M = [\mathbf{B} \quad \mathbf{AB} \quad \mathbf{A^2B}] = \begin{bmatrix} 0 & 0 & 0 \\ 0 & 0 & 0 \\ 1 & -2 & 4 \end{bmatrix} \qquad (12.30)$$

Not only is the determinant of this matrix equal to zero, but so is the determinant of any 2×2 submatrix. Thus the rank of Equation 12.30 is 1. The system is uncontrollable because the rank of C_M is 1, which is less than the order, 3, of the system.

In summary, then, pole-placement design through state-variable feedback is simplified by using the phase-variable form for the plant's state equations. However, controllability, the ability for pole-placement design to succeed, can be visualized best in the parallel form, where the system matrix is diagonal with distinct roots. In any event, the controllability matrix will always tell the designer whether the implementation is viable for state-feedback design. In the next section we will show how to design state-variable feedback for systems not represented in phase-variable form. We will use the controllability matrix as a tool for transforming a system to phase-variable form for the design of state-variable feedback.

12.4 Controller Design for Systems Not Represented in Phase-Variable Form

In Section 12.2 we showed how to design state-variable feedback to yield desired closed-loop poles. We demonstrated this method using systems represented in phase-variable form and saw how simple it was to calculate the feedback gains. Many times the physics of the problem requires feedback from state variables that are not phase variables. For these systems we have some choices for a design methodology.

The first method consists of matching the coefficients of $\det(s\mathbf{I} - (\mathbf{A} - \mathbf{BK}))$ with the coefficients of the desired characteristic equation, which is the same method we used for systems represented in phase variables. This technique, in general, leads to difficult calculations of the feedback gains, especially for higher-order systems not represented with phase variables. Let us illustrate this technique with an example.

Example 12.3

Design state feedback for systems not represented in phase-variable form by matching coefficients of the characteristic equation with the desired characteristic equation.

Problem Given a plant, $\dfrac{Y(s)}{U(s)} = \dfrac{10}{(s+1)(s+2)}$, design state feedback for the plant represented in cascade form to yield a 15% overshoot with a settling time of 0.5 seconds.

Solution The signal-flow diagram for the plant in cascade form is shown in Figure 12.8(*a*). Figure 12.8(*b*) shows the system with state feedback added. Writing the state

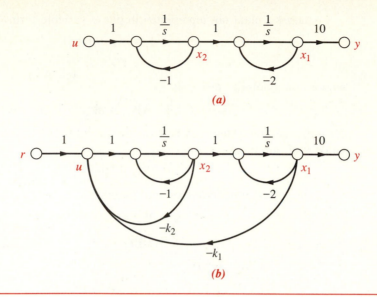

Figure 12.8 (*a*) Signal-Flow Graph in Cascade Form for
$$G(s) = \frac{10}{(s + 1)(s + 2)};$$ (*b*) System with State Feedback Added

equations from Figure 12.8(*b*), we have

$$\dot{\mathbf{x}} = \begin{bmatrix} -2 & 1 \\ -k_1 & -(k_2 + 1) \end{bmatrix} \mathbf{x} + \begin{bmatrix} 0 \\ 1 \end{bmatrix} r \tag{12.31a}$$

$$y = \begin{bmatrix} 10 & 0 \end{bmatrix} \mathbf{x} \tag{12.31b}$$

where the characteristic equation is

$$s^2 + (k_2 + 3)s + (2k_2 + k_1 + 2) = 0 \tag{12.32}$$

Using the transient response requirements stated in the problem, we obtain the desired characteristic equation

$$s^2 + 16s + 239.505 = 0 \tag{12.33}$$

Equating the middle coefficients of Equations 12.32 and 12.33, we find $k_2 = 13$. Equating the last coefficients of these equations along with the result for k_2 yields $k_1 = 211.505$.

The second method consists of transforming the system to phase-variables, designing the feedback gains, and transforming the designed system back to its original state-variable representation.[4] This method requires that we first develop the transformation between a system and its representation in phase-variable form.

---

[4]See the discussions of Ackermann's formula in Franklin (1986) and Ogata (1990), listed in the Bibliography.

Assume a plant not represented in phase-variable form,

$$\dot{\mathbf{z}} = \mathbf{Az} + \mathbf{B}u \tag{12.34a}$$

$$y = \mathbf{Cz} \tag{12.34b}$$

whose controllability matrix is

$$\mathbf{C_{Mz}} = [\mathbf{B} \quad \mathbf{AB} \quad \mathbf{A^2B} \quad \cdots \quad \mathbf{A}^{n-1}\mathbf{B}] \tag{12.35}$$

Assume that the system can be transformed into the phase-variable ($\mathbf{x}$) representation with the transformation

$$\mathbf{z} = \mathbf{Px} \tag{12.36}$$

Substituting this transformation into Equation 12.34, we get

$$\dot{\mathbf{x}} = \mathbf{P^{-1}APx} + \mathbf{P^{-1}B}u \tag{12.37a}$$

$$y = \mathbf{CPx} \tag{12.37b}$$

whose controllability matrix is

$$
\begin{aligned}
\mathbf{C_{Mx}} &= \left[\mathbf{P^{-1}B} \quad (\mathbf{P^{-1}AP})(\mathbf{P^{-1}B}) \quad (\mathbf{P^{-1}AP})^2(\mathbf{P^{-1}B}) \quad \cdots \quad (\mathbf{P^{-1}AP})^{n-1}(\mathbf{P^{-1}B}) \right] \\
&= \left[\mathbf{P^{-1}B} \quad (\mathbf{P^{-1}AP})(\mathbf{P^{-1}B}) \quad (\mathbf{P^{-1}AP})(\mathbf{P^{-1}AP})(\mathbf{P^{-1}B}) \quad \cdots \quad (\mathbf{P^{-1}AP}) \right. \\
&\qquad \left. (\mathbf{P^{-1}AP})(\mathbf{P^{-1}AP}) \quad \cdots \quad (\mathbf{P^{-1}AP})(\mathbf{P^{-1}B}) \right] \\
&= \mathbf{P^{-1}} \left[\mathbf{B} \quad \mathbf{AB} \quad \mathbf{A^2B} \quad \cdots \quad \mathbf{A}^{n-1}\mathbf{B} \right]
\end{aligned} \tag{12.38}
$$

Substituting Equation 12.35 into 12.38 and solving for $\mathbf{P}$, we obtain

$$\mathbf{P} = \mathbf{C_{Mz}}\mathbf{C_{Mx}}^{-1} \tag{12.39}$$

Thus the transformation matrix, $\mathbf{P}$, can be found from the two controllability matrices.

After transforming the system to phase-variables, we design the feedback gains as in Section 12.2. Hence, including both feedback and input, $u = -\mathbf{K_x x} + r$, Equation 12.37 becomes

$$
\begin{aligned}
\dot{\mathbf{x}} &= \mathbf{P^{-1}APx} - \mathbf{P^{-1}BK_x x} + \mathbf{P^{-1}B}r \\
&= (\mathbf{P^{-1}AP} - \mathbf{P^{-1}BK_x})\mathbf{x} + \mathbf{P^{-1}B}r
\end{aligned} \tag{12.40a}
$$

$$y = \mathbf{CPx} \tag{12.40b}$$

Since this equation is in phase-variable form, the zeros of this closed-loop system are determined from the polynomial formed from the elements of $\mathbf{CP}$, as explained in Section 12.2.

Using $\mathbf{x} = \mathbf{P^{-1}z}$, we transform Equation 12.40 from phase variables back to the original representation and get

$$\dot{\mathbf{z}} = \mathbf{Az} - \mathbf{BK_x P^{-1}z} + \mathbf{B}r = (\mathbf{A} - \mathbf{BK_x P^{-1}})\mathbf{z} + \mathbf{B}r \tag{12.41a}$$

$$y = \mathbf{Cz} \tag{12.41b}$$

Comparing Equation 12.41 with 12.3, the state variable feedback gain, $\mathbf{K_z}$, for the original system is

$$\mathbf{K_z} = \mathbf{K_x P^{-1}} \tag{12.42}$$

The transfer function of this closed-loop system is the same as the transfer function for Equation 12.40 since Equations 12.40 and 12.41 represent the same system. Thus the zeros of the closed-loop transfer function are the same as the zeros of the uncompensated plant, based upon the development in Section 12.2. Let us demonstrate with a design example.

Example 12.4   Design a controller for a plant not represented in phase-variable form by first transforming to phase variables.

Problem   Design a state-variable feedback controller to yield a 20.79% overshoot and a settling time of 4 seconds for a plant,

$$G(s) = \frac{(s + 4)}{(s + 1)(s + 2)(s + 5)} \tag{12.43}$$

that is represented in cascade form as shown in Figure 12.9.

Solution   First, find the state equations and the controllability matrix. The state equations written from Figure 12.9 are

$$\dot{\mathbf{z}} = \mathbf{A}_z\mathbf{z} + \mathbf{B}_z u = \begin{bmatrix} -5 & 1 & 0 \\ 0 & -2 & 1 \\ 0 & 0 & -1 \end{bmatrix}\mathbf{z} + \begin{bmatrix} 0 \\ 0 \\ 1 \end{bmatrix}u \tag{12.44}$$

$$y = \mathbf{C}_z\mathbf{z} = [-1 \quad 1 \quad 0]\mathbf{z}$$

from which the controllability matrix is evaluated as

$$\mathbf{C}_{\mathbf{Mz}} = [\mathbf{B}_z \quad \mathbf{A}_z\mathbf{B}_z \quad \mathbf{A}_z^2\mathbf{B}_z] = \begin{bmatrix} 0 & 0 & 1 \\ 0 & 1 & -3 \\ 1 & -1 & 1 \end{bmatrix} \tag{12.45}$$

Since the determinant of $\mathbf{C}_{\mathbf{Mz}}$ is -1, the system is controllable.

We now convert the system to phase variables by first finding the characteristic equation and using this equation to write the phase-variable form. The characteristic equation, $\det(s\mathbf{I} - \mathbf{A}_z)$, is

$$\det(s\mathbf{I} - \mathbf{A}_z) = s^3 + 8s^2 + 17s + 10 = 0 \tag{12.46}$$

Figure 12.9 Signal-Flow Graph for Plant of Example 12.4

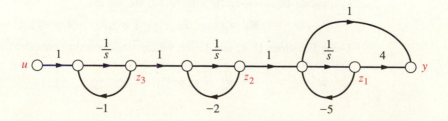

Using the coefficients of Equation 12.46 and our knowledge of the phase-variable form, we write the phase-variable representation of the system as

$$\dot{\mathbf{x}} = \mathbf{A_x x} + \mathbf{B_x}u = \begin{bmatrix} 0 & 1 & 0 \\ 0 & 0 & 1 \\ -10 & -17 & -8 \end{bmatrix} \mathbf{x} + \begin{bmatrix} 0 \\ 0 \\ 1 \end{bmatrix} u \tag{12.47a}$$

$$y = \begin{bmatrix} 4 & 1 & 0 \end{bmatrix} \mathbf{x} \tag{12.47b}$$

The output equation was written using the coefficients of the numerator of Equation 12.43 since the the transfer function must be the same for the two representations. The controllability matrix, $\mathbf{C_{Mx}}$, for the phase-variable system is

$$\mathbf{C_{Mx}} = \begin{bmatrix} \mathbf{B_x} & \mathbf{A_x B_x} & \mathbf{A_x^2 B_x} \end{bmatrix} = \begin{bmatrix} 0 & 0 & 1 \\ 0 & 1 & -8 \\ 1 & -8 & 47 \end{bmatrix} \tag{12.48}$$

Using Equation 12.39, we can now calculate the transformation matrix between the two systems as

$$\mathbf{P} = \mathbf{C_{Mz} C_{Mx}^{-1}} = \begin{bmatrix} 1 & 0 & 0 \\ 5 & 1 & 0 \\ 10 & 7 & 1 \end{bmatrix} \tag{12.49}$$

We now design the controller using the phase-variable representation and then use Equation 12.49 to transform the design back to the original representation. For a 20.79% overshoot and a settling time of 4 seconds, a factor of the characteristic equation of the designed closed-loop system is $s^2 + 2s + 5$. Since the closed-loop zero will be at $s = -4$, we choose the third closed-loop pole to cancel the closed-loop zero. Hence the total characteristic equation of the desired closed-loop system is

$$D(s) = (s + 4)(s^2 + 2s + 5) = s^3 + 6s^2 + 13s + 20 = 0 \tag{12.50}$$

The state equations for the phase-variable form with state variable feedback are

$$\dot{\mathbf{x}} = (\mathbf{A_x} - \mathbf{B_x K_x})\mathbf{x} = \begin{bmatrix} 0 & 1 & 0 \\ 0 & 0 & 1 \\ -(10 + k_{1_x}) & -(17 + k_{2_x}) & -(8 + k_{3_x}) \end{bmatrix} \mathbf{x} \tag{12.51a}$$

$$y = \begin{bmatrix} 4 & 1 & 0 \end{bmatrix} \mathbf{x} \tag{12.51b}$$

The characteristic equation for Equation 12.51 is

$$\det(s\mathbf{I} - (\mathbf{A_x} - \mathbf{B_x K_x})) = s^3 + (8 + k_{3_x})s^2 + (17 + k_{2_x})s + (10 + k_{1_x})$$
$$= 0 \tag{12.52}$$

Comparing Equation 12.50 with 12.52, we see that

$$\mathbf{K_x} = \begin{bmatrix} k_{1_x} & k_{2_x} & k_{3_x} \end{bmatrix} = \begin{bmatrix} 10 & -4 & -2 \end{bmatrix} \tag{12.53}$$

Using Equations 12.42 and 12.49, we can transform the controller back to the original system as

$$\mathbf{K_z} = \mathbf{K_x P^{-1}} = \begin{bmatrix} -20 & 10 & -2 \end{bmatrix} \tag{12.54}$$

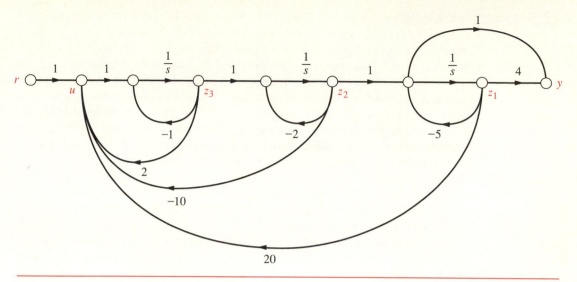

Figure 12.10 Designed System with State-Variable Feedback for Example 12.4

The final closed-loop system with state-variable feedback is shown in Figure 12.10, with the input applied as shown.

Let us now verify our design. The state equations for the designed system shown in Figure 12.10 with input r are

$$\dot{\mathbf{z}} = (\mathbf{A_z} - \mathbf{B_z K_z})\mathbf{z} + \mathbf{B_z}r = \begin{bmatrix} -5 & 1 & 0 \\ 0 & -2 & 1 \\ 20 & -10 & 1 \end{bmatrix}\mathbf{z} + \begin{bmatrix} 0 \\ 0 \\ 1 \end{bmatrix}r \qquad (12.55a)$$

$$y = \mathbf{C_z z} = [-1 \quad 1 \quad 0]\mathbf{z} \qquad (12.55b)$$

Using Equation 3.73 to find the closed-loop transfer function, we obtain

$$T(s) = \frac{(s+4)}{s^3 + 6s^2 + 13s + 20} = \frac{1}{s^2 + 2s + 5} \qquad (12.56)$$

The requirements of our design have been met.

In this section we showed how to design state-variable feedback for plants not represented in phase-variable form. Using controllability matrices, we were able to transform a plant to phase-variable form, design the controller, and finally transform the controller design back to the plant's original representation. The design of the controller relies on the availability of the states for feedback. In the next section we discuss the design of state-variable feedback when some or all of the states are not available.

12.5 Observer Design

Controller design relies upon access to the state variables for feedback through adjustable gains. This access can be provided by hardware. For example, gyros can measure position and velocity on a space vehicle. Sometimes, it is impractical to use this hardware for reasons of cost, accuracy, or availability. For example, in powered flight of space vehicles, inertial measuring units can be used to calculate the acceleration. However, their alignment deteriorates with time; thus, other means of measuring acceleration may be desirable (Rockwell International, 1984). In other applications, some of the state variables may not be available at all, or it is too costly to measure them or send them to the controller. If the state variables are not available because of system configuration or cost, it is possible to estimate the states. Estimated states, rather than actual states, are then fed to the controller. One scheme is shown in Figure 12.11(*a*). An *observer,* sometimes called an *estimator,* is used to calculate state variables that are not accessible from the plant. Here, the observer is a model of the plant. Let us look at the disadvantages of such a configuration.

Assume a plant,

$$\dot{\mathbf{x}} = \mathbf{A}\mathbf{x} + \mathbf{B}u \tag{12.57a}$$

$$y = \mathbf{C}\mathbf{x} \tag{12.57b}$$

and an observer,

$$\dot{\hat{\mathbf{x}}} = \mathbf{A}\hat{\mathbf{x}} + \mathbf{B}u \tag{12.58a}$$

$$\hat{y} = \mathbf{C}\hat{\mathbf{x}} \tag{12.58b}$$

Subtracting Equations 12.58 from 12.57, we obtain

$$\dot{\mathbf{x}} - \dot{\hat{\mathbf{x}}} = \mathbf{A}(\mathbf{x} - \hat{\mathbf{x}}) \tag{12.59a}$$

$$y - \hat{y} = \mathbf{C}(\mathbf{x} - \hat{\mathbf{x}}) \tag{12.59b}$$

Thus the dynamics of the difference between the actual and estimated states is unforced, and if the plant is stable, this difference, due to differences in initial state vectors, approaches zero. However, the speed of convergence between the actual state and the estimated state is the same as the transient response of the plant since the characteristic equation for 12.59 is the same as for 12.57. Since the speed of convergence is too slow, we seek a way to speed up the observer and make its response time much faster than that of the controlled closed-loop system, so that, effectively, the controller will receive the estimated states instantaneously.

To increase the speed of convergence between the actual and estimated states, we use feedback, shown conceptually in Figure 12.11(*b*) and in more detail in Figure 12.11(*c*). The error between the outputs of the plant and the observer is fed back to the derivatives of the observer's states. The system corrects to drive this error to zero. With feedback, we can design a desired transient response into the observer that is much quicker than that of the plant or controlled closed-loop system.

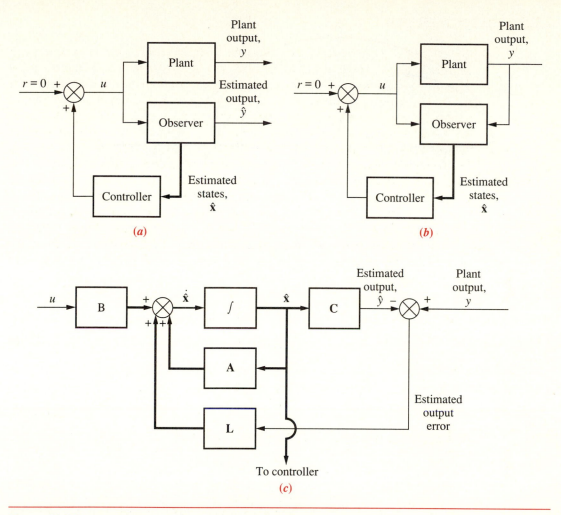

Figure 12.11 State-Feedback Design Using an Observer to Estimate Unavailable State Variables: (**a**) Open-Loop Observer; (**b**) Closed-Loop Observer; (**c**) Exploded View of a Closed-Loop Observer, Showing Feedback Arrangement to Reduce State-Variable Estimation Error

When we implemented the controller, we found that the phase-variable form yielded an easy solution for the controller gains. In designing an observer, it is the dual phase-variable form that yields the easy solution for the observer gains. Figure 12.12(a) shows an example of a third-order plant represented in dual phase-variable form. In Figure 12.12(b), the plant is configured as an observer with the addition of feedback, as previously described.

The design of the observer is separate from the design of the controller. Similar to the design of the controller vector, **K**, the design of the observer consists of evaluating the constant vector, **L**, so that the transient response of the observer is

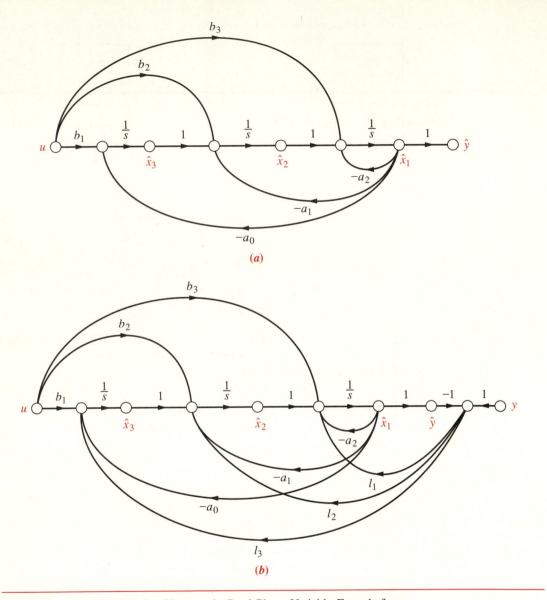

Figure 12.12 (***a***) Third-Order Observer in Dual Phase-Variable Form before Addition of Feedback; (***b***) Third-Order Observer in Dual Phase-Variable Form after Addition of Feedback

faster than the response of the controlled loop in order to yield a rapidly updated estimate of the state vector. We now derive the design methodology.

We will first find the state equations for the error between the actual state vector and the estimated state vector, $(\mathbf{x} - \hat{\mathbf{x}})$. Then we will find the characteristic equation for the error system and evaluate the required $\mathbf{L}$ to meet a rapid transient response for the observer.

Writing the state equations of the observer from Figure 12.11(c), we have

$$\dot{\hat{\mathbf{x}}} = \mathbf{A}\hat{\mathbf{x}} + \mathbf{B}u + \mathbf{L}(y - \hat{y}) \tag{12.60a}$$

$$\hat{y} = \mathbf{C}\hat{\mathbf{x}} \tag{12.60b}$$

But the state equations for the plant are

$$\dot{\mathbf{x}} = \mathbf{A}\mathbf{x} + \mathbf{B}u \tag{12.61a}$$

$$y = \mathbf{C}\mathbf{x} \tag{12.61b}$$

Subtracting Equations 12.60 from 12.61, we obtain

$$(\dot{\mathbf{x}} - \dot{\hat{\mathbf{x}}}) = \mathbf{A}(\mathbf{x} - \hat{\mathbf{x}}) - \mathbf{L}(y - \hat{y}) \tag{12.62a}$$

$$(y - \hat{y}) = \mathbf{C}(\mathbf{x} - \hat{\mathbf{x}}) \tag{12.62b}$$

where $\mathbf{x} - \hat{\mathbf{x}}$ is the error between the actual state vector and the estimated state vector, and $y - \hat{y}$ is the error between the actual output and the estimated output.

Substituting the output equation into the state equation, we obtain the state equation for the error between the estimated state vector and the actual state vector:

$$(\dot{\mathbf{x}} - \dot{\hat{\mathbf{x}}}) = (\mathbf{A} - \mathbf{LC})(\mathbf{x} - \hat{\mathbf{x}}) \tag{12.63a}$$

$$(y - \hat{y}) = \mathbf{C}(\mathbf{x} - \hat{\mathbf{x}}) \tag{12.63b}$$

Letting $\mathbf{e_x} = (\mathbf{x} - \hat{\mathbf{x}})$, we have

$$\dot{\mathbf{e}}_{\mathbf{x}} = (\mathbf{A} - \mathbf{LC})\mathbf{e_x} \tag{12.64a}$$

$$y - \hat{y} = \mathbf{C}\mathbf{e_x} \tag{12.64b}$$

Equation 12.64a is unforced. If the eigenvalues are all negative, then the estimated state vector error, $\mathbf{e_x}$, will decay to zero. The design then consists of solving for the values of $\mathbf{L}$ to yield a desired characteristic equation or response for Equations 12.64. The characteristic equation is found from Equations 12.64 to be

$$\det[\lambda\mathbf{I} - (\mathbf{A} - \mathbf{LC})] = 0 \tag{12.65}$$

Now we select the eigenvalues of the observer to yield stability and a desired transient response that is faster than the controlled closed-loop response. These eigenvalues determine a characteristic equation that we set equal to Equation 12.65 to solve for $\mathbf{L}$.

Let us demonstrate the procedure for an nth-order plant represented in dual phase-variable form. We first evaluate $\mathbf{A} - \mathbf{LC}$. The form of $\mathbf{A}$, $\mathbf{L}$, and $\mathbf{C}$ can be derived by extrapolating the form of these matrices from a third-order plant, which you can derive from Figure 12.12. Thus,

$$
\mathbf{A} - \mathbf{LC} =
\begin{bmatrix}
-a_{n-1} & 1 & 0 & 0 & \cdots & 0 \\
-a_{n-2} & 0 & 1 & 0 & \cdots & 0 \\
\vdots & \vdots & \vdots & \vdots & \vdots & \vdots \\
-a_1 & 0 & 0 & 0 & \cdots & 1 \\
-a_0 & 0 & 0 & 0 & \cdots & 0
\end{bmatrix}
-
\begin{bmatrix}
l_1 \\
l_2 \\
\vdots \\
l_{n-1} \\
l_n
\end{bmatrix}
[1 \ \ 0 \ \ 0 \ \ 0 \ \ \cdots \ \ 0]
$$

$$
=
\begin{bmatrix}
-(a_{n-1} + l_1) & 1 & 0 & 0 & \cdots & 0 \\
-(a_{n-2} + l_2) & 0 & 1 & 0 & \cdots & 0 \\
\vdots & & \vdots & \vdots & \vdots & \vdots \\
-(a_1 + l_{n-1}) & 0 & 0 & 0 & \cdots & 1 \\
-(a_0 + l_n) & 0 & 0 & 0 & \cdots & 0
\end{bmatrix}
\tag{12.66}
$$

The characteristic equation for $\mathbf{A} - \mathbf{LC}$ is

$$
s^n + (a_{n-1} + l_1)s^{n-1} + (a_{n-2} + l_2)s^{n-2} \cdots + (a_1 + l_{n-1})s
$$
$$
+ (a_0 + l_n) = 0 \tag{12.67}
$$

Notice the relationship between Equation 12.67 and the characteristic equation, $\det(s\mathbf{I} - \mathbf{A}) = 0$, for the plant, which is

$$
s^n + a_{n-1}s^{n-1} + a_{n-2}s^{n-2} \cdots + a_1 s + a_0 = 0 \tag{12.68}
$$

Thus, if desired, Equation 12.67 can be written by inspection if the plant is represented in dual phase-variable form. We now equate Equation 12.67 with the desired closed-loop observer characteristic equation, which is chosen on the basis of a desired transient response. Assume the desired characteristic equation is

$$
s^n + d_{n-1}s^{n-1} + d_{n-2}s^{n-2} \cdots + d_1 s + d_0 = 0 \tag{12.69}
$$

We can now solve for the l_i's by equating the coefficients of Equation 12.67 with 12.69. Hence,

$$
l_i = d_{n-i} - a_{n-i} \qquad i = 1, 2, \ldots, n \tag{12.70}
$$

Let us demonstrate the design of an observer using the dual phase-variable form. In subsequent sections we will show how to design the observer for other than dual phase-variable form.

Example 12.5   Design an observer for a plant represented in dual phase-variable form.

Problem   Design an observer for the plant

$$
G(s) = \frac{(s + 4)}{(s + 1)(s + 2)(s + 5)} = \frac{s + 4}{s^3 + 8s^2 + 17s + 10} \tag{12.71}
$$

which is represented in dual phase-variable form. The observer will respond 10 times faster than the controlled loop designed in Example 12.4.

Solution

a. First represent the estimated plant in dual phase-variable form. The result is shown in Figure 12.13(a).

b. Now form the difference between the plant's actual output, y, and the observer's estimated output, $\hat{y}$, and add the feedback paths from this difference to the derivative of each state variable. The result is shown in Figure 12.13(b).

Figure 12.13 (**a**) Signal-Flow Graph of a System Using Dual Phase-Variables; (**b**) Additional Feedback to Create Observer

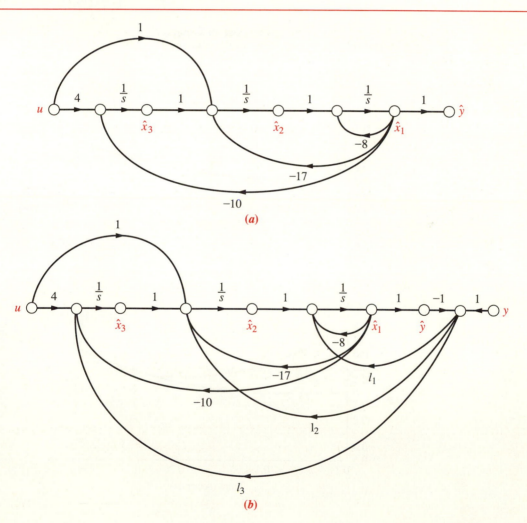

c. Next find the characteristic polynomial. The state equations for the estimated plant shown in Figure 12.13(*a*) are

$$\dot{\mathbf{x}} = \mathbf{A}\hat{\mathbf{x}} + \mathbf{B}u = \begin{bmatrix} -8 & 1 & 0 \\ -17 & 0 & 1 \\ -10 & 0 & 0 \end{bmatrix}\hat{\mathbf{x}} + \begin{bmatrix} 0 \\ 1 \\ 4 \end{bmatrix}u \tag{12.72a}$$

$$\hat{y} = \mathbf{C}\hat{\mathbf{x}} = [1 \quad 0 \quad 0]\hat{\mathbf{x}} \tag{12.72b}$$

From Equations 12.64 and 12.66, the observer error is

$$\dot{\mathbf{e}}_\mathbf{x} = (\mathbf{A} - \mathbf{LC})\mathbf{e}_\mathbf{x} = \begin{bmatrix} -(8 + l_1) & 1 & 0 \\ -(17 + l_2) & 0 & 1 \\ -(10 + l_3) & 0 & 0 \end{bmatrix}\mathbf{e}_\mathbf{x} \tag{12.73}$$

Figure 12.14 Simulation Showing Response of Observer: (**a**) Closed-Loop; (**b**) Open-Loop with Observer Gains Disconnected

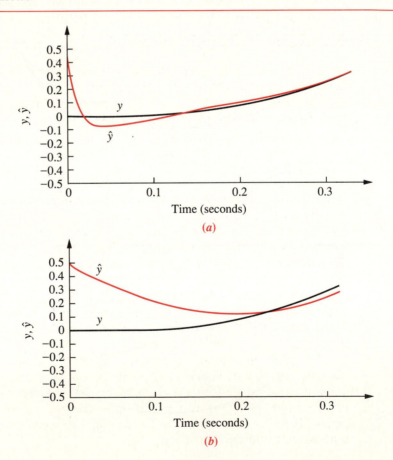

Using Equation 12.65, we obtain the characteristic polynomial

$$s^3 + (8 + l_1)s^2 + (17 + l_2)s + (10 + l_3) \tag{12.74}$$

d. Now evaluate the desired polynomial, set the coefficients equal to those of Equation 12.74, and solve for the gains, l_i. From Equation 12.50, the closed-loop controlled system has dominant second-order poles at $-1 \pm j2$. To make our observer 10 times faster, we design the observer poles to be at $-10 \pm j20$. We select the third pole to be 10 times the real part of the dominant second-order poles, or -100. Hence the desired characteristic polynomial is

$$(s + 100)(s^2 + 20s + 500) = s^3 + 120s^2 + 2500s + 50{,}000 \tag{12.75}$$

Equating Equations 12.74 and 12.75, we find $l_1 = 112$, $l_2 = 2483$, and $l_3 = 49{,}990$.

A simulation of the observer with an input of $r(t) = 100t$ is shown in Figure 12.14. The initial conditions of the plant were all zero, and the initial condition of $\hat{x}_1$ was 0.5. Since the dominant pole of the observer is $-10 \pm j20$, the expected settling time should be about 0.4 sec. It is interesting to note the slower response in Figure 12.14(b), where the observer gains are disconnected, and the observer is simply a copy of the plant with a different initial condition.

In this section we designed an observer in dual phase-variable form that uses the output of a system to estimate the state variables. In the next section we examine the conditions under which an observer cannot be designed. In Section 12.7 we design observers for plants not represented in dual phase-variable form.

12.6 Observability

Recall that the ability to control all of the state variables is a requirement for the design of a controller. Design of state-variable feedback gains cannot be accomplished if any state variable is uncontrollable. Uncontrollability can be viewed best with diagonalized systems. The signal-flow graph showed clearly that the uncontrollable state variable was not connected to the control signal of the system.

A similar concept governs our ability to create a design for an observer. Specifically, we are using the output of a system to deduce the state variables. If any state variable has no effect upon the output, then we cannot evaluate this state variable by observing the output.

The ability to observe a state variable from the output is best seen from the diagonalized system. Figure 12.15(a) shows a system where each state variable can be observed at the output since each is connected to the output. Figure 12.15(b) is an example of a system where all state variables cannot be observed at the output. Here, x_1 is not connected to the output and could not be estimated from a measurement of the output.

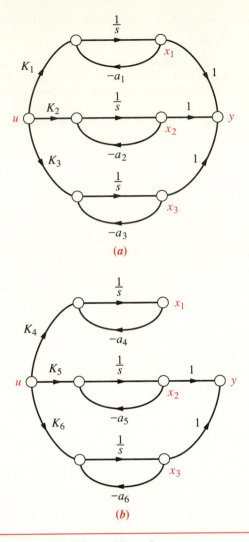

Figure 12.15 Comparison of (*a*) Observable and (*b*) Unobservable Systems

We now make the following definition based upon the previous discussion: If the initial-state vector, $\mathbf{x}(t_0)$, can be found from $u(t)$ and $y(t)$ measured over a finite interval of time from t_0, the system is said to be *observable*; otherwise the system is said to be *unobservable*. Simply stated, observability is the ability to deduce the state variables from a knowledge of the input, $u(t)$, and the output, $y(t)$. Pole-placement for an observer is a viable design technique only for systems that are observable. This section shows us how to determine, a priori, whether or not pole-placement is a viable design technique for an observer.

Observability of Systems Represented in Parallel Form with Distinct Poles

We can also explore observability from the output equation of a diagonalized system. The output equation for the diagonalized system of Figure 12.15(*a*) is

$$y = \mathbf{C}\mathbf{x} = [1 \quad 1 \quad 1]\mathbf{x} \tag{12.76}$$

On the other hand, the output equation for the unobservable system of Figure 12.15(*b*) is

$$y = \mathbf{C}\mathbf{x} = [0 \quad 1 \quad 1]\mathbf{x} \tag{12.77}$$

Notice that the first column of Equation 12.77 is zero. For systems represented in parallel form with distinct eigenvalues, if any column of the output coupling matrix is zero, the diagonal system is not observable.

Observability of Systems Not Represented in Parallel Form: The Observability Matrix

Again, as for controllability, systems represented in other than diagonalized form cannot be reliably evaluated for observability by inspection. In order to determine observability for systems under any representation or choice of state variables, a matrix can be derived that must have a particular property if all state variables are to be observed at the output.[5] We now state the requirements for observability including the form, property, and name of this matrix.

An nth-order plant whose state and output equations are, respectively,

$$\dot{\mathbf{x}} = \mathbf{A}\mathbf{x} + \mathbf{B}\mathbf{u} \tag{12.78a}$$

$$\mathbf{y} = \mathbf{C}\mathbf{x} \tag{12.78b}$$

is observable if the matrix

$$\mathbf{O_M} = \begin{bmatrix} \mathbf{C} \\ \mathbf{CA} \\ \vdots \\ \mathbf{CA}^{n-1} \end{bmatrix} \tag{12.79}$$

is of rank n, where $\mathbf{O_M}$ is called the *observability matrix*. Two examples illustrate the use of the observability matrix.

Example 12.6   Determine the observability of a system.

Problem   Determine if the system of Figure 12.16 is observable.

---

[5] See Ogata (1990: 706–08) for a derivation.

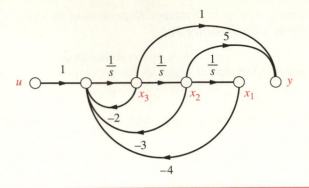

Figure 12.16 System of Example 12.6

Solution The state and output equations for the system are

$$\dot{\mathbf{x}} = \mathbf{Ax} + \mathbf{B}u = \begin{bmatrix} 0 & 1 & 0 \\ 0 & 0 & 1 \\ -4 & -3 & -2 \end{bmatrix} \mathbf{x} + \begin{bmatrix} 0 \\ 0 \\ 1 \end{bmatrix} u \qquad (12.80a)$$

$$y = \mathbf{Cx} = [\,0 \quad 5 \quad 1\,]\mathbf{x} \qquad (12.80b)$$

Thus the observability matrix, $\mathbf{O_M}$, is

$$\mathbf{O_M} = \begin{bmatrix} \mathbf{C} \\ \mathbf{CA} \\ \mathbf{CA}^2 \end{bmatrix} = \begin{bmatrix} 0 & 5 & 1 \\ -4 & -3 & 3 \\ -12 & -13 & -9 \end{bmatrix} \qquad (12.81)$$

Since the determinant of $\mathbf{O_M}$ equals -344, $\mathbf{O_M}$ is of full rank equal to 3. The system is thus observable.

The reader might have been misled and concluded by inspection that the system is unobservable because the state variable x_1 is not fed *directly* to the output. Remember that conclusions about observability by inspection are only valid for diagonalized systems that have distinct eigenvalues.

Example 12.7 Determine the observability of a system.

Problem Determine if the system of Figure 12.17 is observable.

Solution The state and output equations for the system are

$$\dot{\mathbf{x}} = \mathbf{Ax} + \mathbf{B}u = \begin{bmatrix} 0 & 1 \\ -5 & -21/4 \end{bmatrix} \mathbf{x} + \begin{bmatrix} 0 \\ 1 \end{bmatrix} u \qquad (12.82a)$$

$$y = \mathbf{Cx} = [\,5 \quad 4\,]\mathbf{x} \qquad (12.82b)$$

The observability matrix, $\mathbf{O_M}$, for this system is

$$\mathbf{O_M} = \begin{bmatrix} \mathbf{C} \\ \mathbf{CA} \end{bmatrix} = \begin{bmatrix} 5 & 4 \\ -20 & -16 \end{bmatrix} \qquad (12.83)$$

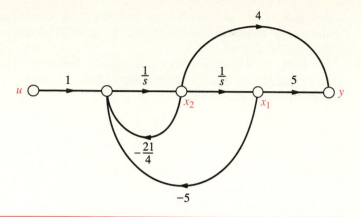

Figure 12.17 System of Example 12.7

The determinant for this observability matrix equals 0. Thus, the observability matrix does not have full rank, and the system is not observable.

Again, the reader might conclude by inspection that the system is observable because all states feed the output. Remember that observability by inspection is only valid for a diagonalized representation of a system with distinct eigenvalues.

Now that we have discussed observability and the observability matrix, we are ready to talk about the design of an observer for a plant not represented in dual phase-variable form.

12.7 Observer Design for Systems Not Represented in Dual Phase-Variable Form

Earlier in the chapter we discussed how to design controllers for systems not represented in phase-variable form. One method is to match the coefficients of $\det[s\mathbf{I} - (\mathbf{A} - \mathbf{BK})]$ with the coefficients of the desired characteristic polynomial. This method can yield difficult calculations for higher-order systems. Another method is to transform the plant to phase-variable form, design the controller, and transfer the design back to its original representation. The transformations were derived from the controllability matrix.

In this section we use a similar idea for the design of observers not represented in dual phase-variable form. One method is to match the coefficients of $\det[s\mathbf{I} - (\mathbf{A} - \mathbf{LC})]$ with the coefficients of the desired characteristic polynomial. Again, this method can yield difficult calculations for higher-order systems. Another method is first to transform the plant to dual phase-variable form so that the design equations are simple, then perform the design in dual phase-variable form, and finally transform the design back to the original representation. Let us pursue this second method.

First, we will derive the transformation between a system representation and its representation in dual phase-variable form. Assume a plant not represented in dual phase-variable form,

$$\dot{\mathbf{z}} = \mathbf{A}\mathbf{z} + \mathbf{B}u \qquad (12.84a)$$

$$y = \mathbf{C}\mathbf{z} \qquad (12.84b)$$

whose observability matrix is

$$\mathbf{O}_{\mathbf{Mz}} = \begin{bmatrix} \mathbf{C} \\ \mathbf{CA} \\ \mathbf{CA}^2 \\ \vdots \\ \mathbf{CA}^{n-2} \\ \mathbf{CA}^{n-1} \end{bmatrix} \qquad (12.85)$$

Now, assume that the system can be transformed to the dual phase-variable form, $\mathbf{x}$, with the transformation

$$\mathbf{z} = \mathbf{P}\mathbf{x} \qquad (12.86)$$

Substituting Equation 12.86 into Equation 12.84 and premultiplying the state equation by $\mathbf{P}^{-1}$, we find that the state equations in dual phase-variable form are

$$\dot{\mathbf{x}} = \mathbf{P}^{-1}\mathbf{A}\mathbf{P}\mathbf{x} + \mathbf{P}^{-1}\mathbf{B}u \qquad (12.87a)$$

$$y = \mathbf{C}\mathbf{P}\mathbf{x} \qquad (12.87b)$$

whose observability matrix, $\mathbf{O}_{\mathbf{Mx}}$, is

$$\mathbf{O}_{\mathbf{Mx}} = \begin{bmatrix} \mathbf{CP} \\ \mathbf{CP}(\mathbf{P}^{-1}\mathbf{AP}) \\ \mathbf{CP}(\mathbf{P}^{-1}\mathbf{AP})(\mathbf{P}^{-1}\mathbf{AP}) \\ \vdots \\ \mathbf{CP}(\mathbf{P}^{-1}\mathbf{AP})(\mathbf{P}^{-1}\mathbf{AP})\cdots(\mathbf{P}^{-1}\mathbf{AP}) \end{bmatrix} = \begin{bmatrix} \mathbf{C} \\ \mathbf{CA} \\ \mathbf{CA}^2 \\ \vdots \\ \mathbf{CA}^n \end{bmatrix}\mathbf{P} \qquad (12.88)$$

Substituting Equation 12.85 into 12.88 and solving for $\mathbf{P}$, we obtain

$$\mathbf{P} = \mathbf{O}_{\mathbf{Mz}}^{-1}\mathbf{O}_{\mathbf{Mx}} \qquad (12.89)$$

Thus the transformation, $\mathbf{P}$, can be found from the two observability matrices.

After transforming the plant to dual phase-variable form, we design the feedback gains, $\mathbf{L_x}$, as in Section 12.5. Using the matrices from Equation 12.87 and the form suggested by Equation 12.64, we have

$$\dot{\mathbf{e}}_{\mathbf{x}} = (\mathbf{P}^{-1}\mathbf{A}\mathbf{P} - \mathbf{L_x}\mathbf{C}\mathbf{P})\mathbf{e}_{\mathbf{x}} \qquad (12.90a)$$

$$y - \hat{y} = \mathbf{C}\mathbf{P}\mathbf{e}_{\mathbf{x}} \qquad (12.90b)$$

Since $\mathbf{x} = \mathbf{P}^{-1}\mathbf{z}$, and $\hat{\mathbf{x}} = \mathbf{P}^{-1}\hat{\mathbf{z}}$, then $\mathbf{e_x} = \mathbf{x} - \hat{\mathbf{x}} = \mathbf{P}^{-1}\mathbf{e_z}$. Substituting $\mathbf{e_x} = \mathbf{P}^{-1}\mathbf{e_z}$ into Equations 12.90 transforms Equations 12.90 back to the original representation. The result is

$$\dot{\mathbf{e}}_z = (\mathbf{A} - \mathbf{PL_xC})\mathbf{e_z} \tag{12.91a}$$

$$y - \hat{y} = \mathbf{Ce_z} \tag{12.91b}$$

Comparing Equation 12.91 to 12.64, we see that the observer gain vector is

$$\mathbf{L_z} = \mathbf{PL_x} \tag{12.92}$$

We now demonstrate the design of an observer for a plant not represented in dual phase-variable form.

Example 12.8   Design an observer for a plant not represented in dual phase-variable form.

Problem   Design an observer for the plant

$$G(s) = \frac{1}{(s+1)(s+2)(s+5)} \tag{12.93}$$

represented in cascade form. The closed-loop performance of the observer is governed by the characteristic polynomial used in Example 12.5: $s^3 + 120s^2 + 2500s + 50{,}000$.

Solution   First represent the plant in its original cascade form.

$$\dot{\mathbf{z}} = \mathbf{Az} + \mathbf{B}u = \begin{bmatrix} -5 & 1 & 0 \\ 0 & -2 & 1 \\ 0 & 0 & -1 \end{bmatrix}\mathbf{z} + \begin{bmatrix} 0 \\ 0 \\ 1 \end{bmatrix}u \tag{12.94a}$$

$$y = \mathbf{Cz} = [1 \quad 0 \quad 0]\mathbf{z} \tag{12.94b}$$

The observability matrix, $\mathbf{O}_{\text{Mz}}$, is,

$$\mathbf{O}_{\text{Mz}} = \begin{bmatrix} \mathbf{C} \\ \mathbf{CA} \\ \mathbf{CA}^2 \end{bmatrix} = \begin{bmatrix} 1 & 0 & 0 \\ -5 & 1 & 0 \\ 25 & -7 & 1 \end{bmatrix} \tag{12.95}$$

whose determinant equals 1. Hence the plant is observable.

The characteristic equation for the plant is

$$\det(s\mathbf{I} - \mathbf{A}) = s^3 + 8s^2 + 17s + 10 = 0 \tag{12.96}$$

We can use the coefficients of this characteristic polynomial to form the dual phase-variable form:

$$\dot{\mathbf{x}} = \mathbf{A_x x} + \mathbf{B_x}u \tag{12.97a}$$

$$y = \mathbf{C_x x} \tag{12.97b}$$

where

$$\mathbf{A_x} = \begin{bmatrix} -8 & 1 & 0 \\ -17 & 0 & 1 \\ -10 & 0 & 0 \end{bmatrix}; \qquad \mathbf{C_x} = [1 \quad 0 \quad 0] \tag{12.98}$$

The observability matrix for the dual phase-variable form is

$$\mathbf{O_{Mx}} = \begin{bmatrix} \mathbf{C_x} \\ \mathbf{C_x A_x} \\ \mathbf{C_x A_x}^2 \end{bmatrix} = \begin{bmatrix} 1 & 0 & 0 \\ -8 & 1 & 0 \\ 47 & -8 & 1 \end{bmatrix} \tag{12.99}$$

We now design the observer for the dual phase-variable form. First, form $(\mathbf{A_x} - \mathbf{L_x C_x})$,

$$\mathbf{A_x} - \mathbf{L_x C_x} = \begin{bmatrix} -8 & 1 & 0 \\ -17 & 0 & 1 \\ -10 & 0 & 0 \end{bmatrix} - \begin{bmatrix} l_1 \\ l_2 \\ l_3 \end{bmatrix} \begin{bmatrix} 1 & 0 & 0 \end{bmatrix}$$

$$= \begin{bmatrix} -(8 + l_1) & 1 & 0 \\ -(17 + l_2) & 0 & 1 \\ -(10 + l_3) & 0 & 0 \end{bmatrix} \tag{12.100}$$

whose characteristic polynomial is

$$\det[s\mathbf{I} - (\mathbf{A_x} - \mathbf{L_x C_x})] = s^3 + (8 + l_1)s^2 + (17 + l_2)s + (10 + l_3) \tag{12.101}$$

Equating this polynomial to the desired closed-loop observer characteristic equation, $s^3 + 120s^2 + 2500s + 50{,}000$, we find

$$\mathbf{L_x} = \begin{bmatrix} 112 \\ 2483 \\ 49{,}990 \end{bmatrix} \tag{12.102}$$

Figure 12.18 Observer Design

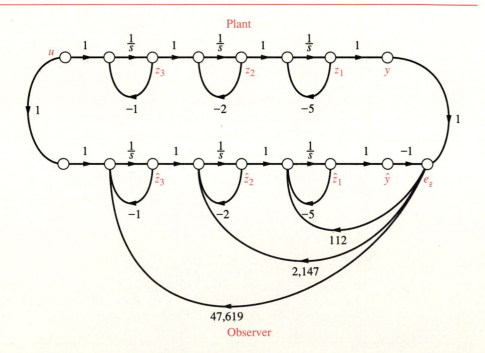

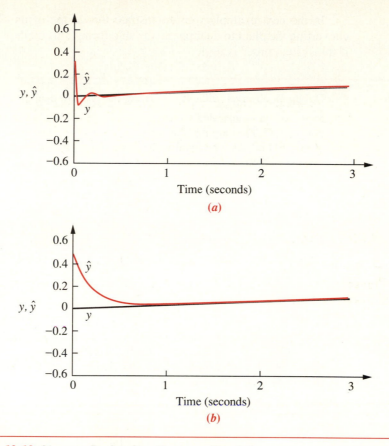

Figure 12.19 Observer Design Step Response Simulation:
(*a*) Closed-Loop Observer; (*b*) Open-Loop Observer with
Observer Gains Disconnected

Now transform the design back to the original representation. Using Equation 12.89, the transformation matrix is

$$\mathbf{P} = \mathbf{O_{Mz}}^{-1}\mathbf{O_{Mx}} = \begin{bmatrix} 1 & 0 & 0 \\ -3 & 1 & 0 \\ 1 & -1 & 1 \end{bmatrix} \quad (12.103)$$

Transforming $\mathbf{L_x}$ to the original representation, we obtain

$$\mathbf{L_z} = \mathbf{PL_x} = \begin{bmatrix} 112 \\ 2147 \\ 47,619 \end{bmatrix} \quad (12.104)$$

The final configuration is shown in Figure 12.18.

A simulation of the observer is shown in Figure 12.19(*a*). To demonstrate the effect of the observer design, Figure 12.19(*b*) shows the reduced speed if the observer is simply a copy of the plant and all observer feedback paths are disconnected.

In the next example we demonstrate the design of an observer without first converting the plant to dual phase-variable form. This method can become difficult if the system order is high.

Example 12.9

Design an observer without converting the plant to dual phase-variable form.

Problem A time-scaled model for the body's blood glucose level is shown in Equation 12.105. The output is the deviation in glucose concentration from its mean value in mg/100 ml, and the input is the intravenous glucose injection rate in gm/kgm/hr (Milhorn, 1966).

$$G(s) = \frac{407(s + 0.916)}{(s + 1.27)(s + 2.69)} \tag{12.105}$$

Design an observer for the phase variables with a transient response described by $\zeta = 0.7$ and $\omega_n = 100$.

Solution We can first model the plant in phase-variable form. The result is shown in Figure 12.20(*a*).

For the plant

$$\mathbf{A} = \begin{bmatrix} 0 & 1 \\ -3.42 & -3.96 \end{bmatrix}; \qquad \mathbf{C} = [\,372.81 \quad 407\,] \tag{12.106}$$

calculation of the observability matrix, $\mathbf{O_M} = [\,\mathbf{C} \quad \mathbf{CA}\,]^T$, shows that the plant is observable and we can proceed with the design. Next, find the characteristic equation of the observer. First, we have

$$
\begin{aligned}
\mathbf{A} - \mathbf{LC} &= \begin{bmatrix} 0 & 1 \\ -3.42 & -3.96 \end{bmatrix} - \begin{bmatrix} l_1 \\ l_2 \end{bmatrix}[\,372.81 \quad 407\,] \\
&= \begin{bmatrix} -372.81l_1 & (1 - 407l_1) \\ -(3.42 + 372.81l_2) & -(3.96 + 407l_2) \end{bmatrix}
\end{aligned} \tag{12.107}
$$

Now evaluate $\det[\lambda\mathbf{I} - (\mathbf{A} - \mathbf{LC})] = 0$ in order to obtain the characteristic equation:

$$
\begin{aligned}
\det[\lambda\mathbf{I} - (\mathbf{A} - \mathbf{LC})] &= \det \begin{bmatrix} (\lambda + 372.81l_1) & -(1 - 407l_1) \\ (3.42 + 372.81l_2) & (\lambda + 3.96 + 407l_2) \end{bmatrix} \\
&= \lambda^2 + (3.96 + 372.81l_1 + 407l_2)\lambda \\
&\quad + (3.42 + 84.39l_1 + 372.81l_2) = 0
\end{aligned} \tag{12.108}
$$

From the problem statement, we want $\zeta = 0.7$ and $\omega_n = 100$. Thus,

$$\lambda^2 + 140\lambda + 10{,}000 = 0 \tag{12.109}$$

Comparing the coefficients of Equations 12.108 and 12.109, we find the values of l_1 and l_2 to be -38.397 and 35.506, respectively. Using Equation 12.60, where

$$\mathbf{A} = \begin{bmatrix} 0 & 1 \\ -3.42 & -3.96 \end{bmatrix}; \quad \mathbf{B} = \begin{bmatrix} 0 \\ 1 \end{bmatrix}; \quad \mathbf{C} = [\,372.81 \quad 407\,]; \quad \mathbf{L} = \begin{bmatrix} -38.397 \\ 35.506 \end{bmatrix} \tag{12.110}$$

the observer is implemented and shown in Figure 12.20(*b*).

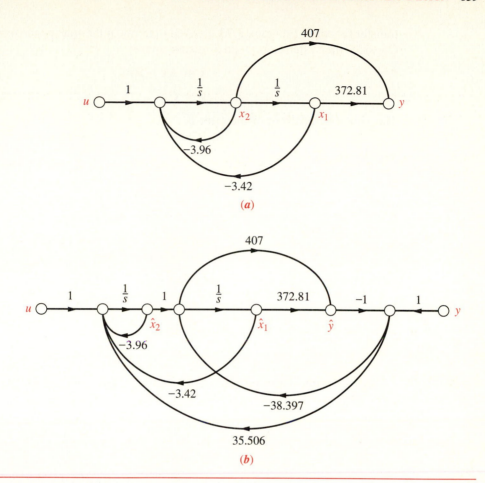

Figure 12.20 (*a*) Plant; (*b*) Designed Observer for Example 12.9

Now that we have explored transient response design using state-space techniques, let us turn our attention to the analysis and design of steady-state error characteristics.

12.8 Steady-State Error Analysis and Design

In this section we discuss steady-state error analysis and design from the state-space viewpoint. We first cover the analysis problem by discussing two different methods and conclude with the design problem.

Analysis via Final Value Theorem

A single-input, single-output system represented in state-space can be analyzed for steady-state error using the final value theorem and the closed-loop transfer function, Equation 3.73, derived in terms of the state-space representation. Consider the system represented in state space:

$$\dot{\mathbf{x}} = \mathbf{A}\mathbf{x} + \mathbf{B}r \qquad (12.111a)$$

$$y = \mathbf{C}\mathbf{x} \qquad (12.111b)$$

The Laplace transform of the error is

$$E(s) = R(s) - Y(s) \qquad (12.112)$$

But

$$Y(s) = R(s)T(s) \qquad (12.113)$$

where $T(s)$ is the closed-loop transfer function. Substituting Equation 12.113 into 12.112, we obtain

$$E(s) = R(s)[1 - T(s)] \qquad (12.114)$$

Using Equation 3.73 for $T(s)$, we find

$$E(s) = R(s)[1 - \mathbf{C}(s\mathbf{I} - \mathbf{A})^{-1}\mathbf{B}] \qquad (12.115)$$

Applying the final value theorem, we have

$$\lim_{s \to 0} sE(s) = \lim_{s \to 0} sR(s)[1 - \mathbf{C}(s\mathbf{I} - \mathbf{A})^{-1}\mathbf{B}] \qquad (12.116)$$

Let us apply the result to an example.

Example 12.10   Evaluate the steady-state error via the final value theorem for systems represented in state-space.

Problem   Evaluate the steady-state error for the controller designed in Example 12.4 for unit step and unit ramp inputs. Use the final value theorem.

Solution   The designed system is given in Equation 12.55. Thus, for Equation 12.116,

$$\mathbf{A} = \begin{bmatrix} -5 & 1 & 0 \\ 0 & -2 & 1 \\ 20 & -10 & 1 \end{bmatrix}; \qquad \mathbf{B} = \begin{bmatrix} 0 \\ 0 \\ 1 \end{bmatrix}; \qquad \mathbf{C} = [-1 \quad 1 \quad 0] \qquad (12.117)$$

Substituting Equations 12.117 into 12.116, we obtain

$$e(\infty) = \lim_{s \to 0} sR(s)\left(1 - \frac{s + 4}{s^3 + 6s^2 + 13s + 20}\right)$$

$$= \lim_{s \to 0} sR(s)\left(\frac{s^3 + 6s^2 + 12s + 16}{s^3 + 6s^2 + 13s + 20}\right) \qquad (12.118)$$

For a unit step, $R(s) = 1/s$, and $e(\infty) = 4/5$. For a unit ramp, $R(s) = 1/s^2$, and $e(\infty) = \infty$.

Analysis via Input Substitution

Another method for steady-state analysis that avoids taking the inverse of $(s\mathbf{I} - \mathbf{A})$ and can be expanded to multiple-input, multiple-output systems substitutes the input along with an assumed solution into the state equations (Hostetter, 1989). We will derive the results for unit step and unit ramp inputs.

Step Inputs

Given the state Equations 12.111, if the input is a unit step where $r = 1$, a steady-state solution, $\mathbf{x}_{ss}$, for $\mathbf{x}$, is

$$\mathbf{x}_{ss} = \begin{bmatrix} V_1 \\ V_2 \\ \vdots \\ V_n \end{bmatrix} = \mathbf{V} \tag{12.119}$$

where V_i is constant. Also,

$$\dot{\mathbf{x}}_{ss} = \mathbf{0} \tag{12.120}$$

Substituting $r = 1$, a unit step, along with Equations 12.119 and 12.120, into Equations 12.111 yields

$$\mathbf{0} = \mathbf{AV} + \mathbf{B} \tag{12.121a}$$

$$y_{ss} = \mathbf{CV} \tag{12.121b}$$

where y_{ss} is the steady-state output. Solving for $\mathbf{V}$ yields

$$\mathbf{V} = -\mathbf{A}^{-1}\mathbf{B} \tag{12.122}$$

But the steady-state error is the difference between the steady-state input and the steady-state output. The final result for the steady-state error for a unit step input into a system represented in state space is

$$e(\infty) = 1 - y_{ss} = 1 - \mathbf{CV} = 1 + \mathbf{CA}^{-1}\mathbf{B} \tag{12.123}$$

Ramp Inputs

For unit ramp inputs, $r = t$, a steady-state solution for $\mathbf{x}$ is

$$\mathbf{x}_{ss} = \begin{bmatrix} V_1 t + W_1 \\ V_2 t + W_2 \\ \vdots \\ V_n t + W_n \end{bmatrix} = \mathbf{V}t + \mathbf{W} \tag{12.124}$$

where V_i and W_i are constants. Hence,

$$\dot{\mathbf{x}}_{ss} = \begin{bmatrix} V_1 \\ V_2 \\ \vdots \\ V_n \end{bmatrix} = \mathbf{V} \tag{12.125}$$

Substituting $r = t$ along with Equations 12.124 and 12.125 into Equations 12.111 yields

$$\mathbf{V} = \mathbf{A}(\mathbf{V}t + \mathbf{W}) + \mathbf{B}t \qquad (12.126a)$$

$$y_{ss} = \mathbf{C}(\mathbf{V}t + \mathbf{W}) \qquad (12.126b)$$

Equating matrix coefficients of t, $\mathbf{AV} = -\mathbf{B}$, or

$$\mathbf{V} = -\mathbf{A}^{-1}\mathbf{B} \qquad (12.127)$$

Equating constant terms, we have $\mathbf{AW} = \mathbf{V}$, or

$$\mathbf{W} = \mathbf{A}^{-1}\mathbf{V} \qquad (12.128)$$

Substituting Equations 12.127 and 12.128 into 12.126b yields

$$y_{ss} = \mathbf{C}[-\mathbf{A}^{-1}\mathbf{B}t + \mathbf{A}^{-1}(-\mathbf{A}^{-1}\mathbf{B})] = -\mathbf{C}[\mathbf{A}^{-1}\mathbf{B}t + (\mathbf{A}^{-1})^2\mathbf{B}] \qquad (12.129)$$

The steady-state error is therefore

$$e(\infty) = \lim_{t \to \infty}(t - y_{ss}) = \lim_{t \to \infty}[(1 + \mathbf{CA}^{-1}\mathbf{B})t + \mathbf{C}(\mathbf{A}^{-1})^2\mathbf{B}] \qquad (12.130)$$

Notice that in order to use this method, $\mathbf{A}^{-1}$ must exist.

We now demonstrate the use of Equations 12.123 and 12.130 to find the steady-state error for step and ramp inputs.

Example 12.11

Evaluate the steady-state error for a system represented in state space, using input substitution.

Problem Evaluate the steady-state error for the controller designed in Example 12.4 for unit step and unit ramp inputs. Use input substitution.

Solution For a unit step input, the steady-state error given by Equation 12.123 is

$$e(\infty) = 1 + \mathbf{CA}^{-1}\mathbf{B} = 1 - 0.2 = 0.8 \qquad (12.131)$$

where $\mathbf{C}$, $\mathbf{A}$, and $\mathbf{B}$ are as follows:

$$\mathbf{A} = \begin{bmatrix} -5 & 1 & 0 \\ 0 & -2 & 1 \\ 20 & -10 & 1 \end{bmatrix}; \quad \mathbf{B} = \begin{bmatrix} 0 \\ 0 \\ 1 \end{bmatrix}; \quad \mathbf{C} = \begin{bmatrix} -1 & 1 & 0 \end{bmatrix} \qquad (12.132)$$

For ramp input, using Equation 12.130, we have

$$e(\infty) = \lim_{t \to \infty}[(1 + \mathbf{CA}^{-1}\mathbf{B})t + \mathbf{C}(\mathbf{A}^{-1})^2\mathbf{B}] = \lim_{t \to \infty}(0.8t + 0.08) = \infty \qquad (12.133)$$

We have covered the steady-state error analysis problem. Now we look at ways of improving the steady-state error for a system represented in state space.

Design with Integral Control

Consider Figure 12.21. The previously designed controller discussed in Section 12.2 is shown inside the dashed box. A feedback path from the output has been added to form the error, e, which is fed forward to the controlled plant via an integrator. The integrator increases the system type and reduces the previous finite error to zero.

We will now derive the form of the state equations for the system of Figure 12.21 and then use that form to design a controller. Thus, we will be able to design a system for zero steady-state error for a step input as well as design the desired transient response.

An additional state variable, x_N, has been added at the output of the leftmost integrator. The error is the derivative of this variable. Now, from Figure 12.21,

$$\dot{x}_N = r - \mathbf{Cx} \tag{12.134}$$

Writing the state equations from Figure 12.21, we have

$$\dot{\mathbf{x}} = \mathbf{Ax} + \mathbf{B}u \tag{12.135a}$$

$$\dot{x}_N = -\mathbf{Cx} + r \tag{12.135b}$$

$$y = \mathbf{Cx} \tag{12.135c}$$

Equations 12.135 can be written as augmented vectors and matrices. Hence,

$$\begin{bmatrix} \dot{\mathbf{x}} \\ \dot{x}_N \end{bmatrix} = \begin{bmatrix} \mathbf{A} & \mathbf{0} \\ -\mathbf{C} & 0 \end{bmatrix} \begin{bmatrix} x \\ x_N \end{bmatrix} + \begin{bmatrix} \mathbf{B} \\ 0 \end{bmatrix} u + \begin{bmatrix} \mathbf{0} \\ 1 \end{bmatrix} r \tag{12.136a}$$

$$y = \begin{bmatrix} \mathbf{C} & 0 \end{bmatrix} \begin{bmatrix} \mathbf{x} \\ x_N \end{bmatrix} \tag{12.136b}$$

Figure 12.21 Integral Control for Steady-State Error Design

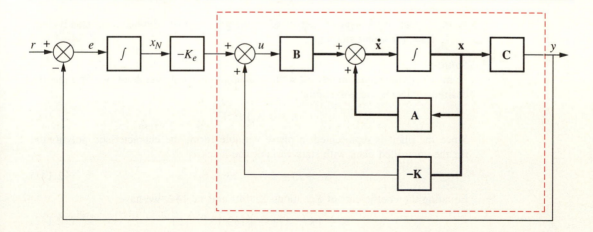

But

$$u = -\mathbf{K}\mathbf{x} - K_e x_N = -[\,\mathbf{K}\quad K_e\,]\begin{bmatrix} \mathbf{x} \\ x_N \end{bmatrix} \tag{12.137}$$

Substituting Equation 12.137 into 12.136a and simplifying, we obtain

$$\begin{bmatrix} \dot{\mathbf{x}} \\ \dot{x}_\mathbf{N} \end{bmatrix} = \begin{bmatrix} (\mathbf{A} - \mathbf{B}\mathbf{K}) & -\mathbf{B}K_e \\ -\mathbf{C} & 0 \end{bmatrix}\begin{bmatrix} \mathbf{x} \\ x_N \end{bmatrix} + \begin{bmatrix} \mathbf{0} \\ 1 \end{bmatrix} r \tag{12.138a}$$

$$y = [\,\mathbf{C}\quad 0\,]\begin{bmatrix} \mathbf{x} \\ x_N \end{bmatrix} \tag{12.138b}$$

Thus, the system type has been increased, and we can use the characteristic equation of Equation 12.138a to design $\mathbf{K}$ and K_e to yield the desired transient response. The effect on the transient response of any closed-loop zeros in the final design must also be taken into consideration. One possible assumption is that the closed-loop zeros will be the same as those of the open-loop plant. This assumption, which of course must be checked, helps place higher-order poles. Let us demonstrate with an example.

Example 12.12   Design integral control for a system represented in state space.

Problem   Consider the plant of Equations 12.139:

$$\dot{\mathbf{x}} = \begin{bmatrix} 0 & 1 \\ -3 & -5 \end{bmatrix}\mathbf{x} + \begin{bmatrix} 0 \\ 1 \end{bmatrix} u \tag{12.139a}$$

$$y = [\,1\quad 0\,]\mathbf{x} \tag{12.139b}$$

a. Design a controller without integral control to yield a 10% overshoot and a settling time of 0.5 seconds. Evaluate the steady-state error for a unit step input.

b. Repeat the design of part *a* using integral control. Evaluate the steady-state error for a unit step input.

Solution

a. Using the requirements for settling time and percent overshoot, we find that the desired characteristic polynomial is

$$s^2 + 16s + 183.137 \tag{12.140}$$

Since the plant is represented in phase-variable form, the characteristic polynomial for the controlled plant with state-variable feedback is

$$s^2 + (5 + k_2)s + (3 + k_1) \tag{12.141}$$

Equating the coefficients of Equations 12.140 and 12.141, we have

$$\mathbf{K} = [\,k_1\quad k_2\,] = [\,180.137\quad 11\,] \tag{12.142}$$

From Equations 12.3, the controlled plant with state-variable feedback represented in phase-variable form is

$$\dot{\mathbf{x}} = (\mathbf{A} - \mathbf{BK})\mathbf{x} + \mathbf{B}r = \begin{bmatrix} 0 & 1 \\ -183.137 & -16 \end{bmatrix}\mathbf{x} + \begin{bmatrix} 0 \\ 1 \end{bmatrix}r \qquad (12.143a)$$

$$y = \mathbf{Cx} = [1 \quad 0] \qquad (12.143b)$$

Using Equation 12.123, we find that the steady-state error for a step input is

$$e(\infty) = 1 + \mathbf{C}(\mathbf{A} - \mathbf{BK})^{-1}\mathbf{B}$$

$$= 1 + [1 \quad 0]\begin{bmatrix} 0 & 1 \\ -183.137 & -16 \end{bmatrix}^{-1}\begin{bmatrix} 0 \\ 1 \end{bmatrix}$$

$$= 0.995 \qquad (12.144)$$

b. We now use Equation 12.138 to represent the integral-controlled plant as follows:

$$\begin{bmatrix} \dot{x}_1 \\ \dot{x}_2 \\ \dot{x}_N \end{bmatrix} = \begin{bmatrix} \begin{bmatrix} 0 & 1 \\ -3 & -5 \end{bmatrix} - \begin{bmatrix} 0 \\ 1 \end{bmatrix}[k_1 \quad k_2] & -\begin{bmatrix} 0 \\ 1 \end{bmatrix}K_e \\ -[1 \quad 0] & 0 \end{bmatrix}\begin{bmatrix} x_1 \\ x_2 \\ x_N \end{bmatrix} + \begin{bmatrix} 0 \\ 0 \\ 1 \end{bmatrix}r$$

$$= \begin{bmatrix} 0 & 1 & 0 \\ -(3 + k_1) & -(5 + k_2) & -K_e \\ -1 & 0 & 0 \end{bmatrix}\begin{bmatrix} x_1 \\ x_2 \\ x_N \end{bmatrix} + \begin{bmatrix} 0 \\ 0 \\ 1 \end{bmatrix}r \qquad (12.145a)$$

$$y = [1 \quad 0 \quad 0]\begin{bmatrix} x_1 \\ x_2 \\ x_N \end{bmatrix} \qquad (12.145b)$$

Using Equation 3.73 and the plant of Equation 12.139, we find that the transfer function of the plant is $G(s) = \dfrac{1}{s^2 + 5s + 3}$. The desired characteristic polynomial for the closed-loop integral-controlled system is shown in Equation 12.140. Since the plant has no zeros, we assume no zeros for the closed-loop system and augment Equation 12.140 with a third pole, $(s + 100)$, which has a real part greater than five times that of the desired dominant second-order poles. The desired third-order closed-loop system characteristic polynomial is

$$(s + 100)(s^2 + 16s + 183.137) = s^3 + 116s^2 + 1783.137s + 18,313.7 \qquad (12.146)$$

The characteristic polynomial for the system of Equation 12.145 is

$$s^3 + (5 + k_2)s^2 + (3 + k_1)s - K_e \qquad (12.147)$$

Matching coefficients from Equations 12.146 and 12.147, we obtain

$$k_1 = 1780.137 \qquad (12.148a)$$

$$k_2 = 111 \qquad (12.148b)$$

$$K_e = -18,313.7 \qquad (12.148c)$$

Substituting these values into Equation 12.145 yields the closed-loop integral-controlled system:

$$
\begin{bmatrix} \dot{x}_1 \\ \dot{x}_2 \\ \dot{x}_N \end{bmatrix} = \begin{bmatrix} 0 & 1 & 0 \\ -1783.137 & -116 & 18{,}313.7 \\ -1 & 0 & 0 \end{bmatrix} \begin{bmatrix} x_1 \\ x_2 \\ x_N \end{bmatrix} + \begin{bmatrix} 0 \\ 0 \\ 1 \end{bmatrix} r \qquad (12.149a)
$$

$$
y = \begin{bmatrix} 1 & 0 & 0 \end{bmatrix} \begin{bmatrix} x_1 \\ x_2 \\ x_N \end{bmatrix} \qquad (12.149b)
$$

In order to check our assumption for the zero, we now apply Equation 3.73 to Equation 12.149 and find the closed-loop transfer function to be

$$
T(s) = \frac{18{,}313.7}{s^3 + 116s^2 + 1783.137s + 18{,}313.7} \qquad (12.150)
$$

Since the transfer function matches our design, we have the desired transient response.

Now let us find the steady-state error for a unit step input. Applying Equation 12.123 to Equation 12.149, we obtain

$$
e(\infty) = 1 + \begin{bmatrix} 1 & 0 & 0 \end{bmatrix} \begin{bmatrix} 0 & 1 & 0 \\ -1783.137 & -116 & 18{,}313.7 \\ -1 & 0 & 0 \end{bmatrix}^{-1} \begin{bmatrix} 0 \\ 0 \\ 1 \end{bmatrix} = 0 \qquad (12.151)
$$

Now that we have designed controllers and observers for transient response and steady-state error, we summarize the chapter with examples demonstrating the combined design of controllers and observers.

12.9 Chapter-Objective Demonstration Problem

In this section we use our ongoing antenna azimuth control system to demonstrate the combined design of a controller and an observer. We will assume that the states are not available and must be estimated from the output. The block diagram of the original system is shown in Figure 10.53. Setting the preamplifier gain to 200, the forward transfer function is simplified to that shown in Figure 12.22.

The design example will specify a transient response for the system and a faster transient response for the observer. The final design configuration will consist of the plant, the observer, and the controller, as shown conceptually in Figure 12.23. The design of the observer and the controller will be separate.

Example 12.13

Design a controller and observer for an antenna azimuth tracking system.

Problem   Using the simplified block diagram of the plant for the antenna azimuth tracking system shown in Figure 12.22, design a controller to yield a 10% overshoot and a settling time of 1 second. Place the third pole 10 times further from the imaginary axis than the second-order dominant pair.

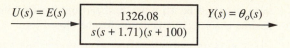

$$U(s) = E(s) \qquad \boxed{\dfrac{1326.08}{s(s + 1.71)(s + 100)}} \qquad Y(s) = \theta_o(s)$$

Figure 12.22 Simplified Block Diagram of Antenna Azimuth Position Control System Shown in Figure 10.53

Assume that the state variables of the plant are not accessible and design an observer to estimate the states. The desired transient response for the observer is a 10% overshoot and a natural frequency 10 times greater than the system response above. As in the case of the controller, place the third pole 10 times further from the imaginary axis than the observer's dominant second-order pair.

Solution

Controller design We first design the controller by finding the desired characteristic equation. A 10% overshoot and a settling time of 1 second yield $\zeta = 0.591$ and $\omega_n = 6.766$. Thus, the characteristic equation for the dominant poles is $s^2 + 8s + 45.779 = 0$, where the dominant poles are located at $-4 \pm j5.457$. The

Figure 12.23 Conceptual State-Space Design Configuration, Showing Plant, Observer, and Controller

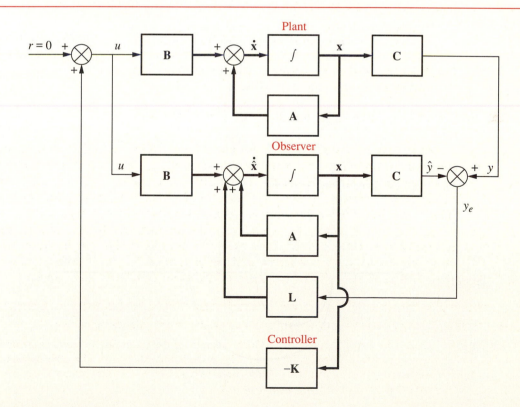

third pole will be 10 times further from the imaginary axis, or at -40. Hence the desired characteristic equation for the closed-loop system is

$$(s^2 + 8s + 45.779)(s + 40) = s^3 + 48s^2 + 365.779s + 1831.16 = 0 \qquad (12.152)$$

Next we find the actual characteristic equation of the closed-loop system. The first step is to model the closed-loop system in state space and then find its characteristic equation. From Figure 12.22, the transfer function of the plant is

$$G(s) = \frac{1326.08}{s(s + 1.71)(s + 100)} = \frac{1326.08}{s(s^2 + 101.71s + 171)} \qquad (12.153)$$

Using phase-variables, this transfer function is converted into the signal-flow graph shown in Figure 12.24, and the state equations are written as follows:

$$\dot{\mathbf{x}} = \begin{bmatrix} 0 & 1 & 0 \\ 0 & 0 & 1 \\ 0 & -171 & -101.71 \end{bmatrix} \mathbf{x} + \begin{bmatrix} 0 \\ 0 \\ 1 \end{bmatrix} u = \mathbf{Ax} + \mathbf{B}u \qquad (12.154a)$$

$$y = [\,1326.08 \quad 0 \quad 0\,]\mathbf{x} = \mathbf{Cx} \qquad (12.154b)$$

We now pause in our design to evaluate the controllability of the system. The controllability matrix, $\mathbf{C_M}$, is

$$\mathbf{C_M} = [\,\mathbf{B} \quad \mathbf{AB} \quad \mathbf{A^2B}\,] = \begin{bmatrix} 0 & 0 & 1 \\ 0 & 1 & -101.71 \\ 1 & -101.71 & 10{,}173.92 \end{bmatrix} \qquad (12.155)$$

The determinant of $\mathbf{C_M}$ is -1; thus, the system is controllable.

Continuing with the design of the controller, we show the controller's configuration with the feedback from all state variables in Figure 12.25. We now find the characteristic equation of the system of Figure 12.25. From Equation 12.7 and Equation 12.154, the system matrix, $\mathbf{A - BK}$, is

$$\mathbf{A - BK} = \begin{bmatrix} 0 & 1 & 0 \\ 0 & 0 & 1 \\ -k_1 & -(171 + k_2) & -(101.71 + k_3) \end{bmatrix} \qquad (12.156)$$

Figure 12.24 Signal-Flow Graph for
$$G(s) = \frac{1326.08}{s(s^2 + 101.71s + 171)}$$

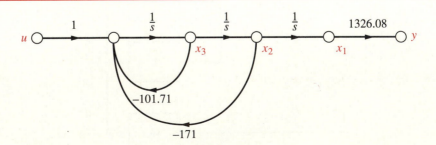

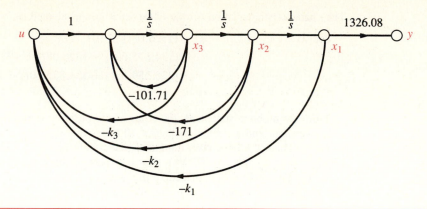

Figure 12.25 Plant with State-Variable Feedback for Controller Design

Thus, the closed-loop system's characteristic equation is

$$\det[s\mathbf{I} - (\mathbf{A} - \mathbf{BK})] = s^3 + (101.71 + k_3)s^2 + (171 + k_2)s + k_1 = 0 \qquad (12.157)$$

Matching the coefficients of Equation 12.152 with those of Equation 12.157, we evaluate the k_i's as follows:

$$k_1 = 1831.16 \qquad (12.158a)$$

$$k_2 = 194.779 \qquad (12.158b)$$

$$k_3 = -53.71 \qquad (12.158c)$$

Observer Design   Before designing the observer, we test the system for observability. Using the $\mathbf{A}$ and $\mathbf{C}$ matrices from Equations 12.154, the observability matrix, $\mathbf{O_M}$, is

$$\mathbf{O_M} = \begin{bmatrix} \mathbf{C} \\ \mathbf{CA} \\ \mathbf{CA}^2 \end{bmatrix} = \begin{bmatrix} 1326.08 & 0 & 0 \\ 0 & 1326.08 & 0 \\ 0 & 0 & 1326.08 \end{bmatrix} \qquad (12.159)$$

The determinant of $\mathbf{O_M}$ is 1326.08^3. Thus $\mathbf{O_M}$ is of rank 3, and the system is observable.

We now proceed to design the observer. Since the order of the system is not high, we will design the observer directly without first converting to dual phase-variable form. From Equation 12.64, we need first to find $\mathbf{A} - \mathbf{LC}$. $\mathbf{A}$ and $\mathbf{C}$ from Equation 12.154 along with

$$\mathbf{L} = \begin{bmatrix} l_1 \\ l_2 \\ l_3 \end{bmatrix} \qquad (12.160)$$

are used to evaluate $\mathbf{A} - \mathbf{LC}$ as follows:

$$\mathbf{A} - \mathbf{LC} = \begin{bmatrix} -1326.08l_1 & 1 & 0 \\ -1326.08l_2 & 0 & 1 \\ -1326.08l_3 & -171 & -101.71 \end{bmatrix} \qquad (12.161)$$

The characteristic equation for the observer is now evaluated as

$$
\begin{aligned}
\det[\lambda \mathbf{I} - (\mathbf{A} - \mathbf{L}\mathbf{C})] = {}& \lambda^3 + (1326.08l_1 + 101.71)\lambda^2 \\
& + (134{,}875.60l_1 + 1326.08l_2 + 171)\lambda \\
& + (226{,}759.68l_1 + 134{,}875.60l_2 + 1326.08l_3) \\
= {}& 0
\end{aligned} \tag{12.162}
$$

From the problem statement, the poles of the observer are to be placed to yield a 10% overshoot and a natural frequency 10 times that of the system's dominant pair of poles. Thus, the observer's dominant poles yield $[s^2 + (2 \times 0.591 \times 67.66)s + 67.66^2] = (s^2 + 80s + 4577.88)$. The real part of the roots of this polynomial is -40. The third pole is then placed 10 times further from the imaginary axis at -400. The composite characteristic equation for the observer is

$$
\begin{aligned}
(s^2 + 80s + 4577.88)(s + 400) = {}& s^3 + 480s^2 + 36{,}577.88s \\
& + 1{,}831{,}152 = 0
\end{aligned} \tag{12.163}
$$

Figure 12.26 Completed State-Space Design for the Antenna Azimuth Control System, Showing Controller and Observer

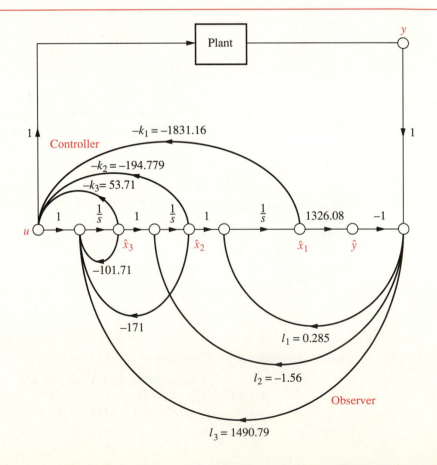

Matching coefficients from Equations 12.162 and 12.163, we solve for the coefficients:

$$l_1 = 0.285 \tag{12.164a}$$

$$l_2 = -1.56 \tag{12.164b}$$

$$l_3 = 1490.79 \tag{12.164c}$$

Figure 12.26, which follows the general configuration of Figure 12.23, shows the completed design, including the controller and the observer.

The results of the design are shown in Figure 12.27. Figure 12.27(*a*) shows the impulse response of the closed-loop system without any difference between the plant

Figure 12.27 State-Space Design Response of Antenna Azimuth Control System: (*a*) Portion of Impulse Response—Plant and Observer the Same, $x_1(0) = \hat{x}_1(0) = 0$; (*b*) Impulse Response—Plant and Observer with Different initial Conditions, $x_1(0) = 0.006$ for the Plant, $\hat{x}_1(0) = 0$ for the Observer

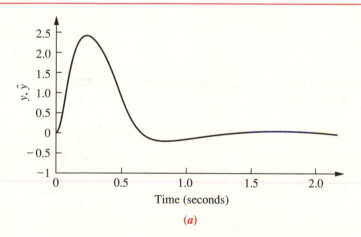

(*a*)

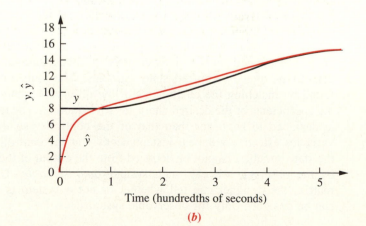

(*b*)

and its modeling as an observer. The undershoot and settling time approximately meet the requirements set forth in the problem statement of 10% and 1 second, respectively. In Figure 12.27(*b*), we see the response designed into the observer. An initial condition of 0.006 was given to x_1 in the plant to make the modeling of the plant and observer different. Notice how the observer's response follows the plant's response by the time 0.04 second is reached.

In this section we demonstrated the concepts of state-space design using our ongoing antenna azimuth control system as an example. Assuming that the states were not available from the plant, our design also included an observer to emulate the plant and provide the state variables for feedback. The controller was designed first to yield the desired transient response. Then the observer was designed to yield a transient response 10 times faster than the controller's so that, effectively, estimated states were available instantaneously.

12.10 Summary

This chapter has followed the path established by Chapters 9 and 11—control system design. Chapter 9 used root locus techniques to design a control system with a desired transient response. Sinusoidal frequency response techniques for design were covered in Chapter 11, and in this chapter, we used state-space design techniques.

State-space design consists of specifying the system's desired pole locations and then designing a controller consisting of state-variable feedback gains to meet these requirements. If the state variables are not available, then an observer is designed to emulate the plant and provide estimated state variables.

Controller design consists of feeding back the state variables to the input, u, of the system through specified gains. The values of these gains are found by matching the coefficients of the system's characteristic equation with the coefficients of the desired characteristic equation. In some cases, the control signal, u, cannot affect one or more state variables. We call such a system *uncontrollable*. For this system, a total design is not possible. Using the controllability matrix, a designer can tell whether or not a system is controllable prior to the design.

Observer design consists of feeding back the error between the actual output and the estimated output. This error is fed back through specified gains to the derivatives of the estimated state variables. The values of these gains are also found by matching the coefficients of the observer's characteristic equation with the coefficients of the desired characteristic equation. The response of the observer is designed to be faster than that of the controller so that the estimated state variables effectively appear instantaneously at the controller. For some systems, the state variables cannot be deduced from the output of the system, as is required by the observer. We call such systems *unobservable*. Using the observability matrix, the designer can tell whether or not a system is observable. Observers can be designed only for observable systems.

Finally, we discussed ways of evaluating and designing the steady-state error performance of systems represented in state-space. The addition of an integration before the controlled plant yields improvement in the steady-state error. In this chapter, this additional integration was incorporated into the controller design.

Several advantages of state-space design are apparent. First, in contrast to the root locus method, all pole locations can be specified to ensure a negligible effect of the nondominant poles upon the transient response. With the root locus, we were forced to justify an assumption that the nondominant poles did not appreciably affect the transient response. We were not always successful in this endeavor. Second, with the use of an observer, we are no longer forced to acquire the actual system variables for feedback. The advantage here is that sometimes the variables cannot be physically accessed, or it may be too expensive to provide that access. Finally, the methods shown lend themselves to design automation using the digital computer.

A disadvantage of the design methods covered in this chapter is the designer's inability to design the location of open or closed-loop zeros that may affect the transient response. In root locus or frequency response design, the zeros of the lag or lead compensator can be specified. Another disadvantage of state-space methods concerns the designer's ability to relate all pole locations to the desired response; this relationship is not always apparent. Also, once the design is completed, we may not be satisfied with the sensitivity to parameter changes. Finally, as previously discussed, state-space techniques do not satisfy our intuition as much as root locus techniques, where the effect of parameter changes can be immediately seen as changes in closed-loop pole locations.

REVIEW QUESTIONS

1. Briefly describe an advantage that state-space techniques have over root locus techniques in the placement of closed-loop poles for transient response design.
2. Briefly describe the design procedure for a controller.
3. Different signal-flow graphs can represent the same system. Which form facilitates the calculation of the variable gains during controller design?
4. In order to effect a complete controller design, a system must be controllable. Describe the physical meaning of controllability.
5. Under what conditions can inspection of the signal-flow graph of a system yield immediate determination of controllability?
6. In order to mathematically determine controllability, the controllability matrix is formed and its rank determined. What is the final step in determining controllability if the controllability matrix is a square matrix?
7. What is an observer?
8. Under what conditions would you use an observer in your state-space design of a control system?

9. Briefly describe the configuration of an observer.

10. What plant representation lends itself to easier design of an observer?

11. Briefly describe the design technique for an observer, given the configuration you described in Question 9.

12. Compare the major difference in the transient response of an observer to that of a controller. Why does this difference exist?

13. From what equation do we find the characteristic equation of the controller-compensated system?

14. From what equation do we find the characteristic equation of the observer?

15. In order to effect a complete observer design, a system must be observable. Describe the physical meaning of observability.

16. Under what conditions can inspection of the signal-flow graph of a system yield immediate determination of observability?

17. In order to mathematically determine observability, the observability matrix is formed and its rank determined. What is the final step in determining observability if the observability matrix is a square matrix?

PROBLEMS

1. Consider the following open-loop transfer functions, where $G(s) = Y(s)/U(s)$, $Y(s)$ is the Laplace transform of the output, and $U(s)$ is the Laplace transform of the input control signal:

   i. $G(s) = \dfrac{(s + 1)}{s(s + 2)}$

   ii. $G(s) = \dfrac{(s + 2)}{(s + 3)(s + 4)}$

   iii. $G(s) = \dfrac{10(s + 2)(s + 3)}{s(s + 4)(s + 5)}$

   iv. $G(s) = \dfrac{10s}{(s + 2)(s + 4)(s + 6)}$

   For each of these transfer functions, do the following:
   a. Draw the signal-flow graph in phase-variable form.
   b. Add state-variable feedback to the signal-flow graph.
   c. For each closed-loop signal-flow graph, write the state equations.
   d. Write, *by inspection*, the closed-loop transfer function, $T(s)$, for your closed-loop signal-flow graphs.
   e. Verify your answers for $T(s)$ by finding the closed-loop transfer functions from the state equations and Equation 3.73.

2. The following open-loop transfer function can be represented by a signal-flow graph in cascade form.

$$G(s) = \frac{20(s + 5)}{(s + 1)(s + 4)(s + 6)}$$

    a. Draw the signal-flow graph and show the state-variable feedback.
    b. Find the closed-loop transfer function with state-variable feedback.

3. The following open-loop transfer function can be represented by a signal-flow graph in parallel form.

$$G(s) = \frac{100(s + 10)(s + 20)}{s(s + 15)(s + 30)}$$

    a. Draw the signal-flow graph and show the state-variable feedback.
    b. Find the closed-loop transfer function with state-variable feedback.

4. Given the following open-loop plant,

$$G(s) = \frac{10}{s(s + 2)(s + 4)}$$

design a controller to yield a 10% overshoot and a settling time of 0.5 second. Place the third pole 10 times further from the imaginary axis than the dominant pole pair. Use the phase variables for state-variable feedback.

5. Given the following open-loop plant,

$$G(s) = \frac{50(s + 1)}{(s + 2)(s + 4)(s + 6)}$$

design a controller to yield a 20% overshoot and a settling time of 1 second. Place the third pole 10 times further from the imaginary axis than the dominant pole pair. Use the phase variables for state-variable feedback.

6. Repeat Problem 4 assuming that the plant is represented in the cascade form. Do not convert to phase-variable form.

7. Repeat Problem 5 assuming that the plant is represented in the parallel form. Do not convert to phase-variable form.

8. For each of the plants represented by signal-flow graphs in Figure P12.1, determine the controllability. If the controllability can be determined by inspection, state that it can and then verify your conclusions using the controllability matrix.

9. Given the plant shown in Figure P12.2, what relationship exists between b_1 and b_2 to make the system uncontrollable?

10. Consider the following transfer function:

$$G(s) = \frac{50(s + 1)}{(s + 2)(s + 4)(s + 6)}$$

If the system is represented in cascade form, design a controller to yield a closed-loop response of 10% overshoot with a settling time of 2 seconds. Design the controller by first transforming the plant to phase variables.

11. Repeat Problem 10 assuming that the plant is represented in parallel form.

Figure P12.1

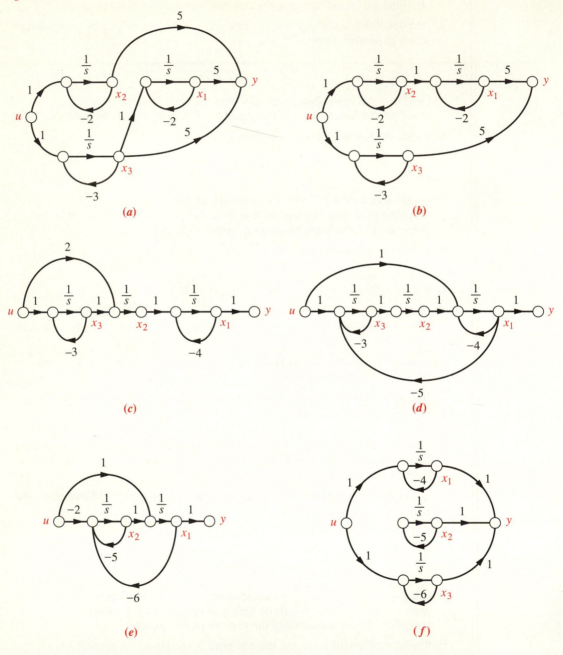

(a)

(b)

(c)

(d)

(e)

(f)

Figure P12.2

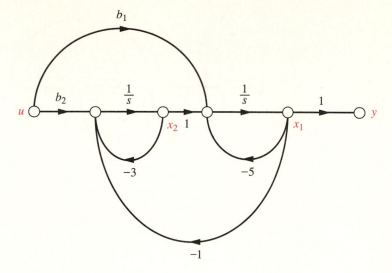

12. Consider the plant

$$G(s) = \frac{1}{s(s + 2)(s + 6)}$$

whose state variables are not available. Design an observer for the dual phase-variables to yield a transient response described by $\zeta = 0.5$ and $\omega_n = 50$. Place the third pole 10 times further from the imaginary axis than the dominant poles.

13. Repeat Problem 12 assuming that the plant is represented in phase-variable form. Do not convert to dual phase-variable form.

14. Consider the plant

$$G(s) = \frac{(s + 1)}{(s + 4)(s + 6)}$$

whose phase variables are not available. Design an observer for the phase variables with a transient response described by $\zeta = 0.7$ and $\omega_n = 100$. Do not convert to dual phase-variable form.

15. Determine whether or not each of the systems shown in Figure P12.1 is observable.

16. Given the plant of Figure P12.3, what relationship must exist between c_1 and c_2 in order for the system to be unobservable?

17. Design an observer for the plant

$$G(s) = \frac{1}{(s + 3)(s + 5)(s + 10)}$$

represented in cascade form. Transform the plant to dual phase-variable form for the design. Then transform the design back to cascade form. The characteristic polynomial for the observer is to be $s^3 + 480s^2 + 36,500s + 1,800,000$.

Figure P12.3

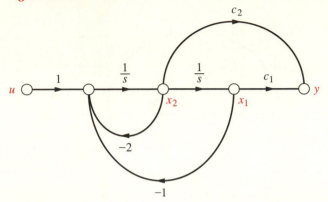

18. Repeat Problem 17 assuming that the plant is represented in parallel form.

19. For each of the following closed-loop systems, find the steady-state error for unit step and unit ramp inputs. Use both the final value theorem and input substitution methods.

$$\textbf{a. } \dot{\mathbf{x}} = \begin{bmatrix} -5 & -4 & -2 \\ -3 & -10 & 0 \\ -1 & 1 & -5 \end{bmatrix}\mathbf{x} + \begin{bmatrix} 1 \\ 1 \\ 0 \end{bmatrix}r; \qquad y = [\,-1 \quad 2 \quad 1\,]\mathbf{x}$$

$$\textbf{b. } \dot{\mathbf{x}} = \begin{bmatrix} 0 & 1 & 0 \\ -5 & -9 & 7 \\ -1 & 0 & 0 \end{bmatrix}\mathbf{x} + \begin{bmatrix} 0 \\ 0 \\ 1 \end{bmatrix}r; \qquad y = [\,1 \quad 0 \quad 0\,]\mathbf{x}$$

$$\textbf{c. } \dot{\mathbf{x}} = \begin{bmatrix} -9 & -5 & -1 \\ 1 & 0 & -2 \\ -3 & -2 & -5 \end{bmatrix}\mathbf{x} + \begin{bmatrix} 2 \\ 3 \\ 5 \end{bmatrix}r; \qquad y = [\,1 \quad -2 \quad 4\,]\mathbf{x}$$

20. Given the plant

$$\dot{\mathbf{x}} = \begin{bmatrix} -1 & 1 \\ 0 & 2 \end{bmatrix}\mathbf{x} + \begin{bmatrix} 0 \\ 1 \end{bmatrix}u; \qquad y = [\,1 \quad 1\,]\mathbf{x}$$

design an integral controller to yield a 10% overshoot, 0.5-second settling time, and zero steady-state error for a step input.

21. Repeat Problem 20 for the following plant:

$$\dot{\mathbf{x}} = \begin{bmatrix} -2 & 1 \\ 0 & -5 \end{bmatrix}\mathbf{x} + \begin{bmatrix} 0 \\ 1 \end{bmatrix}u; \qquad y = [\,1 \quad 1\,]\mathbf{x}$$

22. *Chapter-objective problem:* The conceptual block diagram of a gas-fired heater is shown in Figure P12.4. The commanded fuel pressure is proportional to the desired temperature. The difference between the commanded fuel pressure and a measured pressure related to the output temperature is used to actuate a valve and release fuel to the heater. The rate of fuel flow determines the temperature. When the output temperature equals the equivalent commanded temperature as determined by the commanded fuel pressure, the fuel flow is stopped, and the heater shuts off (Tyner, 1968).

Figure P12.4 Block Diagram of a Gas-Fired Heater

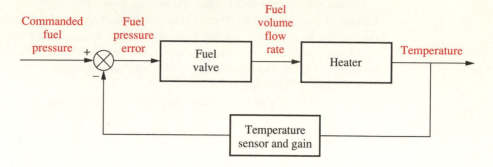

If the transfer function of the heater, $G_H(s)$, is

$$G_H(s) = \frac{1}{(s + 0.4)(s + 0.8)} \frac{\text{degrees F}}{\text{ft}^3/\min}$$

and the transfer function of the fuel valve, $G_v(s)$, is

$$G_v(s) = \frac{5}{s + 5} \frac{\text{ft}^3/\min}{\text{psi}}$$

replace the temperature feedback path with a phase-variable controller that yields a 5% overshoot and a settling time of 10 minutes. Also, design an observer that will respond 10 times faster than the system but with the same percent overshoot.

23. *Chapter-objective problem:* Consider the position control system of Figure P10.9 with a preamplifier gain of $K = 20$.
    a. Design a controller to yield a 15% overshoot and a settling time of 2 seconds. Place the third pole 10 times further from the imaginary axis than the second-order dominant pole-pair.
    b. Redraw Figure P10.9, showing a tachometer that yields rate feedback along with any added gains or attenuators required to implement the state-variable feedback gains.
    c. Assume that the tachometer is not available to provide rate feedback. Design an observer to estimate the states. The observer will respond with a 10% overshoot and a natural frequency 10 times greater than the system response. Place the observer's third pole 10 times further from the imaginary axis than the observer's dominant second-order pole-pair.
    d. Redraw Figure P10.9, showing the implementation of the controller and the observer.

BIBLIOGRAPHY

D'Azzo, J. J., and Houpis, C. H. *Linear Control System Analysis and Design: Conventional and Modern.* 3d ed. McGraw-Hill, New York, 1988.

Franklin, G. F.; Powell, J. D.; and Emami-Naeini, A. *Feedback Control of Dynamic Systems.* Addison-Wesley, Reading, Mass., 1986.

Hostetter, G. H.; Savant, C. J., Jr.; and Stefani, R. T. *Design of Feedback Control Systems*. 2d ed. Saunders College Publishing, New York, 1989.

Kailath, T. *Linear Systems*. Prentice Hall, Englewood Cliffs, N.J., 1980.

Luenberger, D. G. Observing the State of a Linear System, *IEEE Transactions on Military Electronics*, vol. MIL-8, April 1964, pp. 74–80.

Milhorn, H. T., Jr. *The Application of Control Theory to Physiological Systems*. W. B. Saunders, Philadelphia, 1966.

Ogata, K. *Modern Control Engineering*. 2d ed. Prentice Hall, Englewood Cliffs, N.J., 1990.

Ogata, K. *State Space Analysis of Control Systems*. Prentice Hall, Englewood Cliffs, N.J., 1967.

Rockwell International. *Space Shuttle Transportation System*. 1984 (Press Information).

Sinha, N. K. *Control Systems*. Holt, Rinehart & Winston, New York, 1986.

Timothy, L. K., and Bona, B. E. *State Space Analysis: An Introduction*. McGraw-Hill, New York, 1968.

Tyner, M., and May, F. P. *Process Engineering Control*. Ronald Press, New York, 1968.

13

DIGITAL CONTROL SYSTEMS

13.1 Introduction

This chapter is an introduction to digital control systems and will cover only frequency domain analysis and design. Students are encouraged to pursue the study of state-space techniques in an advanced course in sampled-data control systems. In this chapter we introduce analysis and design of stability, steady-state error, and transient response for computer-controlled systems.

With the development of the minicomputer in the mid-1960s and the microcomputer in the mid-1970s, physical systems need no longer be controlled by expensive mainframe computers. For example, milling operations that required mainframe computers in the past can now be controlled by a Macintosh computer.

The digital computer can perform two functions: (1) supervisory—external to the feedback loop; and (2) control—internal to the feedback loop. Examples of supervisory functions consist of scheduling tasks, monitoring parameters and variables for out-of-range values, or initiating safety shut-down. Control functions are of primary interest to us, since a computer that performs within the feedback loop replaces the methods of compensation heretofore discussed. Examples of control functions are lead and lag compensation.

Transfer functions, representing compensators built with analog components, are now replaced with a digital computer that performs calculations that emulate the physical compensator. What advantages are there to replacing analog components with a digital computer?

The use of digital computers in the loop yields the following advantages over analog systems: (1) reduced cost, (2) flexibility in response to design changes, and (3) noise immunity. Modern control systems require control of numerous loops at the same time—pressure, position, velocity, and tension, for example. In the steel industry, a single digital computer can replace numerous analog controllers with a subsequent reduction in cost. Where analog controllers implied numerous adjustments and resulting hardware, digital systems are now installed. Banks of equipment, meters, and knobs are replaced with computer terminals, where information about settings and performance are obtained through menus and screen displays. Digital computers in the loop can yield a degree of flexibility in response

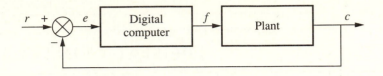

Figure 13.1 Placement of the Digital Computer within the Loop

to changes in design. Any changes or modifications that are required in the future can be implemented with simple software changes rather than expensive hardware modifications. Finally, digital systems exhibit more noise immunity than analog systems by virtue of the methods of implementation.

Where then is the computer placed in the loop? Remember that the digital computer is controlling numerous loops; thus, its position in the loop depends upon the function it performs. Typically, the computer replaces the cascade compensator and is thus positioned at the place shown in Figure 13.1.

The signals *r, e, f,* and *c* shown in Figure 13.1 can take on two forms: (1) digital, or (2) analog. Up to this point we have used analog signals exclusively. Digital signals, which consist of a sequence of binary numbers, can be found in loops containing digital computers.

Loops containing both analog and digital signals must provide a means for conversion from one form to the other as required by each subsystem. A device that converts analog signals to digital signals is called an *analog-to-digital converter*. Conversely, a device that converts digital signals to analog signals is called a *digital-to-analog converter*. For example, in Figure 13.1, if the plant output, *c,* and the system input, *r,* are analog signals, then an analog-to-digital converter must be provided at the input to the digital computer. Also, if the plant input, *f,* is an analog signal, then a digital-to-analog converter must be provided at the output of the digital computer.

Digital-to-Analog Conversion

Digital-to-analog conversion is simple and effectively instantaneous. Properly weighted voltages are summed together to yield the analog output. For example, in Figure 13.2 three weighted voltages are summed. The three-bit binary code is represented by the switches. Thus, if the binary number were 110_2, the center and bottom switches are on, and the analog output is 6 volts. In actual use the switches are electronic and are set by the input binary code.

Analog-to-Digital Conversion

Analog-to-digital conversion, on the other hand, is a two-step process and is not instantaneous. There is a delay between the input analog voltage and the output digital word. In an analog-to-digital converter, the analog signal is first converted to a sampled signal and then converted to a sequence of binary numbers, the digital signal.

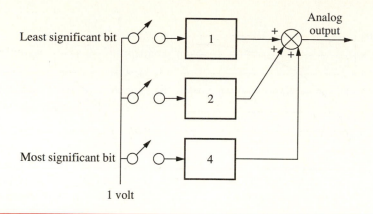

Figure 13.2 Digital-to-Analog Converter

The sampling rate must be at least twice the bandwidth of the signal, or else later recovery of the analog signal will yield distortion. This minimum sampling frequency is called the *Nyquist sampling rate*.[1]

In Figure 13.3(*a*) we start with the analog signal. In Figure 13.3(*b*) we see the analog signal sampled at periodic intervals and held over the sampling interval by a device called a *zero-order sample-and-hold* that yields a staircase approximation to the analog signal. Higher-order holds, such as a first-order hold, generate more complex and more accurate waveshapes between samples. For example, a first-order hold generates a ramp between the samples. Samples are held before being digitized because the analog-to-digital converter converts the voltage to a digital number via a digital counter, which takes time to reach the correct digital number. Hence, the constant analog voltage must be present during the conversion process.

After sampling and holding, the analog-to-digital converter converts the sample to a digital number, which is arrived at in the following manner. The dynamic range of the analog signal's voltage is divided into discrete levels, and each level is assigned a digital number. For example, in Figure 13.3(*b*), the analog signal is divided into eight levels. A three-bit digital number can represent each of the eight levels as shown in the figure. Thus, the difference between quantization levels is $M/8$ volts, where M is the maximum analog voltage. In general for any system, this difference is $M/2^n$ volts, where n is the number of binary bits used for the analog-to-digital conversion.

Looking at Figure 13.3(*b*), we can see that there will be an associated error for each digitized analog value except those voltages at the boundaries such as $M/8$, $2M/8$, and so on. We call this error the *quantization error*. Assuming that the quantization process rounds off the analog voltage to the next higher or lower level, the maximum value of the quantization error is $1/2$ the difference between quantization levels in the range of analog voltages from 0 to $15M/16$. In general for any system using roundoff, the quantization error will be $(1/2)(M/2^n) = M/2^{n+1}$.

---

[1]See Ogata (1987: 170–77) for a detailed discussion.

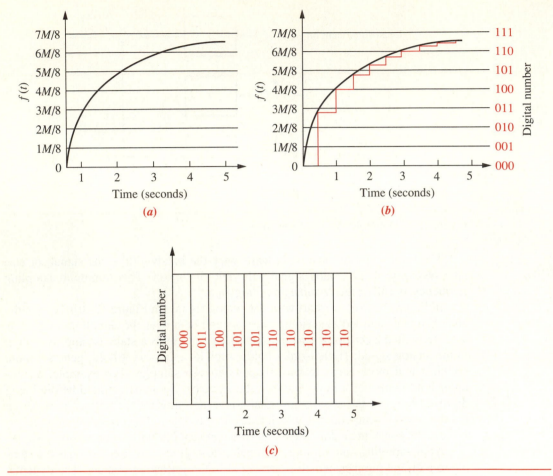

Figure 13.3 Steps in Analog-to-Digital Conversion: (**a**) Analog Signal; (**b**) Analog Signal After Sample-and-Hold; (**c**) Conversion of Samples to Digital Numbers

Chapter Objective

Given the antenna azimuth position control system shown in Figure 13.4(*a*), the student will be able to (1) convert the analog control system to the digital control system of Figure 13.4(*b*) and (2) design the gain to meet a transient response requirement.

Section Summary

In this section we covered the basic concepts of digital systems. We found out why they are used, where the digital computer is placed in the loop, and how to convert from analog to digital signals. Since the computer can replace the compensator, we have to realize that the computer is working with a quantized

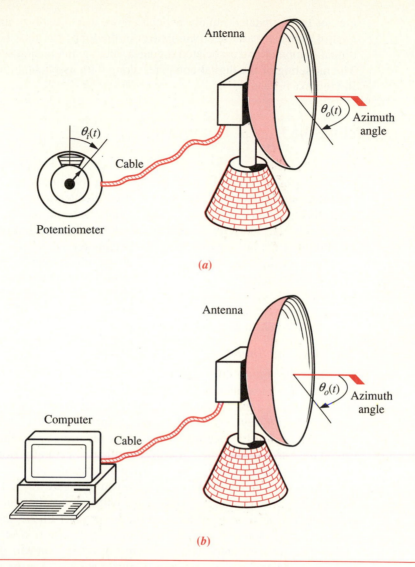

Figure 13.4 Conversion of an (**a**) Analog Antenna Azimuth
Control to a (**b**) Digital Antenna Azimuth Control

amplitude representation of the analog signal formed from values of the analog
signal at discrete intervals of time. Ignoring the quantization error, we see that
the computer performs just as the compensator does, except that signals pass
through the computer only at the sampled intervals of time. We will find that the
sampling of data has an unusual effect upon the performance of a closed-loop
feedback system since stability and transient response are now dependent upon
the sampling rate; if it is too slow, and the system can be unstable since the values

are not being updated rapidly enough. If we are to analyze and design feedback control systems with digital computers in the loop, we must be able to model the digital computer and associated digital-to-analog and analog-to-digital converters. The modeling of the digital computer along with associated converters is covered in the next section.

13.2 Modeling the Digital Computer

If we think about it, the form of the signals in a loop is not as important as what happens to them. For example, if analog-to-digital conversion could happen instantaneously, and time samples occurred at intervals of time that approached zero, there would be no need to differentiate between the digital signals and the analog signals. Thus, previous analysis and design techniques would be valid regardless of the presence of the digital computer.

The fact that signals are sampled at specified intervals and held causes the system performance to change with changes in sampling rate. Basically, then, the computer's effect upon the signal comes from this sampling and holding. Thus, in order to model digital control systems, we must come up with a mathematical representation of this sample-and-hold process.

Modeling the Sampler

Our objective at this point is to derive a mathematical model for the digital computer as represented by a sampler and zero-order hold. Our goal is to represent the computer as a transfer function similar to that for any subsystem. When signals are sampled, however, the Laplace transform that we have dealt with becomes a bit unwieldy. The Laplace transform can be replaced by another related transform called the *z-transform*. The z-transform will arise naturally from our development of the mathematical representation of the computer.

Consider the models for sampling shown in Figure 13.5. The model in Figure 13.5(*a*) is a switch turning on and off at a uniform sampling rate. In Figure 13.5(*b*) sampling can also be considered to be the product of the time waveform to be sampled, $f(t)$, and a sampling function, $s(t)$. If $s(t)$ is a sequence of pulses of width T_W, constant amplitude, and uniform rate as shown, the sampled output, $f^*_{T_W}(t)$, will consist of a sequence of sections of $f(t)$ at regular intervals. This view is equivalent to the switch model of Figure 13.5(*a*).

We can now write the time equation of the sampled waveform, $f^*_{T_W}(t)$. Using the model shown in Figure 13.5(*b*), we have

$$f^*_{T_W}(t) = f(t)s(t) = f(t) \sum_{k=-\infty}^{\infty} u(t - kT) - u(t - kT - T_W) \qquad (13.1)$$

where k is an integer between $+\infty$ and $+\infty$, T is the period of the pulse train, and T_W is the pulse width.

Since Equation 13.1 is the product of two time functions, taking the Laplace transform in order to find a transfer function is not simple. A simplification can be

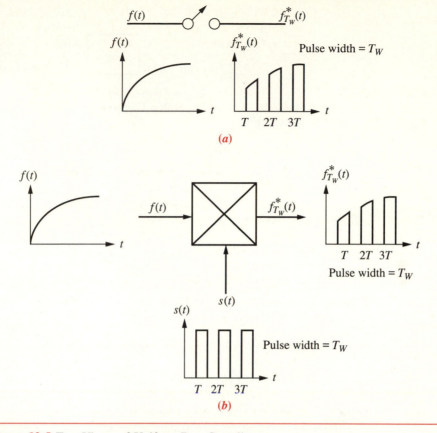

Figure 13.5 Two Views of Uniform-Rate Sampling:
(**a**) Switch Opening and Closing; (**b**) Product of Time
Waveform and Sampling Waveform

made if we assume that the pulse width, T_W, is so small that $f(t)$ remains constant during the sampling interval. Over the sampling interval, then, $f(t) = f(kT)$. Hence,

$$f^*_{T_W}(t) = \sum_{k=-\infty}^{\infty} f(kT)[u(t - kT) - u(t - kT - T_W)] \qquad (13.2)$$

for small T_W.

Equation 13.2 can be further simplified through insight provided by the Laplace transform. Taking the Laplace transform of Equation 13.2, we have

$$F^*_{T_W}(s) = \sum_{k=-\infty}^{\infty} f(kT)\left[\frac{e^{-kTs}}{s} - \frac{e^{-kTs - T_W s}}{s}\right] = \sum_{k=-\infty}^{\infty} f(kT)\left[\frac{1 - e^{-T_W s}}{s}\right] e^{-kTs}$$

$$(13.3)$$

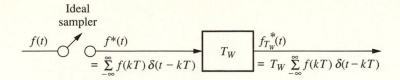

Figure 13.6 A Model of Sampling with a Uniform Rectangular Pulse Train

Replacing $e^{-T_W s}$ with its series expansion, we obtain

$$F_{T_W}^*(s) = \sum_{k=-\infty}^{\infty} f(kT) \left[\frac{1 - \left\{ 1 - T_W s + \frac{(T_W s)^2}{2!} - \cdots \right\}}{s} \right] e^{-kTs} \tag{13.4}$$

For small T_W, Equation 13.4 becomes

$$F_{T_W}^*(s) = \sum_{k=-\infty}^{\infty} f(kT) \left[\frac{T_W s}{s} \right] e^{-kTs} = \sum_{k=-\infty}^{\infty} f(kT) T_W e^{-kTs} \tag{13.5}$$

Finally, converting back to the time domain, we have

$$f_{T_W}^*(t) = T_W \sum_{k=-\infty}^{\infty} f(kT)\delta(t - kT) \tag{13.6}$$

where $\delta(t - kT)$ are Dirac delta functions.

Thus, the result of sampling with rectangular pulses can be thought of as a series of delta functions whose area is the product of the rectangular pulse width and the amplitude of the sampled waveform, or $T_W f(kT)$.

Equation 13.6 is portrayed in Figure 13.6. The sampler is divided into two parts; (1) an ideal sampler described by the portion of Equation 13.6 that is not dependent upon the sampling waveform characteristics,

$$f^*(t) = \sum_{k=-\infty}^{\infty} f(kT)\delta(t - kT) \tag{13.7}$$

and (2) the portion dependent upon the sampling waveform's characteristics, T_W.

Modeling the Zero-Order Hold

The final step in modeling the digital computer is modeling of the zero-order hold (z.o.h.) that follows the sampler. Figure 13.7 summarizes the function of the zero-order hold, which is to hold the last sampled value of $f(t)$. If we assume an ideal sampler (equivalent to setting $T_W = 1$), then $f^*(t)$ is represented by a sequence of delta functions. The zero-order hold yields a staircase approximation to $f(t)$. Hence, the output from the hold is a sequence of step functions whose amplitude is $f(t)$ at the sampling instant, or $f(kT)$. We have previously seen that

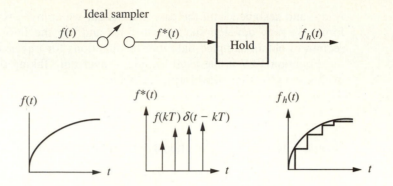

Figure 13.7 Ideal Sampling and the Zero-Order Hold

the transfer function of any linear system is identical to the Laplace transform of the impulse response since the Laplace transform of a unit impulse or delta function input is unity. Since a single impulse from the sampler yields a step over the sampling interval, the Laplace transform of this step, $G_h(s)$, which is the impulse response of the zero-order hold, is the transfer function of the zero-order hold. Using an impulse at zero time, the transform of the resulting step that starts at $t = 0$ and ends at $t = T$ is

$$G_h(s) = \frac{1 - e^{-Ts}}{s} \qquad (13.8)$$

In a physical system, samples of the input time waveform, $f(kT)$, are held over the sampling interval. We can see from Equation 13.8 that the hold circuit integrates the input and holds its value over the sampling interval. Since the area under the delta functions coming from the ideal sampler is $f(kT)$, we can then integrate the ideal sampled waveform and obtain the same result as for the physical system. In other words, if the ideal sampled signal, $f^*(t)$, is followed by a hold, we can use the ideal sampled waveform as the input, rather than $f^*_{T_W}(t)$.

In this section we modeled the digital computer by cascading two elements: (1) an ideal sampler, and (2) a zero-order hold. Together, the model is known as a *zero-order sample-and-hold*. The ideal sampler is modeled by Equation 13.7, and the zero-order hold is modeled by Equation 13.8.

13.3 The z-Transform

The effect of sampling within a system is pronounced. Whereas the stability and transient response of analog systems depend upon gain and component values, sampled-data system stability and transient response depend also upon sampling rate. Our goal is to develop a transform that contains the information of sampling from which sampled-data systems can be modeled with transfer functions, analyzed, and designed with the ease and insight we enjoyed with the Laplace transform. We now develop such a transform and use the information from the last section to obtain sampled-data transfer functions for physical systems.

Equation 13.7 is the ideal sampled waveform. Taking the Laplace transform of this sampled time waveform, we obtain

$$F^*(s) = \sum_{k=0}^{\infty} f(kT)e^{-kTs} \tag{13.9}$$

Now, letting $z = e^{Ts}$, Equation 13.9 can be written as

$$F(z) = \sum_{k=0}^{\infty} f(kT)z^{-k} \tag{13.10}$$

Equation 13.10 defines the *z-transform*. That is, an $F(z)$ can be transformed to $f(kT)$, or an $f(kT)$ can be transformed to $F(z)$. Alternately, we can write

$$f(kT) \Longleftrightarrow F(z) \tag{13.11}$$

Paralleling the development of the Laplace transform, we can form a table relating $f(kT)$, the value of the sampled time function at the sampling instants, to $F(z)$. Let us look at an example.

Example 13.1   Find the *z*-transform of a sampled time function.

Problem   Find the *z*-transform of a sampled unit ramp.

Solution   For a unit ramp, $f(kT) = kT$. Hence the ideal sampled step can be written from Equation 13.7 as

$$f^*(t) = \sum_{k=0}^{\infty} kT\delta(t - kT) \tag{13.12}$$

Taking the Laplace transform, we obtain

$$F^*(s) = \sum_{k=0}^{\infty} kTe^{-kTs} \tag{13.13}$$

Converting to the *z*-transform by letting $e^{-kTs} = z^{-k}$, we have

$$F(z) = \sum_{k=0}^{\infty} kTz^{-k} = T\sum_{k=0}^{\infty} kz^{-k} = T(z^{-1} + 2z^{-2} + 3z^{-3} + \cdots) \tag{13.14}$$

Equation 13.14 can be converted to a closed form by forming the series for $zF(z)$ and subtracting $F(z)$. Multiplying Equation 13.14 by z, we get

$$zF(z) = T(1 + 2z^{-1} + 3z^{-2} + \cdots) \tag{13.15}$$

Subtracting Equation 13.14 from Equation 13.15, we obtain

$$zF(z) - F(z) = (z - 1)F(z) = T(1 + z^{-1} + z^{-2} + \cdots) \tag{13.16}$$

But

$$\frac{1}{1 - z^{-1}} = 1 + z^{-1} + z^{-2} + z^{-3} + \cdots \tag{13.17}$$

which can be verified by performing the indicated division. Substituting Equation 13.17 into 13.16 and solving for $F(z)$ yields

$$F(z) = T\frac{z}{(z-1)^2} \tag{13.18}$$

as the z-transform of $f(kT) = kT$.

The example demonstrates that any function of $s, F^*(s)$, that represents a sampled time waveform can be transformed into a function of $z, F(z)$. The final result, $F(z) = \dfrac{Tz}{(z-1)^2}$ is in a closed form, unlike $F^*(s)$. If this is the case for numerous other sampled time waveforms, then we have the convenient transform that we were looking for. In a similar way, z-transforms for other waveforms can be obtained that parallel the table of Laplace transforms in Chapter 2. A partial table of z-transforms is shown in Table 13.1, and a partial table of z-transform theorems is shown in Table 13.2. For functions not in the table, we must perform an inverse

Table 13.1 Partial Table of z- and s-Transforms

	$f(t)$	$F(s)$	$F(z)$
1.	$u(t)$	$\dfrac{1}{s}$	$\dfrac{z}{z-1}$
2.	$tu(t)$	$\dfrac{1}{s^2}$	$\dfrac{Tz}{(z-1)^2}$
3.	$t^n u(t)$	$\dfrac{n!}{s^{n+1}}$	$\lim\limits_{a\to 0}(-1)^n \dfrac{d^n}{da^n}\left[\dfrac{z}{z-e^{-aT}}\right]$
4.	$e^{-at}u(t)$	$\dfrac{1}{s+a}$	$\dfrac{z}{z-e^{-aT}}$
5.	$t^n e^{-aT}u(t)$	$\dfrac{n!}{(s+a)^{n+1}}$	$(-1)^n \dfrac{d^n}{da^n}\left[\dfrac{z}{z-e^{-aT}}\right]$
6.	$\sin\omega t u(t)$	$\dfrac{\omega}{s^2+\omega^2}$	$\dfrac{z\sin\omega T}{z^2-2z\cos\omega T+1}$
7.	$\cos\omega t u(t)$	$\dfrac{s}{s^2+\omega^2}$	$\dfrac{z(z-\cos\omega T)}{z^2-2z\cos\omega T+1}$
8.	$e^{-at}\sin\omega t u(t)$	$\dfrac{\omega}{(s+a)^2+\omega^2}$	$\dfrac{ze^{-aT}\sin\omega T}{z^2-2ze^{-aT}\cos\omega T+e^{-2aT}}$
9.	$e^{-at}\cos\omega t u(t)$	$\dfrac{s+a}{(s+a)^2+\omega^2}$	$\dfrac{z^2-ze^{-aT}\cos\omega T}{z^2-2ze^{-aT}\cos\omega T+e^{-2aT}}$

Note: kT may be substituted for t in the table when drawing the relationship to the z-transform.

Table 13.2 z-Transform Theorems

	Theorem	Name
1.	$z(af(t)) = aF(z)$	Linearity theorem
2.	$z[f_1(t) + f_2(t)] = F_1(z) + F_2(z)$	Linearity theorem
3.	$z[e^{-at}f(t)] = F(e^{aT}z)$	
4.	$z[tf(t)] = -Tz\dfrac{dF(z)}{dz}$	
5.	$f(0) = \lim\limits_{z\to\infty} F(z)$	Initial value theorem
6.	$f(\infty) = \lim\limits_{z\to 1}(1 - z^{-1})F(z)$	Final value theorem

Note: kT may be substituted for t in the table.

z-transform calculation similar to the inverse Laplace transform by partial fraction expansion. Let us now see how we can work in the reverse direction and find the time function from its z-transform.

The Inverse z-Transform

Two methods for finding the inverse z-transform (the sampled time function from its z-transform) will be described: (1) partial fraction expansion, and (2) the power series method. Regardless of the method used, remember that since the z-transform came from the sampled waveform, the inverse z-transform will yield only the values of the time function at the sampling intervals. Keep this in mind as we proceed, because even as we obtain closed-form time functions as results, they are only valid at sampling intervals.

Inverse z-Transforms via Partial Fraction Expansion

Recall that the inverse Laplace transform consists of a partial fraction that yields a sum of terms leading to exponentials, that is, $A/(s + a)$. Taking this lead and looking at Table 13.1, we find that sampled exponential time functions are related to their z-transforms as follows:

$$e^{-akT} \Longleftrightarrow \frac{z}{z - e^{aT}} \tag{13.19}$$

We thus predict that a partial fraction expansion should be of the following form:

$$F(z) = \frac{Az}{z - z_1} + \frac{Bz}{z - z_2} + \cdots \tag{13.20}$$

Since our partial fraction expansion of $F(s)$ did not contain terms with s in the numerator of the partial fractions, we first form $F(z)/z$ to eliminate the z terms in the numerator, perform a partial fraction expansion of $F(z)/z$, and finally multiply the result by z to replace the z's in the numerator. An example follows.

Example 13.2   Find the inverse z-transform via the partial fraction expansion.

Problem   Given the function in Equation 13.21, find the sampled time function.

$$F(z) = \frac{0.5z}{(z - 0.5)(z - 0.7)} \tag{13.21}$$

Solution   Begin by dividing Equation 13.21 by z and performing a partial fraction expansion.

$$\frac{F(z)}{z} = \frac{0.5}{(z - 0.5)(z - 0.7)} = \frac{A}{z - 0.5} + \frac{B}{z - 0.7} = \frac{-2.5}{z - 0.5} + \frac{2.5}{z - 0.7} \tag{13.22}$$

Next, multiply through by z.

$$F(z) = \frac{0.5z}{(z - 0.5)(z - 0.7)} = \frac{-2.5z}{z - 0.5} + \frac{2.5z}{z - 0.7} \tag{13.23}$$

Using Table 13.1, we find the inverse z-transform of each partial fraction. Hence, the value of the time function at the sampling instants is

$$f(kT) = -2.5(0.5)^k + 2.5(0.7)^k \tag{13.24}$$

Also, from Equations 13.7 and 13.24, the ideal sampled time function is

$$f^*(t) = \sum_{k=-\infty}^{\infty} f(kT)\delta(t - kT) = \sum_{k=-\infty}^{\infty} [-2.5(0.5)^k + 2.5(0.7)^k]\delta(t - kT) \tag{13.25}$$

If we substitute $k = 0, 1, 2$, and 3, we can find the first four samples of the ideal sampled time waveform. Hence,

$$f^*(t) = 0\delta(t) + 0.5\delta(t - T) + 0.6\delta(t - 2T) + 0.545\delta(t - 3T) \tag{13.26}$$

Inverse z-Transform via the Power Series Method

The values of the sampled time waveform can also be found directly from $F(z)$. This method, however, does not yield closed-form expressions for $f(kT)$. The method consists of performing the indicated division, which yields a power series for $F(z)$. The power series can then be easily transformed into $F^*(s)$ and $f^*(t)$.

Example 13.3   Find the inverse z-transform via the power series method.

Problem   Given the function in Equation 13.21, find the sampled time function.

Solution   Begin by converting the numerator and denominator of $F(z)$ to polynomials in z.

$$F(z) = \frac{0.5z}{(z - 0.5)(z - 0.7)} = \frac{0.5z}{z^2 - 1.2z + 0.35} \tag{13.27}$$

Now perform the indicated division.

$$z^2 - 1.2z + 0.35 \overline{)0.5z} \quad \frac{0.5z^{-1} + 0.6z^{-2} + 0.545z^{-3}}{} \qquad (13.28)$$

$$\begin{array}{r} 0.5z - 0.6 + 0.175z^{-1} \\ \hline 0.6 - 0.175z^{-1} \\ 0.6 - 0.720z^{-1} + 0.21 \\ \hline 0.545z^{-1} - 0.21 \end{array}$$

Using the numerator and the definition of z, we obtain

$$F^*(s) = 0.5e^{-Ts} + 0.6e^{-2Ts} + 0.545e^{-3Ts} + \cdots \qquad (13.29)$$

from which

$$f^*(t) = 0.5\delta(t - T) + 0.6\delta(t - 2T) + 0.545\delta(t - 3T) + \cdots \qquad (13.30)$$

The student should compare Equation 13.30 with Equation 13.26, the result obtained via partial expansion.

In this section we learned about the z-transform; in the following section, we apply it.

13.4 The Transfer Function of Sampled-Data Systems

Now that we have established the z-transform, let us apply it to physical systems by finding transfer functions of sampled-data systems. Consider the continuous system shown in Figure 13.8(a). If the input is sampled as shown in Figure 13.8(b), the output is still a continuous signal. If, however, we are satisfied with finding the output at the sampling instants and not in between, the representation of the sampled-data system can be greatly simplified. Our assumption is visually described in Figure 13.8(c), where the output is conceptually sampled in synchronization with the input by a phantom sampler. Using the concept described in Figure 13.8(c), we derive the pulse transfer function of $G(s)$.

Derivation of the Pulse Transfer Function

Using Equation 13.7, we find that the sampled input, $r^*(t)$, to the system of Figure 13.8(c) is

$$r^*(t) = \sum_{n=0}^{\infty} r(nT)\delta(t - nT) \qquad (13.31)$$

which is a sum of impulses. Since the impulse response of a system, $G(s)$, is $g(t)$, we can write the time output of $G(s)$ as the sum of impulse responses generated

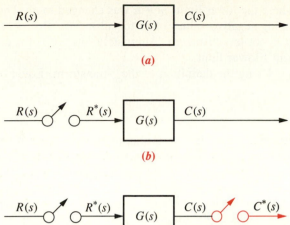

Note: Phantom sampler is shown in color.

Figure 13.8 Sampled-Data Systems: (*a*) Continuous;
(*b*) Sampled Input; (*c*) Sampled Input and Output

by the input, Equation 13.31. Thus,

$$c(t) = \sum_{n=0}^{\infty} r(nT)g(t - nT) \tag{13.32}$$

From Equation 13.10,

$$C(z) = \sum_{k=0}^{\infty} c(kT)z^{-k} \tag{13.33}$$

Using Equation 13.32 with $t = kT$, we obtain

$$c(kT) = \sum_{n=0}^{\infty} r(nT)g(kT - nT) \tag{13.34}$$

Substituting Equation 13.34 into Equation 13.33, we obtain

$$C(z) = \sum_{k=0}^{\infty} \sum_{n=0}^{\infty} r(nT)g\left[(k - n)T\right]z^{-k} \tag{13.35}$$

Letting $m = k - n$, we find

$$C(z) = \sum_{m+n=0}^{\infty} \sum_{n=0}^{\infty} r(nT)g(mT)z^{-(m+n)}$$

$$= \left\{ \sum_{m=0}^{\infty} g(mT)z^{-m} \right\} \left\{ \sum_{n=0}^{\infty} r(nT)z^{-n} \right\} \tag{13.36}$$

where the lower limit, $m + n$ was changed to m. The reasoning is that $m + n = 0$ yields negative values of m for all $n > 0$. But, since $g(mT) = 0$ for all $m < 0$, m is not less than zero. Alternately, $g(t) = 0$ for $t < 0$. Thus, $n = 0$ in the first sum's lower limit.

Using the definition of the z-transform, Equation 13.36 becomes

$$C(z) = \sum_{m=0}^{\infty} g(mT)z^{-m} \sum_{n=0}^{\infty} r(nT)z^{-n} = G(z)R(z) \tag{13.37}$$

Equation 13.37 is a very important result since it shows that the transform of the sampled output is the product of the transforms of the sampled input and the pulse transfer function of the system. Remember that although the output of the system is a continuous function, we had to make an assumption of a sampled output (phantom sampler) in order to arrive at the compact result of Equation 13.37.

One way of finding the pulse transfer function, $G(z)$, is to start with $G(s)$, find $g(t)$, and then use Table 13.1 to find $G(z)$. Let us look at an example.

Example 13.4   Convert $G(s)$ to $G(z)$.

Problem   Given the function in Equation 13.38, find the sampled-data transfer function, $G(z)$, if the sampling time, T, is 0.5 seconds.

$$G(s) = \frac{1 - e^{-Ts}}{s} \frac{(s + 2)}{(s + 1)} = (1 - e^{-Ts}) \frac{(s + 2)}{s(s + 1)} \tag{13.38}$$

Solution   Equation 13.38 represents the system whose transfer function is $\dfrac{(s + 2)}{(s + 1)}$ preceded by a hold, $\dfrac{(1 - e^{-Ts})}{s}$. If we are satisfied with knowing the output of $G(s)$ only at the sampling instants, then we can convert $G(s)$ to $G(z)$. Begin by finding the impulse response (inverse Laplace transform) of $G_1(s) = \dfrac{s + 2}{s(s + 1)}$.

$$G_1(s) = \frac{s + 2}{s(s + 1)} = \frac{A}{s} + \frac{B}{s + 1} = \frac{2}{s} - \frac{1}{s + 1} \tag{13.39}$$

Taking the inverse Laplace transform, we get

$$g_1(t) = 2 - e^{-t} \tag{13.40}$$

from which

$$g_1(kT) = 2 - e^{-kT} \tag{13.41}$$

Using Table 13.1, we find

$$G_1(z) = \frac{2z}{z - 1} - \frac{z}{z - e^{-T}} \tag{13.42}$$

Substituting $T = 0.5$ yields

$$G_1(z) = \frac{2z}{z - 1} - \frac{z}{z - 0.607} = \frac{z^2 - 0.214z}{(z - 1)(z - 0.607)} \tag{13.43}$$

Next, find the z-transform of $1 - e^{-Ts}$. By direct substitution, the z-transform of $1 - e^{-Ts}$ is $1 - z^{-1} = (z-1)/z$. Therefore,

$$G(z) = \frac{z-1}{z}G_1(z) = \frac{z - 0.214}{z - 0.607} \tag{13.44}$$

The major discovery in this section was that once the pulse transfer function, $G(z)$, of a system is obtained, the transform of the sampled output response, $C(z)$, for a given sampled input can be evaluated using the relationship $C(z) = R(z)G(z)$. Finally, the time function can be found by taking the inverse z-transform, as covered in Section 13.3.

13.5 Block Diagram Manipulation of Sampled-Data Systems

Up to this point, we have defined the z-transform, defined sampled-data system transfer functions, and showed how to obtain the sampled response. Basically, we are paralleling discussions that we have seen before for the Laplace transform in Chapters 2 and 4. We now draw a parallel with some of the objectives of Chapter 5, namely, block diagram manipulation. Our objective here is to be able to find the sampled-data transfer function of an arrangement of subsystems that have a computer in the loop. When manipulating block diagrams for sampled-data systems the reader must be careful and remember the definition of the sampled-data system transfer function (derived in the last section) to avoid mistakes. For example, $z\{G_1(s)G_2(s)\} \neq G_1(z)G_2(z)$, where $z\{G_1(s)G_2(s)\}$ denotes the z-transform. The s-domain functions have to be multiplied together first before taking the z-transform. In the ensuing discussion, we will use the notation $G_1G_2(s)$ to denote a single function that is $G_1(s)G_2(s)$ after evaluating the product. Hence, $z\{G_1(s)G_2(s)\} = z\{G_1G_2(s)\} = G_1G_2(z) \neq G_1(z)G_2(z)$.

Let us look at the sampled-data systems shown in Figure 13.9.

The sampled-data systems are shown under the column marked s. Their z-transforms are shown under the column marked z. The standard system that we derived earlier is shown in Figure 13.9(a), where the transform of the output, $C(z)$, is equal to $R(z)G(z)$. This system forms the basis for the other entries in Figure 13.9.

In Figure 13.9(b) there is no sampler between $G_1(s)$ and $G_2(s)$. Thus, we can think of a single function, $G_1(s)G_2(s)$, denoted $G_1G_2(s)$, existing between the two samplers and yielding a single transfer function, as shown in Figure 13.9(a). Hence, the pulse transfer function is $z\{G_1G_2(s)\} = G_1G_2(z)$. The transform of the output, $C(z) = R(z)G_1G_2(z)$.

In Figure 13.9(c) we have the cascaded two subsystems of the type shown in Figure 13.9(a). For this case, then, the z-transform is the product of the

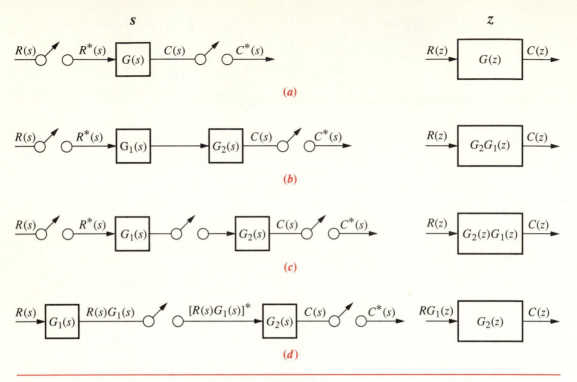

Figure 13.9 Sampled-Data Systems and Their z-Transforms

two z-transforms, or $G_2(z)G_1(z)$. Hence the transform of the output $C(z) = R(z)G_2(z)G_1(z)$.

Finally, in Figure 13.9(d), we see that the continuous signal entering the sampler is $R(s)G_1(s)$. Thus, the model is the same as (a) with $R(s)$ replaced by $R(s)G_1(s)$, and $G_2(s)$ in Figure 13.9(d) replacing $G(s)$ in Figure 13.9(a). The z-transform of the input to $G_2(s)$ is $z\{R(s)G_1(s)\} = z\{RG_1(s)\} = RG_1(z)$. The pulse transfer function for the system $G_2(s)$ is $G_2(z)$. Hence, the output $C(z) = RG_1(z)G_2(z)$.

Using the basic forms shown in Figure 13.9, we now can find the z-transform of feedback control systems. We have already shown that any system, $G(s)$, with sampled input and sampled output, such as that shown in Figure 13.9(a) can be represented as a sampled-data transfer function, $G(z)$. Thus, we want to perform block diagram manipulations that result in subsystems, as well as the entire feedback system, that have sampled inputs and sampled outputs. Then we can make the transformation to sampled-data transfer functions. An example follows.

Example 13.5   Find the z-transform of a feedback system with a computer in the loop.

Problem   Find the z-transform of the system shown in Figure 13.10(a).

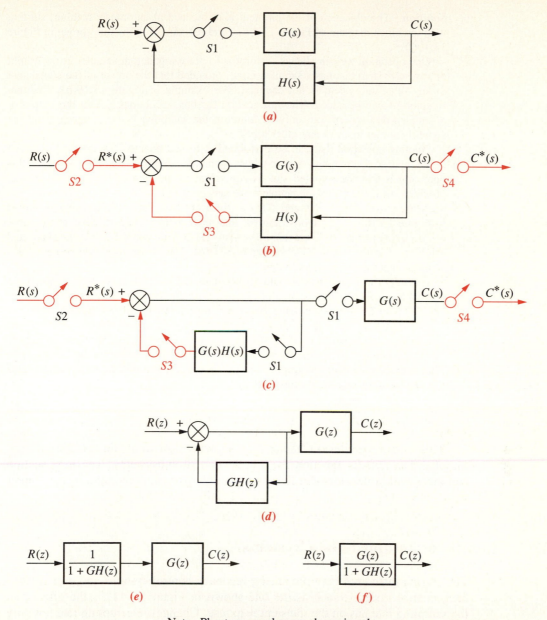

Note: Phantom samplers are shown in color.

Figure 13.10 Steps in the Block Diagram Reduction of a Sampled-Data System

Solution The objective of the problem is to proceed in an orderly fashion, starting with the block diagram of Figure 13.10(*a*) and reducing it to the one shown in Figure 13.10(*f*).

One operation we can always perform is to place a phantom sampler at the output of any subsystem that has a sampled input, provided that the nature of the signal sent to any other subsystem is not changed. For example in Figure 13.10(*b*), phantom sampler $S4$ can be added. The justification for this, of course, is that the output of a sampled-data system can only be found at the sampling instants anyway, and the signal is not an input to any other block.

Another operation that can be performed is to add phantom samplers $S2$ and $S3$ at the input to a summing junction whose output is sampled. The justification for this operation is that the sampled sum is equivalent to the sum of the sampled inputs, provided, of course, that all samplers are synchronized.

Next, move sampler $S1$ and $G(s)$ to the right past the pickoff point, as shown in Figure 13.10(*c*). The motivation for this move is to yield a sampler at the input of $G(s)H(s)$ to match Figure 13.9(*b*). Also, $G(s)$ with sampler $S1$ at the input and sampler $S4$ at the output matches Figure 13.9(*a*). The closed-loop system now has a sampled input and a sampled output.

$G(s)H(s)$ with samplers $S1$ and $S3$ becomes $GH(z)$, and $G(s)$ with samplers $S1$ and $S4$ becomes $G(z)$, as shown in Figure 13.10(*d*). Also, converting $R^*(s)$ to $R(z)$ and $C^*(s)$ to $C(z)$, we now have the system represented totally in the z-domain.

The equations derived in Chapter 5 for transfer functions represented with the Laplace transform can be used for sampled-data transfer functions with only a change in variables from s to z. Thus, using the feedback formula, we obtain the first block of Figure 13.10(*e*). Finally, multiplication of the cascaded sampled-data systems yields the final result shown in Figure 13.10(*f*).

This section paralleled Chapter 5 by showing how to obtain the closed-loop, sampled-data transfer function for a collection of subsystems. The next section continues with a development that parallels the discussion of stability in Chapter 6.

13.6 Stability of Sampled-Data Systems

The glaring difference between analog feedback control systems and digital feedback control systems, such as the one shown in Figure 13.11, is the effect that the sampling rate has on the transient response. Changes in sampling rate not only change the nature of the response from overdamped to underdamped, but they can also turn a stable system into an unstable one. As we proceed with our discussion, the aforementioned effects will become apparent. The reader is encouraged to be on the lookout.

In the s-plane, the region of stability is the left half-plane. If the transfer function, $G(s)$, is transformed into a sampled-data transfer function, $G(z)$, the region of stability on the z-plane can be evaluated from the definition $z = e^{sT}$.

Figure 13.11 Robotic Arms Controlled by a Personal Computer (Courtesy of Rhino Robots, Inc.)

Letting $s = \alpha + j\omega$, we obtain

$$
\begin{aligned}
z = e^{Ts} = e^{T(\alpha + j\omega)} &= e^{\alpha T} e^{j\omega T} \\
&= e^{\alpha T}(\cos \omega T + j \sin \omega T) \\
&= e^{\alpha T} \angle \omega T
\end{aligned}
\tag{13.45}
$$

since $(\cos \omega T + j \sin \omega T) = 1 \angle \omega T$.

Each region of the s-plane can be mapped into a corresponding region on the z-plane (see Figure 13.12). Points that have positive values of α are in the right half of the s-plane, region C. From Equation 13.45, the magnitudes of the mapped points are $e^{\alpha T} > 1$. Thus points in the right half of the s-plane map into points outside the unit circle on the z-plane.

Points on the $j\omega$-axis, region B, have zero values of α and yield points on the z-plane with magnitude = 1, the unit circle. Hence points on the $j\omega$-axis in the s-plane map into points on the unit circle on the z-plane.

Finally, points on the s-plane that yield negative values of α (left half-plane roots, region A) map into the inside of the unit circle on the z-plane.

Thus, a digital control system is (1) stable if all poles of the closed-loop transfer function, $T(z)$, are inside the unit circle on the z-plane, (2) unstable if any pole is outside the unit circle, and (3) marginally stable if poles are on and inside the unit circle.

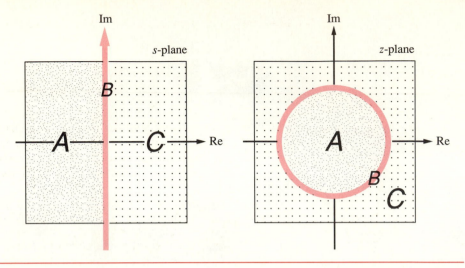

Figure 13.12 Mapping Regions of the *s*-Plane Onto the *z*-Plane

Example 13.6   Model a digital control system and find its stability.

Problem   The missile shown in Figure 13.13(*a*) can be aerodynamically controlled by torques created by the deflection of control surfaces on the missile's body. The commands to deflect these control surfaces come from a computer that uses tracking data along with programmed guidance equations to determine whether the missile is on track. The information from the guidance equations is used to develop flight-control commands for the missile. A simplified model is shown in Figure 13.13(*b*). Here the computer performs the function of controller by using tracking information to develop input commands to the missile. An accelerometer in the missile detects the actual acceleration, which is fed back to the computer. Find the closed-loop digital transfer function for this system and determine if the system is stable for $K = 20$ and $K = 100$ with a sampling interval of $T = 0.1$.

Solution   The input to the control system is an acceleration command developed by the computer. The computer can be modeled by a sample-and-hold. The *s*-plane model is shown in Figure 13.13(*c*). The first step in finding the *z*-plane model is to find $G(z)$, the forward transfer function. From Figure 13.13(*c*),

$$G(s) = \frac{1 - e^{-Ts}}{s} \frac{Ka}{s(s + a)} \tag{13.46}$$

where $a = 27$. The *z*-transform, $G(z)$, is $(1 - z^{-1})z\left\{ \dfrac{Ka}{s^2(s + a)} \right\}$.

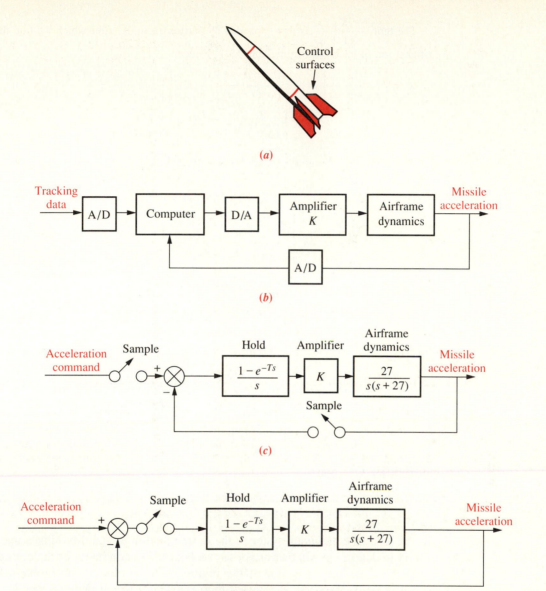

Figure 13.13 Finding Stability of a Missile Control System: (*a*) Missile;
(*b*) Conceptual Block Diagram; (*c*) Block Diagram; (*d*) Block Diagram
with Equivalent Single Sampler

The term $\dfrac{Ka}{s^2(s+a)}$ is first expanded by partial fractions, after which we find the z-transform of each term from Table 13.1. Hence,

$$z\left\{\frac{Ka}{s^2(s+a)}\right\} = Kz\left\{\frac{a}{s^2(s+a)}\right\} = Kz\left\{\frac{1}{s^2} - \frac{1/a}{s} + \frac{1/a}{s+a}\right\}$$

$$= K\left\{\frac{Tz}{(z-1)^2} - \frac{z/a}{z-1} + \frac{z/a}{z-e^{-aT}}\right\}$$

$$= K\left\{\frac{Tz}{(z-1)^2} - \frac{(1-e^{-aT})z}{a(z-1)(z-e^{-aT})}\right\} \tag{13.47}$$

Thus,

$$G(z) = K\left\{\frac{T(z-e^{-aT}) - (z-1)\left(\dfrac{1-e^{-aT}}{a}\right)}{(z-1)(z-e^{-aT})}\right\} \tag{13.48}$$

Letting $T = 0.1$ and $a = 27$, we have

$$G(z) = \frac{K(0.0655z + 0.02783)}{(z-1)(z-0.0672)} \tag{13.49}$$

Finally, we find the closed-loop transfer function, $T(z)$, for a unity feedback system,

$$T(z) = \frac{G(z)}{1+G(z)} = \frac{K(0.0655z + 0.02783)}{z^2 + (0.0655K - 1.0672)z + (0.02783K + 0.0672)} \tag{13.50}$$

The stability of the system is found by finding the roots of the denominator. For $K = 20$, the roots of the denominator are $0.12 \pm j0.78$. The system is thus stable for $K = 20$ since the poles are inside the unit circle. For $K = 100$, the poles are at -0.58 and -4.9. Since one of the poles is outside the unit circle, the system is unstable for $K = 100$.

In the case of continuous systems, the determination of stability hinges upon our ability to determine whether or not the roots of the denominator of the closed-loop transfer function are in the stable region of the s-plane. The problem is complicated by the fact that the closed-loop transfer function denominator is in polynomial form, not factored form. The same problem surfaces with closed-loop sampled-data transfer functions.

Tabular methods for determining stability, such as the Routh-Hurwitz method used for higher-order continuous systems, exist for sampled-data systems. These methods, which will not be covered in this introductory chapter to digital control systems, can be used to determine stability in higher-order digital systems. The reader who wishes to go further into the area of digital system stability is encouraged to look at Raible's tabular method or Jury's stability test for determining the number of a sampled-data system's closed-loop poles that exist outside the unit circle and thus indicate instability.[2]

The following example demonstrates the effect of sampling rate on the stability of a closed-loop feedback control system. All parameters are constant except for the sampling interval, T. We will see that varying T will lead us through regions of stability and instability just as though we were varying the forward path gain, K.

Example 13.7   Find the range of sampling interval that makes a digital feedback control system stable.

Problem   Determine the range of sampling interval, T, that will make the system shown in Figure 13.14 stable, and the range that will make it unstable.

Solution   Since $H(s) = 1$, the z-transform of the closed-loop system, $T(z)$, is found from Figure 13.10 to be

$$T(z) = \frac{G(z)}{1 + G(z)} \tag{13.51}$$

To find $G(z)$, first find the partial fraction expansion of $G(s)$.

$$G(s) = 10\frac{1 - e^{-sT}}{s(s + 1)} = 10(1 - e^{-sT})\left(\frac{1}{s} - \frac{1}{s + 1}\right) \tag{13.52}$$

Taking the z-transform, we obtain

$$G(z) = \frac{10(z - 1)}{z}\left[\frac{z}{z - 1} - \frac{z}{z - e^{-T}}\right] = 10\frac{(1 - e^{-T})}{(z - e^{-T})} \tag{13.53}$$

Substituting Equation 13.53 into 13.51 yields

$$T(z) = \frac{10(1 - e^{-T})}{z - (11e^{-T} - 10)} \tag{13.54}$$

---

[2]A discussion of Raible's tabular method and Jury's stability test can be found in Kuo (1980: 278–86).

Figure 13.14 Digital System for Example 13.7

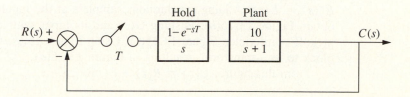

$R(s)$ + $\bigotimes$ $\xrightarrow{\quad T \quad}$ Hold $\dfrac{1 - e^{-sT}}{s}$ Plant $\dfrac{10}{s + 1}$ $C(s)$

The pole of Equation 13.54, $(11e^{-T} - 10)$, monotonically decreases from $+1$ to -1 for $0 < T < 0.2$. For $0.2 < T < \infty$, $(11e^{-T} - 10)$ monotonically decreases from -1 to -10. Thus, the pole of $T(z)$ will be inside the unit circle, and the system will be stable if $0 < T < 0.2$. In terms of frequency, where $f = 1/T$ the system will be stable as long as the sampling frequency is $1/0.2 = 5$ hertz or greater.

In this section we covered the concepts of stability for digital systems. The highlight of the section is that sampling rate (along with system parameters such as gain and component values) helps to determine the stability of a digital system. In general, if the sampling rate is too slow, the closed-loop digital system will be unstable. We now move from stability to steady-state errors, paralleling our previous discussion of steady-state errors in analog systems.

13.7 Steady-State Errors for Sampled-Data Systems

We now examine the effect of sampling upon the steady-state error for digital systems. Any general conclusion for the steady-state error is difficult because of the dependency of those conclusions upon the placement of the sampler in the loop. Remember that the position of the sampler could change the open-loop transfer function. In the discussion of analog systems, there was only one open-loop transfer function, $G(s)$, upon which the general theory of steady-state error was based and from which came the standard definitions of static error constants. For digital systems, however, the placement of the sampler changes the open-loop transfer function and thus precludes any general conclusions. In this section we assume the typical placement of the sampler after the error and in the position of the cascade controller, and we derive our conclusions accordingly about the steady-state error of digital systems.

Consider the digital system in Figure 13.15(a), where the digital computer is represented by the sampler and zero-order hold. The transfer function of the plant is represented by $G_1(s)$ and the transfer function of the z.o.h. by $\dfrac{(1 - e^{-Ts})}{s}$. Letting $G(s)$ equal the product of the z.o.h. and $G_1(s)$ and using the block diagram reduction techniques for sampled-data systems, we can find the sampled error, $E^*(s) = E(z)$. Adding synchronous samplers at the input and the feedback, we obtain Figure 13.15(b). Pushing $G(s)$ and its input sampler to the right past the pickoff point yields Figure 13.15(c). Using Figure 13.9(a), we can convert each block to its z-transform, resulting in Figure 13.15(d).

From this figure, $E(z) = R(z) - E(z)G(z)$, or

$$E(z) = \frac{R(z)}{1 + G(z)} \tag{13.55}$$

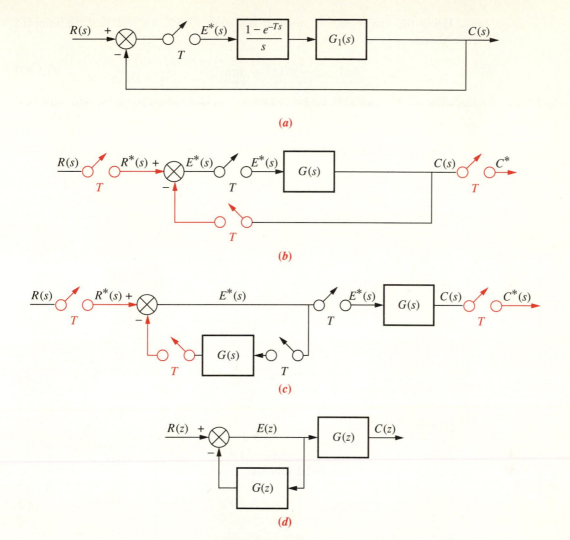

Note: Phantom samplers are shown in color.

Figure 13.15 (*a*) Digital Feedback Control System for Evaluation of Steady-State Errors; (*b*) Phantom Samplers Added; (*c*) Pushing *G(s)* and Its Samplers to the Right Past the Pickoff Point; (*d*) *z*-Transform Equivalent System

A final value theorem for discrete signals can be derived, which states that

$$e^*(\infty) = \lim_{z \to 1}(1 - z^{-1})E(z) \tag{13.56}$$

where $e^*(\infty)$ is the final sampled value of $e(t)$, or (alternately) the final value of $e(kT)$.[3]

---

[3] See Ogata (1987: 59) for a derivation.

Using the final value theorem on Equation 13.55, we find that the sampled steady-state error, $e^*(\infty)$, is

$$e^*(\infty) = \lim_{z \to 1}(1 - z^{-1})E(z) = \lim_{z \to 1}(1 - z^{-1})\frac{R(z)}{1 + G(z)} \tag{13.57}$$

Equation 13.57 must now be evaluated for each input: step, ramp, and parabola.

Unit Step Input

For a unit step input, $R(s) = 1/s$. From Table 13.1,

$$R(z) = \frac{z}{z - 1} \tag{13.58}$$

Substituting Equation 13.58 into Equation 13.57, we have

$$e^*(\infty) = \frac{1}{1 + \lim_{z \to 1} G(z)} \tag{13.59}$$

Defining the static error constant, K_p, as

$$K_p = \lim_{z \to 1} G(z) \tag{13.60}$$

we rewrite Equation 13.59 as

$$e^*(\infty) = \frac{1}{1 + Kp} \tag{13.61}$$

Unit Ramp Input

For a unit ramp input, $R(z) = \dfrac{Tz}{(z - 1)^2}$. Following the procedure for the step input, the reader can derive the fact that

$$e^*(\infty) = \frac{1}{K_v} \tag{13.62}$$

where

$$K_v = \frac{1}{T} \lim_{z \to 1}(z - 1)G(z) \tag{13.63}$$

Unit Parabolic Input

For a unit parabolic input, $R(z) = \dfrac{T^2 z(z + 1)}{2(z - 1)^3}$. Similarly,

$$e^*(\infty) = \frac{1}{K_a} \tag{13.64}$$

where

$$K_a = \frac{1}{T^2} \lim_{z \to 1}(z - 1)^2 G(z) \tag{13.65}$$

Summary of Steady-State Errors

The equations developed above for $e^*(\infty)$, K_p, K_v, and K_a are similar to the equations developed for analog systems. Whereas multiple pole placement at the origin of the s-plane reduced steady-state errors to zero in the analog case, we can see that multiple pole placement at $z = 1$ reduces the steady-state error to zero for digital systems of the type discussed in this section. This conclusion makes sense when one considers that $s = 0$ maps into $z = 1$ under $z = e^{-sT}$.

For example, for a step input, we see that if $G(z)$ in Equation 13.59 has one pole at $z = 1$, the limit will become infinite, and the steady-state error will reduce to zero.

For a ramp input, if $G(z)$ in Equation 13.63 has two poles at $z = 1$, the limit will become infinite, and the error will reduce to zero.

Similar conclusions can be drawn for the parabolic input and Equation 13.65. Here, $G(z)$ needs three poles at $z = 1$ in order for the steady-state error to be zero. Let's look at an example.

Example 13.8   Find the steady-state error of a sampled-data feedback control system.

Problem   For step, ramp, and parabolic inputs, find the steady-state error for the feedback control system shown in Figure 13.15(a) if

$$G_1(s) = \frac{10}{s(s + 1)} \tag{13.66}$$

Solution   First find $G(s)$, the product of the z.o.h. and the plant.

$$G(s) = \frac{10(1 - e^{-sT})}{s^2(s + 1)} = 10(1 - e^{-sT})\left[\frac{1}{s^2} - \frac{1}{s} + \frac{1}{s + 1}\right] \tag{13.67}$$

The z-transform is then

$$G(z) = 10(1 - z^{-1})\left[\frac{Tz}{(z - 1)^2} - \frac{z}{z - 1} + \frac{z}{z - e^{-T}}\right]$$

$$= 10\left[\frac{T}{z - 1} - 1 + \frac{z - 1}{z - e^{-T}}\right] \tag{13.68}$$

For a step input,

$$K_p = \lim_{z \to 1} G(z) = \infty; \qquad e^*(\infty) = \frac{1}{1 + K_p} = 0 \tag{13.69}$$

For a ramp input,

$$K_v = \frac{1}{T}\lim_{z \to 1}(z - 1)G(z) = 10; \qquad e^*(\infty) = \frac{1}{K_v} = 0.1 \tag{13.70}$$

For a parabolic input,

$$K_a = \frac{1}{T^2}\lim_{z \to 1}(z - 1)^2 G(z) = 0; \qquad e^*(\infty) = \frac{1}{K_a} = \infty \tag{13.71}$$

The student will notice that the answers obtained above are the same results obtained for the analog system. However, since stability depends upon the sampling interval, be sure to check the stability of the system after a sampling interval is established before making steady-state error calculations.

In this section we discussed and evaluated the steady-state error of digital systems for step, ramp, and parabolic inputs. The equations for steady-state error parallel those for analog systems. Even the definitions of the static error constants were similar. Poles at the origin of the s-plane for analog systems were replaced with poles at $+1$ on the z-plane to improve the steady-state error. We continue our parallel discussion by moving into a discussion of transient response and the root locus for digital systems.

13.8 Transient Response on the z-Plane

Recall that for analog systems, a transient response requirement was specified by selecting a closed-loop, s-plane pole. In Chapter 8, the closed-loop pole was on the existing root locus, and the design consisted of a simple gain adjustment. If the closed-loop pole was not on the existing root locus, then a cascade compensator was designed to reshape the original root locus to go through the desired closed-loop pole. A gain adjustment then completed the design.

In the next two sections we want to parallel the described analog methods and apply similar techniques to digital systems. For this introductory chapter, we will parallel the discussion through design via gain adjustment. The design of compensation is left to the student to pursue in an advanced course.

Chapter 4 established the relationships between transient response and the s-plane. We saw that vertical lines on the s-plane were lines of constant settling time, horizontal lines were lines of constant peak time, and radial lines were lines of constant percent overshoot. In order to draw equivalent conclusions on the z-plane, we now map those lines through $z = e^{sT}$.

The vertical lines on the s-plane are lines of constant settling time and are characterized by the equation $s = \sigma_1 + j\omega$, where the real part, $\sigma_1 = -4/T_s$, is constant and is in the left half-plane for stability. Substituting this into $z = e^{sT}$, we obtain

$$z = e^{\sigma_1 T} e^{j\omega T} = r_1 e^{j\omega T} \tag{13.72}$$

Equation 13.72 denotes concentric circles of radius r_1. If σ_1 is positive, the circle has a larger radius than the unit circle. On the other hand, if σ_1 is negative, the circle has a smaller radius than the unit circle. The circles of constant settling time, normalized to the sampling interval, are shown in Figure 13.16 with radius

$e^{\sigma_1 T} = e^{-4/(T_s/T)}$. Also, $T_s/T = -4/\ln(r)$, where r is the radius of the circle of constant settling time.

The horizontal lines are lines of constant peak time. The lines are characterized by the equation $s = \sigma + j\omega_1$, where the imaginary part, $\omega_1 = \pi/T_p$, is constant. Substituting this into $z = e^{sT}$, we obtain

$$z = e^{\sigma T}e^{j\omega_1 T} = e^{\sigma T}e^{j\theta_1} \tag{13.73}$$

Equation 13.73 represents radial lines at an angle of θ_1. If σ is negative, that section of the radial line lies inside the unit circle. If σ is positive, that section of the radial line lies outside the unit circle. The lines of constant peak

Figure 13.16 Constant Damping Ratio, Normalized Settling Time, and Normalized Peak Time Plots on the z-Plane

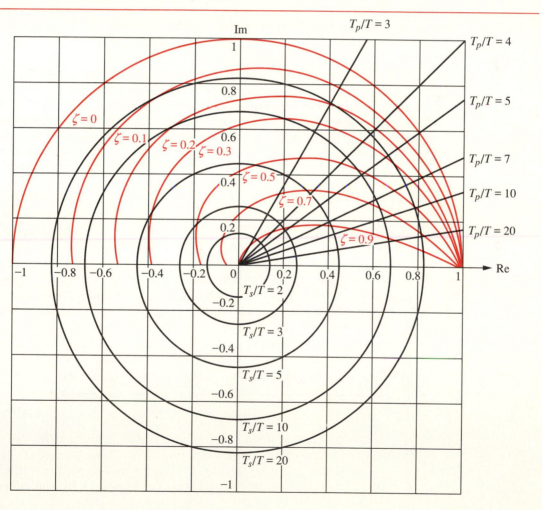

time normalized to the sampling interval are shown in Figure 13.16. The angle of each radial line is $\omega_1 T = \theta_1 = \dfrac{\pi}{T_p/T}$ from which, $\dfrac{T_p}{T} = \dfrac{\pi}{\theta_1}$.

Finally, we map the radial lines of the s-plane onto the z-plane. Remember, these radial lines are lines of constant percent overshoot on the s-plane. From Figure 13.17, these radial lines are represented by

$$\frac{\sigma}{\omega} = -\tan\left(\sin^{-1}\zeta\right) = -\frac{\zeta}{\sqrt{1-\zeta^2}} \tag{13.74}$$

Hence,

$$s = \sigma + j\omega = -\omega\frac{\zeta}{\sqrt{1-\zeta^2}} + j\omega \tag{13.75}$$

Transforming Equation 13.75 to the z-plane yields

$$z = e^{sT} = e^{-\omega T(\zeta/\sqrt{1-\zeta^2})}e^{j\omega T} = e^{-\omega T(\zeta/\sqrt{1-\zeta^2})}\angle\omega T \tag{13.76}$$

Thus, given a desired damping ratio, ζ, Equation 13.76 can be plotted on the z-plane through a range of ωT as shown in Figure 13.16. These curves can then be used as constant percent overshoot curves on the z-plane.

This section has set the stage for the analysis and design of transient response for digital systems. In the next section we apply the results to digital systems using the root locus.

Figure 13.17 The s-Plane Sketch of Constant Percent Overshoot Line

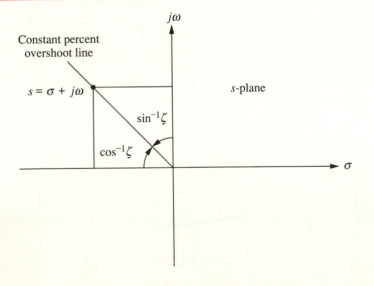

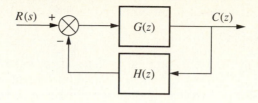

Figure 13.18 Generic Digital Feedback Control System

13.9 The Root Locus on the z-Plane and Design via Gain Adjustment

In this section we will plot root loci and determine the gain required for stability as well as the gain required to meet a transient response requirement. Since the open-loop and closed-loop transfer functions for the generic digital system shown in Figure 13.18 are identical to the continuous system except for a change in variables from s to z, we can use the same rules for plotting a root locus.

However, from our previous discussion, the region of stability on the z-plane is within the unit circle and not the left half-plane. Thus, in order to determine stability, we must search for the intersection of the root locus with the unit circle rather than the imaginary axis.

In the last section we derived the curves of constant settling time, peak time, and damping ratio. In order to design a digital system for transient response, we find the intersection of the root locus with the appropriate curves as they appear on the z-plane in Figure 13.16. Let us look at the following example.

Example 13.9   Sketch the root locus of a digital feedback control system and determine the range of gain for stability from the root locus.

Problem   Sketch the root locus for the system shown in Figure 13.19. Also, determine the range of gain, K, for stability from the root locus plot.

Solution   Treat the system as if z were s, and sketch the root locus. The result is shown in Figure 13.20. Using the root locus program in Appendix C, search along the unit circle for 180°. Identification of the gain, K, at this point yields the range of

Figure 13.19 Digital Feedback Control for Example 13.9

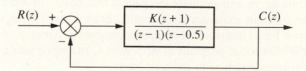

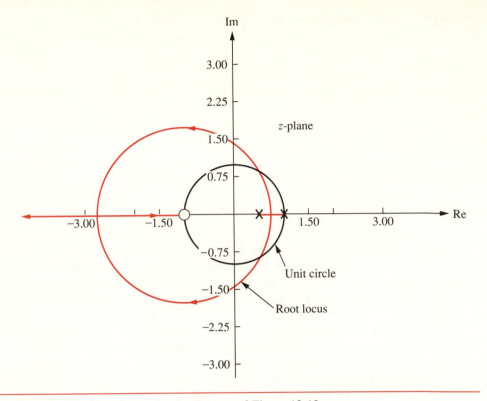

Figure 13.20 Root Locus for the System of Figure 13.19

gain for stability. Using the program, we find that the intersection of the root locus with the unit circle is $1\angle 60°$. The gain at this point is 0.5. Hence, the range of gain for stability is $0 < K < 0.5$.

In the next example, we design the value of gain, K, in Figure 13.19 to meet a transient response specification. The problem is handled similarly to the analog system design, where we found the gain at the point where the root locus crossed the specified damping ratio, settling time, or peak time curve. In digital systems, these curves are as shown in Figure 13.16. In summary, then, draw the root locus of the digital system and superimpose the curves of Figure 13.16. Then find out where the root locus intersects the desired damping ratio, settling time, or peak time curve and evaluate the gain at that point. In order to simplify the calculations and obtain more accurate results, draw a radial line through the point where the root locus intersects the appropriate curve. Measure the angle of this line and use the root locus program in Appendix C to search along this radial line for the point of intersection with the root locus.

Example 13.10   Design a digital system via simple gain adjustment to meet a transient response requirement.

Problem   For the system of Figure 13.19, find the value of gain, K, to yield a damping ratio of 0.7.

Solution   Figure 13.21 shows the constant damping ratio curves superimposed over the root locus for the system as determined from the last example. Draw a radial line from the origin to the intersection of the root locus with the 0.7 damping ratio curve (a 16.62° line). The root locus program in Appendix C can now be used to obtain the gain by searching along a 16.62° line for 180°, the intersection with the root locus. The results of the program show that the gain, K, is 0.0627 at $0.719 + j0.215$, the point where the 0.7 damping ratio curve intersects the root locus.

We can now check our design by finding the unit sampled step response of the system of Figure 13.19 . Using our design for K, 0.0627, along with $R(z) = \dfrac{z}{(z-1)}$, a sampled step input, we find the sampled output to be

$$C(z) = \frac{R(z)G(z)}{1 + G(z)} = \frac{0.0627z^2 + 0.0627z}{z^3 - 2.4373z^2 + 2z - 0.5627} \tag{13.77}$$

Figure 13.21 Root Locus for the System of Figure 13.19 with Constant 0.7 Damping Ratio Curve

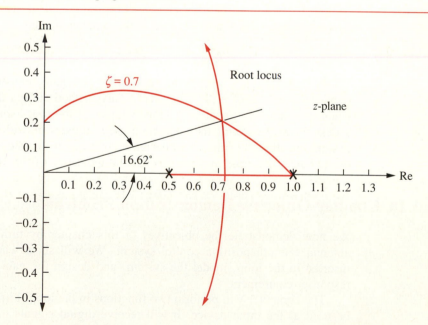

Note: Valid only at integer values of sampling instant.

Figure 13.22 Sampled Step Response of the System of
Figure 13.19 with $K = 0.0627$

Performing the indicated division, we obtain the output valid at the sampling instants, as shown in Figure 13.22. Since the overshoot is approximately 5%, the requirement of a 0.7 damping ratio has been met. The student should remember, however, that the plot is only valid at integer values of the sampling instants.

In this section we used the root locus and gain adjustment to design the transient response of a digital system. This method suffers the same drawbacks when it is applied to analog systems; namely, if the root locus does not intersect a desired design point, then a simple gain adjustment will not accomplish the design objective. Techniques to design compensation for digital systems can then be applied. This topic, however, is beyond the scope of this introductory chapter.

13.10 Chapter-Objective Demonstration Problems

We now demonstrate the objectives of this chapter by turning to our ongoing antenna azimuth position control system. We will show where the computer is inserted in the loop, model the system, and design the gain to meet a transient response requirement.

The computer will perform two functions in the loop. First, the computer will be used as the input device. It will receive digital signals from the keyboard in the form of commands, and digital signals from the output for closed-loop control.

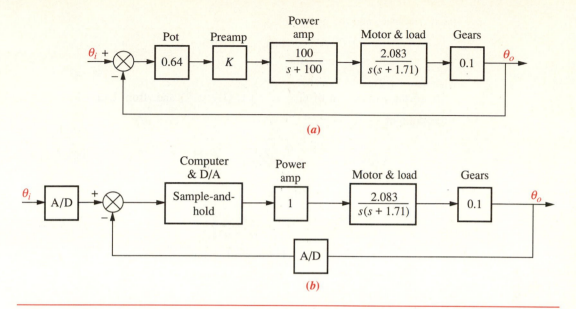

Figure 13.23 Antenna Azimuth Control System: (*a*) Analog
Implementation; (*b*) Digital Implementation

The keyboard will replace the input potentiometer, and an analog-to-digital (A/D) converter will replace the feedback transducer.

Figure 13.23(*a*) shows the original analog system, and Figure 13.23(*b*) shows the system with the computer in the loop. Here the computer is receiving digital signals from two sources, (1) the input via the keyboard or other tracking commands, and (2) from the output via an A/D converter. The plant is receiving signals from the digital computer via a digital-to-analog converter (D/A) and the sample-and-hold.

Figure 13.23(*b*) shows some simplifying assumptions we have made. The power amplifier's pole is assumed to be far enough away from the motor's pole that we can represent the power amplifier as a pure gain equal to its dc gain of unity. Also, we have absorbed any preamplifier and potentiometer gain in the computer and its associated D/A converter.

Example 13.11   Design the gain of a computer controlled system to meet a transient response requirement.

Problem   Design the gain for the antenna azimuth control system shown in Figure 13.23(*b*) to yield a closed-loop damping ratio of 0.5. Assume a sampling interval of $T = 0.1$.

Solution

Modeling the System   Our first objective is to model the system in the z-domain. The forward transfer function, $G(s)$, which includes the sample-and-hold, power amplifier, motor and load, and the gears, is

$$G(s) = \frac{1 - e^{-sT}}{s} \frac{0.2083}{s(s + a)} = \frac{0.2083}{a}(1 - e^{sT})\frac{a}{s^2(s + a)} \tag{13.78}$$

where $a = 1.71$, and $T = 0.1$.

Since the z-transform of $(1 - e^{-sT})$ is $(1 - z^{-1})$ and, from Example 13.6, the z-transform of $\dfrac{a}{s^2(s + a)}$ is

$$z\left\{\frac{a}{s^2(s + a)}\right\} = \left[\frac{Tz}{(z - 1)^2} - \frac{(1 - e^{-aT})z}{a(z - 1)(z - e^{-aT})}\right] \tag{13.79}$$

the z-transform of the plant, $G(z)$, is

$$\begin{aligned}
G(z) &= \frac{0.2083}{a}(1 - z^{-1})z\left\{\frac{a}{s^2(s + a)}\right\} \\
&= \frac{0.2083}{a^2}\left[\frac{[aT - (1 - e^{-aT})]z + [(1 - e^{-aT}) - aTe^{-aT}]}{(z - 1)(z - e^{-aT})}\right]
\end{aligned} \tag{13.80}$$

Substituting the values for a and T, we obtain

$$G(z) = \frac{9.846 \times 10^{-4}(z + 0.945)}{(z - 1)(z - 0.843)} \tag{13.81}$$

Figure 13.24 shows the computer and plant as part of the digital feedback control system.

Designing for Transient Response Now that the modeling in the z-domain is complete, we can begin to design the system for the required transient response. We superimpose the root locus over the constant damping ratio curves in the z-plane, as shown in Figure 13.25. A line drawn from the origin to the intersection forms an 8.58° angle. Searching along this line for 180°, we find the intersection to be $(0.915 + j0.138)$, with a loop gain, $9.846 \times 10^{-4}K$, of 0.0135. Hence $K = 13.71$.

Checking the design by finding the unit sampled step response of the closed-loop system yields the plot of Figure 13.26, which exhibits the 18.5% overshoot expected from a system with $\zeta = 0.473$.

Figure 13.24 Analog Antenna Azimuth Control System Converted to a Digital System

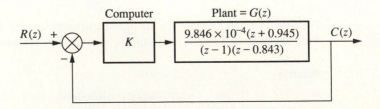

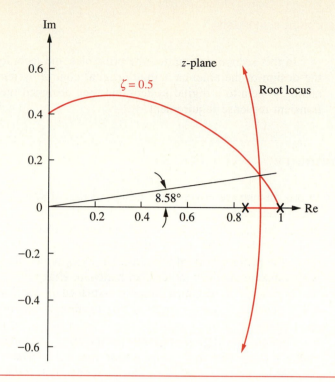

Figure 13.25 Root Locus Superimposed over Constant Damping Ratio Curves

Figure 13.26 Sampled Step Response of the Antenna Azimuth Control System

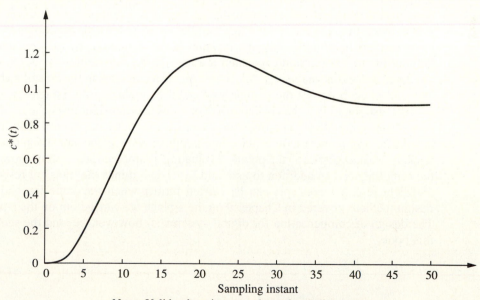

Note: Valid only at integer values of sampling instant.

In this section we applied the principles and techniques of digital systems to the design of the antenna azimuth digital control system. We first converted the analog system to a digital system and then designed the system's gain to meet a transient response requirement.

13.11 Summary

In this chapter we covered the design of digital systems using classical methods. State-space techniques were not covered. The student, however, is encouraged to pursue this topic in a course dedicated to sampled-data control systems.

We looked at the advantages of digital control systems. These systems can control numerous loops at reduced cost. System modifications can be implemented with software changes rather than hardware changes.

Typically, the digital computer is placed in the forward path preceding the plant. Digital-to-analog and analog-to-digital conversion is required within the system to ensure compatibility of the analog and digital signals throughout the system. The digital computer in the loop is modeled as a sample-and-hold network along with any compensation that it performs.

Throughout the chapter we saw direct parallels to the methods used for s-plane analysis of transients, steady-state errors, and the stability of analog systems. The parallel is made possible by the z-transform, which replaces the Laplace transform as the transform of choice for analyzing sampled-data systems. The z-transform allows us to represent sampled waveforms at the sampling instants. We can handle sampled systems as easily as continuous systems, including block diagram reduction, since both signals and systems can be represented in the z-domain and manipulated algebraically. Complex systems can be reduced to a single block through techniques that parallel those used with the s-plane. Time responses can be obtained through division of the numerator by the denominator without the partial fraction expansion required in the s-domain.

Digital systems analysis parallels the s-plane techniques in the area of stability. The unit circle becomes the boundary of stability, replacing the imaginary axis.

We also found that the concepts of root locus and transient response are easily carried into the z-plane. The rules for sketching the root locus do not change. We are able to map points on the s-plane into points on the z-plane and attach transient response characteristics to the points. Evaluating a sampled-data system shows that the sampling rate, in addition to gain and load, determines the transient response. Transient response concepts can be carried farther when the same lag and lead design methods covered in Chapter 9 on the s-plane are carried out on the z-plane. The design of compensation for digital systems is, however, beyond the scope of this book.

REVIEW QUESTIONS

1. Name two functions that the digital computer can perform when used with feedback control systems.
2. Name three advantages of using digital computers in the loop.
3. Name two important considerations in analog-to-digital conversion that yield errors.
4. Of what does the block diagram model for a computer consist?
5. What is the z-transform?
6. What does the inverse z-transform of a time waveform actually yield?
7. Name two methods of finding the inverse z-transform.
8. What method of finding the inverse z-transform yields a closed-form expression for the time function?
9. What method of finding the inverse z-transform immediately yields the values of the time waveform at the sampling instants?
10. In order to find the z-transform of a $G(s)$, what must be true of the input and the output?
11. If input $R(z)$ to system $G(z)$ yields output $C(z)$, what is the nature of $c(t)$?
12. If a time waveform, $c(t)$, at the output of system $G(z)$ is plotted using the inverse z-transform, and a typical second-order response with damping ratio = 0.5 results, can we say that the system is stable?
13. What must exist in order for cascaded sampled-data systems to be represented by the product of their pulse transfer functions, $G(z)$?
14. Where is the region for stability on the z-plane?
15. What methods for finding the stability of digital systems can replace the Routh-Hurwitz criterion for analog systems?
16. To drive steady-state errors in analog systems to zero, a pole can be placed at the origin of the s-plane. Where on the z-plane should a pole be placed to drive the steady-state error of a sampled system to zero?
17. How do the rules for sketching the root locus on the z-plane differ from those for sketching the root locus on the s-plane?
18. Given a point on the z-plane, how can one determine the associated percent overshoot, settling time, and peak time?
19. Given a desired percent overshoot and settling time, how can one tell which point on the z-plane is the design point?

PROBLEMS

1. Derive the z-transforms for the time functions listed below. Do not use any z-transform tables. Follow the following plan: $f(t) \rightarrow f^*(t) \rightarrow F^*(s) \rightarrow F(z)$ followed by converting $F(z)$ into closed form making use of the fact that $\dfrac{1}{1 - z^{-1}} = 1 + z^{-1} + z^{-2} + z^{-3} + \cdots$. Assume ideal sampling.
   a. $e^{-at}u(t)$
   b. $u(t)$

c. $t^2 e^{-at}$

d. $\sin \omega t$

2. For each $F(z)$ below, find $f(kT)$ using partial fraction expansion.

a. $F(z) = \dfrac{z(z+1)(z+2)}{(z-0.5)(z-0.7)(z-0.9)}$

b. $F(z) = \dfrac{(z+0.3)(z+0.5)}{(z-0.4)(z-0.6)(z-0.8)}$

c. $F(z) = \dfrac{(z+1)(z+0.2)}{z(z-0.5)(z-0.6)}$

3. For each $F(z)$ in Problem 2, do the following:
   a. Find $f(kT)$ using the power series expansion.
   b. Check your results against your answers from Problem 2.

4. Using partial fraction expansion and Table 13.1, find the z-transform for each $G(s)$ shown below if $T = 0.5$ seconds.

a. $G(s) = \dfrac{(s+3)}{s(s+1)(s+2)}$

b. $G(s) = \dfrac{(s+1)(s+2)}{(s+3)(s+4)}$

c. $G(s) = \dfrac{27}{(s+2)(s^2+4s+13)}$

d. $G(s) = \dfrac{10}{s(s+2)(s^2+12s+61)}$

5. Find $G(z) = C(z)/R(z)$ for each of the block diagrams shown in Figure P13.1 if $T = 0.1$ second.

Figure P13.1

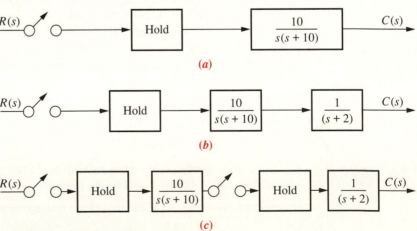

(a)

(b)

(c)

6. Find $T(z) = C(z)/R(z)$ for each of the systems shown in Figure P13.2.

Figure P13.2

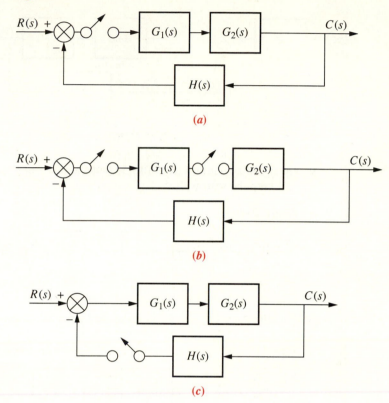

(a)

(b)

(c)

7. Find $C(z)$ in general terms for the digital system shown in Figure P13.3.

Figure P13.3

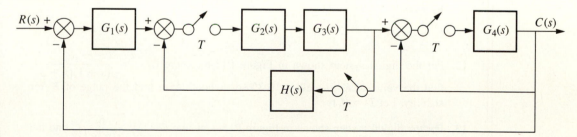

8. Find the closed-loop transfer function, $T(z) = C(z)/R(z)$, for the system shown in Figure P13.4.

Figure P13.4

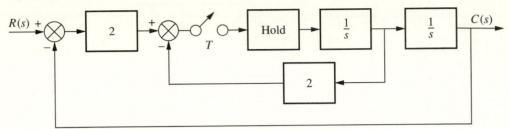

9. Given the system in Figure P13.5, find the range of sampling interval, T, that will keep the system stable.

Figure P13.5

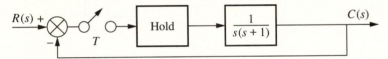

10. Find the range of gain, K, to make the system shown in Figure P13.6 stable.

Figure P13.6

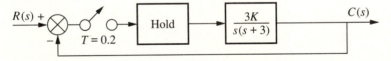

11. Find the static error constants and the steady-state error for each of the digital systems shown in Figure P.13.7 if the inputs are

a. $u(t)$

b. $tu(t)$

c. $\dfrac{1}{2}t^2 u(t)$

12. For the digital system shown in Figure P13.8, where $G(s) = \dfrac{K}{(s+1)(s+3)}$, find the value of K to yield a 16.3% overshoot. Also find the range of K for stability. Let $T = 0.1$.

13. For the digital system shown in Figure P13.8, where $G(s) = \dfrac{K}{s(s+1)}$, find the value of K to yield a peak time of 0.7 seconds if the sampling interval, T, is 0.1 second. Also, find the range of K for stability.

Figure P13.7

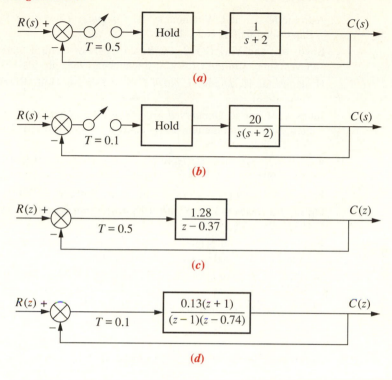

(a)

(b)

(c)

(d)

Figure P13.8

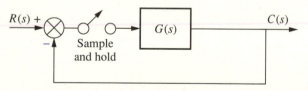

14. **a.** Convert the heading control for the Unmanned Free-Swimming Submersible vehicle shown in Figure P6.8 into a digitally controlled system.
   b. Find the closed-loop pulse transfer function, $T(z)$.
   c. Find the range of heading gain to keep the digital system stable.

15. *Chapter-objective problem:* Convert the antenna azimuth control system shown in Figure P2.36 to a digital system with $T = 0.1$.
   a. Design the gain, K, for a 16.3% overshoot.
   b. For your designed value of gain, find the steady-state error for a unit ramp input.

BIBLIOGRAPHY

Astrom, K. J., and Wittenmark, B. *Computer Controlled Systems*. Prentice Hall, Englewood Cliffs, N.J., 1984.

Boyd, M., and Yingst, J.C. PC-Based Operator Control Station Simplifies Process, Saves Time. *Chilton's I & CS,* September 1988, pp. 99–101.

Hostetter, G. H. *Digital Control System Design*. Holt, Rinehart & Winston, New York, 1988.

Johnson, H., et al. *Unmanned Free-Swimming Submersible (UFSS) System Description*. NRL Memorandum Report 4393. Naval Research Laboratory, Washington, D.C., 1980.

Katz, P. *Digital Control Using Microprocessors*. Prentice Hall, Englewood Cliffs, N.J., 1981.

Kuo, B. C. *Digital Control Systems*. Holt, Rinehart & Winston, New York, 1980.

Ogata, K. *Discrete-Time Control Systems*. Prentice Hall, Englewood Cliffs, N.J., 1987.

Phillips, C. L., and Nagle, H. T., Jr. *Digital Control System Analysis and Design*. Prentice Hall, Englewood Cliffs, N.J., 1984.

Smith, C. L. *Digital Computer Process Control*. Intext Educational Publishers, N.Y., 1972.

Tou, J. *Digital and Sampled-Data Control Systems*. McGraw-Hill, New York, 1959.

APPENDIX A

$\%OS$	Percent overshoot
A	Ampere—unit of electrical current
A	System matrix for state-space representation
B	Input coupling matrix for state-space representation
C	Electrical capacitance
C	Output matrix for state-space representation
$C(s)$	Laplace transform of the output of a system
$c(t)$	Output of a system
$\mathbf{C_M}$	Controllability matrix
D	Feedforward matrix for state-space representation
D	Mechanical rotational coefficient of viscous friction
D_a	Motor armature coefficient of viscous damping
D_m	Total coefficient of viscous friction at the armature of a motor, including armature coefficient of viscous friction and reflected load coefficient of viscous friction
E	Energy
$E(s)$	Laplace transform of the error or actuating signal
$e(t)$	Error or actuating signal
$E_a(s)$	Laplace transform of the motor armature input voltage
$e_a(t)$	Motor armature input voltage
F	Farad—unit of electrical capacitance
$F(s)$	Laplace transform of $f(t)$
$f(t)$	Mechanical force; general time function
f_{Φ_M}	Phase-margin frequency
f_{G_M}	Gain-margin frequency
f_v	Mechanical translational coefficient of viscous friction
g	Acceleration due to gravity
$G(s)$	Forward-path transfer function
$G_c(s)$	Compensator transfer function
G_M	Gain margin
H	Henry—unit of electrical inductance
$H(s)$	Feedback-path transfer function
I	Identity matrix
$i(t)$	Electrical current
J	Moment of inertia
J_a	Motor armature moment of inertia
J_m	Total moment of inertia at the armature of a motor, including armature moment of inertia and reflected load moment of inertia
K	Controller gain matrix

K	Mechanical translational or rotational spring constant; amplifier gain; residue
k	Controller feedback gain; running index
K_a	Acceleration constant
K_b	Back emf constant
kg	Kilogram = newton seconds^2/meter—unit of mass
kg-m^2	Kilogram meters^2 = newton-meters seconds^2/radian—unit of moment of inertia
K_p	Position constant
K_t	Motor torque constant relating developed torque to armature current
K_v	Velocity constant
L	Electrical inductance
L	Observer gain matrix
l	Observer feedback gain
M	Mass; slope of the root locus asymptotes
m	Meter—unit of mechanical translational displacement
$M(\omega)$	Magnitude of a sinusoidal response
m/s	Meters/second—unit of mechanical translational velocity
M_P	Peak magnitude of the sinusoidal magnitude response
N	Newton—unit of mechanical translational force
N-s/m	Newton-seconds/meter—unit of mechanical translational coefficient of viscous friction
n	System type
N/m	Newton/meter—unit of mechanical translational spring constant
N-m	Newton-meter—unit of mechanical torque
N-m s/rad	Newton-meter seconds/radian—unit of mechanical rotational coefficient of viscous friction
N-m/A	Newton-meter/ampere—unit of motor torque constant
N-m/rad	Newton-meter/radian—unit of mechanical rotational spring constant
O$_M$	Observability matrix
P	Similarity transformation matrix
p_c	Compensator pole
Q	Coulomb—unit of electrical charge
$q(t)$	Electrical charge
R	Electrical resistance
$R(s)$	Laplace transform of the input to a system
$r(t)$	Input to a system
R_a	Motor armature resistance
rad	Radian—unit of angular displacement
rad/s	Radian/second—unit of angular velocity
s	Complex variable for the Laplace transform
$S_{F:P}$	Sensitivity of F to a fractional change in P
T	Time constant; sampling interval for digital signals
$T(s)$	Closed-loop transfer function; Laplace transform of mechanical torque
$T(t)$	Mechanical torque
$T_m(t)$	Torque at the armature developed by a motor
$T_m(s)$	Laplace transform of the torque at the armature developed by a motor
T_p	Peak time
T_r	Rise time

T_s	Settling time
T_W	Pulse width
u	Input or control vector for state-space representation
$u(t)$	Input control signal for state-space representation
V-s/rad	Volt-seconds/radian—unit of motor back emf constant
$v(t)$	Mechanical translation velocity; electrical voltage
x	State vector for state-space representation
$x(t)$	Mechanical translation displacement; a state variable
$\dot{x}$	Time derivative of a state variable
$\dot{\mathbf{x}}$	Time derivative of the state vector
y	Output vector for state-space representation
$y(t)$	Output scalar for state-space representation
z	Complex variable for the z-transform
z_c	Compensator zero

α	Pole-scaling factor for a lag compensator, where $\alpha > 1$; angle of attack
β	Pole-scaling factor for a lead compensator, where $\beta < 1$
γ	Pole-scaling factor for a lag-lead compensator, where $\gamma > 1$
δ	Thrust angle
ζ	Damping ratio
θ	Angle of a vector with the positive extension of the real axis
$\theta(t)$	Angular displacement
θ_c	Angular contribution of a compensator on the s-plane
$\theta_m(t)$	Angular displacement of the armature of a motor
λ	Eigenvalue
σ	Real part of the Laplace transform variable, s
σ_0	Real-axis intercept of the root locus asymptotes
Φ_M	Phase margin
$\Phi(t)$	State transition matrix
ϕ	Sinusoidal phase angle; body angle
ϕ_c	Sinusoidal phase angle of a compensator
ϕ_{max}	Maximum sinusoidal phase angle
Ω	Ohm—unit of electrical resistance
ω	Imaginary part of the Laplace transform variable, s
$\omega(t)$	Angular velocity in rad/s
ω_{BW}	Bandwidth in rad/s
ω_d	Damped frequency of oscillation in rad/s
ω_n	Natural frequency in rad/s
ω_P	Peak-magnitude frequency of the magnitude frequency response

APPENDIX B

MATRICES, DETERMINANTS, AND SYSTEMS OF EQUATIONS

B.1 Matrix Definitions and Notations

Matrix

An $m \times n$ *matrix* is a rectangular or square array of elements with m rows and n columns. An example of a matrix is shown in Equation B.1.

$$\mathbf{A} = \begin{bmatrix} a_{11} & a_{12} & \cdots & a_{1n} \\ a_{21} & a_{22} & \cdots & a_{2n} \\ \vdots & \vdots & \vdots & \vdots \\ a_{m1} & a_{m2} & \cdots & a_{mn} \end{bmatrix} \tag{B.1}$$

For each subscript, a_{ij}, $i =$ the row, and $j =$ the column. If $m = n$, the matrix is said to be a *square matrix*.

Vector

If a matrix has just one row, it is called a *row vector*. An example of a row vector follows:

$$\mathbf{B} = \begin{bmatrix} b_{11} & b_{12} & \cdots & b_{1n} \end{bmatrix} \tag{B.2}$$

If a matrix has just one column, it is called a *column vector*. An example of a column vector follows:

$$\mathbf{C} = \begin{bmatrix} c_{11} \\ c_{21} \\ \vdots \\ c_{m1} \end{bmatrix} \tag{B.3}$$

Partitioned Matrix

A matrix can be partitioned into component matrices or vectors. For example, let

$$\mathbf{A} = \begin{bmatrix} a_{11} & a_{12} & a_{13} & a_{14} \\ a_{21} & a_{22} & a_{23} & a_{24} \\ a_{31} & a_{32} & a_{33} & a_{34} \\ a_{41} & a_{42} & a_{43} & a_{44} \end{bmatrix} = \left[\begin{array}{cc|cc} a_{11} & a_{12} & a_{13} & a_{14} \\ a_{21} & a_{22} & a_{23} & a_{24} \\ \hline a_{31} & a_{32} & a_{33} & a_{34} \\ a_{41} & a_{42} & a_{43} & a_{44} \end{array} \right] = \begin{bmatrix} \mathbf{A}_{11} & \mathbf{A}_{12} \\ \mathbf{A}_{21} & \mathbf{A}_{22} \end{bmatrix} \qquad (B.4)$$

where

$$\mathbf{A}_{11} = \begin{bmatrix} a_{11} & a_{12} \\ a_{21} & a_{22} \\ a_{31} & a_{32} \end{bmatrix}; \quad \mathbf{A}_{12} = \begin{bmatrix} a_{13} & a_{14} \\ a_{23} & a_{24} \\ a_{33} & a_{34} \end{bmatrix}$$

$$\mathbf{A}_{21} = \begin{bmatrix} a_{41} & a_{42} \end{bmatrix}; \quad \mathbf{A}_{22} = \begin{bmatrix} a_{43} & a_{44} \end{bmatrix}$$

Null Matrix

A matrix with all elements equal to zero is called the *null matrix*; that is, $a_{ij} = 0$ for all i and j. An example of a null matrix follows:

$$\mathbf{A} = \begin{bmatrix} 0 & 0 & 0 & & \cdots & & 0 \\ 0 & 0 & 0 & & \cdots & & 0 \\ \vdots & \vdots & \vdots & \vdots & \vdots & \vdots & \vdots \\ 0 & 0 & 0 & 0 & 0 & 0 & 0 \end{bmatrix} \qquad (B.5)$$

Diagonal Matrix

A square matrix with all elements off of the diagonal equal to zero is said to be a *diagonal matrix*; that is, $a_{ij} = 0$ for $i \neq j$. An example of a diagonal matrix follows:

$$\mathbf{A} = \begin{bmatrix} a_{11} & 0 & 0 & \cdots & 0 \\ 0 & a_{22} & 0 & \cdots & 0 \\ 0 & 0 & a_{33} & \cdots & 0 \\ \vdots & \vdots & \vdots & \vdots & \vdots \\ 0 & 0 & 0 & \cdots & a_{nn} \end{bmatrix} \qquad (B.6)$$

Identity Matrix

A diagonal matrix with all diagonal elements equal to unity is called an *identity matrix* and is denoted by **I**; that is, $a_{ij} = 1$ for $i = j$, and $a_{ij} = 0$ for $i \neq j$. An example of an identity matrix follows:

$$\mathbf{A} = \begin{bmatrix} 1 & 0 & 0 & \cdots & 0 \\ 0 & 1 & 0 & \cdots & 0 \\ \vdots & \vdots & \vdots & \vdots & \vdots \\ 0 & 0 & 0 & \cdots & 1 \end{bmatrix} \qquad (B.7)$$

Symmetric Matrix

A square matrix for which $a_{ij} = a_{ji}$ is called a *symmetric matrix*. An example of a symmetric matrix follows:

$$\mathbf{A} = \begin{bmatrix} 3 & 8 & 7 \\ 8 & 9 & 2 \\ 7 & 2 & 4 \end{bmatrix} \tag{B.8}$$

Matrix Transpose

The *transpose* of matrix $\mathbf{A}$, designated $\mathbf{A}^T$, is formed by interchanging the rows and columns of $\mathbf{A}$. Thus, if $\mathbf{A}$ is an $m \times n$ matrix with elements a_{ij}, the transpose is an $n \times m$ matrix with elements a_{ji}. An example follows. Given

$$\mathbf{A} = \begin{bmatrix} 1 & 7 & 9 \\ 2 & 6 & -3 \\ 4 & 8 & 5 \\ -1 & 3 & -2 \end{bmatrix} \tag{B.9}$$

then

$$\mathbf{A}^T = \begin{bmatrix} 1 & 2 & 4 & -1 \\ 7 & 6 & 8 & 3 \\ 9 & -3 & 5 & -2 \end{bmatrix} \tag{B.10}$$

Determinant of a Matrix

The *determinant* of a square matrix is denoted by $\det \mathbf{A}$, or

$$\begin{vmatrix} a_{11} & a_{12} & \cdots & a_{1n} \\ a_{21} & a_{22} & \cdots & a_{2n} \\ \vdots & \vdots & \vdots & \vdots \\ a_{m1} & a_{m2} & \cdots & a_{mn} \end{vmatrix} \tag{B.11}$$

The determinant of a 2×2 matrix,

$$\mathbf{A} = \begin{bmatrix} a_{11} & a_{12} \\ a_{21} & a_{22} \end{bmatrix} \tag{B.12}$$

is evaluated as

$$\det \mathbf{A} = \begin{vmatrix} a_{11} & a_{12} \\ a_{21} & a_{22} \end{vmatrix} = a_{11}a_{22} - a_{21}a_{12} \tag{B.13}$$

Minor of an Element

The *minor*, M_{ij}, of element a_{ij} of $\det \mathbf{A}$ is the determinant formed by removing the ith row and the jth column from $\det \mathbf{A}$. As an example, consider the following determinant:

$$\det \mathbf{A} = \begin{vmatrix} 3 & 8 & 7 \\ 6 & 9 & 2 \\ 5 & 1 & 4 \end{vmatrix} \tag{B.14}$$

The minor M_{32} is the determinant formed by removing the third row and the second column from det $\mathbf{A}$. Thus,

$$M_{32} = \begin{vmatrix} 3 & 7 \\ 6 & 2 \end{vmatrix} = -36 \tag{B.15}$$

Cofactor of an Element

The *cofactor*, C_{ij}, of element a_{ij} of det $\mathbf{A}$ is defined to be

$$C_{ij} = (-1)^{(i+j)}M_{ij} \tag{B.16}$$

For example, given the determinant of Equation B.14,

$$C_{21} = (-1)^{(2+1)}M_{21} = (-1)^3 \begin{vmatrix} 8 & 7 \\ 1 & 4 \end{vmatrix} = -25 \tag{B.17}$$

Evaluating the Determinant of a Square Matrix

The determinant of a square matrix can be evaluated by expanding minors along any row or column. Expanding along any row, we find

$$\det \mathbf{A} = \sum_{k=1}^{n} a_{ik}C_{ik} \tag{B.18}$$

where n = number of columns of $\mathbf{A}$; i is the ith row selected to expand by minors; and C_{ik} is the cofactor of a_{ik}. Expanding along any column, we find

$$\det \mathbf{A} = \sum_{k=1}^{m} a_{kj}C_{kj} \tag{B.19}$$

where m = number of rows of $\mathbf{A}$; j is the jth column selected to expand by minors; and C_{kj} is the cofactor of a_{kj}. For example, if

$$\mathbf{A} = \begin{bmatrix} 1 & 3 & 2 \\ -5 & 6 & -7 \\ 8 & 5 & 4 \end{bmatrix} \tag{B.20}$$

then, expanding by minors on the third column, we find

$$\det \mathbf{A} = 2 \begin{vmatrix} -5 & 6 \\ 8 & 5 \end{vmatrix} - (-7) \begin{vmatrix} 1 & 3 \\ 8 & 5 \end{vmatrix} + 4 \begin{vmatrix} 1 & 3 \\ -5 & 6 \end{vmatrix} = -195 \tag{B.21}$$

Expanding by minors on the second row, we find

$$\det \mathbf{A} = -(-5) \begin{vmatrix} 3 & 2 \\ 5 & 4 \end{vmatrix} + 6 \begin{vmatrix} 1 & 2 \\ 8 & 4 \end{vmatrix} - (-7) \begin{vmatrix} 1 & 3 \\ 8 & 5 \end{vmatrix} = -195 \tag{B.22}$$

Singular Matrix

A matrix is *singular* if its determinant equals zero.

Nonsingular Matrix

A matrix is *nonsingular* if its determinant does not equal zero.

Adjoint of a Matrix

The *adjoint* of a square matrix, **A**, written adj **A**, is the matrix formed from the transpose of the matrix **A** after all elements have been replaced by their cofactors. Thus,

$$\text{adj }\mathbf{A} = \begin{bmatrix} C_{11} & C_{12} & \cdots & C_{1n} \\ C_{21} & C_{22} & \cdots & C_{2n} \\ \vdots & \vdots & \vdots & \vdots \\ C_{n1} & C_{n2} & \cdots & C_{nn} \end{bmatrix}^{T} \tag{B.23}$$

For example, consider the following matrix:

$$\mathbf{A} = \begin{bmatrix} 1 & 2 & 3 \\ -1 & 4 & 5 \\ 6 & 8 & 7 \end{bmatrix} \tag{B.24}$$

Hence,

$$\text{adj }\mathbf{A} = \begin{bmatrix} \begin{vmatrix} 4 & 5 \\ 8 & 7 \end{vmatrix} & -\begin{vmatrix} -1 & 5 \\ 6 & 7 \end{vmatrix} & \begin{vmatrix} -1 & 4 \\ 6 & 8 \end{vmatrix} \\ -\begin{vmatrix} 2 & 3 \\ 8 & 7 \end{vmatrix} & \begin{vmatrix} 1 & 3 \\ 6 & 7 \end{vmatrix} & -\begin{vmatrix} 1 & 2 \\ 6 & 8 \end{vmatrix} \\ \begin{vmatrix} 2 & 3 \\ 4 & 5 \end{vmatrix} & -\begin{vmatrix} 1 & 3 \\ -1 & 5 \end{vmatrix} & \begin{vmatrix} 1 & 2 \\ -1 & 4 \end{vmatrix} \end{bmatrix}^{T} = \begin{bmatrix} -12 & 10 & -2 \\ 37 & -11 & -8 \\ -32 & 4 & 6 \end{bmatrix} \tag{B.25}$$

Rank of a Matrix

The *rank* of a matrix, **A**, equals the number of linearly independent rows or columns. The rank can be found by finding the highest-order square submatrix that is nonsingular. For example, consider the following:

$$\mathbf{A} = \begin{bmatrix} 1 & -5 & 2 \\ 4 & 7 & -5 \\ -3 & 15 & -6 \end{bmatrix} \tag{B.26}$$

The determinant of **A** $= 0$. Since the determinant is zero, the 3×3 matrix is singular. Choosing the submatrix

$$\mathbf{A} = \begin{bmatrix} 1 & -5 \\ 4 & 7 \end{bmatrix} \tag{B.27}$$

whose determinant equals 27, we conclude that **A** is of rank 2.

B.2 Matrix Operations

Addition

The sum of two matrices, written $\mathbf{A} + \mathbf{B} = \mathbf{C}$, is defined by $a_{ij} + b_{ij} = c_{ij}$. For example,

$$\begin{bmatrix} 2 & -1 \\ 3 & 5 \end{bmatrix} + \begin{bmatrix} 7 & -5 \\ -4 & 3 \end{bmatrix} = \begin{bmatrix} 9 & -6 \\ -1 & 8 \end{bmatrix} \tag{B.28}$$

Subtraction

The difference between two matrices, written $\mathbf{A} - \mathbf{B} = \mathbf{C}$, is defined by $a_{ij} - b_{ij} = c_{ij}$. For example,

$$\begin{bmatrix} 2 & -1 \\ 3 & 5 \end{bmatrix} - \begin{bmatrix} 7 & -5 \\ -4 & 3 \end{bmatrix} = \begin{bmatrix} -5 & 4 \\ 7 & 2 \end{bmatrix} \tag{B.29}$$

Multiplication

The product of two matrices, written $\mathbf{AB} = \mathbf{C}$, is defined by $c_{ij} = \sum_{k=1}^{n} a_{ik} b_{kj}$. For example, if

$$\mathbf{A} = \begin{bmatrix} a_{11} & a_{12} & a_{13} \\ a_{21} & a_{22} & a_{23} \end{bmatrix}; \quad \mathbf{B} = \begin{bmatrix} b_{11} & b_{12} & b_{13} \\ b_{21} & b_{22} & b_{23} \\ b_{31} & b_{32} & b_{33} \end{bmatrix} \tag{B.30}$$

then

$$\mathbf{C} = \begin{bmatrix} (a_{11}b_{11} + a_{12}b_{21} + a_{13}b_{31}) & (a_{11}b_{12} + a_{12}b_{22} + a_{13}b_{32}) & (a_{11}b_{13} + a_{12}b_{23} + a_{13}b_{33}) \\ (a_{21}b_{11} + a_{22}b_{21} + a_{23}b_{31}) & (a_{21}b_{12} + a_{22}b_{22} + a_{23}b_{32}) & (a_{21}b_{13} + a_{22}b_{23} + a_{23}b_{33}) \end{bmatrix} \tag{B.31}$$

Notice that multiplication is defined only if the number of columns of $\mathbf{A}$ equals the number of rows of $\mathbf{B}$.

Multiplication by a Constant

A matrix can be multiplied by a constant by multiplying every element of the matrix by that constant. For example, if

$$\mathbf{A} = \begin{bmatrix} a_{11} & a_{12} \\ a_{21} & a_{22} \end{bmatrix} \tag{B.32}$$

then

$$k\mathbf{A} = \begin{bmatrix} ka_{11} & ka_{12} \\ ka_{21} & ka_{22} \end{bmatrix} \tag{B.33}$$

Inverse

An $n \times n$ square matrix, $\mathbf{A}$, has an inverse, denoted by $\mathbf{A}^{-1}$, which is defined by

$$\mathbf{AA}^{-1} = \mathbf{I} \tag{B.34}$$

where $\mathbf{I}$ is an $n \times n$ identity matrix. The inverse of $\mathbf{A}$ is given by

$$\mathbf{A}^{-1} = \frac{\text{adj } \mathbf{A}}{\det \mathbf{A}} \tag{B.35}$$

For example, find the inverse of $\mathbf{A}$ in Equation B.24. The adjoint was calculated in Equation B.25. The determinant of $\mathbf{A}$ is

$$\det \mathbf{A} = 1 \begin{vmatrix} 4 & 5 \\ 8 & 7 \end{vmatrix} - (-1) \begin{vmatrix} 2 & 3 \\ 8 & 7 \end{vmatrix} + 6 \begin{vmatrix} 2 & 3 \\ 4 & 5 \end{vmatrix} = -34 \tag{B.36}$$

Hence,

$$\mathbf{A}^{-1} = \frac{\begin{bmatrix} -12 & 10 & -2 \\ 37 & -11 & -8 \\ -32 & 4 & 6 \end{bmatrix}}{-34} = \begin{bmatrix} 0.353 & -0.294 & 0.059 \\ -1.088 & 0.324 & 0.235 \\ 0.941 & -0.118 & -0.176 \end{bmatrix} \tag{B.37}$$

B.3 Matrix and Determinant Identities

The following are identities that apply to matrices and determinants.

Matrix Identities

Commutative Law

$$\mathbf{A} + \mathbf{B} = \mathbf{B} + \mathbf{A} \tag{B.38}$$

$$\mathbf{AB} \neq \mathbf{BA} \tag{B.39}$$

Associative Law

$$\mathbf{A} + (\mathbf{B} + \mathbf{C}) = (\mathbf{A} + \mathbf{B}) + \mathbf{C} \tag{B.40}$$

$$\mathbf{A}(\mathbf{BC}) = (\mathbf{AB})\mathbf{C} \tag{B.41}$$

Transpose of Sum

$$(\mathbf{A} + \mathbf{B})^T = \mathbf{A}^T + \mathbf{B}^T \tag{B.42}$$

Transpose of Product

$$(\mathbf{AB})^T = \mathbf{B}^T \mathbf{A}^T \tag{B.43}$$

Determinant Identities

Multiplication of a Single Row or Single Column of a Matrix, A, by a Constant

If a single row or single column of a matrix, $\mathbf{A}$, is multiplied by a constant, k, forming the matrix, $\tilde{\mathbf{A}}$, then

$$\det \tilde{\mathbf{A}} = k \det \mathbf{A} \tag{B.44}$$

Multiplication of All Elements of an $n \times n$ Matrix, A, by a Constant

$$\det (k\mathbf{A}) = k^n \det \mathbf{A} \tag{B.45}$$

Transpose

$$\det \mathbf{A}^T = \det \mathbf{A} \tag{B.46}$$

Determinant of the Product of Square Matrices

$$\det \mathbf{AB} = \det \mathbf{A} \det \mathbf{B} \tag{B.47}$$

$$\det \mathbf{AB} = \det \mathbf{BA} \tag{B.48}$$

B.4 Systems of Equations

Representation of Systems of Equations

Assume the following system of n linear equations:

$$
\begin{aligned}
a_{11}x_1 + a_{12}x_2 + \cdots + a_{1n} &= b_1 \\
a_{21}x_1 + a_{22}x_2 + \cdots + a_{2n} &= b_2 \\
&\vdots \\
a_{n1}x_1 + a_{n2}x_2 + \cdots + a_{nn} &= b_n
\end{aligned}
\tag{B.49}
$$

This system of equations can be represented in vector-matrix form as

$$\mathbf{Ax} = \mathbf{B} \tag{B.50}$$

where

$$
\mathbf{A} = \begin{bmatrix}
a_{11} & a_{12} & \cdots & a_{1n} \\
a_{21} & a_{22} & \cdots & a_{2n} \\
\vdots & \vdots & \vdots & \vdots \\
a_{n1} & a_{n2} & \cdots & a_{nn}
\end{bmatrix}; \quad
\mathbf{B} = \begin{bmatrix} b_1 \\ b_2 \\ \vdots \\ b_n \end{bmatrix}; \quad
\mathbf{x} = \begin{bmatrix} x_1 \\ x_2 \\ \vdots \\ x_n \end{bmatrix}
$$

For example, the following system of equations,

$$5x_1 + 7x_2 = 3 \tag{B.51a}$$

$$-8x_1 + 4x_2 = -9 \tag{B.51b}$$

can be represented in vector-matrix form as $\mathbf{Ax} = \mathbf{B}$, or

$$
\begin{bmatrix} 5 & 7 \\ -8 & 4 \end{bmatrix}
\begin{bmatrix} x_1 \\ x_2 \end{bmatrix} =
\begin{bmatrix} 3 \\ -9 \end{bmatrix}
\tag{B.52}
$$

Solving Systems of Equations
via Matrix Inverse

If $\mathbf{A}$ is nonsingular, we can premultiply Equation B.50 by $\mathbf{A}^{-1}$, yielding the solution $\mathbf{x}$. Thus,

$$\mathbf{x} = \mathbf{A}^{-1}\mathbf{B} \tag{B.53}$$

For example, premultiplying both sides of Equation B.52 by $\mathbf{A}^{-1}$, where

$$\mathbf{A}^{-1} = \begin{bmatrix} 5 & 7 \\ -8 & 4 \end{bmatrix}^{-1} = \begin{bmatrix} 0.0526 & -0.0921 \\ 0.1053 & 0.0658 \end{bmatrix} \tag{B.54}$$

we solve for $\mathbf{x} = \mathbf{A}^{-1}\mathbf{B}$ as follows:

$$\begin{bmatrix} x_1 \\ x_2 \end{bmatrix} = \begin{bmatrix} 0.0526 & -0.0921 \\ 0.1053 & 0.0658 \end{bmatrix} \begin{bmatrix} 3 \\ -9 \end{bmatrix} = \begin{bmatrix} 0.987 \\ -0.276 \end{bmatrix} \tag{B.55}$$

Solving Systems of Equations via Cramer's Rule

Equation B.53 allows us to solve for all unknowns, x_i, where $i = 1$ to n. If we are interested in a single unknown, x_k, then Cramer's rule can be used. Given Equation B.50, Cramer's rule states that

$$x_k = \frac{\det \mathbf{A}_k}{\det \mathbf{A}} \tag{B.56}$$

where $\mathbf{A_k}$ is a matrix formed by replacing the kth column of $\mathbf{A}$ by $\mathbf{B}$. For example, solve Equation B.52. Using Equation B.56 with

$$\mathbf{A} = \begin{bmatrix} 5 & 7 \\ -8 & 4 \end{bmatrix}; \quad \mathbf{B} = \begin{bmatrix} 3 \\ -9 \end{bmatrix}$$

we find

$$x_1 = \frac{\begin{vmatrix} 3 & 7 \\ -9 & 4 \end{vmatrix}}{\begin{vmatrix} 5 & 7 \\ -8 & 4 \end{vmatrix}} = \frac{75}{76} = 0.987 \tag{B.57}$$

and

$$x_2 = \frac{\begin{vmatrix} 5 & 3 \\ -8 & -9 \end{vmatrix}}{\begin{vmatrix} 5 & 7 \\ -8 & 4 \end{vmatrix}} = \frac{-21}{76} = -0.276 \tag{B.58}$$

BIBLIOGRAPHY

Dorf, R. C. *Matrix Algebra—A Programmed Introduction*. Wiley, New York, 1969.

Kreyszig, E. *Advanced Engineering Mathematics*. 4th ed. Wiley, New York, 1979.

Wylie, C. R., Jr. *Advanced Engineering Mathematics*. 5th ed. McGraw-Hill, New York, 1982.

APPENDIX C

COMPUTER PROGRAMS

C.1 Program to Plot the Step Response of a System Represented in State Space

The following program to plot the step response of a system represented in state space is written in Microsoft® QuickBASIC and meets the following requirements:

1. The user will input whether a hard copy of the parameters is desired.
2. The user will input system order, system matrix, input coupling matrix, and output matrix.
3. The user will input whether a hard copy of the plot is desired.
4. The user will input the initial conditions.
5. The user will input the following plot parameters: iteration interval, plot interval, maximum time, and maximum amplitude.
6. The program will plot the step response and list the response data.
7. The program will replot the step response after allowing the user to change the initial conditions as well as the plot parameters without reentering the system.

This program runs as is on a Macintosh™ computer. The reader should tailor the program for any other form of BASIC that is incompatible with the program as given as well as for a hand-held calculator.

This program plots the step response of a system represented in state space and permits the user to choose an iteration interval. A helpful technique for finding the iteration interval is to run the program with successively diminishing iteration intervals until reaching an iteration interval below which there is no appreciable change in the results.

Another parameter the user can select is the print interval, which allows the user to print at a larger time interval than the iteration interval.

The execution time of the program is also an input parameter. The user should choose a time for which the output has already reached a steady-state value.

The maximum amplitude is a parameter that permits the plotting program to plot the output within the limited space provided. The user should supply the program with the maximum positive or negative amplitude value expected.

A simplified flowchart for the program is shown in Figure C.1. The Microsoft® Quick-BASIC code follows. Unnumbered lines in the code signify a continuation from the previously numbered line and do not imply a new line.

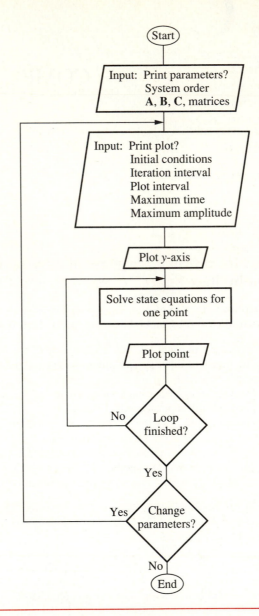

Figure C.1 Flowchart for State-Space Program

```
10 REM ***************Input Parameters************************
20 INPUT"Do you want a hard copy of parameters? (y,n)  >", Prt.Req$
30 PRINT"Enter System Parameters"
40 INPUT"Enter System Order  >", sys.order
50 IF (Prt.Req$="y") THEN LPRINT "System order ="sys.order
60 PRINT "Enter System Matrix"
70 FOR I=1 TO sys.order
80 FOR J=1 TO sys.order
90     PRINT"Enter a(";I;",";J;")  >";
100    INPUT" ", a(I,J)
110    IF (Prt.Req$="y") THEN LPRINT "a(";I;",";J;") ="a(I,J)
120 NEXT J
130 NEXT I
140 PRINT "Enter Input Coupling Matrix"
150 FOR I=1 TO sys.order
160    PRINT "Enter b(";I;") >";
170    INPUT " ",b(I)
180    IF (Prt.Req$="y") THEN LPRINT "b(";I;") ="b(I)
190 NEXT I
200 PRINT "Enter Output Matrix "
210 FOR I=1 TO sys.order
220    PRINT "Enter c(";I;") >";
230    INPUT " ",c(I)
240    IF (Prt.Req$="y") THEN LPRINT "c(";I;") ="c(I)
250 NEXT I
255 INPUT"Do you want a hard copy of plot? (y,n)  >", Prt.Req$
260 PRINT"Enter Initial Conditions "
270 FOR I=1 TO sys.order
280    PRINT"Enter  x(";I;") >";
290    INPUT" ", X(I)
300    IF (Prt.Req$="y") THEN LPRINT "x(";I;") ="X(I)
310 NEXT I
320 PRINT "Enter Plot Parameters"
330 INPUT"Iteration Interval >",DELTAT
340 IF (Prt.Req$="y") THEN LPRINT "Iteration interval ="DELTAT
350 INPUT"Plot Interval >",PRNT.INT
360 IF (Prt.Req$="y") THEN LPRINT "Plot interval ="PRNT.INT
370 INPUT"Maximum Time for Analysis >",MAXTIME
380 IF (Prt.Req$="y") THEN LPRINT "Maximum Time for Analysis ="MAXTIME
390 INPUT"Maximum Amplitude for Plot >",maximum.amplitude
400 IF (Prt.Req$="y") THEN LPRINT "Maximum Amplitude for Plot ="maximum.amplitude
420 LET Y=0
430 REM*******************Plot y Axis************************
440 FOR I=1 TO sys.order
450    Y=Y+c(I)*X(I)                                    'Initial value of output
460 NEXT I
470 Y.SCALED%= (( Y+maximum.amplitude) *20/(maximum.amplitude))
                   'Scaled value of initial output for plotting point on y axis
```

```
480   FOR X=0 TO 40                                 'Do loop index for plotting y axis
490       IF (Y.SCALED% = X) OR (Y.SCALED% < 0 AND X=0) OR (Y.SCALED% > 40 AND X=40)
          THEN PRINT TAB(X+1);"*"; ELSE PRINT TAB(X+1);"-";
500       IF Prt.Req$="y" AND ((Y.SCALED% = X ) OR (Y.SCALED% < 0 AND X=0) OR
          (Y.SCALED% > 40 AND X=40))   THEN
502         LPRINT TAB(X+1);"*";
503         ELSEIF Prt.Req$="y" THEN
505         LPRINT "-";
507       END IF                                    'Plot the y axis OR initial OUTPUT
510   NEXT X                                        'Increment DO LOOP index
520   PRINT TAB(42);"y(0) =";Y                       'Print data for initial value of
                                                                            output
530   IF (Prt.Req$="y") THEN LPRINT TAB(42);"y(0) =";Y
540   REM*****************Start Plot Loop*************************
550   FOR K = 1 TO CINT(MAXTIME/PRNT.INT) STEP 1   'Index for Printing interval
560   REM*****************Start Iteration Loop************************
570   FOR n=1 TO CINT(PRNT.INT/DELTAT) STEP 1      'Index for iteration interval
580   FOR I=1 TO sys.order
590   LET ax=0
600   FOR J=1 TO sys.order
610     ax=ax+a(I,J)*X(J)
620   NEXT J
630     DELTAX(I)=(ax+b(I))*DELTAT                 'Calculate delta X1
640   NEXT I
650   FOR I=1 TO sys.order
660     X(I)=X(I)+DELTAX(I)                        'Calculate next X
670   NEXT I
680   LET cx=0
690   FOR I=1 TO sys.order
700     cx=cx+c(I)*X(I)
710   NEXT I
720     Y=cx                                       'Form output from states
730   NEXT n
740   REM******************End Iteration Loop**********************
750   Y.SCALED% = ((Y+maximum.amplitude)*20/(maximum.amplitude)) 'Scaled output for
                                                                            plot
760   IF Y.SCALED%<20 THEN GOTO 770 ELSE GOTO 800  'Plot output or t axis segment
                                                                            first?
770   PRINT TAB(Y.SCALED%+1);"*";TAB(21);"¦";       'Plot output segment first
780   IF (Prt.Req$="y") THEN LPRINT TAB(Y.SCALED%+1);"*";TAB(21);"¦";
790   GOTO 830
800   IF Y.SCALED%=20 THEN PRINT TAB(21);"*"; ELSE GOTO 810    'Output is on x axis
803   IF (Prt.Req$="y") THEN LPRINT TAB(21);"*";
805   GOTO 830
810   PRINT TAB(21);"¦";                            'Plot x axis segment first
812   IF (Prt.Req$="y") THEN LPRINT TAB(21);"¦";
815   IF (Y.SCALED%+1) < 41 THEN PRINT TAB(Y.SCALED%+1); "*"; ELSE PRINT TAB(41); "*";
825   IF (Prt.Req$="y" AND Y.SCALED%+1 < 41) THEN LPRINT TAB(Y.SCALED%+1); "*";
```

```
827 IF (Prt.Req$="y" AND Y.SCALED%+1 > 40) THEN LPRINT TAB(41); "*";
830 PRINT TAB(42);"y(";K*DELTAT*CINT(PRNT.INT/DELTAT);")=";Y      'Print value
840 IF (Prt.Req$="y") THEN LPRINT TAB(42);"y(";K*DELTAT*CINT(PRNT.INT/DELTAT);")=";Y
850 NEXT K                                      'Set index for next value to be
                                                                     plotted
860 PRINT TAB(21);"t axis"                      'Identify x axis
870 IF (Prt.Req$="y") THEN LPRINT TAB(21);"t axis"
880 REM ******************End Plot Loop**************************
890 PRINT "Hit 'p' 'Enter' to Change Plot Parameters and Replot"
895 PRINT "Hit 'Enter' to Input New System"
900 INPUT" ",q$
905 IF q$="p" THEN GOTO 255 ELSE GOTO 10
```

We will demonstrate the program for the example covered in Section 4.11. The dialog between the user and the computer follows.

```
Do you want a hard copy of parameters? (y,n) >n
Enter System Parameters
Enter System Order > 2
Enter System Matrix
Enter a( 1 , 1 ) > 0
Enter a( 1 , 2 ) > 1
Enter a( 2 , 1 ) > -2
Enter a( 2 , 2 ) > -3
Enter Input Coupling Matrix
Enter b( 1 ) > 0
Enter b( 2 ) > 1
Enter Output Matrix
Enter c( 1 ) > 2
Enter c( 2 ) > 3
Do you want a hard copy of plot? (y,n) > n
Enter Initial Conditions
Enter x( 1 ) > 1
Enter x( 2 ) > -2
Enter Plot Parameters
Iteration Interval > .001
Plot Interval > .1
Maximum Time for Analysis > 2
Maximum Amplitude for Plot > 4
```

The computer plot is shown in Figure C.2

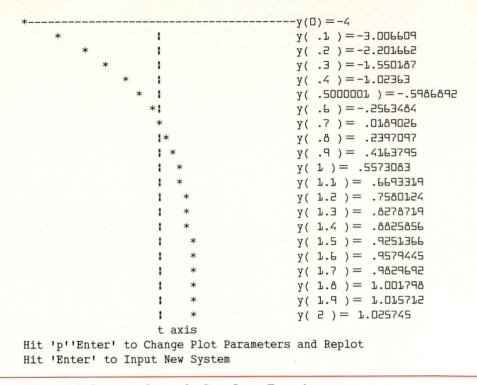

```
*------------------------------------------y(0)=-4
    *                   !               y( .1  )=-3.006609
      *                 !               y( .2  )=-2.201662
        *               !               y( .3  )=-1.550187
          *             !               y( .4  )=-1.02363
            *  !                        y( .5000001 )=-.5986892
              *!                        y( .6  )=-.2563484
               *                        y( .7  )=  .0189026
               !*                       y( .8  )=  .2397097
               ! *                      y( .9  )=  .4163795
               !  *                     y( 1  )=  .5573083
               !  *                     y( 1.1  )=  .6693319
               !   *                    y( 1.2  )=  .7580124
               !   *                    y( 1.3  )=  .8278719
               !   *                    y( 1.4  )=  .8825856
               !    *                   y( 1.5  )=  .9251366
               !    *                   y( 1.6  )=  .9579445
               !    *                   y( 1.7  )=  .9829692
               !    *                   y( 1.8  )=  1.001798
               !    *                   y( 1.9  )=  1.015712
               !    *                   y( 2  )=  1.025745
               t axis
Hit 'p''Enter' to Change Plot Parameters and Replot
Hit 'Enter' to Input New System
```

Figure C.2 Computer Output for State-Space Example

C.2 Root Locus and Frequency Response

The following program to locate points on the root locus and their associated gains as well as yield data to plot a frequency response curve is written in Microsoft® QuickBASIC and meets the following requirements:

1. The user will input the system's open-loop poles and zeros.
2. The user will input any test point on the s-plane in Cartesian or polar coordinates to see if this point lies on the root locus.
3. For each point, the program will calculate the sum of the zero angles minus the sum of the pole angles.
4. For each point the program will calculate the root locus gain, K.

This program runs as is on a Macintosh computer. The reader should tailor the program for any other form of BASIC that is incompatible with the program as given as well as for a hand-held calculator.

The program can also be used to obtain frequency response plots by using test points on the imaginary axis. The gain obtained from the program is the reciprocal of the magnitude, and the angle obtained from the program is the phase angle.

A simplified flowchart for the program is shown in Figure C.3. The Microsoft® Quick-BASIC code follows. Unnumbered lines in the code signify a continuation from the previous numbered line and do not imply a new line.

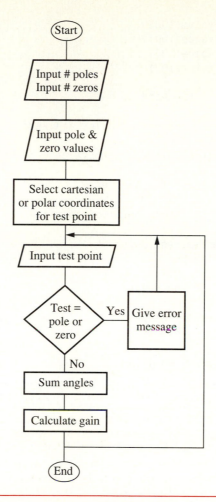

Figure C.3 Flowchart for Program to Find Points on the Root Locus

```
10    DIM pole.real.value(100)
20    DIM pole.imag.value(100)
30    DIM zero.real.value(100)
40    DIM zero.imag.value(100)
50    LET pi=3.14159
60    PRINT"ROOT LOCUS PROGRAM"
70    PRINT
80    REM schedule.enter.number.of.poles.and.zeros
90       GOSUB 250
100   REM schedule.enter.pole.and.zero.values
110      GOSUB 310
120   REM schedule.print.poles.and.zeros
130      GOSUB 470
140   REM schedule.select.test.point.coordinates
150      GOSUB 570
160   REM schedule.enter.test.point
165      LET  error.flag=0
170      GOSUB 610
180   REM schedule.test.the.point
190      GOSUB 770
200   REM schedule.print.test.point
205      IF error.flag=1 GOTO 160
210      GOSUB 1050
220   REM schedule.print.result
230      GOSUB 1090
240   GOTO 160
250   REM ******************Enter.number.of.poles.and.zeros**************
260   INPUT"Enter number of poles >",number.poles
270   INPUT"Enter number of zeros >",number.zeros
280   PRINT
300   RETURN
310   REM ******************Enter.pole.and.zero.values*******************
320   FOR K=1 TO number.poles
330      PRINT"Enter real part of pole #";K;">";
340      INPUT" ", pole.real.value(K)
350      PRINT"Enter imaginary part of pole #";K;">";
360      INPUT" ",pole.imag.value(K)
370      PRINT
380   NEXT K
390   FOR K=1 TO number.zeros
400      PRINT"Enter real part of zero #";K;">";
410      INPUT" ",zero.real.value(K)
420      PRINT"Enter imaginary part of zero #";K;">";
430      INPUT " ",zero.imag.value(K)
440      PRINT
450   NEXT K
460   RETURN
```

```
470   REM *******************Print.poles.and.zeros*******************
480   FOR K=1 TO number.poles
490       PRINT"Open loop pole #";K; " =  (";pole.real.value(K);")
                                        +j(";pole.imag.value(K);")"
500   NEXT K
510   PRINT
520   FOR K=1 TO number.zeros
530       PRINT"Open loop zero #";K; " =  (";zero.real.value(K);")
                                        +j(";zero.imag.value(K);")"
540   NEXT K
550   PRINT
560   RETURN
570   REM *******************Select.test.point.coordinates*******************
580   INPUT"Select test point coord. system (c-cart., p-polar)",test.point.coord$
590   PRINT
600   RETURN
610   REM *******************Enter.test.point*******************
620   IF test.point.coord$="c" THEN GOTO 640
630   IF test.point.coord$="p" THEN GOTO 690
640   REM cartesian
650   INPUT"Enter real part of test point >", test.real.value
660   INPUT"Enter imaginary part of test point >", test.imag.value
670   PRINT
680   GOTO 750
690   REM polar
700   INPUT"Enter magnitude of test point >", test.mag.value
710   INPUT"Enter angle of test point >", test.angle.value
720   PRINT
730   LET test.real.value= test.mag.value*COS(test.angle.value*pi/180)
740   LET test.imag.value= test.mag.value*SIN(test.angle.value*pi/180)
750   REM continue
760   RETURN
770   REM *******************Test.the.point*******************
780   LET magnitude=1
790   LET angle=0
800   FOR K=1 TO number.poles
810       LET deltax=test.real.value-pole.real.value(K)
820       LET deltay=test.imag.value-pole.imag.value(K)
830       IF deltax=0 AND deltay=0 THEN PRINT"ERROR: Test point is the same
                                    as an open-loop pole. Enter new test point."
835       IF deltax=0 AND deltay=0 THEN error.flag=1: GOTO 960
840       GOSUB 970
850       magnitude=magnitude*temp.magnitude
860       angle=angle-temp.angle
870   NEXT K
880   FOR K=1 TO number.zeros
890       LET deltax=test.real.value-zero.real.value(K)
900       LET deltay=test.imag.value-zero.imag.value(K)
```

```
910      IF deltax=0 AND deltay=0 THEN PRINT"ERROR: Test point is the same
                                   as an open-loop zero. Enter new test point."
915      IF deltax=0 AND deltay=0 THEN error.flag=1: GOTO 960
920      GOSUB 970
930      magnitude=magnitude/temp.magnitude
940      angle=angle+temp.angle
950  NEXT K
960  RETURN
970  REM *******************convert.to.magnitude.and.angle******************
980  LET temp.magnitude=SQR(ABS(deltax^2+deltay^2))
990  IF deltax=0 THEN LET temp.angle=pi/2
                            ELSE LET temp.angle =ATN(ABS (deltay)/ABS (deltax))
1000 IF deltay>=0 AND deltax>=0 THEN LET temp.angle=temp.angle
1010 IF deltay>=0 AND deltax<0 THEN LET temp.angle=(pi-temp.angle)
1020 IF deltay<0 AND deltax<=0 THEN LET temp.angle=-(pi-temp.angle)
1030 IF deltay<0 AND deltax>0 THEN LET temp.angle=-temp.angle
1040 RETURN
1050 REM *******************print.test.point************************
1060 IF test.point.coord$="c" THEN PRINT "test point = (";test.real.value;")
                                            +j(";test.imag.value;")"
1070 IF test.point.coord$="p" THEN PRINT "test point = (";test.mag.value;")
                                            < (";test.angle.value;")"
1080 RETURN
1090 REM *******************print.result***********************
1100 angle = angle*180/pi
1110 PRINT "Angle = ";(angle/360-FIX(angle/360))*360    'show angle between +360
                                   and - 360. On BASICs without FIX,
                             SGN(angle/360)*INT(ABS(angle/360)) can be used
1120 PRINT"Gain = ";magnitude
1130 PRINT
1140 RETURN
```

We will demonstrate the program for the root locus shown in Figure C.4, which has open-loop poles at 0, -2, and -4. The locus is first sketched. After the sketch, the program is used to evaluate the exact points at specified locations that may be of interest to us. For example, find the exact point on the root locus that has as its real part -0.25, as shown at point a in Figure C.4. Using the program to search along the line $s = -0.25$, we find that the locus crosses at $-0.25 + j2.3$. The crossing point, a, is where the angles add up to an odd multiple of 180°. The output from the program follows.

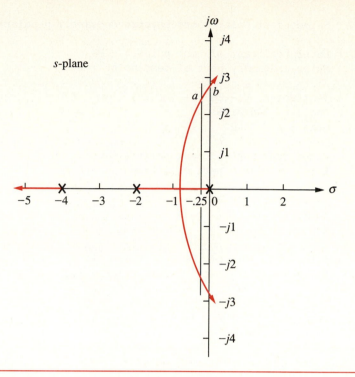

Figure C.4 Root Locus Sketch for the Example

```
ROOT LOCUS PROGRAM

Enter number of poles > 3
Enter number of zeros > 0

Enter real part of pole # 1    >  0
Enter imaginary part of pole # 1 >  0

Enter real part of pole # 2    >  -2
Enter imaginary part of pole # 2 >  0

Enter real part of pole # 3    >  -4
Enter imaginary part of pole # 3 >  0

Open loop pole # 1   =  ( 0 ) +j( 0 )
Open loop pole # 2   =  (-2 ) +j( 0 )
Open loop pole # 3   =  (-4 ) +j( 0 )
```

```
Select test point coord. system (c-cart., p-polar) c

Enter real part of test point > -.25
Enter imaginary part of test point > 1

Test Point = (-.25 ) + j ( 1 )
Angle = -148.7125
Gain = 8.063242

Enter real part of test point > -.25
Enter imaginary part of test point > 2

Test Point = (-.25 ) + j ( 2 )
Angle = -174.0116
Gain = 22.76485

Enter real part of test point > -.25
Enter imaginary part of test point > 2.3

Test Point = (-.25 ) + j ( 2.3 )
Angle = -180.4592
Gain = 29.41407

Enter real part of test point > -.25
Enter imaginary part of test point > 2.5

Test Point = (-.25 ) + j ( 2.5 )
Angle = -184.4086
Gain = 34.55537
```

GLOSSARY

Acceleration constant $\lim_{s \to 0} s^2 G(s)$

Actuating signal The signal that drives the controller. If this signal is the difference between the input and output, it is called the *error*.

Analog-to-digital converter A device that converts analog signals to digital signals.

Armature The rotating member of a dc motor through which a current flows.

Back emf The voltage across the armature of a motor.

Bandwidth The frequency at which the magnitude frequency response is -3 dB below the magnitude at zero frequency.

Basis Linearly independent vectors that define a space.

Block diagram A representation of the interconnection of subsystems that form a system. In a linear system the block diagram consists of blocks representing subsystems, arrows representing signals, summing junctions, and pickoff points.

Bode diagram (plot) A straight-line approximation to a sinusoidal frequency response plot, where the magnitude response is plotted separately from the phase response. The magnitude plot is dB versus $\log \omega$, and the phase plot is phase versus $\log \omega$. In control systems, the Bode plot is usually made for the open-loop transfer function.

Branches Lines that represent subsystems in a signal-flow graph.

Break frequency A frequency where the Bode magnitude plot changes slope.

Breakaway point A point on the real axis of the s-plane where the root locus leaves the real axis and enters the complex plane.

Break-in point A point on the real axis of the s-plane where the root locus enters the real axis from the complex plane.

Characteristic equation The equation formed by setting the characteristic polynomial to zero.

Characteristic polynomial The denominator of a transfer function. Equivalently, the unforced differential equation set equal to zero where the differential operators on the solution are replaced by s or λ.

Classical approach to control systems *See* **frequency domain techniques**.

Closed-loop system A system that monitors its output and corrects for disturbances. It is characterized by feedback paths from the output.

Closed-loop transfer function For a generic feedback system with $G(s)$ in the forward path and $H(s)$ in the feedback path, the closed-loop transfer function, $T(s)$, is $\dfrac{G(s)}{1 \pm G(s)H(s)}$, where the $+$ is for negative feedback, and the $-$ is for positive feedback.

Compensation The addition of a transfer function in the forward path or feedback path for the purpose of improving the transient and steady-state performance of a control system.

Compensator A subsystem inserted into the forward or feedback path for the purpose of improving the transient response or steady-state error.

Constant *M* circles The locus of constant, closed-loop magnitude frequency response for unity feedback systems. It allows the closed-loop magnitude frequency response to be determined from the open-loop magnitude frequency response.

Constant *N* circles The locus of constant, closed-loop phase frequency response for unity feedback systems. It allows the closed-loop phase frequency response to be determined from the open-loop phase frequency response.

Controllability A property of a system by which an input can be found that takes every state variable from a desired initial state to a desired final state.

Controlled variable The output of a plant or process that the system is controlling for the purpose of desired transient response, stability, and steady-state error characteristics.

Controller The subsystem that generates the input to the plant or process.

Critically damped response The step response of a second-order system with a given natural frequency that is characterized by no overshoot and a rise time that is faster than any possible overdamped response with the same natural frequency.

Damped frequency of oscillation The sinusoidal frequency of oscillation of an underdamped response.

Damping ratio The ratio of the exponential decay frequency to the natural frequency.

Decade Frequencies that are separated by a factor of 10.

Decibel (dB) The decibel is defined as $10 \log P_G$, where P_G is the power gain of a signal. Equivalently, the decibel is also $20 \log V_G$, where V_G is the voltage gain of a signal.

Decoupled system A state-space representation in which each state equation is a function of only one state variable. Hence, each differential equation can be solved independently of the other equations.

Digital-to-analog converter A device that converts digital signals to analog signals.

Disturbance An unwanted signal that corrupts the input or output of a plant or process.

Dominant poles The poles that predominantly generate the transient response.

Eigenvalues Any value, λ_i, that satisfies $\mathbf{A}\mathbf{x_i} = \lambda_i \mathbf{x_i}$ for $\mathbf{x_i} \neq 0$. Hence, any value, λ_i, that makes $\mathbf{x_i}$, an eigenvector under the transformation $\mathbf{A}$.

Eigenvector Any vector that is collinear with a new basis vector after a similarity transformation to a diagonal system.

Electrical admittance The inverse of electrical impedance. The ratio of the Laplace transform of the current to the Laplace transform of the voltage.

Electrical impedance The ratio of the Laplace transform of the voltage to the Laplace transform of the current.

Equilibrium The steady-state solution characterized by a constant position or oscillation.

Error The difference between the input and output of a system.

Euler's approximation A method of integration where the area to be integrated is approximated as a sequence of rectangles.

Feedback A path through which a signal flows back to a previous signal in the forward path in order to be added or subtracted.

Feedback compensator A subsystem placed in a feedback path for the purpose of improving the performance of a closed-loop system.

Forced response The part of the total response function due to the input. It is typically of the same form as the input and its derivatives.

Forward-path gain The product of gains found by traversing a path from the input node to the output node of a signal-flow graph.

Frequency domain techniques A method of analyzing and designing linear control systems by using transfer functions and the Laplace transform as well as frequency response techniques.

Frequency response techniques A method of analyzing and designing control systems by using the sinusoidal frequency response characteristics of a system.

Gain margin The amount of additional open-loop gain, expressed in decibels (dB), required at 180° of phase shift to make the closed-loop system unstable.

Gain-margin frequency The frequency at which the phase frequency response plot equals 180°. It is the frequency at which the gain margin is measured.

Homogeneous solution *See* **natural response**.

Ideal derivative compensator *See* **proportional-plus-derivative controller**.

Ideal integral compensator *See* **proportional-plus-integral controller**.

Instability The characteristic of a system defined by a natural response that grows without bounds as time approaches infinity.

Lag compensator A transfer function, characterized by a pole on the negative real axis close to the origin and a zero close and to the left of the pole, that is used for the purpose of improving the steady-state error of a closed-loop system.

Lag-lead compensator A transfer function, characterized by a pole-zero configuration that is the combination of a lag and a lead compensator, that is used for the purpose of improving both the transient response and the steady-state error of a closed-loop system.

Laplace transformation A transformation that transforms linear differential equations into algebraic expressions. The transformation is especially useful for modeling, analyzing, and designing control systems as well as solving linear differential equations.

Lead compensator A transfer function, characterized by a zero on the negative real axis and a pole to the left of the zero, that is used for the purpose of improving the transient response of a closed-loop system.

Linear combination A linear combination of n variables, x_i, for $i = 1$ to n, is given by the following sum, S:

$$S = K_n X_n + K_{n-1} X_{n-1} + \cdots + K_1 X_1,$$

where each K_i is a constant.

Linear independence The variables x_i, for $i = 1$ to n, are said to be linearly independent if their linear combination, S, equals zero *only* if every $K_i = 0$ and *no* $x_i = 0$. Alternately, if the x_i's are linearly independent, then $K_n x_n + K_{n-1} x_{n-1} + \cdots + K_1 x_1 = 0$ cannot be solved for any x_k. Thus, no x_k can be expressed as a linear combination of the other x_i's.

Linear system A system possessing the properties of superposition and homogeneity.

Linearization The process of approximating a nonlinear differential equation with a linear differential equation valid for small excursions about equilibrium.

Loop gain For a signal-flow graph, the product of branch gains found by traversing a path that starts at a node and ends at the same node without passing through any other node more than once, and following the direction of the signal flow.

Major-loop compensation A method of feedback compensation that adds a compensating zero to the open-loop transfer function for the purpose of improving the transient response of the closed-loop system.

Marginal stability The characteristic of a system defined by a natural response that neither decays nor grows, but remains constant or oscillates as time approaches infinity.

Mason's rule A formula from which the transfer function of a system consisting of the interconnection of multiple subsystems can be found.

Mechanical rotational impedance The ratio of the Laplace transform of the torque to the Laplace transform of the angular displacement.

Mechanical translational impedance The ratio of the Laplace transform of the force to the Laplace transform of the linear displacement.

Minor-loop compensation A method of feedback compensation that changes the poles of a forward-path transfer function for the purpose of improving the transient response of the closed-loop system.

Modern approach to control systems *See* **state-space representation**.

Natural frequency The frequency of oscillation of a system if all the damping is removed.

Natural response The part of the total response function due to the system and the way the system acquires or dissipates energy.

Negative feedback The case where a feedback signal is subtracted from a previous signal in the forward path.

Nichols chart The locus of constant closed-loop magnitude and closed-loop phase frequency response plotted on the open-loop dB versus phase-angle plane. It allows the closed-loop frequency response to be determined from the open-loop frequency response.

Nodes Points in a signal-flow diagram that represent signals.

No-load speed The speed produced by a motor with constant input voltage when the torque at the armature is reduced to zero.

Nonminimum-phase system A system whose transfer function has zeros in the right half-plane. The step response is characterized by an initial reversal in direction.

Nontouching-loop gain The product of loop gains from nontouching loops taken two, three, four, and so on at a time.

Nontouching-loops Loops that do not have any nodes in common.

Nyquist criterion If a contour, A, that encircles the entire right half-plane is mapped through $G(s)H(s)$, then the number of closed-loop poles, Z, in the right half-plane equals the number of open-loop poles, P, that are in the right half-plane minus the number of counterclockwise revolutions, N, around -1, of the mapping; that is, $Z = P - N$. The mapping is called the *Nyquist diagram* of $G(s)H(s)$.

Nyquist diagram (plot) A polar frequency response plot, made for the open-loop transfer function.

Nyquist sampling rate The minimum frequency at which an analog signal should be sampled for correct reconstruction. This frequency is twice the bandwidth of the analog signal.

Observability A property of a system by which an initial state vector, $\mathbf{x}(t_0)$, can be found from $u(t)$ and $y(t)$ measured over a finite interval of time from t_0. Simply stated, observability is the property by which the state variables can be found from a knowledge of the input, $u(t)$, and output, $y(t)$.

Observer A system configuration from which inaccessible states can be estimated.

Octave Frequencies that are separated by a factor of two.

Open-loop system A system that does not monitor its output nor correct for disturbances.

Open-loop transfer function For a generic feedback system with $G(s)$ in the forward path and $H(s)$ in the feedback path, the open-loop transfer function is the product of the forward transfer function and the feedback transfer function, or, $G(s)H(s)$.

Operational amplifier An amplifier characterized by a very high input impedance and a very low output impedance that can be used to implement the transfer function of a compensator.

Oscillatory response The step response of a second-order system that is characterized by a pure oscillation.

Output equation The equation that expresses the output variables of a system as linear combinations of the state variables.

Overdamped response The step response of a second-order system that is characterized by no overshoot.

Partial fraction expansion A mathematical equation where a fraction with n factors in its denominator is represented as the sum of simpler fractions.

Particular solution *See* **forced response**.

Passive network A physical network that only stores or dissipates energy. No energy is produced by the network.

Peak time, T_p The time required for the underdamped step response to reach the first, or maximum, peak.

Percent overshoot, $\%OS$ The amount that the underdamped step response overshoots the steady-state, or final, value at the peak time, expressed as a percentage of the steady-state value.

Phase margin The amount of additional open-loop phase shift required at unity gain to make the closed-loop system unstable.

Phase-margin frequency The frequency at which the magnitude frequency response plot equals zero dB. It is the frequency at which the phase margin is measured.

Phase variables State variables such that each subsequent state variable is the derivative of the previous state variable.

Phasor A rotating vector that represents a sinusoid of the form $A \cos(\omega t + \phi)$.

Pickoff point A block diagram symbol that shows the distribution of one signal to multiple subsystems.

Plant or process The subsystem whose output is being controlled by the system.

Poles (1) The values of the Laplace transform variable, s, that cause the transfer function to become infinite, and (2) any roots of factors of the characteristic equation in the denominator that are common to the numerator of the transfer function.

Position constant $\lim_{s \to 0} G(s)$

Positive feedback The case where a feedback signal is added to a previous signal in the forward path.

Proportional-plus-derivative (PD) controller A controller that feeds forward to the plant proportions of the actuating signal plus its derivative for the purpose of improving the transient response of a closed-loop system.

Proportional-plus-integral (PI) controller A controller that feeds forward to the plant proportions of the actuating signal plus its integral for the purpose of improving the steady-state error of a closed-loop system.

Proportional-plus-integral-plus-derivative (PID) controller A controller that feeds forward to the plant proportions of the actuating signal plus its integral plus its derivative for the purpose of improving the transient response and steady-state error of a closed-loop system.

Quantization error The error associated with the digitizing of signals as a result of the finite difference between quantization levels.

Raible's tabular method A tabular method for determining the stability of digital systems that parallels the Routh-Hurwitz method for analog signals.

Rate gyro A device that responds to an angular position input with an output voltage proportional to angular velocity.

Residue The constants in the numerators of the terms in a partial fraction expansion.

Rise time, T_r The time required for the step response to go from 0.1 of the final value to 0.9 of the final value.

Root locus The locus of closed-loop poles as a system parameter is varied. Typically, the parameter is gain. The locus is obtained from the open-loop poles and zeros.

Routh-Hurwitz criterion A method for determining how many roots of a polynomial in s are in the right half of the s-plane, the left half of the s-plane, and on the imaginary axis. Except in some special cases, the Routh-Hurwitz criterion does not yield the coordinates of the roots.

Sensitivity The percent change in a system characteristic for a percent change in a system parameter.

Settling time, T_s The amount of time required for the step response to stay within $\pm 2\%$ of the steady-state value. Strictly speaking, this is the definition of the 2% settling time. Other percentages, for example 5%, also can be used. This book uses the 2% settling time.

Signal-flow graph A representation of the interconnection of subsystems that form a system. It consists of nodes representing signals and lines representing subsystems.

Similarity transformation A transformation from one state-space representation to another state-space representation. Although the state variables are different, each representation is a valid description of the same system and the relationship between the input and output.

Stability The characteristic of a system defined by a natural response that decays to zero as time approaches infinity.

Stall torque The torque produced when a motor's speed is reduced to zero at the armature at a constant input voltage.

State equations A set of n simultaneous, first-order differential equations with n variables, where the n variables to be solved are the state variables.

State-space The n-dimensional space whose axes are the state variables.

State-space representation A mathematical model for a system that consists of simultaneous, first-order differential equations and an output equation.

State-transition matrix The matrix that performs a transformation on $\mathbf{x}(0)$, taking $\mathbf{x}$ from the initial state, $\mathbf{x}(0)$, to the state $\mathbf{x}(t)$ at any time, $t \geq 0$.

State variables The smallest set of linearly independent system variables such that the values of the members of the set at time t_0 along with known forcing functions completely determine the value of all system variables for all $t \geq t_0$.

State vector A vector whose elements are the state variables.

Static error constants The collection of position constant, velocity constant, and acceleration constant.

Steady-state error The difference between the input and output of a system after the transient response has decayed to zero.

Steady-state response *See* **forced response**.

Subsystem A system that is a portion of a larger system.

Summing junction A block diagram symbol that shows the algebraic summation of two or more signals.

System type The number of pure integrations in the forward path.

System variables Any variable that responds to an input or initial conditions in a system.

Tachometer A voltage generator that yields a voltage output proportional to rotational input speed.

Time constant The time for e^{-at} to decay to 37% of its original value at $t = 0$.

Time-domain representation *See* **state-space representation**.

Torque-speed curve The plot that relates a motor's torque to its speed at a constant input voltage.

Transducer A device that converts a signal from one form to another, for example, from a mechanical displacement to an electrical voltage.

Transfer function The ratio of the Laplace transform of the output of a system to the Laplace transform of the input.

Transient response The part of the response curve due to the system and the way the system acquires or dissipates energy. It is the part of the response plot prior to the steady-state response.

Type *See* **system type**.

Underdamped response The step response of a second-order system that is characterized by overshoot.

Velocity constant $\lim_{s \to 0} s G(s)$

z-transformation A transformation related to the Laplace transformation that is used for the representation, analysis, and design of sampled signals and systems.

Zero-input response The part of the response that depends only upon the initial state vector and not the input.

Zero-order sample-and-hold A device that yields a staircase approximation to the analog signal.

Zeros (1) Those values of the Laplace transform variable, s, that cause the transfer function to become zero, and (2) any roots of factors of the numerator that are common to the characteristic equation in the denominator of the transfer function.

Zero-state response That part of the response that depends only upon the input and not the initial state vector.

Chapter 1

9. c. $x(t) = -\dfrac{12}{17}\cos 2t + \dfrac{3}{17}\sin 2t + \left(\dfrac{12}{17}\cos 2t + \dfrac{3}{17}\sin 2t\right)e^{-t}$

10. b. $x(t) = -e^{-t} + 9te^{-t} + 5e^{-2t} + t - 2$

Chapter 2

3. b $x(t) = \dfrac{7}{15} + \dfrac{7}{10}e^{-5t} - \dfrac{7}{6}e^{-3t}$

5. $\dfrac{Y(s)}{X(s)} = \dfrac{s^3 + 2s^2 + 3s + 7}{s^3 + 5s^2 + 7s + 1}$

6. b. $\dfrac{d^3x}{dt^3} + 8\dfrac{d^2x}{dt^2} + 9\dfrac{dx}{dt} + 15x = \dfrac{df}{dt} + 2f(t)$

10. $\dfrac{V_o(s)}{V_i(s)} = \dfrac{s}{2s + 1}$

14. $\dfrac{V_o(s)}{V_i(s)} = \dfrac{s^2 + 3s + 1}{2s^2 + 7s + 2}$

19. $\dfrac{X_1(s)}{F(s)} = \dfrac{f_v s}{M_1 M_2 s^4 + 2f_v(M_1 + M_2)s^3 + (M_1 K_2 + M_2 K_1 + 3f_v^2)s^2 + 2f_v(K_1 + K_2)s + K_1 K_2}$

28. $\dfrac{\theta_2(s)}{T(s)} = \dfrac{3}{20s^2 + 13s + 4}$

29. $\dfrac{\theta_2(s)}{T(s)} = \dfrac{1}{50s^2 + 100s + 100}$

34. $\dfrac{E_o(s)}{T(s)} = \dfrac{\dfrac{5N_1}{\pi J N_2}}{s\left(s + \dfrac{1}{RC}\right)}$

39. $\dfrac{\theta_2(s)}{E_a(s)} = \dfrac{0.0833}{s(s + 0.75)}$

Chapter 3

1. $\dot{\mathbf{x}} = \begin{bmatrix} -\dfrac{3}{2} & -\dfrac{1}{2} & \dfrac{1}{2} \\ \dfrac{1}{2} & -\dfrac{1}{2} & \dfrac{1}{2} \\ \dfrac{1}{2} & -\dfrac{1}{2} & -\dfrac{1}{2} \end{bmatrix} \begin{bmatrix} v_{c_1} \\ i_L \\ v_0 \end{bmatrix} + \begin{bmatrix} 1 \\ 0 \\ 0 \end{bmatrix} v_i$

$y = \begin{bmatrix} 0 & 0 & 1 \end{bmatrix} \begin{bmatrix} v_{c_1} \\ i_L \\ v_0 \end{bmatrix}$

7. $\dot{\mathbf{x}} = \begin{bmatrix} 0 & 1 & 0 & 0 \\ 0 & 0 & 1 & 0 \\ 0 & 0 & 0 & 1 \\ -10 & -5 & -1 & -2 \end{bmatrix} \mathbf{x} + \begin{bmatrix} 0 \\ 0 \\ 0 \\ 1 \end{bmatrix} r(t)$

$c(t) = \begin{bmatrix} 10 & 5 & 0 & 0 \end{bmatrix} \mathbf{x}$

8. $\dfrac{Y(s)}{R(s)} = \dfrac{10}{s^3 + 5s^2 + 2s + 3}$

12. $\dot{\mathbf{x}} = \begin{bmatrix} -\dfrac{D_{eq}}{J_{eq}} & 0 & -\dfrac{K_t}{J_{eq}}\dfrac{N_1}{N_2} \\ 1 & 0 & 0 \\ -\dfrac{K_b}{L_a}\dfrac{N_2}{N_1} & 0 & -\dfrac{R_a}{L_a} \end{bmatrix} \begin{bmatrix} \omega_L \\ \theta_L \\ i_a \end{bmatrix} + \begin{bmatrix} 0 \\ 0 \\ \dfrac{1}{L_a} \end{bmatrix} e_a$

$y = \begin{bmatrix} \dfrac{N_2}{N_1} & 0 & 0 \end{bmatrix} \begin{bmatrix} \omega_L \\ \theta_L \\ i_a \end{bmatrix}$

Chapter 4

6. $x(t) = \dfrac{1}{5}\left[1 - \sqrt{\dfrac{20}{19}}\, e^{-0.5t} \cos\left(\dfrac{\sqrt{19}}{2}t - \arctan\dfrac{1}{\sqrt{19}}\right)\right]$

12. a. $\zeta = 0.548$; $\omega_n = 10.95$ rad/s; $T_s = 0.667$ s; $T_p = 0.343s$; $\%OS = 12.79$

13. a. $s = -8 \pm j10.915$

16. $D = 0.143$ N-m s/rad

19. $R = 912\ \Omega$

23. $s = -1, -4$

24. a. $s^3 - 5s^2 - 8s + 2 = 0;$ **b.** $s = 6.23, 0.221, -1.453$

28. $y = -\dfrac{1}{2}e^{-4t} + \dfrac{1}{2}e^{-2t}$

29. $\Phi(t) = \begin{bmatrix} 1.0455e^{-0.20871t} - 0.045545e^{-4.7913t} & 0.21822e^{-0.20871t} - 0.21822e^{-4.7913t} \\ -0.21822e^{-0.20871t} + 0.21822e^{-4.7913t} & -0.045545e^{-0.20871t} + 1.0455e^{-4.7913t} \end{bmatrix}$

$\mathbf{x}(t) = \begin{bmatrix} 1.0455e^{-0.20871t} - 0.045545e^{-4.7913t} \\ -0.21822e^{-0.20871t} + 0.21822e^{-4.7913t} \end{bmatrix}$

$y = 0.60911e^{-0.20871t} + 0.39089e^{-4.7913t}$

Chapter 5

2. $\dfrac{C(s)}{R(s)} = \dfrac{2s^2 + 6s + 1}{s^5 + 7s^4 + 19s^3 + 26s^2 + 15s + 2}$

4. $\dfrac{C(s)}{R(s)} = \dfrac{50(s - 2)}{s^3 + s^2 + 150s - 100}$

7. $\dfrac{C(s)}{R(s)} = \dfrac{G_1G_5}{1 + G_1G_2 + G_1G_3G_4G_5 + G_1G_3G_5G_6G_7 + G_1G_5G_8}$

12. $\dfrac{C(s)}{R(s)} = \dfrac{G_4G_6 + G_2G_5G_6 + G_3G_5G_6}{1 + G_6 + G_1G_2 + G_1G_3 + G_1G_2G_6 + G_1G_3G_6 + G_4G_6G_7 + G_2G_5G_6G_7 + G_3G_5G_6G_7}$

17. $D = 3560$ N-m s/rad

26. $\dfrac{C(s)}{R(s)} = \dfrac{G_1G_2G_3G_4}{2 + G_2G_3G_4 + 2G_3G_4 + 2G_4}$

27. $\dfrac{C(s)}{R(s)} = \dfrac{G_1G_6G_7(G_2 + G_3)(G_4 + G_5)}{1 - G_6G_7H_3(G_2 + G_3)(G_4 + G_5) - G_6H_1 - G_7H_2 + G_6G_7H_1H_2}$

28. $\dfrac{C(s)}{R(s)} = \dfrac{s^3 + 1}{2s^4 + s^2 + 2s}$

30. a. $\dot{\mathbf{x}} = \begin{bmatrix} -5 & 1 & 0 & 0 \\ 0 & -5 & 0 & 0 \\ 0 & 0 & -7 & 1 \\ 0 & 0 & 0 & -7 \end{bmatrix} \mathbf{x} + \begin{bmatrix} 0 \\ 1 \\ 0 \\ 1 \end{bmatrix} r(t)$

$y = \begin{bmatrix} -\dfrac{3}{4} & 1 & -\dfrac{5}{4} & -1 \end{bmatrix} \mathbf{x}$

35. $\dot{\mathbf{x}} = \begin{bmatrix} 0 & 1 & 0 & 0 \\ -1 & 0 & 1 & 0 \\ 0 & 0 & 0 & 1 \\ 1 & -1 & 0 & 0 \end{bmatrix} \mathbf{x} + \begin{bmatrix} 0 \\ 0 \\ 0 \\ 1 \end{bmatrix} r(t)$

$c = \begin{bmatrix} -1 & 1 & 0 & 0 \end{bmatrix} \mathbf{x}$

Chapter 6

1. 0 rhp, 1 lhp, $4j\omega$

2. 1 rhp, 1 lhp, $4j\omega$

3. 3 rhp, 2 lhp, $0j\omega$

4. 0 rhp, 2 lhp, $2j\omega$

6. Unstable

12. 1 rhp, 2 lhp, $4j\omega$

17. $0 < K < 140.8$

29. a. $-4 < K < 20.41$;   **b.** 1.36 rad/s

31. a. $0 < K < 19.69$;   **b.** $K = 19.69$;   **c.** $s = \pm j1.118, -4.5, -3.5$

32. $-\dfrac{2}{3} < K < 0$

Chapter 7

3. $e(\infty) = 11.25$

7. $\dot{e}(\infty) = 0.9$

10. a. %$OS = 4.32$;   **b.** $T_s = 0.16\,\text{sec}$;   **c.** $e(\infty) = 0$;   **d** $e(\infty) = 0.2$;   **e.** $e(\infty) = \infty$

13. a. $K_p = 1, K_v = 0, K_a = 0$;   **b.** $e(\infty) = 25, \infty, \infty$, respectively;   **c.** Type 0

19. $K = 110,000$

26. $\beta = 1, K = 1.16, \alpha = 7.74$, or $\beta = -1, K = 5.16, \alpha = 1.74$

31. $K = 831,744, \alpha = 831.744$

34. $K_1 = 125,000, K_2 = 0.016$

Chapter 8

14. Breakaway point $= -1.69$; asymptotes: $\sigma_{\text{int}} = -4$; $j\omega$-axis crossing $= \pm\sqrt{32}$

17. b. Asymptotes: $\sigma_{\text{int}} = -\dfrac{8}{3}$;   **c.** $K = 140.8$;   **d.** $K = 13.125$

18. $K = 97; \alpha = 7.29$

20. a. $\sigma_{\text{int}} = -\dfrac{5}{2}$;   **b.** $s = -1.38, -3.62$;   **c.** $0 < K < 126$;   **d.** $K = 10.3$

23. b. $K = 9.4$;   **c.** $T_s = 4.62$ s, $T_p = 1.86$ s;   **d.** $0 < K < 60$;   **e.** $s = -4.27$

27. $\alpha = 9$

30. a. $0 < K < 834.81$;   **b.** $K = 105.1$;   **c.** $K = 61.13$

32. a. $K = 170.2$;   **b.** $K = 16.95$

Chapter 9

1. $G_c(S) = \dfrac{S + 0.1}{S}$; $K \cong 22.5$ for both cases; $K_{p_O} = 1.25$; $K_{p_N} = \infty$; $\%OS_O = \%OS_N = 4.32$;

$T_{s_O} = T_{s_N} = 0.89$ s

5. a. $s = -2.5 \pm j5.67$; **b.** Angle $= 9.78°$; **c.** $p_c = -35.39$; **d.** $K = 1049.41$;
e. $s = -36.33, -1.057$

6. a. $s = -2.4 \pm j4.16$; **b.** $s = -6.06$; **c.** $K = 29.117$; **d.** $s = -1.263$; **f.** $K_a = 4.8$

10. $G_c(s) = \dfrac{s + 5}{s + 15.18}$; $K = 1416.63$

15. $G_c(s) = \dfrac{3.08(s + 6.93)(s + 0.1)}{s(s + 1)(s + 4)}$

16. a. $K_{uc} = 10$; $K_c = 9.95$; **b.** $K_{p_{uc}} = 1.25$; $K_{p_c} = 6.22$; **c.** $\%OS_{uc} = \%OS_c = 4.32$;
d. Closed-loop pole at -0.349 approximately cancels closed-loop zero at -0.5;
e. Approach to final value longer than settling time of uncompensated system;

f. $G_{LLC}(s) = \dfrac{404.067(s + 0.5)(s + 4)}{(s + 2)(s + 4)(s + 0.1)(s + 28.36)}$ yields approximately a 5 times improvement
in speed.

18. Poles $= -0.747 \pm j1.237, -2.51$; zeros—none

Chapter 10

7. a. $0 < K < 810$; **b.** $3.94 < K < \infty$; **c.** $0 < K < 720$ (Answers are from exact frequency response)

8. a. $G_M = 4.19$ dB; $\Phi_M = 14.27°$ (Answers are from exact frequency response)

10. c. $\omega_{BW} = 1.5$ rad/s.

15. $T_s = 2.23$ sec, $T_p = 0.476$ s, $\%OS = 42.62$ (Answers are from exact frequency response)

20. $G_M = 1.17$ dB, $\Phi_M = 6.01°$, $\%OS = 84.78$ (Answers are from exact frequency response)

Chapter 11

1. a. $K = 287.74$ (Answer is from exact frequency response)

2. a. $K = 321.107$ (Answer is from exact frequency response)

3. a. $K = 90.199$ (Answer is from exact frequency response)

7. $G_c(s) = \dfrac{s + 1.124}{s + 4.149}$, $K = 354.24$ (Answer is from exact frequency response)

12. $G_c(s) = \dfrac{(s + 0.042)(s + 1.376)}{s}$, $K = 7.28$ (Answer is from exact frequency response)

Chapter 12

1. e. For function a.: $T(s) = \dfrac{s + 1}{s^2 + (2 + k_2)s + k_1}$

3. b. $T(s) = \dfrac{100(s + 10)(s + 20)}{s^3 + as^2 + bs + c}$

where $a = \left(\dfrac{400}{9}k_{3_z} + \dfrac{100}{9}k_{2_z} + \dfrac{400}{9}k_{1_z} + 45\right)$

$b = \left(2000k_{3_z} + \dfrac{1000}{3}k_{2_z} + \dfrac{2000}{3}k_{1_z} + 450\right)$

$c = 20{,}000k_{3_z}$

and $\mathbf{C} = \begin{bmatrix} 1 & 1 & 1 \end{bmatrix}$; $\mathbf{B} = \dfrac{1}{9}\begin{bmatrix} 400 & 100 & 400 \end{bmatrix}$ was used

8. a. Uncontrollable; **b.** Controllable; **c.** Controllable

10. $\mathbf{K} = \begin{bmatrix} -2.3446 & 0.82892 & -0.14 \end{bmatrix}$ for a characteristic polynomial of $(s + 1)(s^2 + 4s + 11.446)$
$= s^3 + 5s^2 + 15.446s + 11.446$

14. $\mathbf{L} = \begin{bmatrix} -656.4 & 786.4 \end{bmatrix}^T$ for a characteristic polynomial of $s^2 + 140s + 10{,}000$

19. a. Step: $e(\infty) = 1.098$; ramp: $e(\infty) = \infty$

Chapter 13

2. a. $f(kT) = 46.875(0.5)^k - 114.75(0.7)^K + 68.875(0.9)^K$

4. c. $G(z) = 3\left[\dfrac{z}{z - 0.368} - \dfrac{z^2 - 0.026z}{z^2 - 0.052z + 0.135}\right]$

5. b. $G(z) = 0.00125\dfrac{(z + 0.195)(z + 2.821)}{(z - 1)(z - 0.368)(z - 0.819)}$

6. b. $T(z) = \dfrac{G_1(z)G_2(z)}{1 + G_1(z)G_2H(z)}$

10. $0 < K < 11.104$

11. a. $K_p = \dfrac{1}{2}$, $e(\infty) = \dfrac{2}{3}$; $K_v = 0$, $e(\infty) = \infty$; $K_a = 0$, $e(\infty) = \infty$

12. $K = 9.68$ for 16.3% overshoot; $0 < K < 86$ for stability